# Lecture Notes in Mathematics

Volume 2346

**Editors-in-Chief**

Jean-Michel Morel, City University of Hong Kong, Kowloon Tong, China

Bernard Teissier, IMJ-PRG, Paris, France

**Series Editors**

Karin Baur, University of Leeds, Leeds, UK

Michel Brion, UGA, Grenoble, France

Rupert Frank, LMU, Munich, Germany

Annette Huber, Albert Ludwig University, Freiburg, Germany

Davar Khoshnevisan, The University of Utah, Salt Lake City, UT, USA

Ioannis Kontoyiannis, University of Cambridge, Cambridge, UK

Angela Kunoth, University of Cologne, Cologne, Germany

Ariane Mézard, IMJ-PRG, Paris, France

Mark Podolskij, University of Luxembourg, Esch-sur-Alzette, Luxembourg

Mark Policott, Mathematics Institute, University of Warwick, Coventry, UK

László Székelyhidi, MPI for Mathematics in the Sciences, Leipzig, Germany

Gabriele Vezzosi, UniFI, Florence, Italy

Anna Wienhard, MPI for Mathematics in the Sciences, Leipzig, Germany

This series reports on new developments in all areas of mathematics and their applications - quickly, informally and at a high level. Mathematical texts analysing new developments in modelling and numerical simulation are welcome. The type of material considered for publication includes:

1. Research monographs
2. Lectures on a new field or presentations of a new angle in a classical field
3. Summer schools and intensive courses on topics of current research.

Texts which are out of print but still in demand may also be considered if they fall within these categories. The timeliness of a manuscript is sometimes more important than its form, which may be preliminary or tentative. Please visit the LNM Editorial Policy (https://drive.google.com/file/d/1MOg4TbwOSokRnFJ3ZR3ciEeKs9hOnNX_/view?usp=sharing)

*Titles from this series are indexed by Scopus, Web of Science, Mathematical Reviews, and zbMATH.*

Mikko Korhonen

# Maximal Solvable Subgroups of Finite Classical Groups

Springer

Mikko Korhonen
Shenzhen International Center
for Mathematics
Southern University of Science
and Technology
Shenzhen, China

ISSN 0075-8434 ISSN 1617-9692 (electronic)
Lecture Notes in Mathematics
ISBN 978-3-031-62914-3 ISBN 978-3-031-62915-0 (eBook)
https://doi.org/10.1007/978-3-031-62915-0

Mathematics Subject Classification: 20H30, 20F16, 20C99, 20E99, 20B35, 20B15, 20B99

This work was supported by Shenzhen Science and Technology Innovation Program (RCBS20210609104420034).



This Springer imprint is published by the registered company Springer Nature Switzerland AG
The registered company address is: Gewerbestrasse 11, 6330 Cham, Switzerland

# Preface

A subgroup of a group $G$ is said to be *maximal solvable* if it is maximal among the solvable subgroups of $G$. In his 1870 *Traité*, Jordan gave a classification of the maximal solvable subgroups of symmetric groups.

The solution given by Jordan is a massive recursion on the degree of the group $G$. As a first step, he reduces the problem to the case where $G$ is transitive, which in turn is reduced to the case where $G$ is primitive. The primitive case is equivalent to the problem of classifying maximal irreducible solvable subgroups of $\mathrm{GL}_d(p)$, where $p$ is a prime. In $\mathrm{GL}_d(p)$, the problem is reduced to the case of primitive irreducible solvable subgroups. These subgroups are then constructed in terms of maximal irreducible solvable subgroups of general symplectic groups $\mathrm{GSp}_{2\ell}(r)$ ($r$ prime) and orthogonal groups $\mathrm{O}_{2\ell}^{\pm}(2)$.

In this book, we present a proof of Jordan's classification result in modern terms. More generally, we will give a complete classification of the maximal irreducible solvable subgroups of $\mathrm{GL}_n(q)$, $\mathrm{GSp}_{2\ell}(q)$, and $\mathrm{GO}_n^{\pm}(q)$, where $q$ is a power of a prime. We also give similar classifications for maximal irreducible solvable subgroups for $\mathrm{Sp}_{2\ell}(q)$ and $\mathrm{O}_n^{\pm}(q)$. For $q$ even, we classify maximal irreducible solvable subgroups of $\Omega_{2\ell}^{\pm}(q)$.

Along the way, we also obtain some results on irreducible solvable matrix groups, which are of independent interest. For example, for primitive irreducible solvable subgroups $G \leq \mathrm{GL}_n(q)$, we provide upper bounds on the dimension of fixed point spaces of elements of $G$. For irreducible solvable $G \leq \mathrm{GL}_n(q)$, we prove the following result: every abelian subgroup of the affine group $\mathbb{F}_q^n \rtimes G$ has order $\leq q^n$.

This work was partially supported by Shenzhen Science and Technology Program (Grant No. RCBS20210609104420034). The author wishes to thank Tim Burness, Alla Detinko, Dane Flannery, Cai Heng Li, and Alexandre Zalesski for useful discussions, anonymous referees for their useful comments and suggestions, and Fengda Sun for support and many helpful discussions during the preparation of this work.

Shenzhen, China Mikko Korhonen

# Contents

# Chapter 1
# Introduction

## 1.1 Introduction and Historical Background

In his 1870 *Traité des substitutions et des équations algébriques*, one of the questions Camille Jordan considered was the classification of "the general types of equations solvable by radicals" [47, p. VIII]. He interpreted this as equivalent to the solution of the following problem [47, p. 396].

**Problem A** Let $N > 0$ be an integer. Classify the transitive maximal solvable subgroups of the symmetric group $S_N$.

(Here "maximal solvable" means maximal among the solvable subgroups of $S_N$, with respect to inclusion.) The main difficulty in Problem A lies in the classification of primitive maximal solvable subgroups, which Jordan reduced to the following two problems.

**Problem B** Let $n > 0$ be an integer and let $p$ be a prime. Classify the maximal irreducible solvable subgroups of $\mathrm{GL}_n(p)$.

**Problem C** Let $n > 0$ be an integer and let $p$ be a prime. Classify the maximal irreducible solvable subgroups of $\mathrm{GSp}_{2n}(p)$, $\mathrm{O}^+_{2n}(2)$, and $\mathrm{O}^-_{2n}(2)$.

Jordan was able to solve these problems, which he later wrote was "l'objet principal" of his *Traité* [52, p. 263]. He achieves this in the last part of the *Traité*, Livre IV, which takes up nearly 300 pages of the book. The solution given by Jordan is what Dieudonné called a "gigantesque recurrencé" that proceeds by a series of successive reductions [16, p. XXXIV]. Essentially, Jordan identifies the various types of maximal solvable subgroups, and describes them in terms of groups of smaller degree. For example, imprimitive groups are described in terms of primitive groups in smaller degree.

After Jordan, these types of reductions have become a very basic and useful technique—recall for example the O'Nan-Scott theorem on maximal subgroups of

 
M. Korhonen, *Maximal Solvable Subgroups of Finite Classical Groups*, Lecture Notes in Mathematics 2346, https://doi.org/10.1007/978-3-031-62915-0_1

symmetric groups [4, 62, 70]. One can also see the solution of Problems B and C as a predecessor to the classification theorems of Aschbacher, Kleidman–Liebeck, and Bray–Holt–Roney-Dougal on maximal subgroups of classical groups [2, 7, 55].

The results of Livre IV in Jordan's *Traité* seem to have received less attention than some of his other results. Jordan himself wrote later that the proof is "confusing" and contains some mistakes [52, p. 264]. Much later in his life (1908 and 1917), Jordan published two papers where he gives a more clear presentation of his results [51, 52].

One purpose of this book is expository: we give a proof of Jordan's classification in modern terms, for a large part based on the ideas in [51, 52]. More generally, we will use Jordan's methods to classify the maximal irreducible solvable subgroups of $\mathrm{GL}_n(\mathbb{F})$, $\mathrm{GSp}_{2n}(\mathbb{F})$, and $\mathrm{GO}_n^{\epsilon}(\mathbb{F})$ over a finite field $\mathbb{F}$. Previous work in this direction was done by D. A. Suprunenko in the 1950s [76], who studied solvable and nilpotent subgroups of $\mathrm{GL}_n(\mathbb{F})$ over an arbitrary field $\mathbb{F}$. Some of the ideas used by Suprunenko are similar to those of Jordan, but take advantage of the basic methods and language of representation theory.

We will now give an outline of Jordan's classification of maximal solvable subgroups, see also the survey by Dieudonné [16]. In [51, 52] Jordan provides a solution to the following slightly more general problem.

**Problem A'** Let $N > 0$ be an integer. Classify the maximal solvable subgroups of the symmetric group $S_N$.

For the classification of maximal solvable subgroups $G \leq S_N$, the first step is to reduce the problem to the case where $G$ is transitive. If $G$ is intransitive, it is easily seen that $G = G_1 \times \cdots \times G_t$, where $G_i$ is maximal transitive solvable in $S_{n_i}$, and $n_1 + \cdots + n_t = N$. Jordan determines when such a group $G = G_1 \times \cdots \times G_t$ is maximal solvable in $S_N$, thus reducing the problem to the case where $G$ is transitive.

Similarly, if $G$ is transitive and imprimitive, then $G$ is equal to a wreath product $G_1 \wr G_2 \wr \cdots \wr G_t$, where $G_i$ is transitive primitive maximal solvable in $S_{n_i}$, and $N = n_1 n_2 \cdots n_t$ with each $n_i > 1$ equal to some prime power. Jordan proves that such a wreath product is maximal solvable, except when $(n_i, n_{i+1}) = (2, 2)$ for some $i$, which has to be excluded since $S_2 \wr S_2 \lneq S_4$. The problem is thus reduced to the primitive case.

When $G$ is primitive maximal solvable, we have $N = p^n$ for some prime $p$ and $G$ is an affine group of the form $G = \mathbb{F}_p^n \rtimes X$, where $X \leq \mathrm{GL}_n(p)$ is maximal irreducible solvable. It turns out such a group is always maximal solvable in $S_N$, so the following theorems hold.

**Theorem 1.1.1 (Jordan)** *Let $G$ be a maximal solvable subgroup of $S_N$. Then $G$ is of one of the following types:*

*Type (I): $G$ is intransitive, and the following hold:*

- *(a) $N = n_1 + n_2 + \cdots + n_t$, where $t \geq 2$ and $n_i \geq 1$ for all $1 \leq i \leq t$;*
- *(b) $G = G_1 \times G_2 \times \cdots \times G_t$, where $G_i$ is a maximal transitive solvable subgroup of $S_{n_i}$ for all $1 \leq i \leq t$;*

(c) $G_i \not\cong G_j$ *as permutation groups for all* $i \neq j$;
(d) $G_i \not\cong G_j \wr S_2$ *as permutation groups for all* $i \neq j$;
(e) $G_i \not\cong G_j \wr S_3$ *as permutation groups for all* $i \neq j$;

*Type (II):* *G is transitive and imprimitive, and the following hold:*

(a) $N = n_1 n_2 \cdots n_t$, *where* $t \geq 2$ *and* $n_i > 1$ *is a prime power for all* $1 \leq i \leq t$;
(b) $G = G_1 \wr G_2 \wr \cdots \wr G_t$, *where* $G_i$ *is a maximal primitive solvable subgroup of* $S_{n_i}$ *for all* $1 \leq i \leq t$;
(c) $(n_i, n_{i+1}) \neq (2, 2)$ *for all* $1 \leq i < t$.

*Type (III):* *G is primitive, and the following hold:*

(a) $N = p^n$ *for some prime p and integer* $n > 0$;
(b) *G is an affine group* $G = \mathbb{F}_p^n \rtimes X$, *where* $X \leq \mathrm{GL}_n(p)$ *is maximal irreducible solvable.*

**Theorem 1.1.2 (Jordan)** *Let* $G \leq S_N$ *be one of the Types (I)–(III) in Theorem 1.1.1. Then G is maximal solvable in* $S_N$.

We will give a proof of Theorems 1.1.1 and 1.1.2 in Sect. 1.3, following Jordan [52]. Here the proof of Theorem 1.1.1 is straightforward. For Theorem 1.1.2, the key observation is that in a solvable primitive permutation group, a nontrivial element fixes at most half of the points (Lemma 1.3.5).

Problem A' is then reduced to the classification of maximal irreducible solvable subgroups of $\mathrm{GL}_n(p)$ (Problem B), which is where the main difficulties arise.

First one needs to narrow down the structure of maximal irreducible solvable subgroups. This is attained by Jordan in [51], where he describes and constructs the possible candidates for the maximal irreducible solvable subgroups $G$ of $\mathrm{GL}_n(p)$. As a first step, if such a $G$ is imprimitive, there exist integers $k > 1$ and $d > 0$ such that $n = dk$ and $G = G_0 \wr X$, where $G_0 \leq \mathrm{GL}_d(p)$ is primitive maximal irreducible solvable, and $X \leq S_k$ is maximal transitive solvable.

Thus for the construction of maximal irreducible solvable subgroups, it is not difficult to reduce to the case where $G$ is primitive. In this case Jordan first considers a maximal abelian normal subgroup $F \trianglelefteq G$, which he calls the *premier faisceau* of $G$.

In modern terms, it turns out that the $\mathbb{F}_p$-algebra generated by $F$ is a finite field $\mathbb{K} \cong \mathbb{F}_{p^\nu}$, where $n = \mu\nu$ for some integer $\mu \geq 1$. Furthermore, we have $F = \mathbb{K}^\times$, so $F$ is cyclic of order $p^\nu - 1$. If $\mu = 1$, then $F$ is generated by a Singer cycle, and $G$ is equal to the Singer cycle normalizer $\Gamma\mathrm{L}_1(p^n)$ of order $n(p^n - 1)$.

Suppose then that $\mu > 1$. In this case, Jordan shows that $G$ can be constructed in terms of maximal irreducible solvable subgroups of general symplectic groups $\mathrm{GSp}_{2\ell}(r)$ and $\mathrm{O}_{2\ell}^+(2)$. Roughly speaking, Jordan shows that we have a factorization

$$\mu = r_1^{\ell_1} \cdots r_k^{\ell_k},$$

where $r_i$ are primes (with possibly some of the $r_i$ being equal), such that $r_i \mid p^\nu - 1$ for all $1 \le i \le k$, and such that $G$ can be constructed in terms of $X_1, \ldots, X_k$, where:

- $X_i$ is a maximal irreducible solvable subgroup of $\mathrm{GSp}_{2\ell_i}(r_i)$ if $r_i > 2$;
- $X_i$ is a maximal irreducible solvable subgroup of $\mathrm{O}^+_{2\ell_i}(2)$ or $\mathrm{O}^-_{2\ell_i}(2)$ if $r_i = 2$.

To be a bit more specific, one can show that $A = \mathrm{Fit}(C_G(F))$ is an absolutely irreducible subgroup of $C_{\mathrm{GL}_n(q)}(F) = \mathrm{GL}_\mu(\mathbb{K})$, and decomposes as a tensor product $A = R_1 \otimes \cdots \otimes R_k$, where $R_i \trianglelefteq G$ is an extraspecial $r_i$-group of order $r_i^{1+2\ell_i}$ and exponent $r_i \gcd(r_i, 2)$. This provides a homomorphism

$$\pi : N_{\mathrm{GL}_n(p)}(F, R_1, \ldots, R_k) \to \prod_{i=1}^{k} \mathrm{GSp}_{2\ell_i}(r_i)$$

defined by $\pi(g) = (g_1, \ldots, g_k)$, where $g_i$ is the action of $g$ on $R_i/Z(R_i)$. It turns out that $G = \pi^{-1}(X_1 \times \cdots \times X_k)$.

Conversely, for any such factorization $\mu = r_1^{\ell_1} \cdots r_k^{\ell_k}$ and groups $X_1, \ldots, X_k$, one can construct $\pi^{-1}(X_1 \times \cdots \times X_k)$, which turns out to be an irreducible primitive solvable subgroup of $\mathrm{GL}_n(p)$. Thus the main question is reduced to the classification of maximal irreducible solvable subgroups of $\mathrm{GSp}_{2\ell}(r)$ and $\mathrm{O}^\pm_{2\ell}(2)$, in other words, to Problem C.

Using similar reductions—starting again from the imprimitive case—Jordan shows that Problem C can also be reduced to groups of smaller degree. With this the construction of all maximal solvable subgroups of $S_N$ is complete. At the bottom of this massive recursion, we have the normalizers $\Gamma\mathrm{L}_1(p^n)$ of Singer cycles in $\mathrm{GL}_n(p)$, and their analogues in $\mathrm{GSp}_{2\ell}(r)$ and $\mathrm{O}^\pm_{2\ell}(2)$.

In Sects. 2.1–4.3, we will similarly analyze the structure of maximal irreducible solvable subgroups of $\mathrm{GL}_n(q)$, $\mathrm{GSp}_{2\ell}(q)$ and $\mathrm{GO}^\varepsilon_n(q)$, for every prime power $q$. Using the basic approach of Jordan with some adjustments, we will find similarly to Jordan that all such groups are constructed recursively in terms of maximal irreducible solvable subgroups of smaller degree. The base of the recursion again consists of normalizers of Singer cycles in $\mathrm{GL}_n(q)$, and their analogues in symplectic and orthogonal groups.

The recursive construction provides an algorithm to construct a list of groups which contains all maximal irreducible solvable subgroups. It then remains to check which of the subgroups constructed are actually maximal solvable. This question is the topic of [52], where Jordan gives a proof that apart from a few families of examples, the construction always provides maximal solvable subgroups. In Chap. 7, we will prove a similar classification result for $\mathrm{GL}_n(q)$, $\mathrm{GSp}_{2\ell}(q)$, and $\mathrm{GO}^\pm_n(q)$ for every prime power $q$. As a necessary part of this classification, we will also classify maximal irreducible solvable subgroups of $\mathrm{Sp}_{2\ell}(q)$ and $O^\pm_n(q)$ for every prime power $q$. Furthermore, for $q$ even, we will classify maximal irreducible solvable subgroups of $\Omega^\pm_{2\ell}(q)$.

The first step in the classification is to verify some basic properties of the groups given by the construction. For example, one needs to prove that the subgroups

$\pi^{-1}(X_1 \times \cdots \times X_k)$ mentioned earlier are indeed irreducible and primitive. We establish results of this type in Chap. 5. After this, two key results that are needed in the classification are the following.

**Theorem 1.1.3** *Let $G \leq \mathrm{GL}_n(q)$ be primitive irreducible solvable. Then for every $g \in G \setminus \{1\}$, the fixed point space of $g$ on $\mathbb{F}_q^n$ has dimension $\leq 3n/4$.*

**Theorem 1.1.4** *Let $G \leq \mathrm{GL}_n(q)$ be irreducible and solvable. If $D$ is an abelian subgroup of the affine group $\mathbb{F}_q^n \rtimes G$, then $|D| \leq q^n$.*

Jordan proved these results in the case where $q = p$ is a prime, but as we shall see in Chap. 6, the basic idea of his proof works for every prime power $q$.

In fact, for Theorem 1.1.3 Jordan provides a more precise upper bound for elements of prime order, which we will also need. In Jordan's proof, the main argument consists essentially of finding such an upper bound for elements in normalizers of extraspecial groups in $\mathrm{GL}_n(q)$ (Aschbacher class $\mathcal{C}_6$). We will prove the following result in Sect. 6.1.

**Theorem 1.1.5** *Suppose that $R \leq \mathrm{GL}_n(q)$ is absolutely irreducible, where $R$ is an extraspecial $r$-group of exponent $r$ or 4. Let $Z$ be the group of scalar matrices in $\mathrm{GL}_n(q)$.*

*Let $g \in N_{\mathrm{GL}_n(q)}(RZ)$ be an element of prime order $\varpi$ and suppose that $g$ is non-scalar. Let $W$ be a $g$-eigenspace on $\mathbb{F}_q^n$. Then*

$$\dim W \leq \begin{cases} 3n/4, & \text{if } \varpi = 2, \\ 2n/3, & \text{if } \varpi = 3, \\ n/2, & \text{if } \varpi > 3. \end{cases}$$

*Moreover, if $\varpi = 3$ and $n$ is not a multiple of 3, then we also have* $\dim W \leq n/2$.

Some similar bounds were given by Guralnick and Maróti in [31, Section 2], with a different proof that generalizes a result of Hall and Higman [32, Theorem 2.5.1]. Furthermore, Theorem 1.1.3 and the corresponding result for normalizers of extraspecial groups have appeared and have been applied many times in the literature; see for example [26, proof of Proposition 4], [72, Lemma 2.3], [33, p. 452], [10, Lemma 6.3].

Using the results established in Sects. 4.4–6.3, we complete the classification of maximal irreducible solvable subgroups in Chap. 7. More generally in the case of $\mathrm{GSp}_{2\ell}(q)$ and $\mathrm{GO}_n^{\varepsilon}(q)$, in Sect. 7.5 we will also classify *metrically completely reducible* maximal solvable subgroups, where metrically completely reducible means that the group has no nonzero invariant subspaces which are totally isotropic. We will also classify metrically completely reducible maximal solvable subgroups in $\mathrm{Sp}_{2\ell}(q)$, $O_n^{\varepsilon}(q)$, and if $q$ is even, in $\Omega_{2\ell}^{\pm}(q)$.

Finally at the end of this book in Chap. 8, we illustrate our results by providing tables of maximal solvable subgroups in small degrees.

**Remark 1.1.6** From the recursive construction presented in this book, one can extract an efficient algorithm for finding generators for the maximal solvable subgroups of $S_N$, and for the maximal irreducible solvable subgroups of the classical groups that we consider. This could be implemented in a computer algebra system such as GAP or Magma, however we have not written down the algorithm precisely in this book.

Previously an algorithm for constructing maximal irreducible solvable subgroups of $\mathrm{GL}_n(q)$ was proposed in [23, Section 4]. The algorithm in [23] first constructs a list of candidates for maximal irreducible solvable subgroups based on Aschbacher's theorem, and then checks which of these subgroups are maximal solvable in $\mathrm{GL}_n(q)$. However, this approach has issues in certain degrees—see Remark 5.5.16.

**Remark 1.1.7 (Historical Background)** We finish this introduction by providing some more historical background to Jordan's classification of maximal solvable subgroups of symmetric groups.

The origin of group theory is in the study of algebraic equations. In the early 1820s, Niels Henrik Abel (1802–1829) proved that in general, the quintic equation is not solvable by radicals [54, p. 67]. Later in 1826, Abel stated in a letter that he was working on "determining the form of all the algebraic equations which can be solved algebraically", and he believed that he would be able to solve this more general problem [1, p. 256, p. 260] [54, p. 71] [83, p. 101].

Abel died in 1829 and did not have chance to finish his work in this direction, but he did obtain results in some special cases. In modern terms, one of his results is that a polynomial equation $f(X) = 0$ is solvable by radicals if the corresponding Galois group is commutative [54, p. 71].

The general result was given by Évariste Galois (1811–1832), who proved (again in modern terms) that a polynomial equation $f(X) = 0$ over a field of characteristic zero is solvable by radicals if and only if the corresponding Galois group is solvable. This result appeared in the *Premiere mémoire* that Galois had submitted to the Académie des Sciences, but which was rejected [22]. Galois' paper was unclear, and the referee report by Poisson and Lacroix from 1831 suggested that Galois should develop his work further:

> His reasoning is neither clear enough nor well enough developed for us to have been able to judge its correctness [...] The author announces that the proposition which forms the special goal of his memoir is part of a general theory susceptible of many other applications. [. . . ] One may therefore wait until the author will have published his work in its entirely before forming a final opinion; but given the present state of the part that he has submitted to the Academy, we cannot propose to you that you give it your approval. (Translation from [68, IV.2, pp. 148–149])

Galois died in a duel in 1832, and it took a long time before his ideas were fully developed. The *Premiere mémoire* and another unpublished manuscript of Galois were published posthumously in 1846 by Liouville [63]. From here the ideas of Galois spread and were appreciated by famous mathematicians such as Betti, Dedekind, Jordan, Kronecker, and Serret [83, pp. 118–135], [68, I.5]. For example, Betti [6] and Serret [73] wrote texts which contained expositions of Galois' results, correcting mistakes and filling in missing details from the terse and incomplete works of Galois.

The first significant developments on Galois' work were made by Jordan. Starting from his thesis [41], Jordan spent most of 1860–1870 working on topics related to algebraic equations, permutation groups, and solvable groups. This culminated in the publication of the *Traité* in 1870, which greatly expanded on the works of Galois and has had an enormous influence on group theory. As pointed out in [67, pp. 414, 418], Jordan is overly modest in calling all the 667 pages of the *Traité* "just a commentary" on Galois' work [47, p. viii].

In the *Traité*, Jordan was particularly interested in Abel's problem of determining the different types of polynomial equations solvable by radicals, over a field $F$ of characteristic zero [47, p. V, p. VIII]. As a starting point, Jordan takes the result of Galois that $f(X) \in F[X]$ is solvable by radicals if and only if the corresponding Galois group is solvable. To make this criterion more explicit, Jordan was led to consider the construction of maximal transitive solvable subgroups of $S_N$ [43, p. 108] [47, p. 396].

For determining solvability by radicals, what Jordan essentially proposes in [47, p. 396] is the following approach. (This is a similar to how Jordan suggests [47, p. 276] that Galois groups can be computed, see [13, Section 13.3].)

It suffices to consider the irreducible case, so suppose that $f(X) \in F[X]$ is an irreducible polynomial of degree $N$, with $N$ distinct roots $\alpha_1, \ldots, \alpha_N$ in some extension field of $F$. Let $G_1, \ldots, G_t$ be representatives for the conjugacy classes of maximal transitive solvable subgroups of $S_N$. The group corresponding to $f$ is a transitive permutation group $G \leq S_N$ acting on the roots of $f$. Thus by Galois' result, we know that $f$ is solvable by radicals if and only if $G$ is contained in a conjugate of some $G_i$.

To decide whether $G$ is contained in a conjugate of $G_i$, start by finding a polynomial $\psi \in F[X_1, \ldots, X_N]$ such that $G_i$ is the stabilizer of $\psi$ in $S_N$. We consider the resolvent

$$R_\psi(Y, X_1, \ldots, X_N) := \prod_{j=1}^{M} (Y - \psi_j) \in F[Y, X_1, \ldots, X_N],$$

where $\psi_1, \ldots, \psi_M$ is the $S_N$-orbit of $\psi$. Then $R_\psi$ is invariant under the action of $S_N$, so the coefficients of $Y$ can be expressed in terms of elementary symmetric polynomials in $X_1, \ldots, X_N$. Therefore one can compute the specialized resolvent

$$R_\psi^f(Y) := R_\psi(Y, \alpha_1, \ldots, \alpha_N) \in F[Y]$$

using the coefficients of $f$.

At this point Jordan states [47, p. 276] (also in [43, p. 107], [50, footnote, p. 35]) that it suffices to check whether $R_\psi^f(Y)$ has a "rational root", meaning a root in $F$. As pointed out in [13, p. 386], this part of Jordan's argument is missing a detail, due to the possibility of rational roots which are not simple. What is true is the following (see for example [13, Proposition 13.3.2], or [75, Theorem 5]):

- If $G$ is contained in a conjugate of $G_i$, then $R_\psi^f(Y)$ has a rational root.
- If $R_\psi^f(Y)$ has a simple rational root, then $G$ is contained in a conjugate of $G_i$.

Thus if $R_\psi^f(Y)$ has rational roots and all of them occur with multiplicity greater than one, the method is inconclusive. However, we can fix this by modifying $f$ by a suitable Tschirnhaus transformation to get another polynomial $f_0$, such that the Galois groups of $f$ and $f_0$ are isomorphic as permutation groups, and $R_\psi^{f_0}(Y)$ is squarefree [25, Theorem 3, (2)].

Jordan comments that he sees his method describing solvability by radicals "satisfactory from a theoretical point of view", but that in general it leads to computations which are "impractical" [43, pp. 107–108] [47, pp. 276–277]. Indeed, already for small $N$ the degrees of the polynomials involved become too large for calculations by hand. Currently one could implement Jordan's method in a computer algebra system, see for example [24, 37, 75] for computational aspects related to Galois groups and resolvents.

Thus Jordan interprets the question of solvability by radicals as equivalent to the classification of maximal transitive solvable subgroups $S_N$, stating that each maximal transitive solvable subgroup of $S_N$ characterizes a type of equation solvable by radicals [47, p. 396]. Jordan was able to construct and classify maximal solvable subgroups of $S_N$, and later summarized that from his work "the problem of Abel is completely resolved" [50, p. 37].

As mentioned earlier in this introduction (Theorems 1.1.1 and 1.1.2), the classification of maximal transitive solvable subgroups of $S_N$ can be reduced to the case of primitive groups. It was already known to Galois that a solvable primitive permutation group is of prime power degree $p^n$, and can be realized as a group of affine linear transformations over the finite field $\mathbb{F}_p$ of integers modulo $p$ [67]. In modern terms, a solvable primitive permutation group of degree $p^n$ ($p$ prime) is an affine group $(\mathbb{F}_p)^n \rtimes X$, where $X \leq \mathrm{GL}_n(p)$ is irreducible and solvable.

In [63, p. 406], Galois makes a false claim which seems to be equivalent to the following statement: except for $p^n = 3^2$ and $p^n = 5^2$, every irreducible solvable subgroup of $\mathrm{GL}_n(p)$ is conjugate to a subgroup of the semilinear group $\Gamma\mathrm{L}_1(p^n)$. This result would seemingly solve the problem of describing maximal solvable subgroups, but turns out to be completely false, and Jordan points out Galois' mistake in many papers [44, p. 270], [43, p. 108], [46, p. 113], [48, p. 286].

The correct classification of maximal solvable subgroups is considerably more complicated, and takes up the entirety of Livre IV in the *Traité*. Jordan's solution proceeds by a massive recursion, which we have sketched earlier in this introduction. The original proof is somewhat difficult to follow and contains some mistakes, as Jordan himself writes in [52, p. 264].

Jordan also does not provide any concrete examples illustrating his construction in the *Traité*, although in [46] he demonstrates the classification for $\mathrm{GL}_2(p)$. In [48, Table A, p. 288], Jordan gives a table listing the number of maximal irreducible solvable subgroups of $\mathrm{GL}_n(p)$ for $p^n < 10^6$, but the table contains several mistakes and Jordan does not give a list of the groups themselves. For example, as pointed out in [74, p. 94], Jordan claims there are 5 classes of maximal irreducible solvable subgroups in $\mathrm{GL}_4(3)$, while the correct number is 4.

Despite its flaws, in Jordan's proof the key steps and techniques work, and lead to a solution of the problem. From the proof one can also extract intermediate results which are of independent interest, such as Theorem 1.1.4. Here is what Jordan states about mistakes in the *Traité*, in response to some criticism by Netto:

> Nous ne saurions d'ailleurs avoir la prétention de n'avoir laissé se glisser aucune inexactitude dans un ouvrage aussi étendu que le nôtre, et qui traite un sujet nouveau et difficile; mais nous sommes persuadé qu'elles y sont en petit nombre.[1] ([49, p. 258], as quoted in [66])

Perhaps due to its difficulty at the time and a lack of clear exposition, Jordan's classification of maximal solvable subgroups received little attention during his lifetime. One attempt at deciphering Jordan's work is given by Bucht, who in a 96-page paper [9] goes through Jordan's classification in the cases of $\mathrm{GL}_3(p)$ and $\mathrm{GL}_4(p)$. Jordan published papers in 1908 and 1917 [51, 52] which give a more clearly organized and simplified version of the proof. A few gaps, errors, and unconsidered cases still remain in these papers—see for example Remarks 6.3.5, 4.1.8, 7.1.16, and 7.5.6.

Later, maximal solvable subgroups of linear groups have been studied by many authors, most notably Suprunenko in the 1950s and 1960s [76, 77]. Suprunenko studied (among other things) solvable and nilpotent subgroups of $\mathrm{GL}_n(\mathbb{F})$, where $\mathbb{F}$ is an arbitrary field or a division ring. He was certainly familiar with Jordan's results as he mentions in the introduction to [76], which also includes [52] in the bibliography.

In [77, §18–§20], Suprunenko gives a description of the general structure of maximal solvable (not necessarily irreducible) subgroups of $\mathrm{GL}_n(\mathbb{F})$ over an arbitary field, generalizing results of Jordan. For the most part, Suprunenko does not attempt to study when the subgroups given by the construction are maximal solvable, although he does illustrate the results by giving a complete classification of maximal irreducible solvable subgroups of $\mathrm{GL}_r(q)$ for $r$ prime [77, 21.3]. In [78] Suprunenko describes some other special cases, such as the maximal irreducible solvable subgroups of $\mathrm{GL}_4(p)$ for $p$ prime.

Some other work discussing maximal solvable subgroups can be found in [17, 74, 87], and [15]. For more on the history surrounding Abel, Galois, and Jordan, see for example [8, 54, 67, 68, 80, 83].

[1] "We cannot, incidentally, claim that we have not let any inaccuracies slip into a work as extensive as ours, and which deals with a new and difficult subject; but we are convinced that they are there in small numbers."

## 1.2 Basic Notation and Terminology

Let $r$ be a prime and $q$ a power of a prime. Suppose that $G$, $A$, and $B$ are finite groups. We use the following notation:

| | |
|---|---|
| $\mathbb{F}_q$ | Finite field with $q$ elements |
| $\mathbb{F}^\times$ | Multiplicative group of a field $\mathbb{F}$ |
| $O_r(G)$ | Largest normal $r$-subgroup of $G$ |
| $O_{r'}(G)$ | Largest normal $r'$-subgroup of $G$ |
| $A.B$ | Extension of $A$ by $B$ (normal subgroup $A$ with quotient $B$) |
| $A \rtimes B$ | Semidirect product of $A$ by $B$ (normal subgroup $A$ with complement $B$) |
| $A \circ B$ | Central product of $A$ and $B$ |
| $\delta_{i,j}$ | Kronecker delta |
| $S_N$ | Symmetric group of degree $N$ |
| $\mathrm{Sym}(\Omega)$ | Symmetric group on the set $\Omega$ |
| $C_n$ | Cyclic group of order $n$ |
| $\mathrm{Mat}_n(q)$ | Set of $n \times n$ matrices with entries in $\mathbb{F}_q$ |
| $\mathrm{GL}_n(q)$ | General linear group of degree $n$ over $\mathbb{F}_q$ |
| $X^g$ | Fixed point set $\{x \in X : gx = x\}$, for $g \in G$ with $G$ acting on $X$ |
| $X^G$ | Fixed point set $\{x \in X : gx = x$ for all $g \in G\}$, for $G$ acting on $X$ |

With notation such as $A.B$ and $A \rtimes B$ we do not specify the extension, so $G = A.B$ just means that $G$ has a normal subgroup $N \cong A$ with $G/N \cong B$.

Let $a \in \mathbb{Z}$. The Legendre symbol is defined by

$$\left(\frac{a}{r}\right) = \begin{cases} 0, & \text{if } a \equiv 0 \mod r. \\ +1, & \text{if } a \text{ is a square modulo } r \text{ and } a \not\equiv 0 \mod r. \\ -1, & \text{if } a \text{ is not a square modulo } r. \end{cases}$$

We denote by $\nu_r$ the $r$-adic valuation on the integers, so if $a \neq 0$, then $\nu_r(a)$ is the largest integer $k$ such that $r^k$ divides $a$.

By a *form* on a $\mathbb{F}_q$-vector space we will mean a bilinear form, sesquilinear form, or a quadratic form.

If $V$ is a $\mathbb{F}_q$-vector space equipped with a bilinear form $b$, for $W \subseteq V$ we denote by $W^\perp$ the subspace orthogonal to $W$. A subspace $W \subseteq V$ is *totally isotropic* if $W \subseteq W^\perp$, and *non-degenerate* if $W \cap W^\perp = 0$.

If $b$ is an alternating bilinear form, we will also call totally isotropic subspaces *totally singular*. If $V$ is equipped with a quadratic form $Q$, a totally isotropic subspace $W \subseteq V$ is *totally singular* if $Q(W) = 0$.

If $W$ and $W'$ are $\mathbb{F}_q$-vector spaces equipped with bilinear or sesquilinear forms $b$ and $b'$ respectively, then a *similarity* is a bijective linear map $g : W \to W'$ such that for some scalar $\lambda$, we have $b'(gv, gw) = \lambda b(v, w)$ for all $v, w \in W$. If $\lambda = 1$, then $g$ is an *isometry*.

Similarly if $W$ and $W'$ are $\mathbb{F}_q$-vector spaces equipped with quadratic forms $Q$ and $Q'$ respectively, a *similarity* is a bijective linear map $g : W \to W'$ such that for some scalar $\lambda$, we have $Q'(gv) = \lambda Q(v)$ for all $v \in W$. If $\lambda = 1$, then $g$ is an *isometry*.

Let $\kappa$ be a form on a $\mathbb{F}_q$-vector space $V$. Then we denote:

$$\Delta(V, \kappa) = \{g \in \mathrm{GL}(V) : g \text{ is a similarity for } \kappa\}$$

$$I(V, \kappa) = \{g \in \mathrm{GL}(V) : g \text{ is an isometry for } \kappa\}$$

Let $V$ and $V'$ be $\mathbb{F}_q$-vector spaces equipped with forms $\kappa$ and $\kappa'$, respectively. Assume that either both $\kappa$ and $\kappa'$ are bilinear, or that they are both quadratic forms. We say that $(V, \kappa)$ and $(V', \kappa')$ are *similar* if there exists a similarity $V \to V'$. If there exists an isometry $V \to V'$, we say that $(V, \kappa)$ and $(V', \kappa')$ are *isometric*.

Let $H \leq \Delta(V, \kappa)$ and $K \leq \Delta(V', \kappa')$. Then $H$ and $K$ are said to be *similar* if there exists a similarity $g : V \to V'$ such that $gHg^{-1} = K$. We say that $H$ and $K$ are *isometric* if there exists an isometry $g : V \to V'$ such that $gHg^{-1} = K$.

Let $V$ be a finite-dimensional vector space over $\mathbb{F}_q$ with $n = \dim V$. When $n$ is even, there are two types of quadratic forms on $V$ up to isometry, and the two types are distinguished by the dimension of a maximal totally singular subspace. For $n$ odd there is a unique quadratic form on $V$ up to a similarity. (See for example [55, Proposition 2.5.4].)

The type of a quadratic form $Q$ on $V$ is determined by the *signature*, which we define for $n$ even as

$$\mathrm{sgn}(Q) = \begin{cases} +, & \text{if a maximal totally singular subspace has dimension } n/2. \\ -, & \text{if a maximal totally singular subspace has dimension } n/2 - 1. \end{cases}$$

For $n$ odd, we define $\mathrm{sgn}(Q) = \circ$. Usually we will denote $\varepsilon = \mathrm{sgn}(Q)$, where $\varepsilon \in \{\circ, +, -\}$.

A bilinear form $b$ is *reflexive* if $b(v, w) = 0$ implies $b(w, v) = 0$. It is well-known that a reflexive bilinear form must always be symmetric or alternating. Then if $q$ is odd, for a reflexive bilinear form $b$ we define $\mathrm{sgn}(b) = +$ if $b$ is symmetric, and $\mathrm{sgn}(b) = -$ if $b$ is alternating.

If $b$ is a non-degenerate alternating bilinear form on $V$, we will denote

$$\Delta(V, b) = \mathrm{GSp}(V, b) = \mathrm{GSp}_n(q),$$

$$I(V, b) = \mathrm{Sp}(V, b) = \mathrm{Sp}_n(q).$$

If $Q$ is a non-degenerate quadratic form on $V$ with $\mathrm{sgn}(Q) = \varepsilon$, we denote

$$\Delta(V, Q) = \mathrm{GO}(V, Q) = \mathrm{GO}_n^{\varepsilon}(q),$$
$$I(V, Q) = O(V, Q) = O_n^{\varepsilon}(q).$$

In the case where $n$ is odd and $Q$ is a non-degenerate quadratic form on $V$, we will usually denote $\Delta(V, Q) = \mathrm{GO}_n(q)$ and $I(V, Q) = O_n(q)$.

If $b$ is a non-degenerate Hermitian form on $V$, we denote

$$\Delta(V, b) = \Delta U(V, b),$$
$$I(V, b) = \mathrm{GU}(V, b).$$

## 1.3 Reduction to Linear Groups

In this section we will prove Theorems 1.1.1 and 1.1.2, which reduce the classification of maximal solvable subgroups of $S_N$ to the classification of maximal irreducible solvable subgroups of $\mathrm{GL}_n(p)$, where $p$ is a prime and $n > 0$ is an integer.

For the most part, the proofs of these two results only need some basic facts from permutation group theory, as found in standard textbooks such as [81] and [18]. We will use the following terminology.

**Definition 1.3.1** For a transitive permutation group $G \leq \mathrm{Sym}(\Omega)$, a *system of imprimitivity* is a collection $\{B_1, \ldots, B_k\}$ of subsets of $\Omega$ with $k > 1$, such that $\Omega$ is a disjoint union

$$\Omega = B_1 \cup \cdots \cup B_k$$

and $G$ acts on $\{B_1, \ldots, B_k\}$. A system of imprimitivity is *trivial* if $|B_i| = 1$ for all $1 \leq i \leq k$. If $G$ has a nontrivial system of imprimitivity, we say that $G$ is *imprimitive*. Otherwise we say that $G$ is *primitive*.

**Definition 1.3.2** Suppose that $G \leq \mathrm{Sym}(\Omega)$ is imprimitive, with systems of imprimitivity $\Omega = B_1 \cup \cdots \cup B_k$ and $\Omega = C_1 \cup \cdots \cup C_\ell$. We say that $\{B_1, \ldots, B_k\}$ is a *refinement* of $\{C_1, \ldots, C_\ell\}$ if each $C_i$ is a union of some $B_j$'s. If $\{B_1, \ldots, B_k\}$ has no proper nontrivial refinement, we say that $\{B_1, \ldots, B_k\}$ is *nonrefinable*.

The following lemma is probably well known, and is essentially a part of Jordan's proof of Theorem 1.1.2 in [52], see [52, §10].

**Lemma 1.3.3** *Suppose that $G \leq \mathrm{Sym}(\Omega)$ is imprimitive of the form $G = H \wr K$, where $H \leq S_d$ is primitive, $K \leq S_k$ is transitive, and $d, k > 1$. Then $G$ has a unique nonrefinable system of imprimitivity.*

***Proof*** Let $\Omega = B_1 \cup \cdots \cup B_k$ be the system of imprimitivity defining $G$. We have $G = (H_1 \times \cdots \times H_k) \rtimes K$, where $H_i$ acts trivially on $B_j$ for $j \neq i$, and the action of $H_i$ on $B_i$ is isomorphic to $H$ as a permutation group.

Suppose that $\Omega = C_1 \cup \cdots \cup C_\ell$ is another nontrivial system of imprimitivity for $G$. We will show that $\{B_1, \ldots, B_k\}$ is a refinement of $\{C_1, \ldots, C_\ell\}$, which proves the lemma. Each element of $C_1$ is contained in some $B_i$, so without loss of generality we can assume that $C_1 \cap B_1 \neq \emptyset$.

Consider first the case where $C_1 \nsubseteq B_1$. Then there exists $y \in C_1 \cap B_j$ for some $j \neq 1$. For $g \in H_1$ we have $g(y) = y$, so in particular $g(C_1) = C_1$. Thus $H_1$ acts on $C_1$. Because $H_1$ is transitive on $B_1$ and $C_1 \cap B_1 \neq \emptyset$, it follows that $B_1 \subseteq C_1$. By the same argument, we have $B_i \subseteq C_1$ for any $i$ such that $C_1 \cap B_i \neq \emptyset$. Thus $C_1 = B_{i_1} \cup \cdots \cup B_{i_t}$ for some indices $i_1 < \cdots < i_t$, so $\{B_1, \ldots, B_k\}$ is a refinement of $\{C_1, \ldots, C_\ell\}$.

Suppose then that $C_1 \subseteq B_1$. Since $H_1$ acts on $\{C_1, \ldots, C_\ell\}$ and since $H_1$ is transitive on $B_1$, it follows that $B_1 = C_{j_1} \cup \cdots \cup C_{j_s}$ for some indices $j_1 < \cdots < j_s$. Because $\{C_1, \ldots, C_\ell\}$ is a nontrivial system of imprimitivity, by primitivity of $H_1$ we must have $s = 1$ and $B_1 = C_1$. In this case $\{B_1, \ldots, B_k\} = \{C_1, \ldots, C_\ell\}$, as required. □

Recall the following result, which goes back to the work of Galois [67]—see for example [38, II, Satz 3.2] and [18, Theorem 4.7A].

**Proposition 1.3.4** *Let $G \leq \mathrm{Sym}(\Omega)$ be primitive and solvable, where $|\Omega| = N > 1$. Then the following statements hold:*

*(i) $N = p^n$ for some prime $p$ and integer $n > 0$;*
*(ii) $G$ has a unique minimal normal subgroup $K$;*
*(iii) $K$ is transitive and regular, and $K \cong C_p^n$;*
*(iv) $G$ is a semidirect product $G = K \rtimes G_\omega$ for all $\omega \in \Omega$;*
*(v) All complements to $K$ in $G$ are conjugate in $G$;*
*(vi) As a permutation group $G$ is isomorphic to an affine group $V \rtimes X$, where $V = \mathbb{F}_p^n$, and $X \leq \mathrm{GL}_n(p)$ is irreducible and solvable.*

For the proof of Theorem 1.1.2, we will need the following observation from [52, p. 272].

**Lemma 1.3.5** *Let $G$ be a primitive solvable subgroup of $\mathrm{Sym}(\Omega)$. Then for all $g \in G \setminus \{1\}$, we have $|\Omega^g| \leq |\Omega|/2$.*

***Proof*** Since $G$ is primitive solvable, we have $|\Omega| = p^n$ for some prime $p$ and integer $n > 0$ (Proposition 1.3.4). Moreover, as a permutation group $G$ is isomorphic to $V \rtimes X \leq \mathrm{Sym}(V)$, where $V$ is an $n$-dimensional vector space over $\mathbb{F}_p$, and $X \leq \mathrm{GL}(V)$ is irreducible. We can identify $V \rtimes X$ as the set of affine linear transformations

$$\{\varphi_{A,w} : V \to V : A \in X \text{ and } w \in V\} \leq \mathrm{Sym}(V)$$

where $\varphi_{A,w}(v) = Av + w$ for all $v \in V$.

The set of fixed points for $\varphi_{A,w}$ is either empty, or equal to the affine subspace $v_0 + V^A$ for some $v_0 \in V$. We have $|v_0 + V^A| = p^{n'}$ where $n' = \dim(V^A)$, so

$$|\Omega^g| \leq |V^A| \leq p^{n-1} \leq |\Omega|/2$$

if $A \neq 1$. If $A = 1$, then $\varphi_{A,w}$ has no fixed points on $V$ unless $w = 0$, in which case $\varphi_{A,w}$ is the identity. □

***Proof of Theorem 1.1.1*** Let $\Omega = \{1, 2, \ldots, N\}$. Suppose first that $G$ is intransitive. Let

$$\Omega = \Omega_1 \cup \Omega_2 \cup \cdots \cup \Omega_t$$

be the decomposition of $\Omega$ into $G$-orbits, so $t \geq 2$. Set $n_i = |\Omega_i|$, so $N = n_1 + n_2 + \cdots + n_t$, where $n_i \geq 1$ for all $1 \leq i \leq t$. Then $G \leq G_1 \times G_2 \times \cdots \times G_t$ where $G_i \leq S_{n_i}$ is the action of $G$ on $\Omega_i$, so $G = G_1 \times G_2 \times \cdots \times G_t$ since $G$ is maximal solvable. Moreover, each $G_i$ must be maximal transitive solvable since $G$ is maximal solvable.

We next check that conditions (c)–(e) hold for $G$. To this end, suppose that there exists $i \neq j$ such that $G_i \cong G_j \wr S_k$ for some $1 \leq k \leq 3$. Then $G$ is not maximal solvable, since $\left(G_j \wr S_k\right) \times G_j \lneq G_j \wr S_{k+1}$. Therefore (c)–(e) must hold.

Next we consider the case where $G$ is transitive and imprimitive. Let $\Omega = B_1 \cup B_2 \cup \cdots \cup B_r$ be a system of imprimitivity for $G$, where $1 < r < N$. Then $G \leq N_G(B_1) \wr X$, where $N_G(B_1) \leq S_{N/r}$ is the stabilizer of the block $B_1$ in $G$, and $X \leq S_r$ is the action of $G$ on $\{B_1, B_2, \ldots, B_r\}$. Since $G$ is transitive and maximal solvable, we have $G = N_G(B_1) \wr X$ and both $N_G(B_1)$ and $X$ are maximal transitive solvable subgroups of $S_{N/r}$ and $S_r$, respectively.

Repeating this argument with $N_G(B_1)$ and $X$ and applying induction, it follows that $G = G_1 \wr G_2 \wr \cdots \wr G_t$, where $G_i \leq S_{n_i}$ is maximal primitive solvable ($n_i > 1$), and $N = n_1 n_2 \cdots n_t$. Solvable primitive groups only occur in prime power degree, so each $n_i$ is a prime power. Moreover if $(n_i, n_{i+1}) = (2, 2)$ for some $1 \leq i < t$, then $G$ is not maximal solvable, since $S_2 \wr S_2 \lneq S_4$.

It remains to consider the case where $G$ is transitive and primitive. This case follows from Proposition 1.3.4. □

**Remark 1.3.6** In Theorem 1.1.1, for groups of Type (I) it is possible that $n_i = 1$, but by (I.c) this holds for at most one $i$.

We also note that (I.d) excludes $G = S_2 \times S_1 = (S_1 \wr S_2) \times S_1$, which is not maximal solvable in $S_3$. Similarly (I.e) excludes $G = S_3 \times S_1 = (S_1 \wr S_3) \times S_1$, which is not maximal solvable in $S_4$. However, note that for example the point stabilizer $S_4 \times S_1$ is maximal in $S_5$, and thus in particular maximal solvable.

***Proof of Theorem 1.1.2*** By induction on $N = |\Omega|$. There is nothing to prove for $N = 1$, so assume that $N > 1$. If $G$ is not maximal solvable, then by Theorem 1.1.1 we have $G \lneq \overline{G} \leq S_N$, where $\overline{G}$ is of one of the types (I)–(III). We consider the possibilities for the types of $G$ and $\overline{G}$ in turn, and will mostly argue similarly to [52, §9–§14].

**Case 1: $G$ is of type (I)**
In this case $\Omega$ decomposes as a disjoint union of $G$-orbits

$$\Omega = \Omega_1 \cup \Omega_2 \cup \cdots \cup \Omega_t,$$

where $t \geq 2$. Moreover $G = G_1 \times G_2 \times \cdots \times G_t$, where $G_i \leq \mathrm{Sym}(\Omega_i)$ is maximal transitive solvable.

**Case 1.1: $\overline{G}$ is of type (I)**
We have $\overline{G} = \overline{G_1} \times \overline{G_2} \times \cdots \times \overline{G_s}$ and $\Omega = \overline{\Omega_1} \cup \overline{\Omega_2} \cup \cdots \cup \overline{\Omega_s}$ with $\overline{G_i} \leq \mathrm{Sym}(\overline{\Omega_i})$ maximal transitive solvable. Each $\overline{\Omega_i}$ is a union of some $G$-orbits, so there exist indices $1 \leq j_1 < \cdots < j_k \leq t$ such that $G_{j_1} \times G_{j_2} \times \cdots \times G_{j_k} \leq \overline{G_i}$. If $k > 1$, then by induction $G_{j_1} \times G_{j_2} \times \cdots \times G_{j_k}$ is maximal solvable, so $G_{j_1} \times G_{j_2} \times \cdots \times G_{j_k} = \overline{G_i}$, contradicting the transitivity of $\overline{G_i}$. Thus $k = 1$, and so we have $G_{j_1} = \overline{G_i}$, since each $G_j$ is maximal solvable. Since this holds for all $i$, we conclude $G = \overline{G}$, contrary to our assumption $G \lneq \overline{G}$.

**Case 1.2: $\overline{G}$ is of type (II)**
Here $\overline{G}$ can be written as a wreath product $\overline{G} = X \wr \Delta$, where $\Delta \leq S_e$ is primitive solvable, $X \leq S_d$ is transitive solvable, and $N = de$. Let $\Omega = \overline{B_1} \cup \cdots \cup \overline{B_e}$ be the corresponding system of imprimitivity, so now $\overline{G} = (X_1 \times \cdots \times X_e) \rtimes \Delta$, where the $X_i$ are isomorphic to $X$ as permutation groups.

Without loss of generality we may assume that $\Omega_1 \cap \overline{B_1} \neq \emptyset$. Suppose first that $\overline{B_1} \not\subseteq \Omega_1$. Then there exists some $i \neq 1$ such that $\Omega_i \cap \overline{B_1} \neq \emptyset$. Arguing as in the proof of Lemma 1.3.3 (paragraph 3) shows that $\Omega_i \subseteq \overline{B_1}$ for any $i$ such that $\Omega_i \cap \overline{B_1} \neq \emptyset$. Thus $\overline{B_1} = \Omega_{i_1} \cup \cdots \cup \Omega_{i_k}$ for some indices $i_1 < \cdots < i_k$, so $G_{i_1} \times \cdots \times G_{i_k}$ is contained in $X_1$, with $k \geq 2$. But this is a contradiction, since $G_{i_1} \times \cdots \times G_{i_k}$ is maximal solvable by induction, and $X_1$ is transitive solvable.

It follows then that $\overline{B_1} \subseteq \Omega_1$. The group $G$ acts on $\{\overline{B_1}, \ldots, \overline{B_e}\}$ since $\overline{G}$ does, so $\Omega_1 = \overline{B_{j_1}} \cup \cdots \cup \overline{B_{j_{\ell_1}}}$ for some $j_1 < \cdots < j_{\ell_1}$ and $\ell_1 \geq 1$. Since $G_1$ is maximal solvable, it follows that $G_1 = X \wr \Delta_1$ for some maximal transitive solvable $\Delta_1 \leq S_{\ell_1}$. For $i > 1$ the same arguments show that $\overline{B_i}$ is contained in some $\Omega_j$, and $G_i = X \wr \Delta_i$ for some maximal transitive solvable $\Delta_i \leq S_{\ell_i}$, where $\ell_i \geq 1$.

Thus we have $G = X \wr \Delta_1 \times \cdots \times X \wr \Delta_t$. Now $\Delta_1 \times \cdots \times \Delta_t$ is a type (I) subgroup of $S_e$ because $G$ is, so by induction it is maximal solvable. But this is a contradiction, since we have $\Delta_1 \times \cdots \times \Delta_t \leq \Delta$, and $\Delta$ is transitive solvable.

**Case 1.3: $\overline{G}$ is of type (III)**
Now $N = p^n$ for some prime $p$ and integer $n > 0$, and $\overline{G}$ is primitive. By Proposition 1.3.4, the group $\overline{G}$ has a unique minimal normal subgroup $\overline{K}$, with $\overline{K} \cong (C_p)^n$. Moreover for every $\omega \in \Omega$, we have $\overline{G} = \overline{K} \rtimes \overline{G_\omega}$, where $\overline{G_\omega} \leq \mathrm{GL}_n(p)$.

By Lemma 1.3.5, each $g \in \overline{G} \setminus \{1\}$ has at most $N/2$ fixed points. We first use this fact to show that we must have $t = 2$. To this end, note first that by property (I.c) at most one $G_i$ is trivial. Thus if $t > 2$, there exists some $1 \leq i \leq t$ such that $1 < |\Omega_i| < N/2$. But since $G_i$ fixes every point in $\Omega \setminus \Omega_i$, a nontrivial element of $G_i$ has $> N/2$ fixed points, which is a contradiction. Thus $t = 2$, and $G = G_1 \times G_2$.

If $G_i$ is nontrivial, the previous argument shows that $|\Omega_i| \geq N/2$. Thus if $G_1$ and $G_2$ are both nontrivial, we have $|\Omega_1| = |\Omega_2| = N/2$, and every nontrivial element of $G_i$ has no fixed points on $\Omega_i$. Furthermore in this case $N$ is even, so $p = 2$. Nontrivial groups of type (I) and (II) contain nontrivial elements with fixed points, so $G_1$ and $G_2$ are both primitive of type (III). In other words $G_i = (C_2)^{n-1} \rtimes X_i$ with $X_i \leq \mathrm{GL}_{n-1}(2)$ maximal irreducible solvable. But the point stabilizer $X_i$ of $G_i$ is trivial, so $\mathrm{GL}_{n-1}(2)$ is trivial. Therefore $p = n = 2$ and $G_1 \cong G_2 \cong S_2$ as permutation groups, contrary to (I.c).

It remains to consider the case where either $G_1$ or $G_2$ is trivial. Without loss of generality we assume that $G_2 = \{1\}$. Then $G_1$ cannot be maximal solvable of type (I) or (II). Indeed, otherwise $G_1$ would contain a nontrivial element with $\geq |\Omega_1|/2 = (N-1)/2$ fixed points on $\Omega_1$, and thus $G_1$ would contain a nontrivial element with $> N/2$ fixed points on $\Omega$.

Therefore $G_1$ is a primitive solvable subgroup of type (III) in $\mathrm{Sym}(\Omega_1)$, so $p^n - 1 = r^\ell$ for some prime $r$ and integer $\ell > 0$. Since $G_1$ fixes the point in $\Omega_2 = \{\omega\}$, we have $G_1 \leq \overline{G}_\omega \leq \mathrm{GL}_n(p)$. Now $G_1$ contains a minimal normal subgroup $K \cong (C_r)^\ell$ which is transitive and regular on $\Omega_1$ (Proposition 1.3.4).

Thus as a subgroup of $\mathrm{GL}_n(p)$, the minimal normal subgroup $K$ acts transitively on the nonzero vectors in $\mathbb{F}_p^n$. In particular $K$ is irreducible. Since $K$ is abelian, it follows from Schur's lemma that $K$ is a subgroup of the multiplicative group of a finite field, and in particular $K$ is cyclic. Therefore $\ell = 1$, and in this case $G_1 \leq S_r$ is the normalizer of a $r$-cycle and $|G_1| = r(r-1)$.

On the other hand $K$ is an irreducible cyclic subgroup of $\mathrm{GL}_n(p)$ with order $r = p^n - 1$, so it is generated by a Singer cycle; see [38, II, Satz 3.10] and [38, II, Satz 7.3]. Thus as a subgroup of $\mathrm{GL}_n(p)$, the group $G_1$ lies in the normalizer of a Singer cycle. The normalizer of a Singer cycle in $\mathrm{GL}_n(p)$ has order $n(p^n - 1)$ by [38, II, Satz 7.3 (a)], so it follows that $|G_1| = (p^n - 1)(p^n - 2) \leq n(p^n - 1)$ and thus

$$p^n - 2 \leq n.$$

By this inequality, one of the following holds:

- $n = 1$ and $p = 3, r = 2$;
- $n = 2$ and $p = 2, r = 3$.

In the first case $G = S_2 \times S_1$ and in the second case $G = S_3 \times S_1$. But these cases are excluded by (I.d) and (I.e), so we have a contradiction.

**Case 2: $G$ is of type (II)**
By induction, in this case $G$ can be written as a wreath product $G = \Gamma \wr \Delta$, where $\Delta \leq S_t$ is maximal transitive solvable, $\Gamma \leq S_d$ is maximal primitive solvable, and $N = dt$. Let $\Omega = B_1 \cup \cdots \cup B_t$ be the corresponding system of imprimitivity. Then $G = (\Gamma_1 \times \cdots \times \Gamma_t) \rtimes \Delta$, where $\Gamma_i \leq \mathrm{Sym}(B_i)$ is isomorphic to $\Gamma$ as a permutation group for all $1 \leq i \leq t$.

**Case 2.1: $\overline{G}$ is of type (II)**
In this case we can write $\overline{G} = \overline{\Gamma} \wr \overline{\Delta}$ for some $\overline{\Gamma}, \overline{\Delta} \neq \{1\}$ maximal transitive solvable, such that $\overline{\Gamma}$ is primitive. Let $\Omega = \overline{B_1} \cup \cdots \cup \overline{B_s}$ be the corresponding system of imprimitivity for $\overline{G}$, so $\overline{G} = (\overline{\Gamma_1} \times \cdots \times \overline{\Gamma_s}) \rtimes \overline{\Delta}$, where each $\overline{\Gamma_i} \leq \mathrm{Sym}(\overline{B_i})$ is isomorphic as a permutation group to $\overline{\Gamma}$.

Without loss of generality, we can assume that $B_1 \cap \overline{B_1} \neq \emptyset$. It follows from Lemma 1.3.3 that $\{B_1, \ldots, B_t\}$ is a refinement of $\{\overline{B_1}, \ldots, \overline{B_s}\}$. Thus $\overline{B_1} = B_{i_1} \cup \cdots \cup B_{i_k}$ for some $1 = i_1 < \cdots < i_k$, so $\Gamma_{i_1} \times \cdots \times \Gamma_{i_k} \leq \overline{\Gamma_1}$.

Suppose first that $k > 1$. Since $\overline{\Gamma_1}$ is primitive and $\Gamma \neq 1$, by applying Lemma 1.3.5 as in Case 1.3, it follows that $k = 2$ and $\Gamma \cong S_2$. Then $\overline{B_1} = B_{i_1} \cup B_{i_2}$, so $\Delta \leq S_2 \wr \Delta'$, where $\Delta'$ is the action of $G$ on $\{\overline{B_1}, \ldots, \overline{B_s}\}$. Because $\Delta$ is maximal solvable, it follows that $\Delta = S_2 \wr \Delta'$. But then $G = S_2 \wr S_2 \wr \Delta'$, which is excluded by (II.c).

Therefore $k = 1$, in which case $\overline{B_1} = B_1$. Then $\Gamma_1 \leq \overline{\Gamma_1}$, so by maximality of $\Gamma_1$ we have $\Gamma_1 = \overline{\Gamma_1}$. After rearranging the factors if necessary, it follows that $\Gamma_i = \overline{\Gamma_i}$ for all $1 \leq i \leq t$. Then $G = \Gamma \wr \Delta$ and $\overline{G} = \Gamma \wr \overline{\Delta}$, where $\Delta \leq \overline{\Delta}$. Since $\Delta$ is maximal solvable we have $\Delta = \overline{\Delta}$, so in fact $G = \overline{G}$, contrary to our assumption $G \lneq \overline{G}$.

**Case 2.2: $\overline{G}$ is of type (III)**
As in Case 1.3, we must have $t = 2$, as otherwise nontrivial elements of $\Gamma_i$ would have $> N/2$ fixed points on $\Omega$, contradicting Lemma 1.3.5. Thus $G = \Gamma \wr S_2 = (\Gamma_1 \times \Gamma_2) \rtimes S_2$. Since $\Gamma_1$ fixes the $N/2$ points in $B_2$, by Lemma 1.3.5 nontrivial elements of $\Gamma_1$ have no fixed points on $B_1$. As in Case 1.3, it follows that $\Gamma_1$ is primitive and $\Gamma_1 \cong S_2$. But then $G = S_2 \wr S_2$, which is excluded by (II.c).

**Case 3: $G$ is of type (III)**
In this case $N = p^n$ for some prime $p$ and integer $n > 0$. By Proposition 1.3.4, the group $G$ has a unique minimal normal subgroup $K$, which is a transitive regular elementary abelian subgroup isomorphic to $(C_p)^n$. Furthermore $G$ is a semidirect product $G = KG_\omega$ for every $\omega \in \Omega$, and $G_\omega$ is identified as a maximal irreducible solvable subgroup of $\mathrm{Aut}(K) = \mathrm{GL}_n(p)$.

Since $G$ is primitive, the same must be true for $\overline{G}$. Thus similarly by Proposition 1.3.4, the group $\overline{G}$ has a unique minimal normal subgroup $\overline{K}$, and $\overline{G} = \overline{K}\,\overline{G_\omega}$ with $\overline{G_\omega} \leq \mathrm{GL}_n(p)$.

We will show that $K = \overline{K}$. To this end, consider first the possibility that $K \cap \overline{K} = \{1\}$. If this is the case, then we can consider $K$ as a subgroup of $\mathrm{GL}_n(p)$. The number of nonzero vectors in $V = (\mathbb{F}_p)^n$ is $p^n - 1$ which is coprime to $p$, so $K$ fixes some nonzero vector in $V$. Now $G$ acts on $V^K$ since $K$ is a normal subgroup. On the other hand $G$ is primitive, so it must act irreducibly on $V$. Therefore $V^K = V$, which is a contradiction since $K \neq \{1\}$.

Therefore we must have $K \cap \overline{K} \neq \{1\}$, and so $K \leq \overline{K}$ since $K$ is minimal normal. But $|K| = |\overline{K}| = p^n$, so $K = \overline{K}$. In this case $G = KG_\omega$ and $\overline{G} = K\overline{G}_\omega$ with

$$G_\omega \leq \overline{G}_\omega \leq \mathrm{GL}_n(p).$$

Since $G_\omega$ is assumed to be maximal solvable in $\mathrm{GL}_n(p)$, we have $G_\omega = \overline{G}_\omega$. Then $G = \overline{G}$, which is a contradiction. □

We next describe conjugacy among the subgroups of $S_N$ described in Theorem 1.1.1. Since transitivity and primitivity is preserved by conjugacy, it suffices to do this for groups of the same type. Results similar to Propositions 1.3.8 and 1.3.9 below were also observed in [74, Theorem 2.1.4, Theorem 2.1.6].

**Proposition 1.3.7** *Suppose that $G = G_1 \times \cdots \times G_t$ and $\overline{G} = \overline{G_1} \times \cdots \times \overline{G_s}$ are of type (I) as in Theorem 1.1.1. Then $G$ and $\overline{G}$ are conjugate in $S_N$ if and only if all of the following hold:*

*(i) $t = s$;*
*(ii) There exists a permutation $\pi$ of $\{1, \ldots, t\}$ such that $G_i \cong \overline{G_{\pi(i)}}$ as permutation groups for all $1 \leq i \leq t$.*

***Proof*** Sufficiency is clear. For the other direction, suppose that $gGg^{-1} = \overline{G}$ for $g \in S_N$. Let $\{\Omega_1, \ldots, \Omega_t\}$ and $\{\overline{\Omega_1}, \ldots, \overline{\Omega_s}\}$ be the orbits of $G$ and $\overline{G}$, respectively. Then $\{g\Omega_1, \ldots, g\Omega_t\}$ are the orbits of $\overline{G}$, so $t = s$ and there exists a permutation $\pi$ of $\{1, \ldots, t\}$ such that $g\Omega_i = \overline{\Omega_{\pi(i)}}$ for all $1 \leq i \leq t$. It follows that $G_i \cong \overline{G_{\pi(i)}}$ as permutation groups for all $1 \leq i \leq t$. □

**Proposition 1.3.8** *Suppose that $G = G_1 \wr \cdots \wr G_t$ and $\overline{G} = \overline{G_1} \wr \cdots \wr \overline{G_s}$ are of type (II) as in Theorem 1.1.1. Then $G$ and $\overline{G}$ are conjugate in $S_N$ if and only if all of the following hold:*

*(i) $t = s$;*
*(ii) $G_i \cong \overline{G_i}$ as permutation groups for all $1 \leq i \leq t$.*

***Proof*** Sufficiency is straightforward. For the other direction, suppose that $gGg^{-1} = \overline{G}$ for some $g \in S_N$. Write $G = H \wr K$ with $H = G_1$ and $K = G_2 \wr \cdots \wr G_t$, and similarly $\overline{G} = \overline{H} \wr \overline{K}$ with $\overline{H} = \overline{G_1}$ and $\overline{K} = \overline{G_2} \wr \cdots \wr \overline{G_s}$.

Let $\{B_1, \ldots, B_k\}$ be the system of imprimitivity defining $G = H \wr K$, which is nonrefinable since $H$ is primitive. Then $\{gB_1, \ldots, gB_k\}$ is a nonrefinable system of imprimitivity for $\overline{G}$, so by Lemma 1.3.3 it is the system of imprimitivity defining $\overline{G} = \overline{H} \wr \overline{K}$.

The action of $N_G(B_1)$ on $B_1$ is isomorphic to $H$ as a permutation group, while the action of $N_{\overline{G}}(gB_1)$ on $gB_1$ is isomorphic to $\overline{H}$ as a permutation group. On the other hand $gN_G(B_1)g^{-1} = N_{\overline{G}}(gB_1)$, so $H \cong \overline{H}$ as permutation groups.

The action of $K$ on $\{B_1, \ldots, B_k\}$ is faithful and isomorphic as a permutation group to the action of $gKg^{-1}$ on $\{gB_1, \ldots, gB_k\}$. It follows that $K \cong gKg^{-1} \cong \overline{K}$

as permutation groups. If $t = 2$ we are done, and for $t > 2$ the result follows by induction on $t$. □

**Proposition 1.3.9** *Suppose that* $N = p^n$ *and that* $G = (C_p)^n \rtimes X$ *and* $\overline{G} = (C_p)^n \rtimes \overline{X}$ *are of type (III) as in Theorem 1.1.1. Then* $G$ *and* $\overline{G}$ *are conjugate in* $S_N$ *if and only if* $X$ *and* $\overline{X}$ *are conjugate in* $\mathrm{GL}_n(p)$.

***Proof*** Follows from Proposition 1.3.4 (ii) and (v). □

With Theorems 1.1.1–1.1.2 and Propositions 1.3.7–1.3.9, the problem of classifying maximal solvable subgroups of $S_N$ is completely reduced to the problem of classifying maximal irreducible solvable subgroups of $\mathrm{GL}_n(p)$.

**Example 1.3.10** We illustrate Theorems 1.1.1 and 1.1.2 in Table 1.1, where we list all maximal solvable subgroups of $S_n$ up to conjugacy, for $5 \le n \le 10$. In the table, we use the notation

$$\mathrm{AGL}_k(p) := \mathbb{F}_p^k \rtimes \mathrm{GL}_k(p)$$

for the affine permutation group corresponding to $\mathrm{GL}_k(p)$.

The group $\Gamma\mathrm{L}_1(2^3)$ that appears in case $n = 8$ is the normalizer of a Singer cycle in $\mathrm{GL}_3(2)$. It is not too difficult to see that this is the unique maximal irreducible solvable subgroup of $\mathrm{GL}_3(2)$ up to conjugacy; this fact will also follow from results proven in later sections. For maximal transitive solvable subgroups, we give more examples in Sect. 8.1, Table 8.1.

For $1 \le n \le 4$ the symmetric group $S_n$ is solvable, so in these small cases the only maximal solvable subgroup of $S_n$ is the group itself, which is primitive. In terms of the classification in Theorems 1.1.1 and 1.1.2, we have

$$S_2 = \mathbb{F}_2 \rtimes \mathrm{GL}_1(2) = \mathrm{AGL}_1(2),$$

$$S_3 = \mathbb{F}_3 \rtimes \mathrm{GL}_1(3) = \mathrm{AGL}_1(3),$$

$$S_4 = \mathbb{F}_2^2 \rtimes \mathrm{GL}_2(2) = \mathrm{AGL}_2(2),$$

as permutation groups.

**Example 1.3.11** Suppose that $n > 1$ is squarefree, so $n = p_1 \dots p_t$ with $p_1, \dots,$ $p_t$ distinct primes. For each $1 \le i \le t$, the permutation group

$$X_i := \mathbb{F}_{p_i} \rtimes \mathrm{GL}_1(p_i) = \mathrm{AGL}_1(p_i)$$

is the unique maximal primitive solvable subgroup of $S_{p_i}$. By Theorems 1.1.1 and 1.1.2, the maximal transitive solvable subgroups of $S_n$ are precisely the subgroups of the form

$$X_{p_{\pi(1)}} \wr \cdots \wr X_{p_{\pi(t)}},$$

**Table 1.1** Maximal solvable subgroups of $S_n$ for $5 \leq n \leq 10$, up to conjugacy in $S_n$ (See Example 1.3.10)

| $n$ | $X$ | $\|X\|$ | Type |
|---|---|---|---|
| 5 | $AGL_1(5)$ | 20 | Primitive |
| | $S_4 \times S_1$ | 24 | Intransitive |
| | $S_3 \times S_2$ | 12 | Intransitive |
| 6 | $S_3 \wr S_2$ | 72 | Imprimitive |
| | $S_2 \wr S_3$ | 48 | Imprimitive |
| | $AGL_1(5) \times S_1$ | 20 | Intransitive |
| | $S_4 \times S_2$ | 48 | Intransitive |
| 7 | $AGL_1(7)$ | 42 | Primitive |
| | $(S_3 \wr S_2) \times S_1$ | 72 | Intransitive |
| | $(S_2 \wr S_3) \times S_1$ | 48 | Intransitive |
| | $AGL_1(5) \times S_2$ | 40 | Intransitive |
| | $S_4 \times S_3$ | 144 | Intransitive |
| 8 | $\mathbb{F}_2^3 \rtimes \Gamma L_1(2^3)$ | 168 | Primitive |
| | $S_4 \wr S_2$ | 1152 | Imprimitive |
| | $S_2 \wr S_4$ | 384 | Imprimitive |
| | $(S_3 \wr S_2) \times S_2$ | 144 | Intransitive |
| | $AGL_1(5) \times S_3$ | 120 | Intransitive |
| | $AGL_1(7) \times S_1$ | 42 | Intransitive |
| 9 | $AGL_2(3)$ | 432 | Primitive |
| | $S_3 \wr S_3$ | 1296 | Imprimitive |
| | $(\mathbb{F}_2^3 \rtimes \Gamma L_1(2^3)) \times S_1$ | 168 | Intransitive |
| | $(S_4 \wr S_2) \times S_1$ | 1152 | Intransitive |
| | $(S_2 \wr S_4) \times S_1$ | 384 | Intransitive |
| | $AGL_1(7) \times S_2$ | 84 | Intransitive |
| | $(S_2 \wr S_3) \times S_3$ | 288 | Intransitive |
| | $AGL_1(5) \times S_4$ | 480 | Intransitive |
| | $S_4 \times S_3 \times S_2$ | 288 | Intransitive |
| 10 | $AGL_1(5) \wr S_2$ | 800 | Imprimitive |
| | $S_2 \wr AGL_1(5)$ | 640 | Imprimitive |
| | $AGL_2(3) \times S_1$ | 432 | Intransitive |
| | $(S_3 \wr S_3) \times S_1$ | 1296 | Intransitive |
| | $AGL_1(5) \times S_4 \times S_1$ | 480 | Intransitive |
| | $(\mathbb{F}_2^3 \rtimes \Gamma L_1(2^3)) \times S_2$ | 336 | Intransitive |
| | $(S_4 \wr S_2) \times S_2$ | 2304 | Intransitive |
| | $(S_2 \wr S_4) \times S_2$ | 768 | Intransitive |
| | $AGL_1(5) \times S_3 \times S_2$ | 240 | Intransitive |
| | $AGL_1(7) \times S_3$ | 252 | Intransitive |
| | $(S_3 \wr S_2) \times S_4$ | 1728 | Intransitive |
| | $(S_2 \wr S_3) \times S_4$ | 1152 | Intransitive |

where $\pi$ is some permutation of $\{1, \ldots, t\}$. In particular, in this case $S_n$ has exactly $t!$ maximal transitive solvable subgroups, up to conjugacy (Propositions 1.3.8–1.3.9). This example was also observed by Jordan in [48, Table B, p. 288] and by Suprunenko in [76, Example 4(b), pp. 49–50].

**Example 1.3.12** Suppose that $n > 1$ is of the form $n = 4p_1 \cdots p_t$, with $p_1$, ..., $p_t$ distinct odd primes. Then it follows from Theorems 1.1.1, 1.1.2, and Propositions 1.3.8–1.3.9 that $S_n$ has exactly $(t+2)!/2$ maximal transitive solvable subgroups, up to conjugacy. (This is also stated by Jordan in [48, Table B, p. 288].)

For example in the case where $t = 2$ with $n = 4p_1p_2$, representatives for the 12 conjugacy classes of maximal transitive solvable subgroups are given by

$$
\begin{array}{ccc}
S_4 \wr X \wr Y & X \wr S_4 \wr Y & X \wr Y \wr S_4 \\
S_2 \wr X \wr S_2 \wr Y & S_2 \wr X \wr Y \wr S_2 & X \wr S_2 \wr Y \wr S_2
\end{array}
$$

where $\{X, Y\} = \{\mathrm{AGL}_1(p_1), \mathrm{AGL}_1(p_2)\}$.

# Chapter 2
# Basic Structure of Maximal Irreducible Solvable Subgroups

In this chapter, we will begin to consider the classification of maximal irreducible solvable subgroups of the classical groups $\mathrm{GL}_n(q)$, $\mathrm{GSp}_n(q)$, and $\mathrm{GO}_n^{\varepsilon}(q)$, where $q$ is a power of some prime $p$.

We start with Sect. 2.1, where we will fix the basic notation used for the rest of the book, and provide some preliminary results. Then analogously to the case of symmetric groups, we will split the analysis of maximal irreducible solvable subgroups into two cases: the *metrically imprimitive* case, and the *metrically primitive* case. Here metrically imprimitive means that the group preserves some orthogonal decomposition $\mathbb{F}_q^n = W_1 \perp \cdots \perp W_t$ with $t > 1$; otherwise the group is metrically primitive—see the definitions in Sect. 2.2.

In the case of $\mathrm{GL}_n(q)$, the terminology of metrically imprimitive and metrically primitive agrees with the usual meaning of imprimitive and primitive for matrix groups. We will see in the case of $\mathrm{GL}_n(q)$ that an imprimitive maximal irreducible solvable subgroup can be written as a wreath product $H \wr K$, for $H \leq \mathrm{GL}_d(q)$ primitive maximal irreducible solvable and $K \leq S_k$ is maximal transitive solvable, where $n = dk$ with $k > 1$.

Since maximal transitive solvable subgroups of $S_k$ are described in terms of maximal irreducible solvable subgroups of general linear groups of degree $< n$ (Theorem 1.1.1), it follows that maximal irreducible solvable subgroups of $\mathrm{GL}_n(q)$ can be constructed in terms of primitive maximal irreducible solvable subgroups. In the case of $\mathrm{GSp}_n(q)$ and $\mathrm{GO}_n^{\varepsilon}(q)$, we will see similarly that all maximal irreducible solvable subgroups can be constructed in terms of metrically primitive maximal irreducible solvable subgroups.

Not all groups given by these constructions are maximal solvable—for example, $\mathrm{GL}_1(q) \wr S_2$ is maximal irreducible solvable in $\mathrm{GL}_2(q)$ if and only if $q = 4$ or $q > 5$. Later in Chap. 7, we will complete the classification of metrically imprimitive maximal irreducible solvable subgroups.

For metrically primitive maximal irreducible solvable subgroups, we start their analysis in Sect. 2.4 of this chapter by considering a maximal abelian normal



M. Korhonen, *Maximal Solvable Subgroups of Finite Classical Groups*, Lecture Notes in Mathematics 2346, https://doi.org/10.1007/978-3-031-62915-0_2

subgroup $F_0$ of isometries. In Sects. 2.4–2.8, we identify the possible structures of $F_0$, and describe generators for its normalizer in the corresponding classical group. Later in Chap. 4, we will give a more detailed description of the general structure of metrically primitive maximal irreducible solvable subgroups. The classification of metrically primitive maximal irreducible solvable subgroups will be completed in Chap. 7.

## 2.1 Construction of Maximal Irreducible Solvable Groups

In this section, we will establish the basic notation used for the rest of the book, and some preliminary results.

### 2.1.1 General Setup and Notation

We will use notation similar to [55], and the general setup that we use throughout the book is as follows. Let $V$ be a finite-dimensional vector space of dimension $n$ over $\mathbb{F}_q$, equipped with a form $\kappa$. We assume that $\kappa$ is a form of one of the following types.

- $\kappa = 0$ (zero form), in which case $\Delta(V, \kappa) = I(V, \kappa) = \mathrm{GL}_n(q)$.
- $\kappa$ is a non-degenerate alternating bilinear form ($n$ even), in which case $\Delta(V, \kappa) = \mathrm{GSp}_n(q)$ and $I(V, \kappa) = \mathrm{Sp}_n(q)$.
- $\kappa$ is a non-degenerate quadratic form with signature $\varepsilon$, in which case $\Delta(V, \kappa) = \mathrm{GO}_n^{\varepsilon}(q)$ and $I(V, \kappa) = O_n^{\varepsilon}(q)$.

If $\kappa$ is nonzero, we define $e \in \{0, 1\}$ by

$$e = \begin{cases} 0, & \text{if } \kappa \text{ is a quadratic form.} \\ 1, & \text{if } \kappa \text{ is an alternating bilinear form.} \end{cases}$$

For the purposes of this text we are interested in irreducible solvable subgroups of $\Delta(V, \kappa)$, so we will make the following assumptions.

- If $\kappa$ is an alternating bilinear form, we assume that $q$ is odd. For $q$ even, the irreducible solvable subgroups of $\mathrm{GSp}_n(q)$ are contained in $\mathrm{GO}_n^+(q)$ or $\mathrm{GO}_n^-(q)$ by a result of Willems (Theorem 2.1.14). Furthermore, in this case maximal irreducible solvable subgroups of $\mathrm{GO}_n^{\pm}(q)$ are maximal solvable in $\mathrm{GSp}_n(q)$ (Remark 2.3.6).
- If $\kappa$ is a quadratic form and $n$ is odd, we assume that $q$ is odd. If $q$ is even and $n$ is odd, then $\Delta(V, \kappa)$ is not irreducible on $V$, and furthermore $O_n(q) \cong \mathrm{Sp}_{n-1}(q)$ [79, §11, p. 143].

Throughout we will denote by $b$ the bilinear form corresponding to $\kappa$. That is, if $\kappa = 0$ or if $\kappa$ is an alternating bilinear form, we denote $b = \kappa$. In the case where $\kappa$ is a quadratic form, we denote $Q = \kappa$ and denote by $b$ the polarization of $Q$, so

$$b(v, w) = Q(v + w) - Q(v) - Q(w)$$

for all $v, w \in V$.

For $G \leq \Delta(V, \kappa)$, we will use the notation

$$G^\circ := G \cap I(V, \kappa).$$

Note that $G = G^\circ$ if $\kappa$ is the zero form.

If $\kappa \neq 0$, we have the homomorphism $\tau : \Delta(V, \kappa) \to \mathbb{F}_q^\times$, where

$$b(gv, gw) = \tau(g)b(v, w)$$

for all $g \in \Delta(V, \kappa)$ and $v, w \in V$. When $\kappa = Q$ is a quadratic form, we have $Q(gv) = \tau(g)Q(v)$ for all $g \in \Delta(V, \kappa)$ and $v \in V$.

For a quadratic form $\kappa = Q$, we define $\Omega(V, Q) = \Omega_n^\varepsilon(q)$ as follows. If $q$ is odd, then $\Omega(V, Q)$ is defined as the kernel of the spinor norm on $\mathrm{SO}_n^\varepsilon(q)$ [79, §11, p. 163]. In the case where $n$ and $q$ are both even, we have

$$\Omega(V, Q) = \{g \in I(V, Q) : \dim(V^g) \equiv 0 \mod 2\},$$

see [20] or [79, Theorem 11.43, p. 160]. When $\kappa$ is a quadratic form, for $G \leq \Delta(V, \kappa)$ we will use the notation $G^\Omega := G \cap \Omega(V, \kappa)$.

**Definition 2.1.1** Suppose that $\kappa \neq 0$. We say that $G \leq \Delta(V, \kappa)$ is of *multiplier d* if $[\mathbb{F}_q^\times : \tau(G)] = d$.

Note that a maximal irreducible solvable subgroup of $\Delta(V, \kappa)$ has multiplier 1 or 2, since it must contain the group $Z$ of scalar matrices, and $\tau(Z) = \left(\mathbb{F}_q^\times\right)^2$.

For the rest of this section, we will collect some preliminary results which will be needed throughout the text.

### 2.1.2 Generalities on Maximal Solvable Subgroups

Our main results will give a complete classification of maximal irreducible solvable subgroups of $\Delta(V, \kappa)$. In order to do this, it turns out that more generally we need to classify maximal irreducible solvable subgroups of any group $X$ with $I(V, \kappa) \leq X \leq \Delta(V, \kappa)$. By the next two lemmas, it will suffice to consider the cases $X = I(V, \kappa)$ and $X = \Delta(V, \kappa)$.

**Lemma 2.1.2** *Suppose that $Z \leq Z(\mathrm{GL}(V))$, so that $Z$ consists of some scalar matrices in* GL($V$). *Let $X \leq \mathrm{GL}(V)$. Then the following statements hold.*

(i) *If $M$ is a maximal solvable subgroup of $X$, then $MZ$ is a maximal solvable subgroup of $XZ$ and $M = MZ \cap X$.*
(ii) *If $M'$ is a maximal solvable subgroup of $XZ$, then $M' \cap X$ is maximal solvable in $X$ and $M' = (M' \cap X)Z$.*

***Proof*** Straightforward. □

**Lemma 2.1.3** *Suppose that $\kappa \neq 0$, and let $Z \leq \mathrm{GL}(V)$ be the group of scalar matrices. Let $I(V, \kappa) \leq X \leq \Delta(V, \kappa)$. Then the following statements hold.*

(i) *If $\tau(X) \leq (\mathbb{F}_q^\times)^2$, then $M \leq X$ is maximal solvable if and only if $M = M'(X \cap Z)$, where $M'$ is maximal solvable in $I(V, \kappa)$.*
(ii) *If $\tau(X) \not\leq (\mathbb{F}_q^\times)^2$, then $M \leq X$ is maximal solvable if and only if $M = M' \cap X$, where $M'$ is maximal solvable in $\Delta(V, \kappa)$.*

***Proof*** In case (i) we have $X = I(V, \kappa)(Z \cap X)$, and in case (ii) we have $\Delta(V, \kappa) = XZ$. Thus the result follows from Lemma 2.1.2. □

**Lemma 2.1.4** *Let $G$ be a finite group and let $X \leq G$ be a maximal solvable subgroup. If $X$ is abelian, then $X = G$.*

***Proof*** Let $G$ be a minimal counterexample. If $X \neq G$, then there exists a maximal subgroup $M < G$ containing $X$. By minimality of $G$, we must have $X = M$. But then $M$ is a maximal subgroup of $G$ which is abelian, so $G$ is solvable by a result of Herstein [34]. This contradicts the fact that $X$ is maximal solvable, so the lemma follows. □

**Lemma 2.1.5** *Let $G$ be a finite group and let $X \leq G$ be a maximal solvable subgroup. Suppose that $K \trianglelefteq X$ is such that $X/K$ is abelian. Then $X = N_G(K)$.*

***Proof*** Since $X/K$ is abelian and a maximal solvable subgroup of $N_G(K)/K$, it follows from Lemma 2.1.4 that $X = N_G(K)$. □

**Lemma 2.1.6** *Let $G \leq \Delta(V, \kappa)$ be maximal solvable. Then $\langle -I_V \rangle \lneq G^\circ$, except possibly when* dim $V = 1$.

***Proof*** Suppose that $G^\circ = \{\pm I_V\}$. Then $G/G^\circ = G/\{\pm I_V\}$ is cyclic and a maximal solvable subgroup of $\Delta(V, \kappa)/\{\pm I_V\}$, so $G = \Delta(V, \kappa)$ by Lemma 2.1.4. In particular $\Delta(V, \kappa)$ is solvable and $I(V, \kappa) = \{\pm I_V\}$. This can only happen when dim $V = 1$. □

**Lemma 2.1.7** *Let $M \leq I(V, \kappa)$ be maximal solvable. Then the following statements hold:*

(i) *There exists a unique maximal solvable $G \leq \Delta(V, \kappa)$ such that $M \leq G$.*
(ii) *For $G$ as in (i), we have $G^\circ = M$ and $G = N_{\Delta(V,\kappa)}(M)$.*

***Proof*** For (i) the existence part is clear, and uniqueness follows once we prove (ii). For claim (ii), let $G \leq \Delta(V, \kappa)$ be maximal solvable such that $M \leq G$. Then

$M \leq G^\circ$, so by maximality of $M$ we have $M = G^\circ$. Because $G/G^\circ = G/M$ is cyclic, it follows from Lemma 2.1.5 that $G = N_{\Delta(V,\kappa)}(M)$, as claimed. □

**Lemma 2.1.8** *Suppose that $n$ and $q$ are even and that $\kappa$ is a quadratic form. Let $M \leq \Omega(V, \kappa)$ be maximal solvable. Then the following statements hold:*

*(i) There exists a unique maximal solvable $G \leq I(V, \kappa)$ such that $M \leq G$.*
*(ii) For $G$ as in (i), we have $G^\Omega = M$ and $G = N_{I(V,\kappa)}(M)$.*

***Proof*** Since $I(V, \kappa)/\Omega(V, \kappa)$ is cyclic, the result follows with the same proof as Lemma 2.1.7. □

**Remark 2.1.9** By Lemma 2.1.7, every maximal solvable subgroup of $I(V, \kappa)$ is of the form $X^\circ$ for a unique maximal solvable $X \leq \Delta(V, \kappa)$. The converse of this fails. If $X \leq \Delta(V, \kappa)$ is maximal solvable, it is not necessarily the case that $X^\circ$ is maximal solvable in $I(V, \kappa)$. We will give examples of this behaviour in Sect. 5.5.

Similarly when $n$ and $q$ are even and $\kappa$ is a quadratic form, there are examples where $X \leq I(V, \kappa)$ is maximal solvable, but $X^\Omega$ is not maximal solvable in $\Omega(V, \kappa)$ (Remark 5.2.5).

**Lemma 2.1.10** *Let $M \leq \Delta(V, \kappa)$, and denote $\mathscr{C} := \{(xMx^{-1})^\circ : x \in \Delta(V, \kappa)\}$. Then the following statements hold:*

*(i) If $N_{\Delta(V,\kappa)}(M)$ is of multiplier 1, then $\mathscr{C}$ forms a single $I(V, \kappa)$-conjugacy class under conjugation.*
*(ii) If $N_{\Delta(V,\kappa)}(M)$ is of multiplier 2, then $\mathscr{C}$ splits into two $I(V, \kappa)$-conjugacy classes.*
*(iii) Let $x \in \Delta(V, \kappa)$ be such that $\tau(x)$ is nonsquare in $\mathbb{F}_q^\times$. Then in (ii), representatives for the two $I(V, \kappa)$-conjugacy classes are given by $M^\circ$ and $(xMx^{-1})^\circ$.*

***Proof*** Straightforward. □

**Lemma 2.1.11** *Assume that $n$ and $q$ are even, and that $\kappa$ is a quadratic form. Let $M \leq \Delta(V, \kappa)$, and denote $\mathscr{C} := \{(xMx^{-1})^\Omega : x \in \Delta(V, \kappa)\}$. Then the following statements hold:*

*(i) If $N_{I(V,\kappa)}(M) \not\leq \Omega(V, \kappa)$, then $\mathscr{C}$ forms a single $\Omega(V, \kappa)$-conjugacy class under conjugation.*
*(ii) If $N_{I(V,\kappa)}(M) \leq \Omega(V, \kappa)$, then $\mathscr{C}$ splits into two $\Omega(V, \kappa)$-conjugacy classes.*
*(iii) Let $x \in I(V, \kappa) \setminus \Omega(V, \kappa)$. Then in (ii), representatives for the two $\Omega(V, \kappa)$-conjugacy classes are given by $M^\circ$ and $(xMx^{-1})^\circ$.*

***Proof*** Straightforward. □

**Lemma 2.1.12** *Let $G \leq \Delta(V, \kappa)$ be maximal solvable. Suppose that all of the following hold:*

*(i) $G$ is of multiplier 1.*
*(ii) $G^\circ$ is similar to $H^\circ \leq I(Z, \kappa')$.*
*(iii) $H^\circ \leq H$, where $H$ is of multiplier 1.*

*Then $G$ is similar to $H$.*

***Proof*** By (ii), there exists a similarity $\varphi : Z \to W$ such that $\varphi H^\circ \varphi^{-1} = G^\circ$. We have

$$\varphi H \varphi^{-1} \leq \varphi N_{\Delta(Z,\kappa')}(H^\circ)\varphi^{-1} = N_{\Delta(W,\kappa)}(\varphi H^\circ \varphi^{-1}) = N_{\Delta(W,\kappa)}(G^\circ) = G,$$

where the last equality holds by Lemma 2.1.5. Therefore $G^\circ \leq \varphi H \varphi^{-1} \leq G$. Since $H$ is of multiplier 1, the same is true for $\varphi H \varphi^{-1}$, so $\varphi H \varphi^{-1} = G$ and we conclude that $G$ is similar to $H$. □

### 2.1.3 Representation Theory

The following result in Clifford theory is a small generalization of [84, Proposition 5].

**Proposition 2.1.13** *Let $G \leq \Delta(V, \kappa)$ be irreducible and $H \trianglelefteq G$. Then $V \downarrow H$ is completely reducible, and one of the following statements holds:*

*(i) All the homogeneous components in $V \downarrow H$ are non-degenerate subspaces, and:*

- *(a) $V = W_1 \perp \cdots \perp W_k$, where $W_1, \ldots, W_k$ are the homogeneous components of $H$ on $V$;*
- *(b) $G$ acts transitively on $\{W_1, \ldots, W_k\}$.*

*(ii) All the homogeneous components in $V \downarrow H$ are totally isotropic subspaces, and:*

- *(a) $V = (W_1 \oplus W_2) \perp \cdots \perp (W_{2k-1} \oplus W_{2k})$ where $W_1, \ldots, W_{2k}$ are the homogeneous components of $H$ on $V$;*
- *(b) If $H \leq I(V, \kappa)$, then $W_{2i} \cong W_{2i-1}^*$ as $\mathbb{F}_q[H]$-modules for all $1 \leq i \leq k$;*
- *(c) $G$ acts transitively on $\{W_1, \ldots, W_{2k}\}$ and $\{W_{2i-1} \oplus W_{2i} : 1 \leq i \leq k\}$.*

***Proof*** By Clifford's theorem $V \downarrow H$ is completely reducible. Let $V \downarrow H = W_1 \oplus \cdots \oplus W_r$, where $W_1, \ldots, W_r$ are the homogeneous components of $H$ on $V$. Again by Clifford's theorem $G$ acts on $\{W_1, \ldots, W_r\}$ and thus also on the set $\{W_1^\perp, \ldots, W_r^\perp\}$. Hence

$$\bigoplus_{i=1}^{r} W_i \cap W_i^\perp$$

is an $\mathbb{F}_q[G]$-submodule of $V$. Because $G$ is irreducible, either $W_i \cap W_i^\perp = 0$ for all $1 \leq i \leq r$, or $W_i \subseteq W_i^\perp$ for all $1 \leq i \leq r$.

If $W_i \cap W_i^{\perp} = 0$ for all $1 \leq i \leq r$, then all the homogeneous components are non-degenerate subspaces, and $V \downarrow H = W_i \perp W_i^{\perp}$. By uniqueness of the homogeneous components, we have $W_i^{\perp} = \bigoplus_{j \neq i} W_j$. From this it follows that $V \downarrow H = W_1 \perp \cdots \perp W_r$, so (i) holds.

If $W_i \subseteq W_i^{\perp}$ for all $1 \leq i \leq r$, then all the homogeneous components are totally isotropic subspaces. Since $V \downarrow H$ is completely reducible, we have $V \downarrow H = W_i^{\perp} \oplus M$ for some $\mathbb{F}_q[H]$-submodule $M$.

We claim that $M$ is equal to some homogeneous component of $H$. To this end, we have a natural isomorphism of vector spaces $\varphi : V/W_i^{\perp} \to W_i^*$ defined by $\varphi(\overline{v})(w) = b(v, w)$ for all $v \in V$ and $w \in W_i$. It is clear that for all $g \in H$ we have $\varphi(g\overline{v}) = \tau(g)g\varphi(\overline{v})$ for all $\overline{v} \in V/W_i^{\perp}$. Therefore $\varphi$ provides an isomorphism of $\mathbb{F}_q[H]$-modules

$$V/W_i^{\perp} \cong Z \otimes_{\mathbb{F}_q} W_i^*,$$

where $Z$ is the 1-dimensional $\mathbb{F}_q[H]$-module corresponding to the homomorphism $\tau_{|H} : H \to \mathbb{F}_q^{\times}$. Since $W_i$ is homogeneous, it follows then that $V/W_i^{\perp} \cong M$ is a homogeneous $\mathbb{F}_q[H]$-module as well. Because $G$ acts transitively on the homogeneous components, all the homogeneous components have the same dimension—thus from $\dim M = \dim W_i$ we conclude that $M = W_j$ for some $j$.

We have shown that for each homogeneous component $W_i$ there exists a unique homogeneous component $W_j$ such that $b(W_i, W_j) \neq 0$, and in this case $V = W_i^{\perp} \oplus W_j$. Thus the number of homogeneous components is even, say $r = 2k$, and after relabeling we have $V \downarrow H = (W_1 \oplus W_2) \perp \cdots \perp (W_{2k-1} \oplus W_{2k})$. It is clear that $G$ must act on the set of pairs $\{W_{2i-1}, W_{2i}\}_{1 \leq i \leq k}$, and the action is transitive since $G$ is irreducible. Finally if $H \leq I(V, \kappa)$, then $\tau(g) = 1$ for all $g \in H$, so $W_{2i} \cong V/W_{2i-1}^{\perp} \cong Z \otimes_{\mathbb{F}_q} W_{2i-1}^* \cong W_{2i-1}^*$. We have shown that all of the claims in (ii) hold, which completes the proof of the proposition. □

The following theorem is essentially due to Willems.

**Theorem 2.1.14** *Suppose that $q$ is even. Assume that $G \leq \mathrm{GSp}_{2n}(q)$ is irreducible and solvable. Then $G$ is conjugate to a subgroup of $\mathrm{GO}_{2n}^+(q)$ or $\mathrm{GO}_{2n}^-(q)$.*

***Proof*** Let $Z$ be the subgroup of scalar matrices in $\mathrm{GSp}_{2n}(q)$. Since $q$ is even, we have $\tau(Z) = \mathbb{F}_q^{\times}$ and thus $G \leq G^{\circ}Z$. Therefore the subgroup $G^{\circ}$ is irreducible, and it follows from a theorem of Willems [82, Satz 2.8] that $G^{\circ}$ is conjugate to a subgroup of $\mathrm{O}_{2n}^{\varepsilon}(q)$, for some $\varepsilon \in \{+, -\}$. Since $Z \leq \mathrm{GO}_{2n}^{\pm}(q)$, we conclude that $G \leq \mathrm{GO}_{2n}^{\varepsilon}(q)$. □

**Lemma 2.1.15** *Let $G$ be a group and $\mathbb{F}$ a field. Suppose that $V$ is a completely reducible $\mathbb{F}[G]$-module such that $V = W_1 \oplus \cdots \oplus W_k$, where $W_1, \ldots, W_k$ are irreducible and pairwise nonisomorphic $\mathbb{F}[G]$-modules. Then any nonzero $\mathbb{F}[G]$-submodule of $V$ is of the form $W_{i_1} \oplus \cdots \oplus W_{i_\alpha}$, for some $\alpha > 0$ and $1 \leq i_1 < \cdots < i_\alpha \leq k$.*

***Proof*** Well known, see for example [55, Lemma 2.10.11]. □

**Lemma 2.1.16 ([55, Lemma 4.4.3])** *Suppose that* $V = W_1 \otimes W_2$ *and let* $H_i \leq \mathrm{GL}(W_i)$ *for* $1 \leq i \leq 2$*. Then the following hold:*

*(i) If* $H_1$ *and* $H_2$ *are absolutely irreducible, then* $H_1 \otimes H_2 \leq \mathrm{GL}(V)$ *is absolutely irreducible.*
*(ii) If* $H_1$ *and* $H_2$ *are absolutely irreducible, then* $N_{\mathrm{GL}(V)}(H_1 \otimes H_2) = N_{\mathrm{GL}(W_1)}(H_1) \otimes N_{\mathrm{GL}(W_2)}(H_2)$.
*(iii) If* $H_1$ *is absolutely irreducible, then* $N_{\mathrm{GL}(V)}(H_1 \otimes 1) = N_{\mathrm{GL}(W_1)}(H_1) \otimes \mathrm{GL}(W_2)$.
*(iv) If* $H_1$ *is absolutely irreducible, then* $C_{\mathrm{GL}(V)}(H_1 \otimes 1) = 1 \otimes \mathrm{GL}(W_2)$.

**Lemma 2.1.17** *Let* $\mathbb{F}$ *be a field and let* $G$ *be a group. Suppose that* $W$ *is a completely reducible* $\mathbb{F}[G]$*-module that has no trivial submodules. Then*

$$W = \langle (g-1)w : g \in G, w \in W \rangle.$$

***Proof*** It will suffice to prove the lemma in the irreducible case, so suppose that $W$ is a nontrivial irreducible $\mathbb{F}[G]$-module. Then since $W$ is nontrivial, the subspace $\langle (g-1)w : g \in G, w \in W \rangle$ is a nonzero $\mathbb{F}[G]$-submodule of $W$, so by irreducibility $W = \langle (g-1)w : g \in G, w \in W \rangle$. □

### 2.1.4 Number Theory

In this section we will recall some results in elementary number theory. The following lemma on quadratic reciprocity can be found in most basic textbooks on number theory, for example [39, Proposition 5.2.2].

**Lemma 2.1.18** *Let* $r$ *be an odd prime. Then* $\left(\frac{-1}{r}\right)$ *and* $\left(\frac{2}{r}\right)$ *are determined as follows:*

$$\left(\frac{-1}{r}\right) = \begin{cases} +, & \text{if } r \equiv 1 \mod 4 \\ -, & \text{if } r \equiv 3 \mod 4 \end{cases}$$

$$\left(\frac{2}{r}\right) = \begin{cases} +, & \text{if } r \equiv \pm 1 \mod 8 \\ -, & \text{if } r \equiv \pm 3 \mod 8 \end{cases}$$

In the lemmas below, recall that $\nu_2(n)$ denotes the 2-adic valuation of an integer $n$, so $\nu_2(n)$ is the largest integer $k \geq 0$ such that $2^k$ divides $n$.

**Lemma 2.1.19** *Let $n, m \geq 1$ be integers and $x > 1$. Then the following hold:*

*(i)* $\gcd(x^n - 1, x^m - 1) = x^{\gcd(n,m)} - 1$.

*(ii)* $\gcd(x^n + 1, x^m + 1) = \begin{cases} x^{\gcd(n,m)} + 1, & \text{if } \nu_2(n) = \nu_2(m). \\ \gcd(x - 1, 2), & \text{if } \nu_2(n) \neq \nu_2(m). \end{cases}$

*(iii)* $\gcd(x^n + 1, x^m - 1) = \begin{cases} x^{\gcd(n,m)} + 1, & \text{if } \nu_2(n) < \nu_2(m). \\ \gcd(x - 1, 2), & \text{if } \nu_2(n) \geq \nu_2(m). \end{cases}$

***Proof*** Omitted. (The lemma is well known, and can be proved with elementary number theory.) □

**Lemma 2.1.20** *Let $n, m \geq 1$ be integers and $x > 1$. Then the following hold:*

*(i)* $x^n - 1 \mid x^m - 1$ *if and only if* $n \mid m$.
*(ii)* $x^n + 1 \mid x^m - 1$ *if and only if* $n \mid m$ *and* $m/n$ *is even.*
*(iii)* $x^n + 1 \mid x^m + 1$ *if and only if* $n \mid m$ *and* $m/n$ *is odd.*
*(iv)* $x^n - 1 \mid x^m + 1$ *if and only if one of the following holds:*

  *(a)* $x = 2$ *and* $n = 1$.
  *(b)* $x = 3$ *and* $n = 1$.
  *(c)* $x = 2$, $n = 2$, *and* $m$ *is odd.*

***Proof*** A straightforward consequence of Lemma 2.1.19. □

Let $x, n > 1$ be integers. A *primitive prime divisor* of $x^n - 1$ is a prime $r$ such that $r \mid x^n - 1$ and $r \nmid x^{n'} - 1$ for all $0 < n' < n$. We will need the following theorem of Zsigmondy [88] on existence of primitive prime divisors.

**Theorem 2.1.21** *Let $x, n > 1$ be integers. Then one of the following holds:*

*(i)* $x^n - 1$ *has a primitive prime divisor.*
*(ii)* $x + 1$ *is a power of* $2$ *and* $n = 2$.
*(iii)* $x = 2$ *and* $n = 6$.

The following lemma is a special case of Catalan's conjecture (Mihăilescu's theorem [65]).

**Lemma 2.1.22** *Let $q$ be a prime power, and suppose that $q^\nu \pm 1$ is a power of two. Then one of the following holds:*

*(i)* $q$ *is a prime,* $\nu = 1$, *and* $q \pm 1$ *is a power of two;*
*(ii)* $q = 9$ *and* $\nu = 1$.
*(iii)* $q = 3$ *and* $\nu = 2$.

***Proof*** We prove the lemma in the case where $q$ is a prime, from which the general result follows. Consider first the case where $q$ is a prime and $q^\nu - 1$ is a power of two. If $\nu = 1$ then (i) holds, so suppose that $\nu > 1$. Then $q^\nu - 1$ has no primitive prime divisors and $q$ is odd, so by Theorem 2.1.21 we have $\nu = 2$ and $q + 1$ is a power of two. Since $q^2 - 1 = (q - 1)(q + 1)$ is a power of two and $\gcd(q - 1, q + 1) = 2$, it follows that $q = 3$, so (iii) holds.

Next we consider the case where $q$ is a prime and $q^\nu + 1$ is a power of two. Suppose that $\nu > 1$. Since $q$ must be odd, it follows from Theorem 2.1.21 that $q^{2\nu} - 1$ has a primitive prime divisor $r$, so $r \nmid q^{\nu'} - 1$ for all $0 < \nu' < 2\nu$. On the other hand $q^{2\nu} - 1 = (q^\nu - 1)(q^\nu + 1)$, so $r \mid q^\nu + 1$. This is a contradiction, since $q^\nu + 1$ is a power of two, but as a primitive prime divisor $r$ must be odd. Therefore the only possibility is that $\nu = 1$, so (i) holds. □

## 2.2 Metrically Imprimitive Subgroups

**Definition 2.2.1** Let $G \leq \Delta(V, \kappa)$. We say that $G$ is *metrically imprimitive* if there exists an orthogonal decomposition

$$V = W_1 \perp \cdots \perp W_k,$$

where $G$ acts on $\{W_1, \ldots, W_k\}$ and $k > 1$. We call $\{W_1, \ldots, W_k\}$ an *orthogonal system of imprimitivity*.

If no such decomposition exists, we say that $G$ is *metrically primitive*. In the case where $\kappa = 0$, we will also simply call $G$ *imprimitive* or *primitive*, and $\{W_1, \ldots, W_k\}$ is called a *system of imprimitivity*.

**Definition 2.2.2** Let $G \leq \Delta(V, \kappa)$ with orthogonal systems of imprimitivity $\Gamma = \{W_1, \ldots, W_k\}$ and $\Gamma' = \{Z_1, \ldots, Z_\ell\}$. We say that $\Gamma'$ is a *refinement* of $\Gamma$ if each $W_i$ is a direct sum of some $Z_j$'s, for all $1 \leq i \leq k$. If $\Gamma$ has no proper refinement, we say that $\Gamma$ is *nonrefinable*.

**Remark 2.2.3** In the context of metrically imprimitive and metrically primitive subgroups, we will mostly be concerned with the case where $G$ is irreducible. However, in a few occasions (e.g. Lemma 5.6.4) it will be useful for us to consider (orthogonal) systems of imprimitivity for groups which are not necessarily irreducible, so we do not require irreducibility in Definition 2.2.1.

(We have taken the terminology in Definition 2.2.1 from [84].) Similarly to Theorem 1.1.1, we will see that it is natural to split the classification of maximal irreducible solvable subgroups of $\Delta(V, \kappa)$ to the metrically imprimitive case and the metrically primitive case.

In this section, we shall describe the general structure of maximal irreducible solvable subgroups of $\Delta(V, \kappa)$ which are metrically imprimitive. For the most part, these subgroups are determined by two groups $H \leq \Delta(W, \kappa')$ and $K \leq S_k$, where $H$ is metrically primitive maximal irreducible solvable with $\dim W < \dim V$, and $K \leq S_k$ is maximal transitive solvable with $1 < k \leq \dim V$.

We give an explicit construction in Examples 2.2.5 and 2.2.7 below. For example, an imprimitive maximal irreducible solvable subgroup of $\mathrm{GL}_n(q)$ is of the form $H \wr K$, where $H \leq \mathrm{GL}_d(q)$ is primitive maximal irreducible solvable, $K \leq S_k$ is maximal transitive solvable, and $n = dk$ with $k > 1$.

**Lemma 2.2.4** *Let $G \leq \mathrm{GL}(V)$ be irreducible, and suppose that $V = W_1 \oplus \cdots \oplus W_k$ is a system of imprimitivity for $G$ on $V$. Then the following statements hold:*

*(i) $G$ acts transitively on $\{W_1, \ldots, W_k\}$.*
*(ii) $N_G(W_i)$ acts irreducibly on $W_i$ for all $1 \leq i \leq k$.*
*(iii) Let $G_i$ be the image of $N_G(W_i)$ in $\mathrm{GL}(W_i)$. Then $G_i$ is similar to $G_j$ for all $1 \leq i, j \leq k$.*

***Proof*** Well-known, see for example [77, Theorem 15.1]. □

**Example 2.2.5 (Isometric Imprimitive Subgroups)** Let $n = dk$, where $k > 1$. Let $K$ be a subgroup of the symmetric group $S_k$, and let $H \leq \mathrm{GL}_d(q) = \mathrm{GL}(W)$.

Furthermore, if $b$ is an alternating bilinear form, we assume that $d$ is even and $H \leq \mathrm{GSp}_d(q) = \Delta(W, \kappa')$, where $\kappa'$ is a non-degenerate alternating bilinear form on $W$.

If $\kappa$ is a quadratic form with $\mathrm{sgn}(\kappa) = \varepsilon$, we assume that $H \leq \mathrm{GO}_d^{\varepsilon'}(q) = \Delta(W, \kappa')$, where $\kappa'$ is a non-degenerate quadratic form on $W$ and

$$\varepsilon = \begin{cases} (\varepsilon')^k, & \text{if } d \text{ is even.} \\ (-1)^{k(q-1)/4}, & \text{if } d \text{ is odd and } k \text{ is even.} \\ \circ, & \text{if } d \text{ and } k \text{ are odd.} \end{cases}$$

By [55, Proposition 2.5.11, Proposition 2.5.13] we can find an orthogonal decomposition

$$V = W_1 \perp \cdots \perp W_k,$$

where $(W_i, \kappa)$ is isometric to $(W, \kappa')$ for all $1 \leq i \leq k$. Choose a basis $e_1, \ldots, e_d$ for $W$. Then for each $W_i$ we can find a basis $e_1^{(i)}, \ldots, e_d^{(i)}$ such that $b'(e_t, e_s) = b(e_t^{(i)}, e_s^{(i)})$ for all $1 \leq t, s \leq d$.

For each $\sigma \in K$, we have a linear map $\tilde{\sigma} : V \to V$ defined by $e_t^{(i)} \mapsto e_t^{(\sigma(i))}$ for all $1 \leq i \leq k$ and $1 \leq t \leq d$. Then $\tilde{\sigma} \in I(V, \kappa)$, and with the homomorphism $\sigma \mapsto \tilde{\sigma}$ we identify $K$ as a subgroup of $I(V, \kappa)$.

For $1 \leq i \leq k$, let $\psi_i : (W, \kappa') \to (W_i, \kappa)$ be the isometry defined by $e_t \mapsto e_t^{(i)}$. Define $H_i = \psi_i H \psi_i^{-1}$, so that $H_i$ is the subgroup corresponding to $H$ in $\Delta(W_i, \kappa)$. We extend the action of $H_i$ to $V$ by taking it to act trivially on $W_j$ for all $j \neq i$. In this manner, we can consider the wreath product

$$H \wr K = (H_1 \times \cdots \times H_k) \rtimes K$$

as a subgroup of $\mathrm{GL}(V)$.

We will call $G = (H \wr K) \cap \Delta(V, \kappa)$ the *isometric imprimitive subgroup* corresponding to $H$ and $K$ in $\Delta(V, \kappa)$. (These groups are called *décomposable* by Jordan.) Let $H = H^\circ \langle h \rangle$, where $\tau(h)$ generates $\tau(H)$ in $\mathbb{F}_q^\times$. Denote $h_i = \psi_i h \psi_i^{-1}$, so that $H_i = H_i^\circ \langle h_i \rangle$. Then it is easy to see that

$$G = \langle H^\circ \wr K, h_1 \cdots h_k \rangle$$

and $G^\circ = H^\circ \wr K$, so $G$ is of the same multiplier as $H$. Note that if $\kappa = 0$, then $G = H \wr K$.

**Lemma 2.2.6** *Let $H \leq \mathrm{GL}(W)$ and $K \leq S_k$, where $\dim W = d$ and $k > 1$. Let $H \wr K$ be the wreath product subgroup of $\mathrm{GL}(V)$ as in Example 2.2.5, where $\dim V = dk$. Then $H \wr K$ is irreducible if and only if $H$ is nontrivial irreducible and $K$ is transitive.*

***Proof*** Follows from [77, Theorem 15.1, Lemma 15.4]. □

**Example 2.2.7 (Semiprimary Subgroups)** (These groups are called *semi-primaire* by Jordan.) Assume that $\kappa$ is a nonzero form, $q$ is odd, and that $n$ is even. Write $n = 2d$, and let $W$ be an $\mathbb{F}_q$-vector space with $\dim W = d$.

If $\kappa$ is an alternating bilinear form, we assume that $d$ is even and $H \leq \mathrm{Sp}_d(q) = I(W, \kappa')$, where $\kappa'$ is a non-degenerate alternating bilinear form on $W$.

Furthermore, if $\kappa$ is a quadratic form with signature $\varepsilon$, we assume that $H \leq O_d^{\varepsilon'}(q) = I(W, \kappa')$, where $\kappa'$ is a non-degenerate quadratic form on $W$ and

$$\varepsilon = \begin{cases} -1, & \text{if } d \text{ is odd and } q \equiv 1 \mod 4. \\ +1, & \text{otherwise.} \end{cases}$$

Let $\zeta \in \mathbb{F}_q^\times$ be a primitive element. It follows from [55, Proposition 2.5.10, Proposition 2.5.11] that we can find an orthogonal decomposition

$$V = W_1 \perp W_2$$

such that $(W_1, \kappa)$ is isometric to $(W, \kappa)$, and $(W_2, \kappa)$ is isometric to $(W, \zeta\kappa')$. Note that in this case if $d$ is even, then $(W_2, \kappa)$ is isometric to $(W, \kappa)$, and for $d$ odd $(W_2, \kappa)$ is similar but nonisometric to $(W, \kappa')$.

Fix a basis $e_1, \ldots, e_d$ of $W$. We can find a basis $e_1^{(1)}, \ldots, e_d^{(1)}$ of $W_1$ such that $b(e_i^{(1)}, e_j^{(1)}) = b'(e_i, e_j)$ for all $1 \leq i, j \leq d$. Moreover, there is a basis $e_1^{(2)}, \ldots, e_d^{(2)}$ of $W_2$ such that $b(e_i^{(2)}, e_j^{(2)}) = \zeta b'(e_i, e_j)$ for all $1 \leq i, j \leq d$.

For $i = 1, 2$, let $\psi_i : (W, \kappa') \to (W_i, \kappa)$ be the similarity defined by $e_j \mapsto e_j^{(i)}$. Define $H_i = \psi_i H \psi_i^{-1}$, so that $H_i$ is the subgroup corresponding to $H$ in $I(W_i, \kappa)$. We extend the action of $H_i$ to $V$ by taking it to act trivially on $W_j$ for all $j \neq i$.

We then have $H_1 \times H_2$ as a subgroup of $I(V, \kappa)$. Define $\varphi : V \to V$ by

$$\varphi\left(e_i^{(1)}\right) = e_i^{(2)}$$

$$\varphi\left(e_i^{(2)}\right) = \zeta e_i^{(1)}$$

for all $1 \leq i \leq d$. Then $\varphi \in \Delta(V, \kappa)$, and $\varphi H_1 \varphi^{-1} = H_2$ and $\varphi H_2 \varphi^{-1} = H_1$, so $\varphi$ normalizes the subgroup $H_1 \times H_2$. We define

$$\operatorname{semiwr}(H) = \langle H_1 \times H_2, \varphi \rangle,$$

which we call the *semiprimary imprimitive subgroup* corresponding to $H$ in $\Delta(V, \kappa)$.

Note that $\operatorname{semiwr}(H)^\circ = H_1 \times H_2$, and moreover $\tau(\varphi) = \zeta$, so $\operatorname{semiwr}(H)$ is of multiplier 1 in $\Delta(V, \kappa)$.

**Lemma 2.2.8** *Assume that $q$ is odd. Let $H \leq I(W, \kappa')$ and $G = \operatorname{semiwr}(H)$ be as in Example 2.2.7. Then $G$ is irreducible if and only if $H$ is irreducible.*

***Proof*** We have $V = W_1 \perp W_2$ and $G = \langle H_1 \times H_2, \varphi \rangle$ as in Example 2.2.7.

Suppose first that $H$ is irreducible. If $H$ is trivial, then by irreducibility $\dim W = 1$, and thus $\dim V = 2$. We have $\varphi^2 = \zeta I_V$ where $\zeta \in \mathbb{F}_q$ is a primitive element, so $\varphi$ has no eigenvalues on $V$. Thus $G$ is irreducible since $\langle \varphi \rangle$ is.

Suppose then that $H$ is nontrivial irreducible. Now $H_1$ and $H_2$ are similar to $H$ and $H$ is nontrivial, so $W_1$ and $W_2$ are nonisomorphic irreducible $\mathbb{F}_q[H_1 \times H_2]$-modules. Thus it follows from Lemma 2.1.15 that every proper nonzero $\mathbb{F}_q[G]$-submodule of $V$ would have to be equal to $W_1$ or $W_2$. However, the map $\varphi$ swaps $W_1$ and $W_2$. Therefore $W_1$ and $W_2$ are not $G$-invariant, so $G$ must be irreducible.

Conversely, suppose that $G$ is irreducible. We have $N_G(W_1) = \langle H_1 \times H_2, \varphi^2 \rangle$, where $\varphi^2 = \zeta I_V$. Therefore the action of $N_G(W_1)$ on $W_1$ is equal to $H_1 Z$, where $Z \leq \mathrm{GL}(W_1)$ is the group of scalar matrices. It follows from Lemma 2.2.4 (ii) that $H_1 Z$ is irreducible, so $H_1$ is irreducible. Since $H_1$ is similar to $H$, we conclude that $H$ is irreducible. □

**Lemma 2.2.9** *Assume that $q$ is odd. Let $Z$ be the group of scalar matrices in $\Delta(W, \kappa')$. Suppose that $H \leq \Delta(W, \kappa')$ is such that $Z \leq H$ and $H$ is of multiplier 1. Then*

$$\operatorname{semiwr}(H^\circ) \lneq (H \wr S_2) \cap \Delta(V, \kappa),$$

*where $\operatorname{semiwr}(H^\circ)$ and $(H \wr S_2) \cap \Delta(V, \kappa)$ are as in Example 2.2.7 and Example 2.2.5, respectively.*

***Proof*** We have $\operatorname{semiwr}(H^\circ) = \langle H_1^\circ \times H_2^\circ, \varphi \rangle$ with $V = W_1 \perp W_2$, and $\varphi$ defined with respect to a basis $(e_j^{(i)})$ as in Example 2.2.7. Here $H_i = \psi_i H \psi_i^{-1}$ with $\psi_i$ defined as in Example 2.2.7. Since $H$ is of multiplier 1, there exists $h \in H$ such that $\tau(h) = \zeta$, where $\zeta \in \mathbb{F}_q^\times$ is the primitive element used in the construction of Example 2.2.7.

Denote $h_i = \psi_i h \psi_i^{-1}$ for $i = 1, 2$; so $h_i$ is the element corresponding to $h$ in $\Delta(W_i, \kappa)$. Define $g = h_1 h_2$, so $g \in (H_1 \times H_2) \cap \Delta(V, \kappa)$ and $\tau(g) = \zeta$. Next define $\sigma : V \to V$ by

$$\sigma\left(e_i^{(1)}\right) = h_2^{-1} e_i^{(2)}$$

$$\sigma\left(e_i^{(2)}\right) = h_1 e_i^{(1)}$$

for all $1 \leq i \leq d$. Then $\sigma \in I(V, \kappa)$.

Define $f_j^{(1)} := e_j^{(1)}$ and $f_j^{(2)} := h_2^{-1} e_j^{(2)}$ for all $1 \leq j \leq d$. Note that with respect to the basis $(f_j^{(i)})$ of $V$, the map $\sigma \in I(V, \kappa)$ corresponds to the permutation that swaps the summands $W_1$ and $W_2$, as in the construction of Example 2.2.5. Then it is readily seen that we have

$$(H \wr S_2) \cap \Delta(V, \kappa) = \langle H_1^\circ \times H_2^\circ, g, \sigma \rangle,$$

with $(H \wr S_2) \cap \Delta(V, \kappa)$ as in Example 2.2.5.

We next verify that $\varphi \in (H \wr S_2) \cap \Delta(V, \kappa)$. Denote the matrix of $h$ with respect to the basis $(e_j)$ by $A$. Then with respect to the basis $(f_j^{(i)})$, we have

$$\begin{aligned} \sigma\varphi &= \begin{pmatrix} 0 & I_d \\ I_d & 0 \end{pmatrix} \begin{pmatrix} 0 & \zeta A^{-1} \\ A & 0 \end{pmatrix} \\ &= \begin{pmatrix} A & 0 \\ 0 & \zeta A^{-1} \end{pmatrix} \\ &= g \begin{pmatrix} I_d & 0 \\ 0 & \zeta A^{-2} \end{pmatrix}. \end{aligned}$$

Note that here $\zeta A^{-2} \in H_2^\circ$ since $H$ contains scalar matrices, so $\sigma\varphi \in (H \wr S_2) \cap \Delta(V, \kappa)$, giving $\varphi \in (H \wr S_2) \cap \Delta(V, \kappa)$. Thus $\mathrm{semiwr}(H^\circ) \leq (H \wr S_2) \cap \Delta(V, \kappa)$, and the inclusion is proper since $\sigma \notin \mathrm{semiwr}(H^\circ) \cap I(V, \kappa) = H_1^\circ \times H_2^\circ$. □

We will next illustrate the constructions of Examples 2.2.5 and 2.2.7 in the case where $n = 2$.

**Example 2.2.10** Suppose that $n = 2$ and $\kappa = 0$.

In $\Delta(V, \kappa) = \mathrm{GL}_2(q)$, the construction of Example 2.2.5 provides an imprimitive subgroup $G = \mathrm{GL}_1(q) \wr S_2$, which has generators

$$\begin{pmatrix} \zeta & 0 \\ 0 & 1 \end{pmatrix}, \begin{pmatrix} 1 & 0 \\ 0 & \zeta \end{pmatrix}, \begin{pmatrix} 0 & 1 \\ 1 & 0 \end{pmatrix},$$

where $\zeta \in \mathbb{F}_q^\times$ is a primitive element. Thus $G$ consists of matrices of the form

$$\begin{pmatrix} \alpha & 0 \\ 0 & \beta \end{pmatrix}, \begin{pmatrix} 0 & \alpha \\ \beta & 0 \end{pmatrix}$$

with $\alpha, \beta \in \mathbb{F}_q^\times$.

**Example 2.2.11** Suppose that $n = 2$ and $\kappa \neq 0$. If $\Delta(V, \kappa)$ contains a metrically imprimitive subgroup, the orthogonal system of imprimitivity consists of two 1-spaces $V = W_1 \perp W_2$ with $W_i = \langle e_i \rangle$. Thus $\kappa$ is a quadratic form and $q$ is odd, with $\Delta(V, \kappa) = \mathrm{GO}_2^\varepsilon(q)$ and $I(V, \kappa) = O_2^\varepsilon(q)$.

Here the construction of isometric imprimitive subgroups (Example 2.2.5) applies when $\varepsilon = (-1)^{(q-1)/2}$. In this case we can arrange $Q(e_1) = Q(e_2) = 1$, so the construction of Example 2.2.5 gives $G = \langle O_1(q) \wr S_2, h_1h_2 \rangle$, where

$$O_1(q) = \{\pm 1\},$$
$$O_1(q) \wr S_2 = \left\langle \begin{pmatrix} -1 & 0 \\ 0 & 1 \end{pmatrix}, \begin{pmatrix} 1 & 0 \\ 0 & -1 \end{pmatrix} \begin{pmatrix} 0 & 1 \\ 1 & 0 \end{pmatrix} \right\rangle,$$
$$h_1h_2 = \begin{pmatrix} \zeta & 0 \\ 0 & \zeta \end{pmatrix}.$$

Here $|G| = 4(q-1)$, $|G^\circ| = 8$, and $\zeta \in \mathbb{F}_q^\times$ is a primitive element.

For the construction of semiprimary subgroups (Example 2.2.7), we need to assume that $\varepsilon = (-1)^{(q+1)/2}$. In this case we can arrange $Q(e_1) = 1$, $Q(e_2) = \zeta$, so the construction of Example 2.2.7 gives $G = \mathrm{semiwr}(O_1(q)) = \langle O_1(q) \times O_1(q), \varphi \rangle$, where

$$O_1(q) \times O_1(q) = \left\langle \begin{pmatrix} -1 & 0 \\ 0 & 1 \end{pmatrix}, \begin{pmatrix} 1 & 0 \\ 0 & -1 \end{pmatrix} \right\rangle,$$
$$\varphi = \begin{pmatrix} 0 & \zeta \\ 1 & 0 \end{pmatrix}.$$

Here $|G| = 4(q-1)$ and $|G^\circ| = 4$.

**Remark 2.2.12** In Example 2.2.11, the system of imprimitivity used for $G^\circ = O_1(q) \wr S_2$ is $\{\langle e_1 \rangle, \langle e_2 \rangle\}$. It is straightforward to check that for $\lambda \in \mathbb{F}_q$ with $\lambda^2 = \pm 1$, the decomposition

$$V = \langle e_1 + \lambda e_2 \rangle \oplus \langle e_1 - \lambda e_2 \rangle$$

defines another system of imprimitivity for $O_1(q) \wr S_2$. Furthermore one can show that every system of imprimitivity for $G^\circ$ is of this form, see [57, Remark 3.3] or [12, Remark 2.8].

Note that $\langle e_1 + \lambda e_2\rangle$ and $\langle e_1 - \lambda e_2\rangle$ are orthogonal if and only if $\lambda^2 = 1$. Thus in total $G^\circ$ has exactly two orthogonal systems of imprimitivity, $\{\langle e_1\rangle, \langle e_2\rangle\}$ and $\{\langle e_1 + e_2\rangle, \langle e_1 - e_2\rangle\}$. Additionally if $q \equiv 1 \mod 4$, there is a non-orthogonal system of imprimitivity $\{\langle e_1 + \lambda e_2\rangle, \langle e_1 - \lambda e_2\rangle\}$, where $\lambda^2 = -1$.

**Remark 2.2.13** The group $\mathrm{GO}_2^\varepsilon(q)$ is solvable, so in most cases the imprimitive subgroups $G \leq \mathrm{GO}_2^\varepsilon(q)$ described in Example 2.2.11 are not maximal solvable. The few exceptions occur when $q$ is small, and we list them below:

- $q = 3$, semiprimary case: $\mathrm{GO}_2^+(3) = G$ and $O_2^+(3) = O_1(3) \times O_1(3)$.
- $q = 3$, isometric imprimitive case: $\mathrm{GO}_2^-(3) \gneqq G$ and $O_2^-(3) = O_1(3) \wr S_2$.
- $q = 5$, isometric imprimitive case: $\mathrm{GO}_2^+(5) \gneqq G$ and $O_2^+(5) = O_1(5) \wr S_2$.

In all other cases of Example 2.2.11, we have $G \lneqq \mathrm{GO}_2^\varepsilon(q)$ and $G^\circ \lneqq O_2^\varepsilon(q)$. Thus for $q$ odd, we can conclude that $\mathrm{GO}_2^\varepsilon(q)$ is metrically primitive if and only if $q > 3$ or $(q, \varepsilon) = (3, -)$, and $O_2^\varepsilon(q)$ is metrically primitive if and only if $q > 5$ or $(q, \varepsilon) = (5, -)$.

We will now prove that all metrically imprimitive maximal irreducible solvable subgroup of $\Delta(V, \kappa)$ arise from the constructions of Examples 2.2.5 and 2.2.7. This is essentially what is proved by Jordan in [51, §III] assuming that $q$ is a prime, but the same proof works in general. In the case where $\Delta(V, \kappa) = \mathrm{GL}(V)$, Suprunenko gives a proof in [77, Theorem 15.4].

**Theorem 2.2.14** *Let $G$ be a metrically imprimitive maximal irreducible solvable subgroup of $\Delta(V, \kappa)$. Choose an orthogonal system of imprimitivity*

$$V = W_1 \perp \cdots \perp W_k$$

*for $G$ such that $k$ is as large as possible. Let $H \leq \Delta(W_1, \kappa)$ be the image of $N_G(W_1)$ in $\Delta(W_1, \kappa)$, and let $K \leq S_k$ be the image of $G$ in $S_k$. Then the following statements hold:*

*(i) If $G^\circ$ acts transitively on $\{W_1, \ldots, W_k\}$, then:*

- *$H$ is metrically primitive maximal irreducible solvable in $\Delta(W_1, \kappa)$;*
- *$q > 2$ if $\dim W_1 = 1$;*
- *$K$ is maximal transitive solvable in $S_k$;*
- *$G = (H \wr K) \cap \Delta(V, \kappa)$, as constructed in Example 2.2.5.*

*(ii) If $G^\circ$ does not act transitively on $\{W_1, \ldots, W_k\}$, then:*

- *$\kappa \neq 0$, $q$ is odd, and $k$ is even;*
- *$G^\circ$ has exactly two orbits $\{W_1^{(1)}, \ldots, W_{k/2}^{(1)}\}$ and $\{W_1^{(2)}, \ldots, W_{k/2}^{(2)}\}$ on $\{W_1, \ldots, W_k\}$;*
- *$V = Z_1 \perp Z_2$, where $Z_i := W_1^{(i)} \perp \cdots \perp W_{k/2}^{(i)}$ for $1 \leq i \leq 2$.*
- *Let $T$ be the image of $N_G(Z_1)$ in $\Delta(Z_1, \kappa)$. Then $T$ is maximal irreducible solvable in $\Delta(Z_1, \kappa)$ and of multiplier 2;*
- *$G = \mathrm{semiwr}(T^\circ)$ as constructed in Example 2.2.7.*

***Proof*** First note that since $G$ is solvable, so are $H$ and $K$. Moreover, since $G$ is irreducible, $K$ must be transitive and $H$ must be irreducible (Lemma 2.2.4). Since $k$ was chosen to be as large as possible, a straightforward argument shows that $H$ must be metrically primitive on $W_1$ (see for example [77, Lemma 15.2]). The assumption $q > 2$ when $\dim W_1 = 1$ in case (i) is necessary, since otherwise $H \wr K$ is not irreducible (Lemma 2.2.6)

For the remaining claims, we consider the two possibilities in turn.

**Case 1: $G^\circ$ acts transitively on** $\{W_1, \ldots, W_k\}$
Let $e_1^{(1)}, \ldots, e_d^{(1)}$ be a basis of $W_1$. For all $1 < i \leq k$ let $g_i \in G^\circ$ be such that $g_i(W_1) = W_i$. We define $e_j^{(i)} = g_i e_j^{(1)}$ for all $1 \leq j \leq d$, so $e_1^{(i)}, \ldots, e_d^{(i)}$ is a basis of $W_i$.

Let $H_1 = H$, and extend the action of $H_1$ to $V$ be having it act trivially on $W_i$ for all $1 < i \leq k$. Define $H_i = g_i H_1 g_i^{-1}$, so now $H_i$ acts trivially on $W_j$ for $j \neq i$ and corresponds to the image of $N_G(W_i)$ in $\Delta(W_i, \kappa)$. Consider the subgroup $B = H_1 \times \cdots \times H_k$ of $\mathrm{GL}(V)$. Arguing as in [77, Proof of Lemma 15.5], we find that $G$ normalizes $B$ and that $GB = H \wr K$. Since $G$ is maximal solvable in $\Delta(V, \kappa)$, it follows that $G = (H \wr K) \cap \Delta(V, \kappa)$, as in Example 2.2.5. Since $G$ is maximal solvable, it is clear that $H$ must be maximal solvable in $\Delta(W_1, \kappa)$ and $K$ must be maximal solvable in $S_k$.

**Case 2: $G^\circ$ does not act transitively on** $\{W_1, \ldots, W_k\}$
Since $G$ acts transitively on $\{W_1, \ldots, W_k\}$, it is clear that $\kappa \neq 0$ in this case. Let $Z \leq \Delta(V, \kappa)$ be the group formed by scalar matrices, which must be contained in $G$ since $G$ is maximal solvable. Now the subgroup $G^\circ Z$ acts on the summands $W_i$ and has index at most 2 in $G$. Thus since $G$ acts transitively on $\{W_1, \ldots, W_k\}$, we conclude that $G^\circ$ has exactly two orbits on $\{W_1, \ldots, W_k\}$, and that $G^\circ Z$ is a subgroup of index 2 in $G$. Certainly this implies that $q$ is odd, as otherwise $\tau(Z) = \mathbb{F}_q^\times$ and $G = G^\circ Z$. Because $G$ acts on $G^\circ$-orbits, the two $G^\circ$-orbits are of equal size, and in particular $k$ is even.

We have $[G : G^\circ Z] = 2$, so $G$ is of multiplier 1. Let $\zeta \in \mathbb{F}_q^\times$ be a primitive element, and choose $g \in G$ such that $\tau(g) = \zeta$.

Let $\left\{W_1^{(1)}, \ldots, W_{k/2}^{(1)}\right\}$ be the $G^\circ$-orbit of $W_1$. We have $g \notin G^\circ Z$, so $g$ does not act on $\left\{W_1^{(1)}, \ldots, W_{k/2}^{(1)}\right\}$, and thus $g(W_1)$ is not conjugate to $W_1$ under the action of $G^\circ$. It follows that the other $G^\circ$-orbit is equal to $\left\{W_1^{(2)}, \ldots, W_{k/2}^{(2)}\right\}$, where $W_j^{(2)} := g(W_j^{(1)})$ for all $1 \leq j \leq k/2$. Define

$$Z_i := W_1^{(i)} \perp \cdots \perp W_{k/2}^{(i)}$$

for $1 \leq i \leq 2$, so $V = Z_1 \perp Z_2$.

Let $T_1$ be the action of $T = N_G(Z_1)$ on $Z_1$, and consider $T_1 \leq \Delta(V, \kappa)$ by having $T_1$ act trivially on $Z_2$. Set $T_2 = g T_1 g^{-1}$. As in the previous case, arguing as in [77, Proof of Lemma 15.5] we see that $G$ normalizes $T_1 \times T_2$. Thus $G$ normalizes

$T_1^\circ \times T_2^\circ$, so $T_1^\circ \times T_2^\circ \leq G$ by maximality of $G$. Since $G^\circ$ acts on $Z_1$ and $Z_2$ we have $G^\circ \leq T_1 \times T_2$, so we conclude that $G^\circ = T_1^\circ \times T_2^\circ$. Therefore $G = \langle T_1^\circ \times T_2^\circ, g \rangle$, and it is straightforward to see that $G = \mathrm{semiwr}(T^\circ)$ as in Example 2.2.7.

If $T$ is of multiplier 1, it follows from Lemma 2.2.9 that $G$ is not maximal solvable. Therefore $T$ must be of multiplier 2. We check that $T$ must be maximal solvable in $\Delta(Z_1, \kappa)$. To this end, suppose that $T \leq X \leq \Delta(Z_1, \kappa)$ with $X$ maximal solvable. We must have $T^\circ = X^\circ$, as otherwise $T^\circ \lneqq X^\circ$ and $G \lneqq \mathrm{semiwr}(X^\circ)$. Then $X$ must be of multiplier 2, as otherwise $G = \mathrm{semiwr}(T^\circ) = \mathrm{semiwr}(X^\circ)$ is not maximal solvable by Lemma 2.2.9. Hence $X = X^\circ Z = T^\circ Z = T$, where $Z \leq \Delta(Z_1, \kappa)$ is the group of scalar matrices. This proves that $T$ is maximal solvable. □

With Theorem 2.2.14, we have a construction of the potential maximal irreducible solvable subgroups of $\Delta(V, \kappa)$ which are metrically imprimitive. Conversely, with a few exceptions the groups in Theorem 2.2.14 (i) and (ii) are maximal irreducible solvable, a result which we will prove later in Sect. 7.4. For example, it turns out in $\mathrm{GL}_2(q)$ the subgroup $\mathrm{GL}_1(q) \wr S_2$ is maximal irreducible solvable if and only if $q = 4$ or $q > 5$.

We will finish this section by proving that the groups described in Theorem 2.2.14 are irreducible and solvable.

**Lemma 2.2.15** *Let $G \leq \Delta(V, \kappa)$ be metrically primitive maximal irreducible solvable. Then $G^\circ$ is an irreducible subgroup of $I(V, \kappa)$.*

***Proof*** We first show that $V \downarrow G^\circ$ must be homogeneous.

If $V \downarrow G^\circ$ is not homogeneous, then it follows from Proposition 2.1.13 that

$$V \downarrow G^\circ = W_1 \oplus W_2,$$

where $W_1$ and $W_2$ are the homogeneous components of $G^\circ$, and $W_i$ are both totally isotropic as subspaces. Let $\zeta \in \mathbb{F}_q^\times$ be a primitive element, and consider the map

$$\varphi = \begin{pmatrix} I_{W_1} & 0 \\ 0 & \zeta I_{W_2} \end{pmatrix}.$$

Then $\varphi \in \Delta(V, \kappa)$. Moreover $\varphi$ centralizes $G^\circ$ and $\tau(\varphi) = \zeta$, so it follows from Lemma 2.1.5 that $G = G^\circ \langle \varphi \rangle$. However, now $G^\circ$ and $\varphi$ both leave the subspaces $W_1$ and $W_2$ invariant, which contradicts the fact that $G$ is irreducible. Therefore $V \downarrow G^\circ$ is homogeneous.

Since $G/G^\circ$ is cyclic, by Clifford theory the homogeneous components of $V \downarrow G^\circ$ are irreducible [11, Proposition 2.6.2]. Therefore we conclude that $G^\circ$ is irreducible. □

**Remark 2.2.16** In Lemma 2.2.15, it is not necessarily the case that $G^\circ$ is metrically primitive. For example $G = \mathrm{GO}_2^-(3)$ is metrically primitive, but $G^\circ = O_2^-(3) = O_1(3) \wr S_2$ is not (Remark 2.2.13). For metrically primitive maximal irreducible solvable subgroups $G \leq \Delta(V, \kappa)$, we will later classify the cases where $G^\circ$ is metrically imprimitive (Corollary 5.6.13).

**Lemma 2.2.17** *Assume that $n$ and $q$ are even, and that $\kappa$ is a quadratic form. Let $G \leq \Delta(V, \kappa)$ be metrically primitive irreducible. Then either $G^\Omega$ is irreducible, or all of the following hold:*

*(i) We have $V \downarrow G^\Omega = W_1 \oplus W_2$, where $W_1$ and $W_2$ are nonisomorphic irreducible $\mathbb{F}_q[G^\Omega]$-modules.*
*(ii) $W_1$ and $W_2$ are totally isotropic subspaces and $W_2 \cong W_1^*$ as $\mathbb{F}_q[G^\Omega]$-modules.*

***Proof*** Since $q$ is even, we have $G = G^\circ Z$, where $Z \leq \mathrm{GL}(V)$ is the group of scalar matrices. Therefore $G^\circ$ is irreducible and metrically primitive. Suppose that $V \downarrow G^\Omega$ is not irreducible.

Because $G^\circ / G^\Omega$ is cyclic, by Clifford theory the homogeneous components of $V \downarrow G^\Omega$ are irreducible [11, Proposition 2.6.2], so there are at least two of them. Furthermore $G^\circ$ is metrically primitive, so we conclude from Proposition 2.1.13 we have

$$V \downarrow G^\Omega = W_1 \oplus W_2,$$

where $W_1$ and $W_2$ are totally isotropic subspaces and nonisomorphic irreducible $\mathbb{F}_q[G^\Omega]$-modules. The fact that $W_2 \cong W_1^*$ as $\mathbb{F}_q[G^\Omega]$-modules follows Proposition 2.1.13 (ii)(b). □

**Theorem 2.2.18** *Let $G = (H \wr K) \cap \Delta(V, \kappa)$ as in the construction of Example 2.2.5. Suppose that all of the following hold:*

- *$H$ is metrically primitive maximal irreducible solvable in $\Delta(W, \kappa')$;*
- *$K \leq S_k$ is maximal transitive solvable, where $k > 1$;*
- *$q > 2$ if $\dim W = 1$.*

*Then $G$ is irreducible and solvable, and $G^\circ$ is also irreducible. Furthermore, $G$ is of the same multiplier as $H$ if $\kappa \neq 0$.*

***Proof*** Since $H$ and $K$ are solvable, it is immediate that $G$ is solvable as well. By Lemmas 2.1.6 and 2.2.15, we have $H^\circ \neq 1$ and $H^\circ$ is irreducible. It follows from Lemma 2.2.6 that $G^\circ = H^\circ \wr K$ is irreducible, so $G$ must also be irreducible. The fact that $G$ is of the same multiplier as $H$ was noted in Example 2.2.5. □

**Theorem 2.2.19** *Let $G$ be the semiprimary imprimitive subgroup corresponding to $H \leq \Delta(W, \kappa')$, as constructed in Example 2.2.7. Suppose that $H$ is maximal irreducible solvable in $\Delta(W, \kappa')$. Then $G$ is irreducible, solvable, and of multiplier 1.*

***Proof*** Since $H$ is solvable, it is immediate that $G = \langle H_1^\circ \times H_2^\circ, \varphi \rangle$ is solvable as well. It follows from Lemma 2.2.8 that $G$ is irreducible, and $G$ is of multiplier 1 since $\tau(\varphi) \in \mathbb{F}_q^\times$ is a primitive element. □

**Lemma 2.2.20** *Assume that $n$ and $q$ are even, and that $\kappa$ is a quadratic form. Let $G = (H \wr K) \cap \Delta(V, \kappa)$ be an irreducible isometric imprimitive subgroup (Example 2.2.5). Then all of the following statements hold:*

*(i)* $G^{\Omega} = Y \rtimes K$, *where*

$$Y := \{(h_1, \dots, h_k) : \text{the number of } i \text{ such that } h_i \notin H_i^{\Omega} \text{ is even}\}.$$

*(ii)* $H^{\Omega} \wr K \leq G^{\Omega}$.

*(iii)* $G^{\circ} \leq \Omega(V, \kappa)$ *if and only if* $H \leq \Omega(W, \kappa)$.

***Proof*** Let $V = W_1 \perp \cdots \perp W_k$ be the decomposition corresponding to $G$ as in Example 2.2.5, where $\dim W_i = d$ for all $1 \leq i \leq k$. Note that $d$ must be even for $H$ to be irreducible.

We will first prove that $K \leq \Omega(V, \kappa)$. Every $\sigma \in K$ is a permutation matrix, so $\dim V^{\sigma}$ is equal to the number of orbits of $\sigma$ on the basis vectors. Then it is clear from the definition that $\dim V^{\sigma} = dt$, where $t$ is the number of orbits of $\sigma$ on $\{1, \dots, k\}$. Because $d$ is even, we have $\dim V^{\sigma} \equiv 0 \mod 2$, so $\sigma \in \Omega(V, \kappa)$.

Therefore $G^{\Omega} = Y \rtimes K$, where $Y = (H_1^{\circ} \times \cdots \times H_k^{\circ}) \cap \Omega(V, \kappa)$. For $(h_1, \dots, h_k) \in H_1^{\circ} \times \cdots \times H_k^{\circ}$, we have

$$\dim V^{(h_1, \dots, h_k)} = \dim W_1^{h_1} + \cdots + \dim W_k^{h_k},$$

so it follows that $Y = \{(h_1, \dots, h_k) : \text{the number of } i \text{ such that } h_i \notin H_i^{\Omega} \text{ is even}\}$. This completes the proof of (i). Claims (ii) and (iii) are immediate from (i). □

**Lemma 2.2.21** *Assume that $n$ and $q$ are even, and that $\kappa$ is a quadratic form. Let $G = (H \wr K) \cap \Delta(V, \kappa)$ be an irreducible isometric imprimitive subgroup (Example 2.2.5), where $H \leq \Delta(V, \kappa)$ is metrically primitive irreducible and $K \leq S_k$ is transitive for $k > 1$. Then $G^{\Omega}$ is irreducible.*

***Proof*** Let $V = W_1 \perp \cdots \perp W_k$ be the decomposition corresponding to $G$ as in Example 2.2.5, where $\dim W_i = d$ for all $1 \leq i \leq k$. Then $G^{\circ} = (H_1^{\circ} \times \cdots \times H_k^{\circ}) \rtimes K$, where $H_i$ acts irreducibly on $W_i$ and trivially on $W_j$ for $j \neq i$. Note that since $q$ is even, here $d$ must be even.

As seen from Lemma 2.2.20, we have $H^{\Omega} \wr K \leq G^{\Omega}$. Thus if $H^{\Omega}$ is irreducible, we know that $G^{\Omega}$ is irreducible (Lemma 2.2.6).

Suppose then that $H^{\Omega}$ is reducible. It follows from Lemma 2.2.17 that we have a decomposition into totally isotropic subspaces $W = W' \oplus W''$, where $W'$ and $W''$ are nonisomorphic irreducible $\mathbb{F}_q[H^{\Omega}]$-modules. In this case $H$ acts on $\{W', W''\}$. Thus $W_i = W_i' \oplus W_i''$, where $W_i'$ and $W_i''$ are nonisomorphic irreducible $\mathbb{F}_q[H_i^{\Omega}]$-modules, and $H_i$ acts on $\{W_i', W_i''\}$.

We conclude that the action of $H^{\Omega} \wr K$ on $V$ decomposes as $V = V' \oplus V''$, where $V' = W_1' \oplus \cdots \oplus W_k'$ and $V'' = W_1'' \oplus \cdots \oplus W_k''$ are nonisomorphic irreducible $\mathbb{F}_q[H^{\Omega} \wr K]$-modules by Lemma 2.2.6.

Thus by Lemma 2.1.15, every proper nonzero $\mathbb{F}_q[H^\Omega \wr K]$-submodule of $V$ must be equal to $V'$ or $V''$. We will show that $V'$ and $V''$ are not $G^\Omega$-invariant, which will prove that $G^\Omega$ is irreducible. To this end, pick $h_i \in H_i^\circ \setminus H_i^\Omega$ for all $1 \leq i \leq t$. Then for $i \neq j$ we have $g = h_i h_j \in G^\Omega$. Here $h_i$ maps $W_i'$ to $W_i''$, so $g(V') = V''$ and $g(V'') = V'$. Therefore $V'$ and $V''$ are not $G^\Omega$-invariant, so we conclude that $G^\Omega$ is irreducible on $V$. □

## 2.3 Metrically Completely Reducible Subgroups

We will use the following terminology, which is taken from [87, Definition 0.2].

**Definition 2.3.1** Let $G \leq \Delta(V, \kappa)$. We say that $G$ is *metrically completely reducible* if $\kappa \neq 0$ and every $G$-invariant subspace of $V$ is non-degenerate. (Equivalently, the only totally isotropic $G$-invariant subspace is the zero subspace.)

As we shall see later (for example Sect. 4.2), the structure of a metrically primitive maximal solvable subgroup is mostly determined by certain extraspecial groups $R_1, \ldots, R_k$, and groups $X_1, \ldots, X_k$, where $X_i \leq \mathrm{GSp}_{2\ell_i}(r_i)$ is maximal among the metrically completely reducible solvable subgroups of $\mathrm{GSp}_{2\ell_i}(r_i)$.

In this section, we will describe some basic results on metrically completely reducible maximal solvable subgroups of $\Delta(V, \kappa)$ and $I(V, \kappa)$. In the case where $n$ and $q$ are even and $\kappa$ is a quadratic form, we will also consider metrically completely reducible subgroups of $\Omega(V, \kappa)$. Much later in Sect. 7.5, we will obtain a complete classification of metrically completely reducible maximal solvable subgroups.

**Lemma 2.3.2** *Suppose that $G \leq \Delta(V, \kappa)$ is metrically completely reducible. Then $V = W_1 \perp \cdots \perp W_t$ for some irreducible $G$-submodules $W_1, \ldots, W_t$.*

***Proof*** Let $Z \subseteq V$ be an irreducible $G$-submodule of $V$. Then $Z$ is non-degenerate, so $V = Z \perp Z^\perp$. Proceeding by induction on dim $V$, the result follows by applying induction on $Z^\perp$. □

**Lemma 2.3.3** *Suppose that $q$ is even. Assume that $G \leq \mathrm{GSp}_{2n}(q)$ is metrically completely reducible and solvable. Then $G$ is conjugate to a subgroup of $\mathrm{GO}^+_{2n}(q)$ or $\mathrm{GO}^-_{2n}(q)$.*

***Proof*** By Lemma 2.3.2, it suffices to consider the case where $G$ is irreducible. In the irreducible case, the lemma is Theorem 2.1.14. □

**Lemma 2.3.4** *Assume that $n$ and $q$ are even, and that $\kappa = Q$ is a quadratic form with polarization $b$. Let $X \leq I(V, Q)$ and suppose that as an $\mathbb{F}_q[X]$-module $V$ has no trivial composition factors. Then $Q$ is the unique $X$-invariant quadratic form with polarization equal to $b$.*

***Proof*** Let $Q' : V \to \mathbb{F}_q$ be an $X$-invariant quadratic form with polarization $b$, so $Q'(v+w) = Q'(v)+Q'(w)+b(v, w)$ for all $v, w \in V$. Then $\rho : V \to \mathbb{F}_q$ defined by $\rho = Q + Q'$ satisfies $\rho(v + w) = \rho(v) + \rho(w)$ for all $v, w \in V$.

Because $q$ is even, we can define $\rho' : V \to \mathbb{F}_q$ by $\rho'(v) = \sqrt{\rho(v)}$ for all $v \in V$. Then $\rho'$ is a morphism of $\mathbb{F}_q[X]$-modules, where we consider $\mathbb{F}_q$ as the trivial 1-dimensional module. Since $X$ has no trivial composition factors on $V$, we conclude that $\rho' = 0$, which implies $Q = Q'$. □

**Lemma 2.3.5** *Assume that $n$ and $q$ are even, and that $Q$ is a non-degenerate quadratic form with polarization $b$. Let $X \leq O(V, Q) < \mathrm{Sp}(V, b)$ be metrically completely reducible maximal solvable in $O(V, Q)$. Then $X$ is maximal solvable in $\mathrm{Sp}(V, b)$.*

***Proof*** Suppose that $X \leq Y \leq \mathrm{Sp}(V, b)$ with $Y$ solvable. Then $Y$ is metrically completely reducible since $X$ is, so by Lemma 2.3.3 there is a quadratic form $Q'$ on $V$ with polarization $b$ such that $Y \leq O(V, Q')$. Because $X$ is metrically completely reducible, it has no trivial composition factors on $V$, so by Lemma 2.3.4 we have $Q = Q'$. Then $X = Y$ because $X$ is maximal solvable in $O(V, Q)$. □

**Remark 2.3.6** Suppose that $n$ and $q$ are even. By Lemmas 2.3.3 and 2.3.5, the classification of metrically completely reducible maximal solvable subgroups of $\mathrm{Sp}_n(q)$ is equivalent to the classification of metrically completely reducible maximal solvable subgroups of $O_n^+(q)$ and $O_n^-(q)$. In particular, for maximal irreducible solvable subgroups of $\mathrm{Sp}_n(q)$, it suffices to consider those in $O_n^+(q)$ and $O_n^-(q)$.

**Lemma 2.3.7** *Suppose that $L \leq I(V, \kappa)$ is metrically completely reducible maximal solvable. Then there is a decomposition $V = W_1 \perp \cdots \perp W_t$ such that all of the following hold:*

*(i) $L = H_1^\circ \times \cdots \times H_t^\circ$ with respect to $V = W_1 \perp \cdots \perp W_t$.*
*(ii) $H_i$ is maximal irreducible solvable in $\Delta(W_i, \kappa)$ for all $1 \leq i \leq t$.*
*(iii) $H_i^\circ$ is maximal irreducible solvable in $I(W_i, \kappa)$ for all $1 \leq i \leq t$.*

***Proof*** Follows from Lemma 2.3.2. □

**Lemma 2.3.8** *Let $L = H_1^\circ \times \cdots \times H_t^\circ$ with respect to $V = W_1 \perp \cdots \perp W_t$, where $H_i^\circ \leq I(W_i, \kappa)$ is maximal irreducible solvable for all $1 \leq i \leq t$. Then the following statements hold:*

*(i) If $\kappa \neq 0$, then $G$ is metrically completely reducible.*
*(ii) Every $L$-submodule of $V$ is of the form $W_{i_1} \perp \cdots \perp W_{i_r}$ for some indices $1 \leq i_1 < \cdots < i_r \leq t$.*

***Proof*** Claim (ii) is immediate from Lemma 2.1.15, and (i) follows from (ii). □

**Lemma 2.3.9** *Suppose that $L \leq \Delta(V, \kappa)$ is metrically completely reducible maximal solvable. Then there is a decomposition $V = W_1 \perp \cdots \perp W_t$ such that all of the following hold:*

(i) *$L = (H_1 \times \cdots \times H_t) \cap \Delta(V, \kappa)$ with respect to $V = W_1 \perp \cdots \perp W_t$.*
(ii) *$H_i \leq \Delta(W_i, \kappa)$ is maximal irreducible solvable for all $1 \leq i \leq t$.*
(iii) *If $L$ is of multiplier 2, then $H_i^\circ \leq I(W_i, \kappa)$ is maximal irreducible solvable for all $1 \leq i \leq t$.*

***Proof*** Claims (i) and (ii) follow from Lemma 2.3.2. For (iii), note that if $L$ is of multiplier 2 and maximal solvable, then $L = L^\circ Z$ where $Z \leq \Delta(V, \kappa)$ is the group of scalar matrices. Hence $L^\circ$ is maximal solvable in $I(V, \kappa)$, so clearly $H_i^\circ$ is maximal solvable in $I(W_i, \kappa)$ for all $1 \leq i \leq t$. □

**Lemma 2.3.10** *Let $L = (H_1 \times \cdots \times H_t) \cap \Delta(V, \kappa)$ with respect to $V = W_1 \perp \cdots \perp W_t$, where $H_i \leq \Delta(W_i, \kappa)$ is maximal irreducible solvable and $H_i \neq 1$ for all $1 \leq i \leq t$. Then the following statements hold:*

(i) *If $\kappa \neq 0$, then $L^\circ$ is metrically completely reducible.*
(ii) *If $L$ is of multiplier 1, then every $L$-submodule of $V$ is of the form $W_{i_1} \perp \cdots \perp W_{i_r}$ for some indices $1 \leq i_1 < \cdots < i_r \leq t$.*

***Proof*** For claim (i), we have $L^\circ = H_1^\circ \times \cdots \times H_t^\circ$. If $H_i$ is metrically primitive, it follows from Lemma 2.2.15 that $H_i^\circ$ acts irreducibly on $W_i$. In the case where $H_i$ is metrically imprimitive, it follows from Theorems 2.2.14 and 2.2.18 that $H_i^\circ$ is irreducible, except when $H_i$ is of semiprimary type. In the latter case we have $H_i^\circ = X_i^\circ \times Y_i^\circ$ with respect to a decomposition $W_i = W_i' \perp W_i''$, where $W_i'$ is an irreducible $\mathbb{F}_q[X_i^\circ]$-module, and $W_i''$ is an irreducible $\mathbb{F}_q[Y_i^\circ]$-module.

Thus by Lemma 2.1.15, every $\mathbb{F}_q[L^\circ]$-submodule of $V$ is of the form $Z_{i_1} \perp \cdots \perp Z_{i_r}$, where $Z_{i_s} = W_{i_s}$ or $H_{i_s}$ is semiprimary and $Z_{i_s} \in \{W_{i_s}', W_{i_s}''\}$. We conclude that $L^\circ$ is metrically completely reducible.

Next we consider claim (ii). Because $L$ is of multiplier 1, for all $1 \leq i \leq t$ the group $H_i$ is of multiplier 1 in $\Delta(W_i, \kappa)$. Thus for all $1 \leq i \leq t$, the action of $L$ on $W_i$ is equal to $H_i$. In this case the $W_i$ are irreducible $\mathbb{F}_q[L]$-modules and pairwise nonisomorphic, so (ii) follows from Lemma 2.1.15. □

**Remark 2.3.11** The assumption that $L$ is of multiplier 1 is necessary in Lemma 2.3.10 (ii), due to the fact that some of the factors $H_i$ could be metrically imprimitive of semiprimary type (Example 2.2.7). In this case $H_i^\circ$ is not irreducible, and $W_i = W_i' \perp W_i''$ where $W_i'$ and $W_i''$ are nonisomorphic irreducible $\mathbb{F}_q[H_i^\circ]$-modules.

**Lemma 2.3.12** *Assume that $n$ and $q$ are even, and that $\kappa$ is a quadratic form. Let $L = (H_1 \times \cdots \times H_t) \cap \Delta(V, \kappa)$ with respect to $V = W_1 \perp \cdots \perp W_t$, where $H_i \leq \Delta(W_i, \kappa)$ is maximal irreducible solvable and $H_i \neq 1$ for all $1 \leq i \leq t$. Then* *all of the following statements hold:*

(i) *The action of $L^\Omega$ on $W_i$ is equal to $H_i^\circ$ for all $1 \leq i \leq t$, except when:*

(a) *There exists $i$ such that $H_i^\circ \not\leq \Omega(W_i, \kappa)$, and $H_j^\circ \leq \Omega(W_j, \kappa)$ for all $j \neq i$.*

(ii) *$L^\Omega$ is metrically completely reducible, except when:*

(a) *There exists $i$ such that $H_i^\Omega$ is not irreducible, and $H_j^\circ \leq \Omega(W_j, \kappa)$ for all $j \neq i$.*

(iii) *If (ii)(a) holds, then $L^\Omega$ is not metrically completely reducible. Furthermore, every $L^\Omega$-invariant totally isotropic subspace of $V$ is contained in $W_i$.*

(iv) *If $L^\Omega$ is metrically completely reducible, then every $L$-submodule of $V$ is of the form $W_{i_1} \perp \cdots \perp W_{i_r}$ for some indices $1 \leq i_1 < \cdots < i_r \leq t$.*

(v) *Every non-degenerate $L$-submodule of $V$ is of the form $W_{i_1} \perp \cdots \perp W_{i_r}$ for some indices $1 \leq i_1 < \cdots < i_r \leq t$.*

***Proof*** We begin with the proof of (i). Suppose that (i)(a) does not hold, and let $1 \leq i \leq t$. If $H_i^\circ \leq \Omega(W_i, \kappa)$ for all $1 \leq i \leq t$, then the action of $L^\Omega$ on $W_i$ is equal to $H_i^\Omega = H_i^\circ$. Suppose then that $H_i^\circ \not\leq \Omega(W_i, \kappa)$. Because (i)(a) does not hold, there exists $j \neq i$ such that $H_j^\circ \not\leq \Omega(W_j, \kappa)$. It is clear that the action of $L^\Omega$ on $W_i$ contains $H_i^\Omega$. For $h_i \in H_i^\circ \setminus H_i^\Omega$, choose $h_j \in H_j^\circ \setminus H_j^\Omega$, so $h_i h_j \in L^\Omega$. Therefore the action of $L^\Omega$ on $W_i$ contains every $h_i \in H_i^\circ \setminus H_i^\Omega$, and consequently all of $H_i^\circ$.

Before proving (ii), we consider claim (iii). Suppose that (ii)(a) holds. We can assume without loss of generality that $H_1^\Omega$ is not irreducible and $H_j^\circ \leq \Omega(W_j, \kappa)$ for all $j \neq 1$. In this case $H_1$ is metrically primitive by Lemma 2.2.21. Then by Lemma 2.2.17 we have a decomposition into totally isotropic subspaces $W_1 = W_1' \oplus W_1''$, where $W_1'$ and $W_1''$ are nonisomorphic irreducible $\mathbb{F}_q[H_1^\Omega]$-modules. Therefore $L^\Omega$ is not metrically completely reducible. Furthermore, it follows from Lemma 2.1.15 that every $L^\Omega$-submodule of $V$ is of the form

$$Z \perp W_{i_1} \perp \cdots \perp W_{i_r}$$

for some $2 \leq i_2 < \cdots < i_r \leq t$ and $Z \in \{0, W_1', W_1'', W_1\}$. Thus every $L^\Omega$-invariant totally isotropic subspace is equal to 0, $W_1'$, or $W_1''$. Furthermore, this also proves (v) in the case where (ii)(a) holds.

For claim (ii), suppose that (ii)(a) does not hold. We will first prove that the action of $L^\Omega$ on $W_i$ is irreducible for all $1 \leq i \leq t$. If (i)(a) does not hold, by (i) the action of $L^\Omega$ on $W_i$ is equal to $H_i^\circ$ for all $1 \leq i \leq t$, which is irreducible. Consider then the case where (i)(a) holds, so $H_i^\circ \not\leq \Omega(W_i, \kappa)$ and $H_j^\circ \leq \Omega(W_j, \kappa)$ for all $j \neq i$. In this case the action of $L^\Omega$ on $W_i$ must be irreducible since (ii)(a) does not hold, and $L^\Omega$ acts irreducibly on $W_j$ for $j \neq i$ since $H_j^\Omega = H_j^\circ$ does.

It follows that the $W_i$ are pairwise nonisomorphic irreducible $\mathbb{F}_q[L^\Omega]$-modules, so by Lemma 2.1.15 every $L$-submodule of $V$ is of the form $W_{i_1} \perp \cdots \perp W_{i_r}$ for some indices $1 \leq i_1 < \cdots < i_r \leq t$. This implies that $L^\Omega$ is metrically completely reducible, and at the same time proves that (iv) holds.

Claim (v) follows from (iv) when $L^\Omega$ is metrically completely reducible. When $L$ is not metrically completely reducible, we know that (ii)(a) holds, and in this case we have seen in the proof (iii) that (v) holds. □

We end this section with a description of when two subgroups of the form in Lemma 2.3.9 are conjugate in $\Delta(V, \kappa)$, and similarly for Lemma 2.3.7 and conjugacy in $I(V, \kappa)$.

**Lemma 2.3.13** *Let $L = (H_1 \times \cdots \times H_t) \cap \Delta(V, \kappa)$ and $\overline{L} = (\overline{H_1} \times \cdots \times \overline{H_s}) \cap \Delta(V, \kappa)$ be as in Lemma 2.3.9. Assume that $L$ is of multiplier 1.*

*Then $L$ and $\overline{L}$ are conjugate in $\Delta(V, \kappa)$ if and only if all of the following hold:*

*(i) $t = s$;*
*(ii) There exists a permutation $\pi$ of $\{1, \ldots, t\}$ such that $H_i$ is similar to $\overline{H_{\pi(i)}}$ for all $1 \leq i \leq t$.*

***Proof*** Let $V = W_1 \perp \cdots \perp W_t$ be the decomposition defining $L$, so $H_i \leq \Delta(W_i, \kappa)$ is maximal irreducible solvable and $H_i \neq 1$ for all $1 \leq i \leq t$. Similarly let $V = Z_1 \perp \cdots \perp Z_s$ define $\overline{L}$, so $\overline{H_i} \leq \Delta(Z_i, \kappa)$ is maximal irreducible solvable and $\overline{H_i} \neq 1$ for all $1 \leq i \leq s$.

Suppose first that $L$ and $\overline{L}$ are conjugate in $\Delta(V, \kappa)$, say $gLg^{-1} = \overline{L}$ for $g \in \Delta(V, \kappa)$. It follows from Lemma 2.3.10 that $t$ is the number of irreducible $L$-submodules of $V$, and $s$ is the number of irreducible $\overline{L}$-submodules of $V$. Therefore $t = s$.

Let $1 \leq i \leq t$. Since $W_i$ is an irreducible $L$-module, it follows that $gW_i$ is an irreducible $\overline{L}$-module, and so $gW_i = Z_j$ for some $1 \leq j \leq t$ by Lemma 2.3.10. Thus there exists a permutation $\pi$ of $\{1, \ldots, t\}$ such $gW_i = Z_{\pi(i)}$ for all $1 \leq i \leq t$, in which case $gH_ig^{-1} = \overline{H_{\pi(i)}}$.

Conversely, suppose that (i) and (ii) hold. For $1 \leq i \leq t$, let $g_i : W_i \to Z_{\pi(i)}$ be a similarity such that $g_iH_ig_i^{-1} = \overline{H_{\pi(i)}}$. Because each $H_i$ is of multiplier 1, by multiplying $g_i$ with a suitable element of $H_i$ we can assume that $\tau(g_i) = 1$. Then $g = g_1 \cdots g_t \in \Delta(V, \kappa)$ and $gLg^{-1} = \overline{L}$. □

**Lemma 2.3.14** *Let $L^\circ = H_1^\circ \times \cdots \times H_t^\circ$ and $\overline{L} = \overline{H_1^\circ} \times \cdots \times \overline{H_s^\circ}$ be as in Lemma 2.3.7.*

*Then $L$ and $\overline{L}$ are conjugate in $I(V, \kappa)$ if and only if all of the following hold:*

*(i) $t = s$;*
*(ii) There exists a permutation $\pi$ of $\{1, \ldots, t\}$ such that $H_i^\circ$ is isometric to $\overline{H_{\pi(i)}^\circ}$ for all $1 \leq i \leq t$.*

***Proof*** By applying Lemma 2.3.8, the result follows with the same proof as Lemma 2.3.13. □

## 2.4 Metrically Primitive Maximal Irreducible Solvable Subgroups

Let $G \leq \Delta(V, \kappa)$ be maximal irreducible solvable, and suppose that $G$ is metrically primitive. We have $G = \Delta(V, \kappa)$ if $\dim V = 1$, so we assume for the rest of this section that $n = \dim V > 1$.

As a first step towards understanding the structure of $G$, choose a subgroup $F_0 \leq G$ with the following properties:

(F1) $F_0$ is abelian and normal in $G$;
(F2) $F_0 \leq I(V, \kappa)$.
(F3) $F_0$ is maximal among the subgroups of $G$ satisfying (F1) and (F2).

(If the form $\kappa$ is nonzero, it is not necessarily true that $F_0$ is maximal among the abelian normal subgroups of $G$.) Note that $F_0 \neq 1$ by Lemma 2.1.6. We will see later that $F_0$ is uniquely determined (Corollary 5.7.6).

**Remark 2.4.1** A subgroup satisfying properties (F1)–(F3) is what Jordan calls a *premier faisceau* of $G$ in [51, §IV, p. 12].

Our first aim is to describe the structure of $F_0$ and $N_{\Delta(V,\kappa)}(F_0)$ explicitly.

**Lemma 2.4.2** *Suppose that $A \leq \mathrm{GL}(V)$ is abelian and irreducible. Then $A$ is cyclic and $p \nmid |A|$.*

***Proof*** Since $A$ is irreducible, by Schur's lemma $C_{\mathrm{End}(V)}(A)$ is a division ring, and therefore a field since we are working over $\mathbb{F}_q$. We have $A \leq C_{\mathrm{GL}(V)}(A)$, and the multiplicative group of a finite field is cyclic of order $p^\nu - 1$, so $A$ must be cyclic with order coprime to $p$. □

**Lemma 2.4.3** *Suppose that $G$ is metrically primitive irreducible, and let $N$ be an abelian normal subgroup of $G$ such that $N \leq G^\circ$. Then $N$ is cyclic, $p \nmid |N|$, and $V^g = 0$ for all $g \in N \setminus \{1\}$.*

***Proof*** Since $G$ is metrically primitive, by Proposition 2.1.13 either $V \downarrow N$ is homogeneous, or $V \downarrow N \cong W \oplus W^*$, where $W$ is a homogeneous $\mathbb{F}_q[N]$-module. Thus it follows from Lemma 2.4.2 that $N$ is cyclic and $p \nmid |N|$.

Since $N$ is cyclic, every $H \leq N$ is normal in $G$, and thus $G$ acts on $V^H$. Because $G$ is irreducible, we have either $V^H = V$ or $V^H = \{0\}$. If $V^H = V$, then clearly $H = \{1\}$. Therefore $V^H = 0$ for every $1 \neq H \leq N$, which proves that $V^g = 0$ for all $g \in N \setminus \{1\}$. □

In particular Lemma 2.4.3 applies for $F_0$, so let $f$ be a generator for $F_0$. Let $\mathbb{K}/\mathbb{F}_q$ be a splitting field for the representation of $F_0$ on $V$, in other words, the splitting field for the characteristic polynomial of $f$. Then $\mathbb{K} = \mathbb{F}_{q^\nu}$ for some integer $\nu \geq 1$.

We define $V' = \mathbb{K} \otimes_{\mathbb{F}_q} V$. Then $V'$ is a $n$-dimensional $\mathbb{K}$-vector space, and $V$ can be identified as the subspace spanned by vectors of the form $1 \otimes v$. Throughout we will denote $\lambda \otimes v$ by $\lambda v$ for all $\lambda \in \mathbb{K}$ and $v \in V$.

Then $\mathrm{GL}(V) \leq \mathrm{GL}(V')$. By extending the form $\kappa$ to $V'$, we get $\Delta(V,\kappa) \leq \Delta(V',\kappa)$ and $I(V,\kappa) \leq I(V',\kappa)$.

Let $s : V' \to V'$ be the map defined by $s = s_0 \otimes I_V$, where $s_0 : \mathbb{K} \to \mathbb{K}$ is the Frobenius automorphism $\xi \mapsto \xi^q$ that generates $\mathrm{Gal}(\mathbb{K}/\mathbb{F}_q)$. Then $s$ is a semilinear map, with

$$s(v_1 + v_2) = s(v_1) + s(v_2) \qquad \text{for all } v_1, v_2 \in V'.$$
$$s(\lambda v) = \lambda^q s(v) \qquad \text{for all } v \in V' \text{ and } \lambda \in \mathbb{K}.$$

Furthermore, we have

$$\begin{aligned} V &= (V')^s, \\ \mathrm{GL}(V) &= C_{\mathrm{GL}(V')}(s). \end{aligned} \tag{2.1}$$

For the bilinear form $b$ extended to $V'$, we have $b(s(v), s(w)) = b(v, w)^q$ for all $v, w \in V'$.

**Lemma 2.4.4 ([38, V, Hilfssatz 13.2])** *Let $W' \subseteq V'$ be a $\mathbb{K}$-subspace of $V'$. Then $W' = \mathbb{K} \otimes_{\mathbb{F}_q} W$ for some $\mathbb{F}_q$-subspace $W \subseteq V$ if and only if $s(W') \subseteq W'$.*

Let $W'_\lambda$ be the $f$-eigenspace corresponding to $\lambda \in \mathbb{K}$. Then $s(W'_\lambda) = W'_{\lambda^q}$, so the eigenvalues of $f$ are invariant under the Frobenius automorphism $\xi \mapsto \xi^q$. For $\lambda \in \mathbb{K}$, we define

$$Q'_\lambda = \sum_{0 \leq k < [\mathbb{K}:\mathbb{F}_q]} s^k(W'_\lambda).$$

Then $Q'_\lambda$ is $s$-invariant, so by Lemma 2.4.4 there exists an $F_0$-invariant subspace $Q_\lambda \subseteq V$ such that $Q'_\lambda = \mathbb{K} \otimes_{\mathbb{F}_q} Q_\lambda$.

**Lemma 2.4.5** *Two $f$-eigenspaces $W'_\alpha$ and $W'_\beta$ are orthogonal to each other if $\alpha\beta \neq 1$. Furthermore, an $f$-eigenspace $W'_\alpha$ is totally singular if $\alpha \neq \pm 1$.*

***Proof*** Follows from the fact that the form $\kappa$ on $V'$ is $f$-invariant. □

By Lemma 2.4.5, we can find $\lambda_1, \ldots, \lambda_r \in \mathbb{K}$ such that

$$V' = Q'_{\lambda_1} \perp \cdots \perp Q'_{\lambda_t} \perp (Q'_{\lambda_{t+1}} \oplus Q'_{\lambda_{t+1}^{-1}}) \perp \cdots \perp (Q'_{\lambda_r} \oplus Q'_{\lambda_r^{-1}}).$$

This gives an orthogonal decomposition

$$V = Q_{\lambda_1} \perp \cdots \perp Q_{\lambda_t} \perp (Q_{\lambda_{t+1}} \oplus Q_{\lambda_{t+1}^{-1}}) \perp \cdots \perp (Q_{\lambda_r} \oplus Q_{\lambda_r^{-1}}).$$

Since $G$ is metrically primitive and permutes the subspaces $Q_{\lambda_i}$, we conclude that there exists $\lambda \in \mathbb{K}$ such that one of the following holds:

- $V = Q_\lambda$.
- The form $\kappa$ is nonzero, and $V = Q_\lambda \oplus Q_{\lambda^{-1}}$.

This gives us four distinct types of $G$, which we name as follows.

**Definition 2.4.6** We say that $G$ is of type:

- $\mathcal{B}_0$, if $\kappa = 0$.
- $\mathcal{B}_1$, if $\kappa \neq 0$ and $V = Q_\lambda \oplus Q_{\lambda^{-1}}$, with $\lambda$ and $\lambda^{-1}$ on different orbits under the Frobenius automorphism.
- $\mathcal{B}_2$, if $\kappa \neq 0$, $V = Q_\lambda$, and $W'_\lambda$ is totally singular.
- $\mathcal{B}_3$, if $\kappa \neq 0$, $V = Q_\lambda$, and $W'_\lambda$ is not totally singular.

**Remark 2.4.7** Groups of type $\mathcal{B}_1$, $\mathcal{B}_2$, and $\mathcal{B}_3$ in Definition 2.4.6 correspond to what Jordan calls groups of *Premiére catègorie*, *Deuxiéme catègorie*, and *Troisiéme catègorie*, respectively [51, §V], [47, 614, p. 462].

**Remark 2.4.8** In the case where $G$ is of type $\mathcal{B}_1$, we have $G$ metrically primitive, but imprimitive since $G$ acts on $\{Q_\lambda, Q_{\lambda^{-1}}\}$. For groups of types $\mathcal{B}_2$ and $\mathcal{B}_3$, we will see later (Theorem 5.6.9) that the group $G$ is primitive.

Denote $\mu = \dim_{\mathbb{K}}(W'_\lambda)$. Because $\mathbb{K}$ is a splitting field for $f$, the eigenvalues

$$\lambda, \lambda^q, \ldots, \lambda^{q^{[\mathbb{K}:\mathbb{F}_q]-1}}$$

must be distinct. Therefore $n = \mu[\mathbb{K} : \mathbb{F}_q]$ if $V = Q_\lambda$, and $n = 2\mu[\mathbb{K} : \mathbb{F}_q]$ if $V = Q_\lambda \oplus Q_{\lambda^{-1}}$.

**Lemma 2.4.9** *Let $\alpha \in \mathbb{K}$ be an eigenvalue of $f$ on $V'$. Then* $\dim_{\mathbb{K}}(W'_\alpha) = \mu$.

***Proof*** Because $F_0 \trianglelefteq G$, the group $G$ acts on the $f$-eigenspaces on $V'$. When $G$ is of type $\mathcal{B}_1$, by irreducibility $G$ acts transitively on $\{Q_\lambda, Q_{\lambda^{-1}}\}$ (Lemma 2.2.4 (i)). Thus by replacing $W'_\alpha$ with a $G$-conjugate if necessary, we can assume that $\alpha$ is conjugate to $\lambda$ under the Frobenius automorphism $s_0 : \mathbb{K} \to \mathbb{K}$, $\xi \mapsto \xi^q$.

In this case $\alpha = \lambda^{q^k}$ for some $0 \leq k < [\mathbb{K} : \mathbb{F}_q]$. Then $W'_\alpha = s^k(W'_\lambda)$, where $s = s_0 \otimes I_V$, so $\dim_{\mathbb{K}}(W'_\alpha) = \dim_{\mathbb{K}}(W'_\lambda) = \mu$. □

In Sects. 2.5–2.8 that follow, we will describe the structure of $F_0$ and $N_{\Delta(V,\kappa)}(F_0)$ for all types $\mathcal{B}_0$–$\mathcal{B}_3$, in similar lines to the discussion by Jordan in [51, §V, pp. 14–21]. We summarize some of the basic properties of $F_0$ in Table 2.1. In Sect. 2.9, we will discuss the case $\mu = 1$ in more detail.

**Table 2.1** Summary of the different possibilities for $F_0$

| | |
|---|---|
| Type $\mathcal{B}_0$ | $\kappa = 0$, and $V' = Q'_\lambda$ |
| | $\mathbb{K} = \mathbb{F}_{q^\nu}$ for some $\nu \geq 1$ |
| | $q^\nu > 2$ |
| | $\lvert F_0 \rvert = q^\nu - 1$ |
| Type $\mathcal{B}_1$ | $\kappa \neq 0$, and $V' = Q'_\lambda \oplus Q'_{\lambda^{-1}}$ |
| | $\mathbb{K} = \mathbb{F}_{q^\nu}$ for some $\nu \geq 1$ |
| | $q^\nu > 4$ or $(q, \nu) = (4, 1)$ |
| | $\mathrm{sgn}(\kappa) = +$ if $\kappa$ is a quadratic form |
| | $\lvert F_0 \rvert = q^\nu - 1$ |
| Type $\mathcal{B}_2$ | $\kappa \neq 0$, $V' = Q'_\lambda$, and $W'_\lambda$ is totally singular |
| | $\mathbb{K} = \mathbb{F}_{q^{2\nu}}$ for some $\nu \geq 1$ |
| | $\mathrm{sgn}(\kappa) = (-1)^\mu$ if $\kappa$ is a quadratic form |
| | $\lvert F_0 \rvert = q^\nu + 1$ |
| Type $\mathcal{B}_3$ | $\kappa \neq 0$, $V' = Q'_\lambda$, and $W'_\lambda$ is not totally singular |
| | $\mathbb{K} = \mathbb{F}_q$, and $q$ is odd |
| | $\lambda = -1$, and $f = -I_V$, so $\lvert F_0 \rvert = 2$ |

## 2.5 Groups of Type $\mathcal{B}_0$

We continue with the setup of the previous section. We will consider the case where $G$ is of type $\mathcal{B}_0$, so $\kappa = 0$ and $\Delta(V, \kappa) = \mathrm{GL}(V)$. Denote $\nu = [\mathbb{K} : \mathbb{F}_q]$. Then $\mathbb{K} \cong \mathbb{F}_{q^\nu}$ and

$$V' = Q'_\lambda = W'_\lambda \oplus W'_{\lambda^q} \oplus \cdots \oplus W'_{\lambda^{q^{\nu-1}}}, \tag{2.2}$$

so $n = \mu\nu$.

Fix a $\mathbb{K}$-basis $x_1, \ldots, x_\mu$ of $W'_\lambda$. Then we have a $\mathbb{K}$-basis $(x_i^{(k)})$ of $V'$ by defining $x_i^{(k)} = s^k(x_i)$ for all $1 \leq i \leq \mu$ and $0 \leq k < \nu$. For convenience, we denote $x_i^{(\nu)} = x_i^{(0)}$.

**Lemma 2.5.1** *For $G$ of type $\mathcal{B}_0$, we have $q^\nu > 2$.*

***Proof*** If $q^\nu = 2$, then $\mathbb{K} = \mathbb{F}_2$ and $F_0$ must be trivial. But this is a contradiction, since $F_0 \neq 1$ by Lemma 2.1.6. □

Since $g \in \mathrm{GL}(V)$ if and only if $g$ centralizes $s$ by (2.1), the action of $g \in \mathrm{GL}(V)$ on $V'$ is determined by its action on $W_\lambda$. Indeed, for $g \in \mathrm{GL}(V)$ we have $gx_i^{(k)} = s^k g(x_i)$. We will now define generators for $N_{\mathrm{GL}(V)}(F_0)$, which turns out to be a semilinear group $\Gamma\mathrm{L}_\mu(q^\nu)$.

Let $A \in \mathrm{GL}_\mu(\mathbb{K})$. We define $\mathbb{K}$-linear maps $\psi, g_A : V' \to V'$ by

$$\psi(x_i^{(k)}) = x_i^{(k+1)}$$

$$g_A(x_i^{(k)}) = \sum_{j=1}^{\mu} A_{ji}^{q^k} x_j^{(k)}$$

for all $1 \le i \le \mu$ and $0 \le k < \nu$. We will now describe the structure of $N_{\mathrm{GL}(V)}(F_0)$ and $F_0$.

**Lemma 2.5.2** *The following statements hold:*

*(i)* $N_{\mathrm{GL}(V)}(F_0) = C_{\mathrm{GL}(V)}(F_0) \rtimes \langle\psi\rangle$, *where* $|\psi| = \nu$.
*(ii)* $C_{\mathrm{GL}(V)}(F_0) = \{g_A : A \in \mathrm{GL}_\mu(\mathbb{K})\} \cong \mathrm{GL}_\mu(\mathbb{K})$.

*In particular* $N_{\mathrm{GL}(V)}(F_0)$ *is generated by* $\psi$ *and the set of* $g_A$ *with* $A \in \mathrm{GL}_\mu(\mathbb{K})$.

***Proof*** For claim (i), one checks that $\psi$ centralizes $s$ and normalizes $F_0$, so $\psi \in N_{\mathrm{GL}(V)}(F_0)$. Let $g \in N_{\mathrm{GL}(V)}(F_0)$. Then $g$ acts on the $f$-eigenspaces on $V'$, so there exists $0 \le \delta < \nu$ such that $g(W'_\lambda) = W'_{\lambda^{q^\delta}}$. Then $g\psi^{-\delta}$ stabilizes every $f$-eigenspace, so $g\psi^{-\delta} \in C_{\mathrm{GL}(V)}(F_0)$. From this $N_{\mathrm{GL}(V)}(F_0) = C_{\mathrm{GL}(V)}(F_0) \rtimes \langle\psi\rangle$ follows, and it is clear that $|\psi| = \nu$.

For (ii), note that $g \in \mathrm{GL}(V)$ centralizes $F_0$ if and only if it stabilizes the eigenspaces of $F_0$. Therefore

$$C_{\mathrm{GL}(V)}(F_0) = \{g_A : A \in \mathrm{GL}_\mu(\mathbb{K})\} \cong \mathrm{GL}_\mu(\mathbb{K})$$

as claimed. □

**Lemma 2.5.3** *The eigenvalue* $\lambda$ *is a primitive element of* $\mathbb{K}$, *and* $|F_0| = q^\nu - 1$.

***Proof*** Consider the subgroup $F \le \mathrm{GL}(V)$ generated by the linear map defined by $x_i^{(k)} \mapsto \zeta^{q^k} x_i^{(k)}$, where $\zeta \in \mathbb{K}^\times$ is a primitive element. Then $F$ is normalized by $\psi$ and $F$ centralizes $F_0$, so $F \trianglelefteq N_{\mathrm{GL}(V)}(F_0)$ by Lemma 2.5.2 (i). In particular $F$ is normalized by $G$, so $GF$ is solvable and so $F \trianglelefteq G$ since $G$ is maximal solvable. We have $F_0 \le F$ and $F$ is abelian, so $F_0 = F$ by the maximality of $F_0$. Thus $|F_0| = q^\nu - 1$ and $\lambda$ is a primitive element. □

We have the following relations between the generators of $N_{\mathrm{GL}(V)}(F_0)$:

$$\psi^{-1} f \psi = f^q$$

$$\psi^{-1} g_A \psi = g_{\overline{A}}$$

where $\overline{A}$ is defined by $\overline{A}_{ij} = A_{ij}^q$ for all $1 \le i, j \le \mu$.

**Remark 2.5.4** Conversely, suppose that $n = \mu\nu$ for integers $\mu$ and $\nu$. Then we can construct $f \in \mathrm{GL}(V)$ with $|f| = q^\nu - 1$, such that the splitting field of the characteristic polynomial of $f$ is $\mathbb{K} = \mathbb{F}_{q^\nu}$, and $V' := \mathbb{K} \otimes_{\mathbb{F}_q} V$ decomposes with eigenspaces as in (2.2).

First in $\mathrm{GL}(W) = \mathrm{GL}_\nu(q)$, one can construct such an $f_0$ as a Singer cycle, and we will describe this construction later in Remark 2.9.9. Then taking the action of $f_0$ on $V = W \oplus \cdots \oplus W$ ($\mu$ times), we get the desired element $f$. Alternatively, let $f_0 \in \mathrm{GL}_n(q)$ be a Singer cycle, so $|f_0| = q^n - 1$. Then $q^\nu - 1 \mid q^n - 1$ (Lemma 2.1.20), and

$$f = f_0^{\frac{q^n-1}{q^\nu-1}}$$

has the required properties.

**Remark 2.5.5** In the type $\mathcal{B}_0$ case, we now have a complete description of the possible structures for $F_0$ in $G$, and the normalizer $N_{\mathrm{GL}(V)}(F_0)$ in all cases (Lemma 2.5.2). For $\mu = 1$ we also have a precise description of $G$, since in this case $N_{\mathrm{GL}(V)}(F_0)$ is solvable and thus $G = N_{\mathrm{GL}(V)}(F_0) = \langle f \rangle \rtimes \langle \psi \rangle$. Here $|G| = (q^n - 1)n$, and this subgroup is known as the semilinear group $G = \Gamma\mathrm{L}_1(q^n)$, or the normalizer of a Singer cycle in $\mathrm{GL}_n(q)$. (We will provide more details about the $\mu = 1$ case for all types in Sect. 2.9, see e.g. Remark 2.9.9.)

For $\mu > 1$, we will make the structure of $G$ more precise later in Chap. 4. At this point, by Lemma 2.5.2 we know that there exists $A \in \mathrm{GL}_\mu(\mathbb{K})$ and $\delta \mid \nu$ such that $G$ is generated by $C_G(F_0)$ and $g_A\psi^\delta$. Later we will see that we can arrange $\delta \in \{1, 2\}$ (Lemma 4.3.5 (ii)).

**Remark 2.5.6** An alternative approach for describing the structure of $F_0$, more in the lines of [77, §19], is as follows. One finds that the $\mathbb{F}_q$-algebra generated by $F_0$ in $\mathrm{End}(V)$ is equal to $\mathbb{K} = \mathbb{F}_{q^\nu}$, where $\nu$ divides $n$ and $F_0$ is equal to the multiplicative group $F_0 = \mathbb{K}^\times$. This makes $V$ into a $\mathbb{K}$-vector space of dimension $\mu = n/\nu$. Having fixed a $\mathbb{K}$-vector space basis $x_1, \ldots, x_\mu$ of $V$, we can define $\psi : V \to V$ by

$$\psi\left(\sum_{i=1}^{\mu} \alpha_i x_i\right) = \alpha_i^q x_i$$

for all $\alpha_1, \ldots, \alpha_\mu \in \mathbb{K}$. Then $\psi \in N_{\mathrm{GL}(V)}(F_0)$, and

$$N_{\mathrm{GL}(V)}(F_0) = C_{\mathrm{GL}(V)}(F_0) \rtimes \langle \psi \rangle,$$

where

$$C_G(F_0) = \{g \in \mathrm{GL}(V) : g \text{ is } \mathbb{K}\text{-linear}\}.$$

In this setup $g_A : V \to V$ for $A \in \mathrm{GL}_\mu(\mathbb{K})$ is the linear map with matrix $A$ with respect to the basis $x_1, \ldots, x_\mu$. Furthermore, we have $\psi g_A \psi^{-1} = g_{\overline{A}}$, where $\overline{A}_{ij} = A_{ij}^q$ for all $1 \leq i, j \leq \mu$. In particular $\psi f \psi^{-1} = f^q$.

## 2.6 Groups of Type $\mathcal{B}_1$

We continue with the setup of Sect. 2.4, and consider the case where $G$ is of type $\mathcal{B}_1$. In this case $\kappa$ is a nonzero form, and $V = Q_\lambda \oplus Q_{\lambda^{-1}}$, with $\lambda$ and $\lambda^{-1}$ on different orbits under the Frobenius automorphism $\xi \mapsto \xi^q$. Let $\zeta \in \mathbb{K}^\times$ be a primitive element, and denote $\nu = [\mathbb{K} : \mathbb{F}_q]$, so $\mathbb{K} \cong \mathbb{F}_{q^\nu}$.

**Lemma 2.6.1** *For $G$ of type $\mathcal{B}_1$, we have $q^\nu > 4$ or $(q, \nu) = (4, 1)$.*

***Proof*** If $(q, \nu)$ equals $(2, 1)$, $(2, 2)$, or $(3, 1)$, we have $\lambda^q = \lambda^{-1}$, contradicting the assumption that $\lambda$ and $\lambda^{-1}$ are in different orbits under the Frobenius automorphism $\xi \mapsto \xi^q$ of $\mathbb{K}$. □

**Lemma 2.6.2** *For $G$ of type $\mathcal{B}_1$, each $f$-eigenspace $W'_\alpha$ is totally singular. Furthermore, the subspaces $Q_\lambda$ and $Q_{\lambda^{-1}}$ are totally singular.*

***Proof*** Let $\alpha \in \mathbb{K}$ be an eigenvalue of $f$ on $V'$. Then $\alpha$ and $\alpha^{-1}$ are in different orbits under the Frobenius automorphism, so $\alpha \neq \pm 1$ and $W'_\alpha$ is totally singular (Lemma 2.4.5). Because $W'_\alpha$ orthogonal to $W'_\beta$ if $\beta \neq \alpha^{-1}$ (Lemma 2.4.5), we conclude that $Q'_\lambda$ and $Q'_{\lambda^{-1}}$ are totally singular. Therefore $Q_\lambda$ and $Q_{\lambda^{-1}}$ are totally singular. □

**Lemma 2.6.3** *Suppose that $\kappa$ is a quadratic form. Then for $G$ of type $\mathcal{B}_1$, we have $\varepsilon = +$.*

***Proof*** By Lemma 2.6.2 we have a decomposition $V = Q_\lambda \oplus Q_{\lambda^{-1}}$ into totally singular subspaces, so $\varepsilon = +$. □

Fix a $\mathbb{K}$-basis $x_1, \ldots, x_\mu$ of $W'_\lambda$. It follows from Lemma 2.4.5 that $W'_\lambda \oplus W'_{\lambda^{-1}}$ is non-degenerate and the subspaces $W'_{\lambda^{\pm 1}}$ are totally singular, so there exists a $\mathbb{K}$-basis $y_1, \ldots, y_\mu$ of $W_{\lambda^{-1}}$ such that $b(x_i, y_j) = \delta_{i,j}$. Then we have a $\mathbb{K}$-basis $(x_i^{(k)}, y_i^{(k)})$ of $V'$ by defining $x_i^{(k)} = s^k(x_i)$ and $y_i^{(k)} = s^k(y_i)$ for all $1 \leq i \leq \mu$ and $0 \leq k < \nu$. Note that

$$b(x_i^{(k)}, y_j^{(\ell)}) = \delta_{i,j}\delta_{k,\ell}$$

for all $1 \leq i, j \leq \mu$ and $0 \leq k, \ell < \nu$. For the generator $f$ of $F_0$, we have $f(x_i^{(k)}) = \lambda^{q^k} x_i^{(k)}$ and $f(y_i^{(k)}) = \lambda^{-q^k} y_i^{(k)}$ for all $1 \leq i, j \leq \mu$ and $0 \leq k, \ell < \nu$.

We will now describe generators for $N_{\Delta(V,\kappa)}(F_0)$. Recall that we denote $e = 0$ if $\kappa$ is a quadratic form, and $e = 1$ if $\kappa$ is an alternating bilinear form. Let $A \in \mathrm{GL}_\mu(\mathbb{K})$. Define $\mathbb{K}$-linear maps $\eta, \varphi, \psi, g_A : V' \to V'$ by

$$\eta(x_i^{(k)}) = x_i^{(k)} \qquad \eta(y_i^{(k)}) = \zeta^{\frac{q^\nu - 1}{q-1}} y_i^{(k)}$$

$$\varphi(x_i^{(k)}) = y_i^{(k)} \qquad \varphi(y_i^{(k)}) = (-1)^e x_i^{(k)}$$

$$\psi(x_i^{(k)}) = x_i^{(k+1)} \qquad \psi(y_i^{(k)}) = y_i^{(k+1)}$$

$$g_A(x_i^{(k)}) = \sum_{j=1}^{\mu} A_{ji}^{q^k} x_j^{(k)} \qquad g_A(y_i^{(k)}) = \sum_{j=1}^{\mu} B_{ji}^{q^k} y_j^{(k)}$$

for all $1 \leq i \leq \mu$ and $0 \leq k < \nu$, where $B = A^{-T}$ (inverse transpose).

**Lemma 2.6.4** *The following statements hold:*

*(i)* $N_{\Delta(V,\kappa)}(F_0) = N_{I(V,\kappa)}(F_0) \rtimes \langle \eta \rangle$, *where* $|\eta| = q - 1$.
*(ii)* $N_{I(V,\kappa)}(F_0) = \langle C_{I(V,\kappa)}(F_0), \varphi, \psi \rangle$, *where* $|\psi| = \nu$ *and* $\varphi^2 = (-1)^e I_V$.
*(iii)* $N_{I(V,\kappa)}(F_0)/C_{I(V,\kappa)}(F_0) \cong C_2 \times C_\nu$.
*(iv)* $C_{I(V,\kappa)}(F_0) = \{g_A : A \in \mathrm{GL}_\mu(\mathbb{K})\} \cong \mathrm{GL}_\mu(\mathbb{K})$.
*(v)* $C_{\Delta(V,\kappa)}(F_0) = C_{I(V,\kappa)}(F_0) \rtimes \langle \eta \rangle$.

*In particular* $N_{\Delta(V,\kappa)}(F_0)$ *is generated by* $\eta$, $\psi$, $\varphi$, *and the set of* $g_A$ *with* $A \in \mathrm{GL}_\mu(\mathbb{K})$.

***Proof*** One checks that $\eta \in \Delta(V, \kappa)$ and $\eta$ centralizes $F_0$, so $\eta \in N_{\Delta(V,\kappa)}(F_0)$. Now $\tau(\eta) = \zeta^{\frac{q^\nu - 1}{q-1}}$ is a primitive element of $\mathbb{F}_q^\times$ with $|\eta| = q - 1$, so

$$N_{\Delta(V,\kappa)}(F_0) = N_{I(V,\kappa)}(F_0) \rtimes \langle \eta \rangle$$

and (i) holds.

For (ii), note that $\varphi$ centralizes $s$, normalizes $F_0$, and leaves the form $\kappa$ invariant, so $\varphi \in N_{I(V,\kappa)}(F_0)$. Moreover $\varphi$ swaps the two subspaces $Q'_\lambda$ and $Q'_{\lambda^{-1}}$, and $\varphi^2 = (-1)^e I_{V'}$. Similarly $\psi \in N_{I(V,\kappa)}(F_0)$, with $|\psi| = \nu$, and $\varphi\psi = \psi\varphi$.

Let $g \in N_{I(V,\kappa)}(F_0)$. Then $g$ acts on $\{Q'_\lambda, Q'_{\lambda^{-1}}\}$, so there exists $c_1 \in \{0, 1\}$ such that $g\varphi^{-c_1}$ leaves the subspaces $Q'_\lambda$ and $Q'_{\lambda^{-1}}$ invariant. Then $g\varphi^{-c_1}$ maps $W'_\lambda$ to $W'_{\lambda q^{c_2}}$ for some $0 \leq c_2 < \nu$. Then $g\varphi^{-c_1}\psi^{-c_2}$ leaves all of the eigenspaces $W'_{\lambda^{\pm q^i}}$ invariant, in other words, $g\varphi^{-c_1}\psi^{-c_2} \in C_{I(V,\kappa)}(F_0)$. Therefore (ii) holds, and it is clear that

$$\frac{N_{I(V,\kappa)}(F_0)}{C_{I(V,\kappa)}(F_0)} = \langle \overline{\varphi}, \overline{\psi} \rangle \cong C_2 \times C_\nu,$$

as claimed by (iii).

We have $g_A \in C_{I(V,\kappa)}(F_0)$. Since $g \in \mathrm{GL}(V)$ centralizes $F_0$ if and only if it stabilizes the eigenspaces of $F_0$, it follows that

$$C_{I(V,\kappa)}(F_0) = \{g_A : A \in \mathrm{GL}_\mu(\mathbb{K})\} \cong \mathrm{GL}_\mu(\mathbb{K})$$
$$C_{\Delta(V,\kappa)}(F_0) = C_{I(V,\kappa)}(F_0) \rtimes \langle \eta \rangle.$$

as claimed by (iv) and (v). □

**Lemma 2.6.5** *The eigenvalue* $\lambda$ *is a primitive element of* $\mathbb{K}$, *and* $|F_0| = q^\nu - 1$.

***Proof*** Let $F$ be the subgroup generated by the $\mathbb{K}$-linear map $V' \to V'$ defined by $x_i^{(k)} \mapsto \zeta^{q^k} x_i^{(k)}$ and $y_i^{(k)} \mapsto \zeta^{-q^k} y_i^{(k)}$. We have $F \le C_{I(V,\kappa)}(F_0)$. It is straightforward to see that $F$ is normalized by $\varphi$ and $\psi$, and that $F$ is centralized by $\eta$ and $g_A$ for all $A \in \mathrm{GL}_\mu(\mathbb{K})$. Thus it follows from Lemma 2.6.4 that $F \trianglelefteq N_{\Delta(V,\kappa)}(F_0)$.

Since $G$ is maximal solvable, we must have $F \le G$. On the other hand $F_0 \le F \le I(V,\kappa)$, so $F_0 = F$ by the maximality of $F_0$. In particular $|F_0| = |\zeta| = q^\nu - 1$ and $\lambda$ must be a primitive element of $\mathbb{K}$. □

By Lemma 2.6.5, we can assume without loss of generality that $\lambda = \zeta$. Then we have the following relations among the generators of $N_{\Delta(V,\kappa)}(F_0)$:

$$\begin{aligned} \psi\eta &= \eta\psi & \varphi\psi &= \psi\varphi \\ \psi^{-1} f \psi &= f^q & \varphi f \varphi^{-1} &= f^{-1} \\ \varphi\eta\varphi^{-1} &= f^{\frac{q^\nu-1}{q-1}}\eta & \varphi g_A \varphi^{-1} &= g_{A^{-T}} \\ \psi^{-1} g_A \psi &= g_{\overline{A}} \end{aligned} \tag{2.3}$$

where $\overline{A}$ is defined by $\overline{A}_{ij} = A_{ij}^q$ for all $1 \le i, j \le \mu$.

**Lemma 2.6.6** *The group* $G$ *is of multiplier* 1 *and* $G = G^\circ \rtimes \langle\eta\rangle$.

***Proof*** It follows from the relations in (2.3) that the generators of $N_{\Delta(V,\kappa)}(F_0)$ normalize the abelian subgroup $\langle F_0, \eta\rangle$. In particular $G$ normalizes $\langle F_0, \eta\rangle$, so $\langle F_0, \eta\rangle \le G$ since $G$ is maximal solvable in $\Delta(V,\kappa)$.

Hence $\eta \in G$. Now $\tau(\eta) = \zeta^{\frac{q^\nu-1}{q-1}}$ is a primitive element of $\mathbb{F}_q^\times$, so $G$ is of multiplier 1 and $G = G^\circ \rtimes \langle\eta\rangle$. □

**Remark 2.6.7 (Cf. Remark 2.5.4)** Conversely, suppose that $n = 2\mu\nu$ for integers $\mu, \nu \ge 1$. Suppose that $\kappa$ is an alternating bilinear form, or a quadratic form with $\mathrm{sgn}(\kappa) = +$ (Lemma 2.6.3).

Then we can construct $f \in I(V,\kappa)$ with $|f| = q^\nu - 1$ such that $f$ has splitting field $\mathbb{K} = \mathbb{F}_{q^\nu}$ and $V' = \mathbb{K} \otimes_{\mathbb{F}_q} V$ decomposes into $f$-eigenspaces

$$V' = Q'_\lambda \oplus Q'_{\lambda^{-1}} = W'_\lambda \oplus \cdots \oplus W'_{\lambda^{q^{\nu-1}}} \oplus W'_{\lambda^{-1}} \oplus \cdots \oplus W'_{\lambda^{-q^{\nu-1}}}$$

where $\dim_{\mathbb{K}} W'_{\lambda^{\pm q^i}} = \mu$ for all $0 \le i < \nu$.

First in the $\mu = 1$ case, this can be done with a Singer cycle construction, which will be described later in Remark 2.9.10. It follows from the construction given there that we can find $f_0$ with the desired properties in $I(W,\kappa')$, where $\dim W = 2\nu$ and $\kappa'$ is a form of the same type as $\kappa$. Then identifying $V = W \perp \cdots \perp W$ ($\mu$ times), the action of $f_0$ on $V$ provides $f$ with the desired properties.

Alternatively, let $f_0 \in I(V, \kappa)$ be the element constructed in the case $\mu = 1$, so $|f_0| = q^{n/2} - 1$. Then $q^\nu - 1 \mid q^{n/2} - 1$ (Lemma 2.1.20), and one can check that

$$f = f_0^{\frac{q^{n/2}-1}{q^\nu - 1}}$$

has all the required properties.

**Remark 2.6.8** With the results of this section, we have a complete description of the possibilities for $F_0 \leq I(V, \kappa)$ when $G$ is of type $\mathcal{B}_1$. In the case $\mu = 1$ it follows from Lemma 2.6.4 that $C_{I(V,\kappa)}(F_0) = F_0$ and $N_{\Delta(V,\kappa)}(F_0)$ is solvable, so

$$G = N_{\Delta(V,\kappa)}(F_0) = \langle f, \psi, \varphi, \eta \rangle.$$

Therefore for $\mu = 1$, we have $|G^\circ| = n(q^\nu - 1)$ and $|G| = n(q^\nu - 1)(q - 1)$, where $n = 2\nu$. (For more on the $\mu = 1$ case, see Sect. 2.9 and Remark 2.9.10.)

For $\mu > 1$, we will describe $G$ more precisely in Chap. 4. At this point we can note the following, analogously to Remark 2.5.5. Since $G$ is irreducible, there exists some $g \in G$ that swaps the two subspaces $Q'_\lambda$ and $Q'_{\lambda^{-1}}$. Because $\eta$ leaves these subspaces invariant and $G = G^\circ \rtimes \langle \eta \rangle$ (Lemma 2.6.6), we can arrange $g \in G^\circ$, in which case $g = g_A \varphi \psi^\alpha$ for some $\alpha \geq 0$ (Lemma 2.6.4). Therefore the image of $G^\circ / C_{G^\circ}(F_0)$ in $N_{I(V,\kappa)}(F_0)/C_{I(V,\kappa)}(F_0)$ is either a cyclic subgroup $\langle \overline{\varphi \psi^\delta} \rangle$ or a noncyclic subgroup $\langle \overline{\varphi}, \overline{\psi^\delta} \rangle$, for some $\delta \mid \nu$.

Therefore one of the following holds:

- There exists $A \in \mathrm{GL}_\mu(\mathbb{K})$ and $\delta \mid \nu$ such that $G^\circ$ is generated by $C_{G^\circ}(F_0)$ and $g_A \varphi \psi^\delta$;
- There exist $A, B \in \mathrm{GL}_\mu(\mathbb{K})$ and $\delta \mid \nu$ such that $G^\circ$ is generated by $C_{G^\circ}(F_0)$, $g_A \varphi$, and $g_B \psi^\delta$.

Later we will see that we can arrange $\delta \in \{1, 2\}$ (Lemma 4.3.13 (iii)).

## 2.7 Groups of Type $\mathcal{B}_2$

We continue with the setup of Sect. 2.4, and consider the case where $G$ is of type $\mathcal{B}_2$. In this case $\kappa$ is a nonzero form, and $W'_\lambda$ is totally singular. Then all of the eigenspaces $s^k(W'_\lambda) = W'_{\lambda q^k}$ are totally singular.

Since $\kappa$ is non-degenerate, it follows that for each $0 \leq k < [\mathbb{K} : \mathbb{F}_q]$ there exists a unique $0 \leq \ell < [\mathbb{K} : \mathbb{F}_q]$ such that $k \neq l$ and $s^k(W'_\lambda) \oplus s^\ell(W'_\lambda)$ is non-degenerate (Lemma 2.4.5). In particular, there exists a unique $0 < \nu < [\mathbb{K} : \mathbb{F}_q]$ such that $W'_\lambda \oplus s^\nu(W'_\lambda)$ is non-degenerate.

**Lemma 2.7.1** *We have* $\lambda^{1+q^\nu} = 1$ *and* $[\mathbb{K} : \mathbb{F}_q] = 2\nu$.

***Proof*** For $v \in W'_\lambda$ and $w \in s^\nu(W'_\lambda)$, we have

$$b(v, w) = b(fv, fw) = \lambda^{1+q^\nu} b(v, w).$$

Because the bilinear form $b$ is non-degenerate on $W'_\lambda \oplus s^\nu(W'_\lambda)$, it follows that $\lambda^{1+q^\nu} = 1$.

This implies $\lambda^{q^{2\nu}-1} = 1$, so $\lambda$ lies in the subfield of $\mathbb{K}$ isomorphic to $\mathbb{F}_{q^\delta}$, where $\delta = \gcd(2\nu, [\mathbb{K} : \mathbb{F}_q])$. Since $\mathbb{K}$ is the splitting field for $F_0$, we have $\delta = [\mathbb{K} : \mathbb{F}_q]$. Then $[\mathbb{K} : \mathbb{F}_q]$ divides $2\nu$, which combined with $0 < \nu < [\mathbb{K} : \mathbb{F}_q]$ implies that $[\mathbb{K} : \mathbb{F}_q] = 2\nu$. □

Let $\zeta \in \mathbb{K}^\times$ be a primitive element. We define a form $\widehat{b_0} : W'_\lambda \times W'_\lambda \to \mathbb{K}$ by $\widehat{b_0}(v, w) = b(v, s^\nu(w))$ for all $v, w \in W'_\lambda$. Since

$$b(s^\nu(w), v)^{q^\nu} = b(s^{2\nu}(w), s^\nu(v)) = b(w, s^\nu(v))$$

for all $v, w \in V'$, it follows that

$$\widehat{b_0}(v, w) = (-1)^e \widehat{b_0}(w, v)^{q^\nu}$$

for all $v, w \in V'$.

Thus according to whether $e = 0$ or $e = 1$, the form $\widehat{b_0}$ is Hermitian or skew-Hermitian with respect to the automorphism $\xi \mapsto \xi^{q^\nu}$ of $\mathbb{K}$. Furthermore, since $W'_\lambda \oplus W'_{\lambda^{q^\nu}}$ is a non-degenerate subspace with respect to $b$, it follows that $\widehat{b_0}$ is non-degenerate.

Multiplying a skew-Hermitian form with $\xi = \zeta^{\frac{q^\nu+1}{2}}$ makes the form Hermitian, since $\xi^{q^\nu} = -\xi$. We define $\widehat{b} := \xi^{-e}\widehat{b_0}$. Then $\widehat{b} : W'_\lambda \times W'_\lambda \to \mathbb{K}$ is a non-degenerate Hermitian form such that

$$b(v, s^\nu(w)) = \begin{cases} \widehat{b}(v, w), & \text{if } e = 0. \\ \zeta^{\frac{q^\nu+1}{2}} \widehat{b}(v, w), & \text{if } e = 1. \end{cases}$$

for all $v, w \in W'_\lambda$.

Let $x_1, \ldots, x_\mu$ be a basis of $W'_\lambda$. Then we have a $\mathbb{K}$-basis $(x_i^{(k)})$ of $V'$ by defining $x_i^{(k)} = s^k(x_i)$ for all $1 \le i \le \mu$ and $0 \le k < 2\nu$. Then

$$b(x_i^{(k)}, x_j^{(\ell)}) = \begin{cases} \delta_{i,j}\delta_{\ell,k+\nu}\widehat{b}(x_i, x_j)^{q^k}, & \text{if } e = 0. \\ \zeta^{\frac{q^k(q^\nu+1)}{2}} \delta_{i,j}\delta_{\ell,k+\nu}\widehat{b}(x_i, x_j)^{q^k}, & \text{if } e = 1. \end{cases} \tag{2.4}$$

for all $1 \le i, j \le \mu$ and $0 \le k \le \ell < 2\nu$.

Often we will take $x_1, \ldots, x_\mu$ to be an orthonormal basis of $(W'_\lambda, \widehat{b})$. (Recall that any non-degenerate Hermitian space over $\mathbb{K}$ admits an orthonormal basis.) In this case

$$b(x_i^{(k)}, x_j^{(\ell)}) = \begin{cases} \delta_{i,j}\delta_{\ell,k+\nu}, & \text{if } e = 0. \\ \zeta^{\frac{q^k(q^\nu+1)}{2}}\delta_{i,j}\delta_{\ell,k+\nu}, & \text{if } e = 1. \end{cases}$$

for all $1 \le i, j \le \mu$ and $0 \le k \le \ell < 2\nu$.

Occasionally when $\mu = 2\mu'$ is even, we will use a "hyperbolic" basis $x_1, \ldots, x_{\mu'}, x_{\mu'+1}, \ldots, x_\mu$ such that $\widehat{b}(x_i, x_{i+\mu'}) \neq 0$ for $1 \le i \le \mu'$, with rest of the products equal to zero. Furthermore, with the bases that we use, there exist integers $k_1, \ldots, k_{\mu'}$ such that

$$\widehat{b}(x_i, x_{i+\mu'}) = (-1)^{k_i}\zeta^{\frac{q^\nu+1}{2}} \tag{2.5}$$

for all $1 \le i \le \mu'$. Note that then $\widehat{b}(x_{i+\mu'}, x_i) = -\widehat{b}(x_i, x_{i+\mu'})$.

Next in the case where $e = 0$, we will determine the type of the quadratic form $Q$ on $V$. Note that since $Q'_\lambda$ is a totally singular subspace of $V'$ with dimension $n/2$, the quadratic form $Q$ on $V'$ is of type $(+)$. However, for the type of the quadratic form on $V$, this is not necessarily the case and we shall prove the following.

**Lemma 2.7.2** *Suppose that $\kappa$ is a quadratic form. Then for $G$ of type $\mathcal{B}_2$, we have $\varepsilon = (-1)^\mu$.*

***Proof*** We argue similarly to [51, pp. 18–19]. Choose an orthonormal basis $x_1, \ldots, x_\mu$ of $(W'_\lambda, \widehat{b})$. For $1 \le i \le \mu$, let $Z'_i$ be the $\mathbb{K}$-subspace of $V$ with basis

$$\{x_i^{(k)} : 0 \le k < 2\nu\} = \{s^k(x_i) : 0 \le k < 2\nu\}.$$

Then $Z'_i$ is $s$-invariant, so by Lemma 2.4.4 we have $Z'_i = \mathbb{K} \otimes_{\mathbb{F}_q} Z_i$ for some $\mathbb{F}_q$-subspace $Z_i$ of $V$. Thus

$$V = Z_1 \perp \cdots \perp Z_\mu,$$

with $\dim Z_i = 2\nu$, so by [55, Proposition 2.5.11] it will suffice to prove that the quadratic form on $(Z, \kappa) := (Z_i, \kappa)$ is of type $(-)$.

For this we shall count the number of solutions to $Q(v) = 1$ in $(Z, \kappa)$. Let $1 \le i \le \mu$ and denote $x = x_i$, so $(s^k(x))_{0 \le k < 2\nu}$ is a basis of $Z' := Z'_i$. For $v \in Z'$, we have $v \in Z$ if and only if $s(v) = v$, which is equivalent to

$$v = \sum_{0 \le i < 2\nu} \delta^{q^i} s^i(x)$$

for some $\delta \in \mathbb{K}$. For such a $v \in Z$, we have

$$Q(v) = \sum_{0 \le i < \nu} \delta^{q^i} \delta^{q^{i+\nu}} = \sum_{0 \le i < \nu} (\delta^{q^\nu + 1})^{q^i}.$$

Denote $z = \delta^{q^\nu+1}$. Then $z^{q^\nu - 1} = 1$, so $z$ is contained in the subfield $\mathbb{F}_{q^\nu}$ of $\mathbb{K}$, and $Q(v)$ is the trace $\mathrm{Tr}_{\mathbb{F}_{q^\nu}/\mathbb{F}_q}(z)$. Since the trace is a surjective $\mathbb{F}_q$-linear map, there are a total of $q^{\nu-1}$ solutions to $\mathrm{Tr}_{\mathbb{F}_{q^\nu}/\mathbb{F}_q}(z) = 1$ in $\mathbb{F}_{q^\nu}$.

For each of the solutions $z$, there are $q^\nu + 1$ solutions to $\delta^{q^\nu+1} = z$ in $\mathbb{K}$. Thus there are a total of $q^{\nu-1}(q^\nu + 1) = q^{2\nu-1} + q^{\nu+1}$ vectors $v \in Z$ such that $Q(v) = 1$. For a quadratic form of type (+), the number of such vectors would be $q^{2\nu-1} - q^{\nu+1}$ [59, Theorem 6.26, Theorem 6.32], so we conclude that $(Z_i, \kappa)$ is of type (−). □

With the bilinear form $b$ completely defined, we can now describe generators for $N_{\Delta(V,\kappa)}(F_0)$. Let $A \in \mathrm{GL}_\mu(\mathbb{K})$. We define $\mathbb{K}$-linear maps $\eta, \psi, g_A : V' \to V'$ by

$$\begin{aligned}
\eta(x_i^{(k)}) &= \left(\zeta^{\frac{q^\nu-1}{q-1}}\right)^{q^k} x_i^{(k)} \\
\psi(x_i^{(k)}) &= \alpha^{q^k} x_i^{(k+1)} \\
g_A(x_i^{(k)}) &= \sum_{j=1}^{\mu} A_{ji}^{q^k} x_j^{(k)}
\end{aligned} \tag{2.6}$$

for all $1 \le i \le \mu$ and $0 \le k < 2\nu$, where $\alpha \in \mathbb{K}$ is a scalar that depends on the value of $e \in \{0, 1\}$ and the basis of $W'_\lambda$ that we are using. We choose

$$\alpha = \begin{cases} 1, & \text{if } (x_i) \text{ is an orthonormal basis and } e = 0. \\ \zeta^{-\frac{q-1}{2}}, & \text{if } (x_i) \text{ is an orthonormal basis and } e = 1. \\ \zeta^{-\frac{(q-1)(e+1)}{2}} & \text{if } (x_i) \text{ is a hyperbolic basis with products as in (2.5).} \end{cases}$$

**Lemma 2.7.3** *The following statements hold:*

*(i)* $N_{\Delta(V,\kappa)}(F_0) = \langle N_{I(V,\kappa)}(F_0), \eta \rangle$.
*(ii)* $N_{I(V,\kappa)}(F_0) = \langle C_{I(V,\kappa)}(F_0), \psi \rangle$, *where* $\psi^{2\nu} = (-1)^e I_V$.
*(iii)* $N_{I(V,\kappa)}(F_0)/C_{I(V,\kappa)}(F_0) \cong C_{2\nu}$.
*(iv)* $C_{I(V,\kappa)}(F_0) = \{g_A : A \in I(W'_\lambda, \widehat{b})\} \cong I(W'_\lambda, \widehat{b}) \cong \mathrm{GU}_\mu(\mathbb{K})$.
*(v)* $C_{\Delta(V,\kappa)}(F_0) = \{g_A : A \in \Delta(W'_\lambda, \widehat{b})\} \cong \Delta\mathrm{U}_\mu(\mathbb{K})$.
*(vi)* $C_{\Delta(V,\kappa)}(F_0) = \langle C_{I(V,\kappa)}(F_0), \eta \rangle$.

*In particular* $N_{\Delta(V,\kappa)}(F_0)$ *is generated by* $\eta$, $\psi$, *and the set of* $g_A$ *with* $A \in I(W'_\lambda, \widehat{b})$.

***Proof*** It is clear that $\eta$ centralizes $s$ and $F_0$, and that $\eta \in \Delta(V, \kappa)$. Here $\tau(\eta) = \zeta^{\frac{q^{2\nu}-1}{q-1}}$ is a primitive element of $\mathbb{F}_q$, so $N_{\Delta(V,\kappa)}(F_0) = \langle N_{I(V,\kappa)}(F_0), \eta\rangle$, as claimed by (i).

It is readily seen that $\psi$ centralizes $s$ and normalizes $F_0$, no matter what the scalar $\alpha \in \mathbb{K}$ is, so $\psi \in N_{\mathrm{GL}(V)}(F_0)$. If $e = 0$, it follows from (2.4) that $\psi \in I(V, \kappa)$ if and only if

$$\alpha^{q^\nu+1}\widehat{b}(x_i, x_j)^q = \widehat{b}(x_i, x_j)$$

for all of the basis vectors $x_1, \ldots, x_\mu$ of $W'_\lambda$. Similarly if $e = 1$, by (2.4) we have $\psi \in I(V, \kappa)$ if and only if

$$\alpha^{q^\nu+1}\zeta^{\frac{q(q^\nu+1)}{2}}\widehat{b}(x_i, x_j)^q = \zeta^{\frac{q^\nu+1}{2}}\widehat{b}(x_i, x_j)$$

for all $1 \le i, j \le \mu$.

If $x_1, \ldots, x_\mu$ is an orthonormal basis, we have $\widehat{b}(x_i, x_j)^q = \widehat{b}(x_i, x_j) = \delta_{i,j}$, so our choice $\alpha = \zeta^{-\frac{e(q-1)}{2}}$ implies $\psi \in I(V, \kappa)$. In the case of a hyperbolic basis with products as in (2.5), we have $\alpha = \zeta^{-\frac{(q-1)(e+1)}{2}}$, and one verifies that $\psi \in I(V, \kappa)$.

Thus we conclude that $\psi \in N_{I(V,\kappa)}(F_0)$. Now arguing as in the proof of Lemma 2.5.2 (i), we find that $N_{I(V,\kappa)}(F_0) = \langle C_{I(V,\kappa)}(F_0), \psi\rangle$. Furthermore, a calculation shows that $\psi^{2\nu} = (-1)^e I_V$, so (ii) holds.

It follows from (ii) that

$$\frac{N_{I(V,\kappa)}(F_0)}{C_{I(V,\kappa)}(F_0)} = \langle\overline{\psi}\rangle \cong C_{2\nu},$$

as claimed by (iii).

It remains to prove claims (iv)–(vi) about the centralizer of $F_0$. As in the proof of Lemma 2.5.2 (ii), we see that $C_{\mathrm{GL}(V)}(F_0)$ are precisely the maps of the form $g_A \in C_{\mathrm{GL}(V)}(F_0)$ for $A \in \mathrm{GL}_\mu(\mathbb{K})$. It is clear that $g_A \in I(V, \kappa)$ if and only if the action of $A$ on $W'_\lambda$ is contained in $I(W'_\lambda, \widehat{b})$, where $\widehat{b}$ is the Hermitian or skew-Hermitian form defined earlier in this section. Hence

$$C_{I(V,\kappa)}(F_0) = \{g_A : A \in I(W'_\lambda, \widehat{b})\} \cong I(W'_\lambda, \widehat{b}) \cong \mathrm{GU}_\mu(\mathbb{K}).$$

Similarly

$$C_{\Delta(V,\kappa)}(F_0) = \{g_A : A \in \Delta(W'_\lambda, \widehat{b})\} \cong \Delta\mathrm{U}_\mu(\mathbb{K}).$$

The fact that $C_{\Delta(V,\kappa)}(F_0) = \langle C_{I(V,\kappa)}(F_0), \eta\rangle$ follows since $\eta \in C_{\Delta(V,\kappa)}(F_0)$ and $\tau(\eta)$ is a primitive element of $\mathbb{F}_q$. □

**Lemma 2.7.4** *The group $G$ is of multiplier* 1 *and* $G = \langle G^\circ, \eta\rangle$.

***Proof*** It is readily checked that all the generators of $N_{\Delta(V,\kappa)}(F_0)$ (Lemma 2.7.3) normalize $\langle \eta \rangle$. Indeed $g_A$ centralizes $\eta$ for all $A \in \mathrm{GL}_\mu(\mathbb{K})$, and $\psi^{-1}\eta\psi = \eta^q$. Therefore $G$ normalizes $\langle \eta \rangle$, so $\eta \in G$ since $G$ is maximal solvable. Since $\tau(\eta) = \zeta^{\frac{q^{2\nu}-1}{q-1}}$ is a primitive element of $\mathbb{F}_q$, it follows that $G$ is of multiplier 1 and $G = \langle G^\circ, \eta \rangle$. □

**Lemma 2.7.5** $F_0 = \langle \eta^{q-1} \rangle$ *and* $|F_0| = q^\nu + 1$.

***Proof*** We have $\lambda^{q^\nu+1} = 1$, so $\lambda$ is contained in the subgroup of $\mathbb{K}^\times$ generated by $\zeta^{q^\nu-1}$. Therefore $F_0 \leq \langle \eta^{q-1} \rangle \leq I(V, \kappa)$. As noted in Lemma 2.7.4, we have $\langle \eta \rangle \trianglelefteq G$, so by maximality of $F_0$ we have $F_0 = \langle \eta^{q-1} \rangle$. In particular $|F_0| = q^\nu + 1$. □

In view of Lemma 2.7.5, we can assume without loss of generality that $f = \eta^{q-1}$. In this case $\lambda = \zeta^{q^\nu-1}$. We have the following relations among the generators of $N_{\Delta(V,\kappa)}(F_0)$.

$$\begin{aligned} \eta^{q-1} &= f \\ \psi^{-1} f \psi &= f^q \\ \psi^{-1} g_A \psi &= g_{\overline{A}} \end{aligned} \tag{2.7}$$

where $\overline{A}$ is defined by $\overline{A}_{ij} = A_{ij}^q$ for all $1 \leq i, j \leq \mu$.

**Remark 2.7.6 (Cf. Remarks 2.5.4, 2.6.7)** Conversely, suppose that $n = 2\mu\nu$ for integers $\mu, \nu \geq 1$. Suppose that $\kappa$ is an alternating bilinear form, or a quadratic form with $\mathrm{sgn}(\kappa) = (-1)^\mu$ (Lemma 2.7.2).

Then we can construct $f \in I(V, \kappa)$ with $|f| = q^\nu + 1$ such that $f$ has splitting field $\mathbb{K} = \mathbb{F}_{q^{2\nu}}$ and $V' = \mathbb{K} \otimes_{\mathbb{F}_q} V$ decomposes into $f$-eigenspaces

$$V' = W'_\lambda \oplus \cdots \oplus W'_{\lambda^{q^{2\nu-1}}},$$

where $\dim_{\mathbb{K}} W'_{\lambda^{q^i}} = \mu$ for all $0 \leq i < 2\nu$.

First the $\mu = 1$ case, this follows from a construction which we describe later in Remark 2.9.11. It follows from the construction given there that we can find $f_0$ with the desired properties in $I(W, \kappa')$, where $\dim W = 2\nu$ and $\kappa'$ is a form of the same type as $\kappa$. Note that here $\mathrm{sgn}(\kappa') = -$ if $\kappa'$ is a quadratic form (Lemma 2.7.2). Thus we can identify $V = W \perp \cdots \perp W$ ($\mu$ times), and the action of $f_0$ on $V$ provides $f$ with the desired properties.

An alternative construction proceeds as follows. If $\mu$ is odd, let $f_0 \in I(V, \kappa)$ be the element constructed in the case $\mu = 1$, so $|f_0| = q^{n/2}+1$. Then $q^\nu+1 \mid q^{n/2}+1$ (Lemma 2.1.20), and one can check that

$$f = f_0^{\frac{q^{n/2}+1}{q^\nu+1}}$$

has all the required properties.

In the case where $\mu$ is even, let $f_0 \in I(V, \kappa)$ be the element constructed given by the type $\mathcal{B}_1$ construction (Remark 2.6.7) in the case $\mu = 1$, so $|f_0| = q^{n/2} - 1$. Then $q^\nu + 1 \mid q^{n/2} - 1$ (Lemma 2.1.20), and one can check that

$$f = f_0^{\frac{q^{n/2}-1}{q^\nu+1}}$$

has all the required properties.

**Remark 2.7.7 (Cf. Remarks 2.5.5, 2.6.8)** With the results of this section, we have described the possible structures of $F_0 \leq I(V, \kappa)$ when $G$ is of type $\mathcal{B}_2$.

In the case where $\mu = 1$, we also have a complete description of $G$. Indeed, for $\mu = 1$ it follows from Lemma 2.7.3 that $N_{\Delta(V,\kappa)}(F_0)$ is solvable, with generators $\eta$ and $\psi$. Therefore $G = N_{\Delta(V,\kappa)}(F_0) = \langle \eta, \psi \rangle$ and $G^\circ = \langle f, \psi \rangle$. In this case we have $|G| = (q^\nu + 1)(q - 1)n$ and $|G^\circ| = (q^\nu + 1)n$, with $n = 2\nu$. (For more on the $\mu = 1$ case, see Sect. 2.9 and Remark 2.9.11.)

If $\mu > 1$, the structure of $G$ is described in more detail in Chap. 4. Similarly to Remark 2.5.5, at this point we can note that by Lemma 2.7.3, there exists $A \in \mathrm{GL}_\mu(\mathbb{K})$ and an integer $\delta \mid 2\nu$ such that $G^\circ$ is generated by $C_{G^\circ}(F_0)$ and $g_A\psi^\delta$. Later in Lemma 4.3.22 (iii), we will prove that we can arrange $\delta \in \{1, 2\}$.

## 2.8 Groups of Type $\mathcal{B}_3$

We continue from the setup of Sect. 2.4, and consider the case where $G$ is of type $\mathcal{B}_3$. In this case $\kappa$ is a nonzero form, and $V = Q_\lambda$ with $W_\lambda'$ is not totally singular. It follows from Lemma 2.4.5 that $\lambda = \pm 1$, so $Q_\lambda' = W_\lambda'$ and $\lambda$ is the only eigenvalue of $f$. Since $F_0 \neq 1$, we must have $\lambda = -1$ and $F_0 = \langle -I_V \rangle$. We conclude the following.

**Lemma 2.8.1** *For $G$ of type $\mathcal{B}_3$, we have $q$ odd and $\mathbb{K} = \mathbb{F}_q$. Furthermore $n = \mu$, $|F_0| = 2$, and $F_0 = \langle -I_V \rangle$.*

If $\mu = 1$, then $G = \mathrm{GO}_1(q) = \mathbb{F}_q^\times$ and $G^\circ = O_1(q) = \{\pm 1\}$. Thus $|G| = (q - 1)$, $|G^\circ| = 2$, and $G$ is of multiplier 2. For $\mu > 1$, we will make the structure of $G$ more precise later in Chap. 4, and will prove for example that we must have $\mu = 2^\ell$ for some $\ell \geq 1$ (Lemma 4.1.19).

## 2.9 Groups of Type $\mathcal{B}_i$ with $\mu = 1$

We continue with the notation from Sects. 2.4–2.8, where we consider a metrically primitive maximal irreducible solvable subgroup $G$ in $\Delta(V, \kappa)$. As before, we denote by $F_0$ a subgroup satisfying properties (F1)–(F3) defined in Sect. 2.4. We have $F_0 = \langle f \rangle$. As in Sect. 2.4 we denote by $\mu$ the dimension of an $f$-eigenspace on $V' = \mathbb{K} \otimes_{\mathbb{F}_q} V$ (Lemma 2.4.9), where $\mathbb{K}$ is a splitting field for $F_0$.

Following the notation of Sects. 2.5–2.8, we define

$$\nu = \begin{cases} [\mathbb{K} : \mathbb{F}_q]/2, & \text{if } G \text{ is of type } \mathcal{B}_2. \\ [\mathbb{K} : \mathbb{F}_q], & \text{otherwise.} \end{cases}$$

Note that $\nu$ is an integer by Lemma 2.7.1.

In this section, we will summarize the properties of $G$ in the case where $\mu = 1$. We begin with two lemmas which do not assume anything about $\mu$.

**Lemma 2.9.1** *Suppose that $G \leq \Delta(V, \kappa)$ is metrically primitive maximal irreducible solvable. Then $N_{\Delta(V,\kappa)}(F_0)/C_{I(V,\kappa)}(F_0)$ is abelian.*

***Proof*** As seen in Sects. 2.5–2.8, we have isomorphisms

$$\begin{aligned}
&\text{Type } \mathcal{B}_0 : && N_{\Delta(V,\kappa)}(F_0)/C_{I(V,\kappa)}(F_0) = \langle \overline{\psi} \rangle \cong C_\nu, \\
&\text{Type } \mathcal{B}_1 : && N_{\Delta(V,\kappa)}(F_0)/C_{I(V,\kappa)}(F_0) = \langle \overline{\eta}, \overline{\varphi}, \overline{\psi} \rangle \cong C_{q-1} \times C_2 \times C_\nu, \\
&\text{Type } \mathcal{B}_2 : && N_{\Delta(V,\kappa)}(F_0)/C_{I(V,\kappa)}(F_0) = \langle \overline{\eta}, \overline{\psi} \rangle \cong C_{q-1} \times C_{2\nu}, \\
&\text{Type } \mathcal{B}_3 : && N_{\Delta(V,\kappa)}(F_0)/C_{I(V,\kappa)}(F_0) = \Delta(V, \kappa)/I(V, \kappa) \cong C_{q-1}.
\end{aligned}$$

from which the lemma follows. □

**Lemma 2.9.2** *Suppose that $G \leq \Delta(V, \kappa)$ is metrically primitive maximal irreducible solvable. Then $G/C_{G^\circ}(F_0)$ is abelian, and $G = N_{\Delta(V,\kappa)}(C_{G^\circ}(F_0))$.*

***Proof*** The quotient $G/C_{G^\circ}(F_0)$ embeds into $N_{\Delta(V,\kappa)}(F_0)/C_{I(V,\kappa)}(F_0)$, so $G/C_{G^\circ}(F_0)$ is abelian by Lemma 2.9.1. Then $G = N_{\Delta(V,\kappa)}(C_{G^\circ}(F_0))$ follows from Lemma 2.1.5. □

**Lemma 2.9.3** *Suppose that $G \leq \Delta(V, \kappa)$ is metrically primitive maximal irreducible solvable. If $\mu = 1$, then $F_0 = C_{G^\circ}(F_0)$.*

***Proof*** If $\mu = 1$, then it follows from the description of $C_{I(V,\kappa)}(F_0)$ (Lemma 2.5.2 (ii), Lemma 2.6.4 (iv), Lemma 2.7.3 (iv), Lemma 2.8.1) that $C_{I(V,\kappa)}(F_0) = F_0$, from which the lemma follows. □

The following result is proved by Jordan in [51, §VI, Théorème 21, pp. 21–24] when $q$ is a prime. As is expected the same proof works for any prime power $q$; we will instead argue using Lemma 2.1.4.

**Theorem 2.9.4** *Suppose that $G \leq \Delta(V, \kappa)$ is metrically primitive maximal irreducible solvable. If $\mu > 1$, then $F_0 \lneqq C_{G^\circ}(F_0)$.*

***Proof*** Suppose that $F_0 = C_{G^\circ}(F_0)$. It follows then from Lemma 2.9.2 that $G = N_{\Delta(V,\kappa)}(F_0)$. Hence $F_0 = C_{G^\circ}(F_0) = C_{I(V,\kappa)}(F_0)$. We have

$$
\begin{array}{lll}
\text{Type } \mathcal{B}_0: & C_{I(V,\kappa)}(F_0) \cong \mathrm{GL}_\mu(\mathbb{K}) & \text{by Lemma 2.5.2 (ii)},\\
\text{Type } \mathcal{B}_1: & C_{I(V,\kappa)}(F_0) \cong \mathrm{GL}_\mu(\mathbb{K}) & \text{by Lemma 2.6.4 (iv)},\\
\text{Type } \mathcal{B}_2: & C_{I(V,\kappa)}(F_0) \cong \mathrm{GU}_\mu(\mathbb{K}) & \text{by Lemma 2.7.3 (iv)},\\
\text{Type } \mathcal{B}_3: & C_{I(V,\kappa)}(F_0) = I(V,\kappa) & \text{by Lemma 2.8.1.}
\end{array}
$$

If $\mu > 1$, this is clearly in contradiction with the fact that $F_0 = C_{I(V,\kappa)}(F_0)$. □

**Definition 2.9.5** Let $0 \leq i \leq 3$. Let $F_0 \leq \Delta(V, \kappa)$ and $\kappa$ be as described for groups of type $\mathcal{B}_i$ in one of the Sects. 2.5–2.8 in the case $\mu = 1$. We define $G_{1,\nu}^{\mathcal{B}_i} := N_{\Delta(V,\kappa)}(F_0)$.

**Theorem 2.9.6** *Suppose that $G \leq \Delta(V, \kappa)$ is metrically primitive maximal irreducible solvable of type $\mathcal{B}_i$, with $\mu = 1$. Then G is conjugate to $G_{1,\nu}^{\mathcal{B}_i}$ in $\Delta(V, \kappa)$.*

***Proof*** It follows from Lemma 2.9.3 that $C_{G^\circ}(F_0) = F_0$, so by Lemma 2.9.2 we have $G = G_{1,\nu}^{\mathcal{B}_i}$. □

**Remark 2.9.7** Conversely to Theorem 2.9.6, as a part of our main results, we will see that in most cases $G_{1,\nu}^{\mathcal{B}_i}$ is metrically primitive maximal irreducible solvable in $\Delta(V, \kappa)$ (Theorem 7.1.15). For example in the case $\kappa = 0$, we have $G_{1,\nu}^{\mathcal{B}_0} = \Gamma\mathrm{L}_1(q^n)$. It turns out that $\Gamma\mathrm{L}_1(q^n)$ is maximal solvable in $\mathrm{GL}_n(q)$, except for $(n, q) = (2, 3)$. This particular case could also be deduced from a result of Kantor [53], which describes subgroups of $\mathrm{GL}_n(q)$ containing a Singer cycle.

**Remark 2.9.8** Let $\mu$ have prime factorization $\mu = r_1^{\ell_1} \cdots r_k^{\ell_k}$. In later sections, we will generalize the definition of $G_{1,\nu}^{\mathcal{B}_i}$ to certain groups

$$G_{\mu,\nu}^{\mathcal{B}_i}(X_1, \ldots, X_k) \leq \Delta(V, \kappa),$$

where $X_j$ is maximal among the metrically completely reducible solvable subgroups of $\mathrm{GSp}_{2\ell_j}(r_j)$ for all $1 \leq j \leq k$.

Generalizing Theorem 2.9.6, we will see later (Proposition 4.2.4) that every metrically primitive maximal irreducible solvable $G \leq \Delta(V, \kappa)$ of type $\mathcal{B}_i$ is conjugate to $G_{\mu,\nu}^{\mathcal{B}_i}(X_1, \ldots, X_k)$ for some $X_1, \ldots, X_k$. This will complete the recursive construction of the candidates for metrically primitive maximal irreducible solvable subgroups of $\Delta(V, \kappa)$.

**Table 2.2** Structure of metrically primitive maximal irreducible solvable groups with $\mu = 1$

| Type | $\mathbb{K}$ | $G^\circ$ | $G$ | $\lvert G^\circ\rvert$ | $\lvert G\rvert$ |
|---|---|---|---|---|---|
| $\mathcal{B}_0$ | $\mathbb{F}_{q^n}$ | $\langle f\rangle \rtimes \langle\psi\rangle$ | $\langle f\rangle \rtimes \langle\psi\rangle$ | $(q^n-1)n$ | $(q^n-1)n$ |
| $\mathcal{B}_1$ | $\mathbb{F}_{q^{n/2}}$ | $\langle f, \psi, \varphi\rangle$ | $G^\circ \rtimes \langle\eta\rangle$ | $(q^{n/2}-1)n$ | $(q-1)(q^{n/2}-1)n$ |
| $\mathcal{B}_2$ | $\mathbb{F}_{q^n}$ | $\langle f, \psi\rangle$ | $\langle G^\circ, \eta\rangle$ | $(q^{n/2}+1)n$ | $(q-1)(q^{n/2}+1)n$ |
| $\mathcal{B}_3$ | $\mathbb{F}_q$, $q$ odd | $O_1(q)$ | $\mathrm{GO}_1(q)$ | 2 | $q-1$ |

Furthermore, for groups of the form $G = G^{\mathcal{B}_i}_{\mu,\nu}(X_1, \ldots, X_k)$, we will see that in most cases $G$ and $G^\circ$ are metrically primitive maximal irreducible solvable in $\Delta(V, \kappa)$ and $I(V, \kappa)$, respectively (Theorems 7.1.12 and 7.1.13).

We end this section with several remarks, which explain how the groups in Definition 2.9.5 can be constructed, and how one can compute explicit generating matrices for them. Furthermore, in Table 2.2 we list some of their basic properties, which follow from the results of Sects. 2.5–2.8.

**Remark 2.9.9** The groups of type $\mathcal{B}_0$ in Definition 2.9.5 can be constructed explicitly as follows. In this case we have $\kappa = 0$ and $\mathbb{K} = \mathbb{F}_{q^n}$. Let $\lambda \in \mathbb{K}^\times$ be a primitive element.

We can identify $V = \mathbb{K}$ as $\mathbb{F}_q$-vector spaces. Then we define $f : V \to V$ by $f(v) = \lambda v$ for all $v \in V$, and $\phi : V \to V$ by $\phi(v) = v^q$ for all $v \in V$. The minimal polynomial of $f$ over $\mathbb{F}_q$ has roots $\lambda, \lambda^q, \ldots, \lambda^{q^{n-1}}$, so $\mathbb{K}$ is a splitting field for $f$. Then $V' = \mathbb{K} \otimes_{\mathbb{F}_q} V$ decomposes into $f$-eigenspaces

$$V' = W'_\lambda \oplus \cdots \oplus W'_{\lambda^{q^{n-1}}},$$

where $f(v) = \alpha v$ for all $v \in W'_\alpha$ and $\alpha \in \mathbb{K}$.

Set $\psi = \phi^{-1}$. Then $\psi^{-1} f \psi = f^q$, so $\psi(W'_\alpha) = W'_{\alpha^q}$ for all $\alpha \in \mathbb{K}$. Choose a nonzero $v_0 \in W'_\lambda$, and define $v_i := \psi^i v_0$ for all $0 \le i < n$. Then $v_0, \ldots, v_{n-1}$ is a basis of $V'$, and

$$f(v_i) = \lambda^{q^i} v_i$$

$$\psi(v_i) = v_{i+1}$$

for all $0 \le i < n$, where $v_{n+1} = v_0$.

Therefore by Definition 2.9.5 and Lemma 2.5.2, with $F_0 = \langle f\rangle$ we have

$$G^{\mathcal{B}_0}_{1,n} = N_{\mathrm{GL}(V)}(F_0) = \langle f, \psi\rangle \le \mathrm{GL}(V).$$

Here $f$ is a Singer cycle in $\mathrm{GL}(V)$, and $G^{\mathcal{B}_0}_{1,n}$ is the normalizer of the Singer cyclic subgroup $F_0$, which is sometimes denoted by $\Gamma\mathrm{L}_1(q^n)$.

We can get explicit matrices over $\mathbb{F}_q$ for $f$ and $\phi$ by the following well-known computation, see for example [35, Lemma 6.1]. Having identified $V = \mathbb{K}$, we have a basis

$$B = \{1, \lambda, \dots, \lambda^{n-1}\}$$

for $V$ since $\lambda$ is a primitive element. Let $p(x) \in \mathbb{F}_q[x]$ be the minimal polynomial of $\lambda$ over $\mathbb{F}_q$. Then the matrix of $f$ with respect to the basis $B$ is the companion matrix of $p(x)$. By computing $x^{q^i}$ modulo $p(x)$, we can express $\lambda^{q^i}$ as a linear combination of $1, \lambda, \dots, \lambda^{n-1}$ over $\mathbb{F}_q$. This gives us the matrix of $\phi$ with respect to $B$, and thus the matrix of $\psi = \phi^{-1}$.

**Remark 2.9.10** The groups of type $\mathcal{B}_1$ in Definition 2.9.5 can be constructed as follows. Let $n = 2\nu$ be even and $\mathbb{K} = \mathbb{F}_{q^\nu}$. Let $\lambda \in \mathbb{K}^\times$ be a primitive element.

As a $\mathbb{F}_q$-vector space we can identify $V = \mathbb{K} \oplus \mathbb{K}$. We define $\mathbb{F}_q$-linear maps $f, \phi, \varphi, \eta : V \to V$ by

$$\begin{aligned} f(x, y) &= (\lambda x, \lambda^{-1} y) \\ \phi(x, y) &= (x^q, y^q) \\ \varphi(x, y) &= ((-1)^e y, x) \\ \eta(x, y) &= (x, \lambda^{\frac{q^\nu - 1}{q-1}} y) \end{aligned}$$

for all $x, y \in \mathbb{K}$. Set $\psi = \phi^{-1}$.

Define a bilinear form $b : V \times V \to \mathbb{F}_q$ by

$$b((x, y), (x', y')) = \mathrm{Tr}_{\mathbb{K}/\mathbb{F}_q}(xy' + (-1)^e x'y)$$

for all $x, y, x', y' \in \mathbb{K}$. Then $b$ is a non-degenerate symmetric bilinear form if $e = 0$, and a non-degenerate alternating bilinear form if $e = 1$. If $e = 0$, we also define a quadratic form $Q : V \to \mathbb{F}_q$ by

$$Q(x, y) = \mathrm{Tr}_{\mathbb{K}/\mathbb{F}_q}(xy)$$

for all $x, y \in \mathbb{K}$. Then the polarization of $Q$ is equal to $b$, and $\mathrm{sgn}(Q) = +$ since $V = \mathbb{K} \oplus \mathbb{K}$ is a totally singular decomposition.

Set $\kappa = Q$ if $e = 0$, and $\kappa = b$ if $e = 1$. Then $\kappa$ is a non-degenerate form, and a straightforward check shows that $f, \phi, \varphi \in I(V, \kappa)$ and $\eta \in \Delta(V, \kappa)$ with $\tau(\eta) = \lambda^{\frac{q^\nu - 1}{q-1}}$.

The action of $f$ on the summand $\mathbb{K} \oplus 0$ has minimal polynomial over $\mathbb{F}_q$ with roots $\lambda, \lambda^q, \dots, \lambda^{q^{\nu-1}}$, while the action of $f$ on $0 \oplus \mathbb{K}$ has minimal polynomial over $\mathbb{F}_q$ with roots $\lambda^{-1}, \lambda^{-q}, \dots, \lambda^{-q^{\nu-1}}$. Therefore $\mathbb{K}$ is a splitting field for the action of

$f$ on $V$, so $V' = \mathbb{K} \otimes_{\mathbb{F}_q} V$ decomposes into $f$-eigenspaces

$$V' = W'_\lambda \oplus \cdots \oplus W'_{\lambda^{q^{\nu-1}}} \oplus W'_{\lambda^{-1}} \oplus \cdots \oplus W'_{\lambda^{-q^{\nu-1}}}$$

where $f(v) = \alpha v$ for all $v \in W_\alpha$ and $\alpha \in \mathbb{K}$.

Choose a nonzero vector $v_0 \in W'_\lambda$. Define $v_i := \psi^i v_0$ for all $0 < i < \nu$. We have $\psi^{-1} f \psi = f^q$, so $v_i \in W'_{\lambda^{q^i}}$ for all $0 \leq i < \nu$. Next define $w_i := \varphi(v_i)$ for all $0 \leq i < \nu$. We have $\varphi f \varphi^{-1} = f^{-1}$, so $w_i \in W'_{\lambda^{-q^i}}$ for all $0 \leq i < \nu$.

Thus $v_0, \ldots, v_{\nu-1}, w_0, \ldots, w_{\nu-1}$ is a basis of $V'$, and we have

$$\begin{aligned} f(v_i) &= \lambda^{q^i} v_i & f(w_i) &= \lambda^{-q^i} w_i \\ \psi(v_i) &= v_{i+1} & \psi(w_i) &= w_{i+1} \\ \varphi(v_i) &= w_i & \varphi(w_i) &= (-1)^e v_i \\ \eta(v_i) &= v_i & \eta(w_i) &= \lambda^{\frac{q^\nu - 1}{q-1}} w_i \end{aligned}$$

for all $0 \leq i < \nu$, where $v_\nu = v_0$ and $w_\nu = w_0$.

Therefore by Definition 2.9.5 and Lemma 2.6.4, with $F_0 = \langle f \rangle$ we have

$$G_{1,\nu}^{\mathcal{B}_1} = N_{\Delta(V,\kappa)}(F_0) = \langle f, \psi, \varphi, \eta \rangle \leq \Delta(V, \kappa).$$

Since $\lambda$ is a primitive element, the elements $1, \lambda, \ldots, \lambda^{n-1}$ form an $\mathbb{F}_q$-basis of $\mathbb{K}$. Thus we can take

$$B = \left\{(1, 0), (\lambda, 0), \ldots, (\lambda^{n-1}, 0), (0, 1), (0, \lambda), \ldots, (0, \lambda^{n-1})\right\}$$

as a basis of $V = \mathbb{K} \oplus \mathbb{K}$. As in Remark 2.9.9, one can compute the matrix of $f$ and $\psi$ with respect to the basis $B$, using the minimal polynomial of $\lambda$ over $\mathbb{F}_q$. Furthermore, we have

$$\varphi = \begin{pmatrix} 0 & (-1)^e I_\nu \\ I_\nu & 0 \end{pmatrix}, \qquad \eta = \begin{pmatrix} I_\nu & 0 \\ 0 & \lambda^{\frac{q^\nu - 1}{q-1}} I_\nu \end{pmatrix}$$

with respect to the basis $B$. Thus we have explicit matrices representing the generators $f, \psi, \varphi, \eta$ of $G_{1,\nu}^{\mathcal{B}_1}$. Furthermore, the bilinear form $b$ is represented by a matrix

$$b = \begin{pmatrix} 0 & (-1)^e \Omega \\ \Omega & 0 \end{pmatrix},$$

where $\Omega_{ij} := \mathrm{Tr}_{\mathbb{K}/\mathbb{F}_q}(\lambda^{i+j-2})$ for all $1 \leq i, j \leq \nu$.

**Remark 2.9.11** The groups of type $\mathcal{B}_2$ in Definition 2.9.5 can be constructed as follows. Let $n = 2\nu$ be even and $\mathbb{K} = \mathbb{F}_{q^{2\nu}}$. Let $\zeta \in \mathbb{K}^\times$ be a primitive element, and define $\lambda = \zeta^{q^\nu - 1}$.

We identify $V = \mathbb{K}$ as an $\mathbb{F}_q$-vector space. Define linear maps $f, \phi, \eta : V \to V$ by $f(v) = \lambda v$, $\phi(v) = \zeta^{e(\frac{q-1}{2})} v^q$, and $\eta(v) = \zeta^{\frac{q^\nu - 1}{q-1}} v$ for all $v \in V$. We define a bilinear form $b : V \times V \to \mathbb{F}_q$ by

$$b(x, y) = \mathrm{Tr}_{\mathbb{F}_{q^\nu}/\mathbb{F}_q}\left(\zeta^{e(\frac{q^\nu+1}{2})}(xy^{q^\nu} + (-1)^e x^{q^\nu} y)\right) = \mathrm{Tr}_{\mathbb{K}/\mathbb{F}_q}\left(\zeta^{e(\frac{q^\nu+1}{2})} x y^{q^\nu}\right)$$

for all $x, y \in V$. If $e = 0$, we also define a quadratic form $Q : V \to \mathbb{F}_q$ by

$$Q(x) = \mathrm{Tr}_{\mathbb{F}_{q^\nu}/\mathbb{F}_q}(x^{q^\nu+1})$$

for all $x \in V$. Then the polarization of $Q$ is equal to $b$, and by the calculation in the proof of Lemma 2.7.2 we have $\mathrm{sgn}(Q) = -$.

The bilinear form $b$ is non-degenerate since $(x, y) \mapsto \mathrm{Tr}_{\mathbb{K}/\mathbb{F}_q}(xy)$ defines a nondegenerate bilinear form on $\mathbb{K}$. In the case $e = 0$ it is clear that $b$ is symmetric, and for $e = 1$ the bilinear form $b$ is alternating since for $\alpha = \zeta^{\frac{q^\nu+1}{2}}$ we have $\alpha^{q^\nu} = -\alpha$. We define $\kappa = Q$ if $e = 0$, and $\kappa = b$ if $e = 1$.

A straightforward calculation shows that $f, \phi \in I(V, \kappa)$ and $\eta \in \Delta(V, \kappa)$ with $\tau(\eta) = \zeta^{\frac{q^{2\nu}-1}{q-1}}$. The minimal polynomial of $f$ on $V$ has roots $\lambda, \lambda^q, \ldots, \lambda^{q^{2\nu-1}}$, so $\mathbb{K}$ is a splitting field for $f$. Then $V' = \mathbb{K} \otimes_{\mathbb{F}_q} V$ decomposes into $f$-eigenspaces

$$V' = W'_\lambda \oplus \cdots \oplus W'_{\lambda^{q^{2\nu-1}}},$$

where $f(v) = \alpha v$ for all $v \in W'_\alpha$ and $\alpha \in \mathbb{K}$.

Set $\psi = \phi^{-1}$. Then $\psi^{-1} f \psi = f^q$, so $\psi(W'_\alpha) = W'_{\alpha^q}$ for all $\alpha \in \mathbb{K}$. Choose a nonzero $v_0 \in W'_\lambda$, and define $v_i := \psi^i v_0$ for all $0 \leq i < 2\nu$. Then $v_0, \ldots, v_{2\nu-1}$ is a basis of $V'$, and

$$\begin{aligned} f(v_i) &= \lambda^{q^i} v_i \\ \psi(v_i) &= v_{i+1} \\ \eta(v_i) &= \left(\zeta^{\frac{q^\nu-1}{q-1}}\right)^{q^i} v_i \end{aligned}$$

for all $0 \leq i < 2\nu$, where $v_{2\nu} = v_0$.

It follows from Definition 2.9.5 and Lemma 2.7.3 that with $F_0 = \langle f \rangle$, we have

$$G^{\mathcal{B}_2}_{1,\nu} = N_{\Delta(V,\kappa)}(F_0) = \langle f, \psi, \eta \rangle \leq \Delta(V, \kappa).$$

The $\mathbb{F}_q$-algebra generated by $\lambda$ is equal to all of $\mathbb{K}$, so the minimal polynomial of $\lambda$ over $\mathbb{F}_q$ has degree $2\nu$. Thus the elements

$$B = \{1, \lambda, \dots, \lambda^{2\nu-1}\}$$

form a basis of $V$. As in Remark 2.9.9, we can compute the matrix of $f$, $\psi$, and $\eta$ with respect to the basis $B$ by using the minimal polynomial of $\lambda$ over $\mathbb{F}_q$. With respect to the basis $B$, the bilinear form $b$ is represented by the matrix $\Omega$, where

$$\Omega_{ij} = \mathrm{Tr}_{\mathbb{K}/\mathbb{F}_q}\left(\zeta^{e(\frac{q^\nu+1}{2})}\lambda^{i-1+q^\nu(j-1)}\right)$$

for all $1 \le i, j \le n$.

**Remark 2.9.12** For groups of type $\mathcal{B}_3$ we always have $\mathbb{K} = \mathbb{F}_q$ and $\nu = 1$. Then $n = 1$, so $\kappa$ is a quadratic form, $q$ is odd, and

$$G_{1,\nu}^{\mathcal{B}_3} = G_{1,1}^{\mathcal{B}_3} = \mathrm{GO}_1(q) = \mathrm{GL}_1(q).$$

# Chapter 3
# Extraspecial Groups

Throughout this chapter, let $r$ be a prime number. A finite $r$-group $R$ is said to be *extraspecial* if $Z(R)$ is cyclic of order $r$ and $R/Z(R)$ is a nontrivial elementary abelian group.

In Chap. 4, we will find that metrically primitive maximal irreducible solvable subgroups normalize certain extraspecial $r$-groups which have exponent $r$ or 4. For their construction, we need some basic results on extraspecial groups and their irreducible representations; these will be provided in this chapter.

## 3.1 Extraspecial Groups

We refer to [19, Chapter A, Section 20] for the basic properties of extraspecial $r$-groups. Recall that any extraspecial $r$-group has order $r^{1+2\ell}$ for some integer $\ell > 0$, and there are two extraspecial groups of order $r^{1+2\ell}$ up to isomorphism, the group of plus type denoted by $r_+^{1+2\ell}$, and the group of minus type denoted by $r_-^{1+2\ell}$.

Let $z \in R$ be a generator of $Z(R)$. Considering $\overline{R} := R/Z(R)$ as an $\mathbb{F}_r$-vector space, we have a non-degenerate alternating bilinear form $\xi : \overline{R} \times \overline{R} \to \mathbb{F}_r$ defined by

$$[x, y] = z^{\xi(\overline{x}, \overline{y})}$$

for all $x, y \in R$. This observation goes back to Jordan [51, §VII, p. 24], who calls $\xi(\overline{x}, \overline{y})$ the *exposant d'échange* of $x, y \in R$.

In the case where $r = 2$, we define a quadratic form $\vartheta : \overline{R} \to \mathbb{F}_2$ by

$$x^2 = z^{\vartheta(\overline{x})}$$



M. Korhonen, *Maximal Solvable Subgroups of Finite Classical Groups*,
Lecture Notes in Mathematics 2346, https://doi.org/10.1007/978-3-031-62915-0_3

for all $x \in R$. Then $\vartheta$ is non-degenerate with polarization equal to $\xi$. This definition also goes back to Jordan [51, §VII, p. 25], who calls $\vartheta(\overline{x})$ the *caractère* of $x \in R$. We have $R \cong 2_+^{1+2\ell}$ if $\mathrm{sgn}(\vartheta) = +$, and $R \cong 2_-^{1+2\ell}$ if $\mathrm{sgn}(\vartheta) = -$.

We can find generators $R = \langle x_1, y_1, \ldots, x_\ell, y_\ell \rangle$ such that the images $\overline{x_i}, \overline{y_i}$ form a symplectic basis for $\overline{R}$. In other words, we have $[x_i, y_i] = z$ and $[x_i, y_j] = 1$ for $i \neq j$, and $[x_i, x_j] = 1 = [y_i, y_j]$ for all $1 \leq i, j \leq \ell$. If $R \cong r_+^{1+2\ell}$, we can arrange $x_i^r = y_i^r = 1$ for all $1 \leq i \leq \ell$. In the case where $R \cong r_-^{1+2\ell}$, we can find generators such that $x_1^r = y_1^r = z$, and $x_i^r = y_i^r = 1$ for all $2 \leq i \leq \ell$.

Every automorphism of $R$ acts on $R/Z(R)$ and preserves the form $\xi$ up to a scalar, so we have a homomorphism $\phi : \mathrm{Aut}(R) \to \mathrm{GSp}(W, \xi)$, where $\mathrm{GSp}(W, \xi) = \mathrm{GSp}_{2\ell}(r)$. Note that if $r = 2$, then the action of every automorphism leaves the quadratic form $\vartheta$ invariant, so $\mathrm{Im}\,\phi \leq O(W, \vartheta)$. The restriction of $\phi'$ to $C_{\mathrm{Aut}(R)}(Z(R))$ is a homomorphism

$$\phi' : C_{\mathrm{Aut}(R)}(Z(R)) \to \mathrm{Sp}(W, \xi).$$

We note the following for later reference, see for example [19, A, Theorem 20.8].

**Lemma 3.1.1** $\mathrm{Ker}\,\phi' = \mathrm{Inn}(R)$.

In other words, the automorphisms of $R$ that act trivially on $R/Z(R)$ and $Z(R)$ are precisely the inner automorphisms of $R$. We also recall the following result for later use.

**Lemma 3.1.2 ([38, III, Satz 13.7, Satz 13.8])** *The following hold:*

(i) *Every maximal abelian subgroup of $R$ has order $r^{1+\ell}$.*
(ii) *If $r > 2$ and $R$ is of plus type, then every maximal abelian subgroup of $R$ is isomorphic to $C_r^{\ell+1}$.*
(iii) *If $r = 2$ and $R$ is of plus type, then $R$ has maximal abelian subgroups isomorphic to both $C_2^{\ell+1}$ and $C_4 \times C_2^{\ell-1}$.*
(iv) *If $r = 2$ and $R$ is of minus type, then every maximal abelian subgroup of $R$ is isomorphic to $C_4 \times C_2^{\ell-1}$.*

## 3.2 Absolutely Irreducible Representations of Extraspecial Groups

First recall the following result which is well known, see for example [19, Chapter B, Theorem 9.16–9.17].

**Theorem 3.2.1** *Let $R$ be an extraspecial group of order $r^{1+2\ell}$, and let $\mathbb{K}$ be a field that contains a primitive $r$th root of unity. Then the following statements hold:*

(i) *Every irreducible $\mathbb{K}[R]$-module is absolutely irreducible;*
(ii) *Every irreducible $\mathbb{K}[R]$-module of dimension $> 1$ is faithful of dimension $r^\ell$;*

*(iii) There exist exactly $r - 1$ irreducible $\mathbb{K}[R]$-modules of dimension $r^\ell$, up to isomorphism;*
*(iv) If $V$ and $W$ are irreducible $\mathbb{K}[R]$-modules of dimension $r^\ell$, then $V \cong W$ if and only if $V \downarrow Z(R) \cong W \downarrow Z(R)$.*

Let $R$ be an extraspecial group of order $r^{1+2\ell}$, and let $\mathbb{K}$ be a field that contains a primitive $r$th root of unity. It follows from Theorem 3.2.1 and [19, Chapter A, Theorem 20.8] that faithful absolutely irreducible representations are quasiequivalent, which means that they correspond to conjugate subgroups of $\mathrm{GL}_{r^\ell}(\mathbb{K})$.

We will be concerned with groups $R$ of type $r_+^{1+2\ell}$ for $r > 2$, which have exponent $r$; and groups of type $2_\pm^{1+2\ell}$ which have exponent 4. In what follows, we will construct the embedding $R \leq \mathrm{GL}_{r^\ell}(q)$ and provide generators for $N_{\mathrm{GL}_{r^\ell}(q)}(R)$. We will give a similar description for their embeddings and normalizers in unitary, symplectic and orthogonal groups over $\mathbb{F}_q$ as well. The generators are originally due to Jordan [51, §X]—similar constructions appear also for example in [35, Section 9], [36, Section 9], [55, Section 4.6], [77, Theorem 21.4].

### 3.2.1 Absolutely Irreducible Representation of $r_+^{1+2\ell}$

We begin with the construction for groups of type $r_+^{1+2\ell}$. Suppose that $\mathbb{K} = \mathbb{F}_q$, where $q$ is a power of a prime such that $r \mid q - 1$. Let $\theta \in \mathbb{K}^\times$ be a primitive $r$th root of unity.

Let $V$ be an $r$-dimensional $\mathbb{K}$-vector space with basis $v_0, v_1, \ldots, v_{r-1}$. We define linear maps $A, B, C, E : V \to V$ by

$$
\begin{aligned}
Av_\xi &= \theta^\xi v_\xi \\
Bv_\xi &= v_{\xi+1} \\
Cv_\xi &= \sum_{0 \leq i < r} \theta^{i\xi} v_i \\
Ev_\xi &= \theta^{\frac{\xi(\xi-1)}{2}} v_\xi
\end{aligned}
$$

for all $0 \leq \xi < r$, where the indices are interpreted modulo $r$. It is clear that $A, B, E$ are invertible, and $C$ is also invertible by the Vandermonde determinant. Note that $E = 1$ if $r = 2$.

Then $A^r = B^r = 1$ and $[A, B] = \theta I_V$, and it is readily seen that $R_0 = \langle A, B \rangle \cong r_+^{1+2}$ is absolutely irreducible. We have $CAC^{-1} = B^{-1}$ and $CBC^{-1} = A$, so $C$ normalizes $R_0$. If $r > 2$, then $EAE^{-1} = A$ and $EBE^{-1} = BA$, so $E$ normalizes $R_0$ as well.

Since $R_0$ is absolutely irreducible, by Lemma 2.1.16 (i) the tensor product $R = R_0 \otimes \cdots \otimes R_0$ ($\ell$ times) is an absolutely irreducible subgroup of $\mathrm{GL}(W)$, where

$W = V \otimes \cdots \otimes V$ ($\ell$ times). Here $R$ is a central product of $\ell$ copies of $r_+^{1+2}$, so $R \cong r_+^{1+2\ell}$. As a basis for $W$, we take the vectors $v_{\xi_1,\ldots,\xi_\ell} := v_{\xi_1} \otimes \cdots \otimes v_{\xi_\ell}$ for $0 \le \xi_1, \ldots, \xi_\ell < r$. We define linear maps $A_t, B_t, C_t, D_{st}, E_t : W \to W$ for all $1 \le t \le \ell$ and $1 \le s < t \le \ell$ by

$$\begin{aligned}
A_t v_{\xi_1,\ldots,\xi_\ell} &= \theta^{\xi_t} v_{\xi_1,\ldots,\xi_\ell}\\
B_t v_{\xi_1,\ldots,\xi_\ell} &= v_{\xi_1,\ldots,\xi_{t-1},\xi_t+1,\xi_{t+1},\ldots,\xi_\ell}\\
C_t v_{\xi_1,\ldots,\xi_\ell} &= \sum_{0\le i<r} \theta^{i\xi_t} v_{\xi_1,\ldots,\xi_{t-1},i,\xi_{t+1},\ldots,\xi_\ell}\\
D_{st} v_{\xi_1,\ldots,\xi_\ell} &= \theta^{\xi_s\xi_t} v_{\xi_1,\ldots,\xi_\ell}\\
E_t v_{\xi_1,\ldots,\xi_\ell} &= \theta^{\frac{\xi_t(\xi_t-1)}{2}} v_{\xi_1,\ldots,\xi_\ell}
\end{aligned}$$

for all $0 \le \xi_1, \ldots, \xi_\ell < r$, where as before the indices are interpreted modulo $r$. Note that $A_t = I_{r^{t-1}} \otimes A \otimes I_{r^{\ell-t}}$, $B_t = I_{r^{t-1}} \otimes B \otimes I_{r^{\ell-t}}$, $C_t = I_{r^{t-1}} \otimes C \otimes I_{r^{\ell-t}}$, $E_t = I_{r^{t-1}} \otimes E \otimes I_{r^{\ell-t}}$ for all $1 \le t \le \ell$.

Then $R = \langle A_1, B_1, \ldots, A_\ell, B_\ell \rangle$. The linear maps $C_t, D_{st}, E_t$ centralize the generators of $R$, except for the following ones:

$$\begin{aligned}
C_t A_t C_t^{-1} &= B_t^{-1}\\
C_t B_t C_t^{-1} &= A_t\\
D_{st} B_s D_{st}^{-1} &= B_s A_t\\
D_{st} B_t D_{st}^{-1} &= B_t A_s\\
E_t B_t E_t^{-1} &= B_t A_t \text{ (if } r > 2)
\end{aligned} \tag{3.1}$$

Hence $C_t, D_{st}, E_t$ normalize $R$. We have the following well-known result, which is for the most part due to Jordan.

**Theorem 3.2.2** *Let $Z$ be the group of scalar matrices in* $\mathrm{GL}(W)$. *Then the following statements hold:*

*(i)* $N_{\mathrm{GL}(W)}(R)$ *is generated by $Z$ together with the linear maps* $A_t, B_t, C_t, D_{st}, E_t$;
*(ii)* $N_{\mathrm{GL}(W)}(R)/RZ \cong \mathrm{Sp}_{2\ell}(r)$ *if* $r > 2$;
*(iii)* $N_{\mathrm{GL}(W)}(R)/RZ \cong O_{2\ell}^+(2)$ *if* $r = 2$.

***Proof*** The action of $N_{\mathrm{GL}(W)}(R)$ on $R/Z(R)$ induces a map $\pi : N_{\mathrm{GL}(W)}(R) \to \mathrm{Sp}_{2\ell}(r)$, with $\mathrm{Im}\,\pi \le O_{2\ell}^+(2)$ if $r = 2$. Since $\mathrm{Ker}\,\pi$ acts trivially on $R/Z(R)$ and $Z(R)$, it follows from Lemma 3.1.1 that $\mathrm{Ker}\,\pi$ acts on $R$ by inner automorphisms. Thus $\mathrm{Ker}\,\pi = RC_{\mathrm{GL}(W)}(R) = RZ$, since $R$ is absolutely irreducible.

Therefore it will suffice to check that the images of $C_t, D_{st}, E_t$ generate $\mathrm{Sp}_{2\ell}(r)$ if $r > 2$, and $O_{2\ell}^+(2)$ if $r = 2$.

Let $G$ be the group generated by $\pi(C_t)$, $\pi(D_{st})$, $\pi(E_t)$. Denote the images of $A_i$ and $B_i$ in $\overline{R} := R/Z(R)$ by $e_i := \overline{A_i}$ and $f_i := \overline{B_i}$. Then with respect to the alternating bilinear form $\xi$ on $\overline{R}$, the vectors $e_1, f_1, \ldots, e_\ell, f_\ell$ form a symplectic basis of $R/Z(R)$, with $\xi(e_i, f_j) = \delta_{i,j}$. Moreover in the case where $r = 2$, we have $\vartheta(e_i) = 0 = \vartheta(f_i)$ for all $1 \leq i \leq \ell$.

The images of $C_t$, $D_{st}$, $E_t$ act on $\overline{R}$ as

$$\pi(C_t) : \begin{cases} e_t \mapsto -f_t \\ f_t \mapsto e_t \end{cases} \qquad \pi(D_{st}) : \begin{cases} f_t \mapsto f_t + e_s \\ f_s \mapsto f_s + e_t \end{cases}$$

$$\pi(E_t) : f_t \mapsto f_t + e_t \text{ (if } r > 2)$$

with rest of the basis vectors fixed.

We will consider the case $r > 2$ first and prove that $G = \mathrm{Sp}_{2\ell}(r)$. If $\ell = 1$, then

$$G = \langle \pi(C_1), \pi(E_1) \rangle = \left\langle \begin{pmatrix} 0 & -1 \\ 1 & 0 \end{pmatrix}, \begin{pmatrix} 1 & 1 \\ 0 & 1 \end{pmatrix} \right\rangle = \mathrm{Sp}_2(r),$$

as required.

Suppose then that $\ell > 1$ and proceed by induction on $\ell$. Let $H$ be the group generated by $\pi(C_t)$, $\pi(E_t)$ for $2 \leq t \leq \ell$ and $\pi(D_{st})$ for $2 \leq s < t \leq \ell$. By induction, the action of $H$ on $\langle e_i, f_j : 2 \leq i, j \leq \ell \rangle$ generates all of $\mathrm{Sp}_{2\ell-2}(2)$.

We will first show that $G$ acts transitively on nonzero vectors in $\overline{R}$. To this end, suppose that $v \in \overline{R}$ is nonzero, and write $v = v' + v''$, where $v' \in \langle e_1, f_1 \rangle$ and $v'' \in \langle e_i, f_j : 2 \leq i, j \leq \ell \rangle$. We will show that $v$ is on the same $G$-orbit as $e_1$. If $v'' = 0$, then by the $\ell = 1$ case there exists $g \in \langle \pi(C_1), \pi(E_1) \rangle = \mathrm{Sp}_2(r)$ such that $gv = e_1$.

Thus we can assume that $v'' \neq 0$, and then by induction there exists $g \in H$ such that $gv'' = e_2$. Therefore we reduce to the case $v = v' + e_2$. Since $D_{12}C_2(e_2) = -f_2 - e_1$, we can assume that $v' \neq 0$. By the $\ell = 1$ case, there exists $g \in \langle \pi(C_1), \pi(E_1) \rangle$ such that $gv' = f_1$, so $gv = f_1 + e_2$. Now $C_1 D_{12}^{-1}(f_1 + e_2) = e_1$, so we conclude that $G$ is transitive on nonzero vectors in $\overline{R}$.

Next we will show that $G$ acts transitively on hyperbolic pairs, i.e., pairs $(e, f)$ such that $\xi(e, f) = 1$. It will suffice to show that $G$ can transform any hyperbolic pair $(e, f)$ to $(e_1, f_1)$. By transitivity on nonzero vectors we can assume that $e = e_1$, in which case $f = \lambda e_1 + f_1 + v''$ for some scalar $\lambda$ and $v'' \in \langle e_i, f_j : 2 \leq i, j \leq \ell \rangle$. Suppose first that $v'' \neq 0$. By induction there exists $g \in H$ such that $gv'' = e_2$. Then $D_{12}^{-1}g$ transforms $(e, f)$ to $(e_1, \lambda e_1 + f_1)$. Thus we can assume that $v'' = 0$, in which case $\pi(E_1)^{-\lambda}$ transforms $(e, f)$ to $(e_1, f_1)$.

Finally we show that $G$ can transform $\{e_1, f_1, \ldots, e_\ell, f_\ell\}$ to any other symplectic basis $\{e_1', f_1', \ldots, e_\ell', f_\ell'\}$ of $\overline{R}$, where $\xi(e_i', f_j') = \delta_{i,j}$ for all $1 \leq i, j \leq \ell$. First note that since $G$ acts transitively on hyperbolic pairs, we can assume that $e_1' = e_1$ and $f_1' = f_1$. In this case $\langle e_i, f_j : 2 \leq i, j \leq \ell \rangle = \langle e_i', f_j' : 2 \leq i, j \leq \ell \rangle$ since both subspaces are equal to the orthogonal complement of $\langle e_1, f_1 \rangle$. By induction, there exists $g \in H$ such that $g(e_i) = e_i'$ and $g(f_i) = f_i'$ for all $2 \leq i \leq \ell$, which proves the claim. Since any element of $\mathrm{Sp}_{2\ell}(r)$ maps a symplectic basis to a symplectic basis, we conclude that $G = \mathrm{Sp}_{2\ell}(r)$.

It remains to prove that $G = O_{2\ell}^+(2)$ in the case $r = 2$. Here one can argue with a similar proof by induction, so we will only sketch the argument. In the case $\ell = 1$ it is clear that $G = \langle \pi(C_1) \rangle = O_2^+(2)$. For $\ell > 1$, proceed by induction and first show that $G$ acts transitively on the set of nonzero vectors $e \in \overline{R}$ with $\vartheta(e) = 0$. Then one proves that $G$ acts transitively on the set of pairs $(e, f)$ satisfying $\xi(e, f) = 1$ and $\vartheta(e) = \vartheta(f) = 0$. Finally, arguing as in the previous paragraph we conclude that $G = O_{2\ell}^+(2)$. □

### 3.2.2 Unitary Representation of $r_+^{1+2\ell}$

Consider the construction from the previous subsection in the case where $\mathbb{K} = \mathbb{F}_{q^2}$, where $q$ is a power of a prime such that $r \mid q + 1$. With respect to the automorphism $\xi \mapsto \xi^q$ of $\mathbb{K}$, we define a non-degenerate Hermitian form $\widehat{b} : W \times W \to \mathbb{K}$ by taking $\{v_{\xi_1,\ldots,\xi_\ell} : 0 \leq \xi_1, \ldots, \xi_\ell < r\}$ as an orthonormal basis.

It is readily checked that $A_t, B_t, D_{st}, E_t \in \mathrm{GU}(W, \widehat{b})$ for all $1 \leq t \leq \ell$ and $1 \leq s < t \leq \ell$. We have $\widehat{b}(C_t w, C_t w') = r\widehat{b}(w, w')$ for all $w, w' \in W$ and $1 \leq t \leq \ell$. Thus we choose $c \in \mathbb{K}^\times$ such that $rc^{q+1} = 1$, so $cC_t \in \mathrm{GU}(W, \widehat{b})$ for all $1 \leq t \leq \ell$. Then we have the following result.

**Theorem 3.2.3** *Let $Z$ be the group of scalar matrices in $\mathrm{GL}(W)$ and denote $Z' = Z \cap \mathrm{GU}(W, \widehat{b})$. Then the following statements hold:*

*(i)* $N_{\Delta U(W,\widehat{b})}(R) = N_{\mathrm{GL}(W)}(R)$*;*
*(ii)* $N_{\mathrm{GU}(W,\widehat{b})}(R)$ *is generated by $Z'$ together with the linear maps $A_t$, $B_t$, $cC_t$, $D_{st}$, $E_t$;*
*(iii)* $N_{\mathrm{GU}(W,\widehat{b})}(R)/RZ' \cong \mathrm{Sp}_{2\ell}(r)$ *if $r > 2$;*
*(iv)* $N_{\mathrm{GU}(W,\widehat{b})}(R)/RZ' \cong O_{2\ell}^+(2)$ *if $r = 2$.*

***Proof*** Since $A_t, B_t, C_t, D_{st}, E_t \in \Delta U(W, \widehat{b})$, claim (i) follows from Theorem 3.2.2. Let $\pi : N_{\mathrm{GL}(W)}(R) \to \mathrm{Sp}_{2\ell}(r)$ be the map as in the proof of Theorem 3.2.2, and denote the restriction of $\pi$ to $\mathrm{GU}(W, \widehat{b})$ by $\pi'$.

We have $\pi(C_t) = \pi(cC_t)$ for all $1 \leq t \leq \ell$, so $\mathrm{Im}\,\pi = \mathrm{Im}\,\pi'$ and claim (ii) follows from Theorem 3.2.2. Then claims (iii)–(iv) follow from Theorem 3.2.2 (ii)–(iii) and the fact that $\mathrm{Ker}\,\pi' = RZ'$. □

### 3.2.3 Orthogonal Representation of $2_+^{1+2\ell}$

Consider the construction of Sect. 3.2.1 in the case where $r = 2$, so $\mathbb{K} = \mathbb{F}_q$ with $q$ odd. We can define a non-degenerate quadratic form $Q : W \to \mathbb{K}$ by taking

$\{v_{\xi_1,\ldots,\xi_\ell} : 0 \leq \xi_1, \ldots, \xi_\ell \leq 1\}$ to be an orthonormal basis. Then by [55, Proposition 2.5.13] we have $\mathrm{sgn}(Q) = +$, unless $\ell = 1$ and $q \equiv 3 \mod 4$, in which case $\mathrm{sgn}(Q) = -$.

It is straightforward to see that $A_t, B_t, D_{st} \in O(W, Q)$ for all $1 \leq t \leq \ell$ and $1 \leq s < t \leq \ell$. Furthermore $C_t \in \mathrm{GO}(W, Q)$, with $Q(C_t v) = 2Q(v)$ for all $1 \leq t \leq \ell$ and $v \in W$. The following result describes the normalizer of $R$ in $O(W, Q)$ and $\mathrm{GO}(W, Q)$.

**Theorem 3.2.4** *Let $Z$ be the group of scalar matrices in $\mathrm{GL}(W)$, and let $\pi : N_{\mathrm{GL}(W)}(R) \to O^+_{2\ell}(2)$ be the homomorphism corresponding to the action of $N_{\mathrm{GL}(W)}(R)$ on $R/Z(R)$. Then the following statements hold:*

*(i)* $N_{\mathrm{GO}(W,Q)}(R) = N_{\mathrm{GL}(W)}(R)$*;*
*(ii)* $N_{\mathrm{GO}(W,Q)}(R)$ *is generated by $Z$ together with the linear maps $A_t, B_t, C_t, D_{st}$;*
*(iii)* $N_{\mathrm{GO}(W,Q)}(R)/RZ \cong O^+_{2\ell}(2)$*;*
*(iv)* $N_{O(W,Q)}(R)/R \cong \begin{cases} O^+_{2\ell}(2), & \text{if } 2 \text{ is a square in } \mathbb{K}, \\ \Omega^+_{2\ell}(2), & \text{otherwise.} \end{cases}$
*(v) Suppose that 2 is not a square in $\mathbb{K}$, and let $g \in N_{\mathrm{GO}(W,Q)}(R)$. Then $\pi(g) \in \Omega^+_{2\ell}(2)$ if and only if $\tau(g)$ is a square in $\mathbb{K}$.*
*(vi) Suppose that 2 is a square in $\mathbb{K}$. Then $\tau(g)$ is a square in $\mathbb{K}$ for all $g \in N_{\mathrm{GO}(W,Q)}(R)$.*

***Proof*** It follows from Theorem 3.2.2 (i) that $N_{\mathrm{GL}(W)}(R)$ is generated by $Z$ together with the linear maps $A_t, B_t, C_t, D_{st}$; therefore $N_{\mathrm{GO}(W,Q)}(R) = N_{\mathrm{GL}(W)}(R)$ so claim (i) holds. Now claims (ii) and (iii) follow from Theorem 3.2.2.

We have $Q(C_t v) = 2Q(v)$ for all $v \in W$ and $1 \leq t \leq \ell$. Thus if 2 is a square in $\mathbb{K}$, then there exists a scalar $c \in \mathbb{K}$ such that $cC_t \in O(W, Q)$ for all $1 \leq t \leq \ell$. Then $A_t, B_t, cC_t, D_{st} \in N_{O(W,Q)}(R)$, so it follows from (ii) that $\pi(N_{O(W,Q)}(R)) = \pi(N_{\mathrm{GO}(W,Q)}(R)) = O^+_{2\ell}(2)$ and $N_{O(W,Q)}(R)/R \cong O^+_{2\ell}(2)$.

Next we consider (iv) in the case where 2 is not a square in $\mathbb{K}$. For this, note first that $\pi(N_{O(W,Q)}(R)) = \pi(ZN_{O(W,Q)}(R))$ has index $\leq 2$ in $\pi(N_{\mathrm{GO}(W,Q)}(R)) = O^+_{2\ell}(2)$. Since 2 is not a square in $\mathbb{K}$, we have $cC_t \notin O(W, Q)$ for all scalars $c \in \mathbb{K}$. It follows that $\pi(C_t) \notin \pi(N_{O(W,Q)}(R))$ for all $1 \leq t \leq \ell$, and so $\pi(N_{O(W,Q)}(R))$ has index 2 in $O^+_{2\ell}(2)$.

Since $\mathrm{rank}(\pi(C_t) - 1) = 1$ and $\mathrm{rank}(\pi(D_{st}) - 1) = 2$, we have $\pi(C_t) \notin \Omega^+_{2\ell}(2)$ for all $1 \leq t \leq \ell$ and $\pi(D_{st}) \in \Omega^+_{2\ell}(2)$ for all $1 \leq s < t \leq \ell$. Thus we have shown

$$\begin{aligned} &\pi(C_t) \notin \pi(N_{O(W,Q)}(R)) && \pi(C_t) \notin \Omega^+_{2\ell}(2) \\ &\pi(D_{st}) \in \pi(N_{O(W,Q)}(R)) && \pi(D_{st}) \in \Omega^+_{2\ell}(2) \end{aligned}$$

for all $1 \leq t \leq \ell$ and $1 \leq s < t \leq \ell$. Since $\pi(N_{O(W,Q)}(R))$ and $\Omega^+_{2\ell}(2)$ both have index 2 in $O^+_{2\ell}(2)$, we conclude that $\pi(N_{O(W,Q)}(R)) = \Omega^+_{2\ell}(2)$.

Claim (v) follows from (iv), since $O(W, Q)Z$ is precisely the set of $g \in \mathrm{GO}(W, Q)$ such that $\tau(g)$ is a square in $\mathbb{K}$. Claim (vi) follows (ii), using the fact that $\tau(A_t) = \tau(B_t) = 1$ and $\tau(C_t) = 2$ for all $1 \le t \le \ell$, and $\tau(D_{st}) = 1$ for all $1 \le s < t \le \ell$. □

**Remark 3.2.1** In the proof of Theorem 3.2.4 (iv), we could also have used the fact that $O_{2\ell}^+(2)$ has a unique subgroup of index 2 if $\ell \neq 2$ [3, 22.7, 22.9]. However $O_4^+(2)$ has three different subgroups of index 2 (described in [55, Proposition 2.5.9]), so the case $\ell = 2$ still needs a separate argument.

### 3.2.4 Absolutely Irreducible Representation of $2_-^{1+2\ell}$

Suppose that $\mathbb{K} = \mathbb{F}_q$, where $q$ is odd. Let $V$ be a 2-dimensional $\mathbb{K}$-vector space with basis $v_0, v_1$ and define linear maps $A, B, C : V \to V$ as in Sect. 3.2.1, so

$$A = \begin{pmatrix} 1 & 0 \\ 0 & -1 \end{pmatrix}, \quad B = \begin{pmatrix} 0 & 1 \\ 1 & 0 \end{pmatrix}, \quad C = \begin{pmatrix} 1 & 1 \\ 1 & -1 \end{pmatrix}.$$

As in Sect. 3.2.1 we denote $R_0 = \langle A, B \rangle$, so $R_0$ is absolutely irreducible and $R_0 \cong 2_+^{1+2}$.

Next we choose $a, b \in \mathbb{K}$ such that $a^2 + b^2 = -1$ and define $A', B' : V \to V$ by

$$A' = \begin{pmatrix} a & b \\ b & -a \end{pmatrix}, \quad B' = \begin{pmatrix} 0 & -1 \\ 1 & 0 \end{pmatrix}.$$

Furthermore denote

$$C' = A' + B' = \begin{pmatrix} a & b-1 \\ b+1 & -a \end{pmatrix}, \quad E' = A' + I_V = \begin{pmatrix} a+1 & b \\ b & -a+1 \end{pmatrix}.$$

Then $(A')^2 = (B')^2 = -1$ and $[A', B'] = \theta I_V$, so $R_0' = \langle A', B' \rangle \cong 2_-^{1+2}$. It is straightforward to see that $R_0'$ is an absolutely irreducible subgroup of $\mathrm{GL}(V)$. We have $C'A'(C')^{-1} = B'$ and $C'B'(C')^{-1} = A'$, so $C'$ normalizes $R_0'$. Furthermore $E'A'(E')^{-1} = A'$ and $E'B'(E')^{-1} = A'B'$, so $E'$ also normalizes $R_0'$.

Next we define $R = R_0' \otimes R_0 \otimes \cdots \otimes R_0$, where $R_0$ occurs $\ell - 1$ times as a factor. As in Sect. 3.2.1, it follows from Lemma 2.1.16 (i) that $R$ is an absolutely irreducible subgroup of $\mathrm{GL}(W)$, where $W = V \otimes \cdots \otimes V$ ($\ell$ times). Moreover $R$ is a central product of $2_-^{1+2}$ and $\ell - 1$ copies of $2_+^{1+2}$, so $R \cong 2_-^{1+2\ell}$.

We take as a basis of $W$ the vectors $v_{\xi_1,\ldots,\xi_\ell} := v_{\xi_1} \otimes \cdots \otimes v_{\xi_\ell}$, where $0 \le \xi_1, \ldots, \xi_\ell \le 1$. We define maps $A_t, B_t, C_t, D_{st} : W \to W$ for $2 \le t \le \ell$ and $2 \le s < t \le \ell$ as in Sect. 3.2.1.

In addition, we define $A_1', C_1', E_1' : W \to W$ by $A_1' = A' \otimes I_{2^{\ell-1}}$, $C_1' = C' \otimes I_{2^{\ell-1}}$, and $E_1' = E' \otimes I_{2^{\ell-1}}$. Furthermore, for $2 \leq t \leq \ell$ we define $D_{1t}' : W \to W$ by

$$D_{1t}' v_{\xi_1,\ldots,\xi_\ell} = (A_1')^{\delta_{1,\xi_t}} v_{\xi_1,\ldots,\xi_\ell}$$

for all $0 \leq \xi_1, \ldots, \xi_\ell \leq 1$. Thus

$$D_{1t}' v_{\xi_1,\ldots,\xi_\ell} = \begin{cases} v_{\xi_1,\ldots,\xi_\ell}, & \text{if } \xi_t = 0. \\ (-1)^{\xi_1} a v_{\xi_1,\xi_2,\ldots,\xi_\ell} + b v_{\xi_1+1,\xi_2,\ldots,\xi_\ell}, & \text{if } \xi_t = 1. \end{cases}$$

We have $R = \langle A_1', B_1', A_2, B_2, \ldots, A_t, B_t \rangle$. For $C_t$ with $2 \leq t \leq \ell$ and $D_{st}$ with $2 \leq s < t \leq \ell$ the same relations as in (3.1) hold. The maps $E_1'$ and $D_{1t}'$ centralize the generators of $R$, except for the following ones:

$$\begin{aligned} C_1' A_1' (C_1')^{-1} &= B_1' \\ C_1' B_1' (C_1')^{-1} &= A_1' \\ D_{1t}' B_1' (D_{1t}')^{-1} &= B_1' A_t \\ D_{1t}' B_t (D_{1t}')^{-1} &= B_t A_1' A_t \\ E_1' B_1' (E_1')^{-1} &= A_1' B_1' \end{aligned} \tag{3.2}$$

**Theorem 3.2.5** *Let $Z$ be the group of scalar matrices in* $\mathrm{GL}(W)$. *Then the following statements hold:*

*(i)* $N_{\mathrm{GL}(W)}(R)$ *is generated by $Z$ together with the linear maps* $A_1'$, $B_1'$, $C_1'$, $E_1'$, $A_t$, $B_t$, $C_t$, $D_{1t}'$, $D_{st}$ *for $2 \leq t \leq \ell$ and $2 \leq s < t \leq \ell$;*

*(ii)* $N_{\mathrm{GL}(W)}(R)/RZ \cong O_{2\ell}^-(2)$.

***Proof*** As in the proof of Theorem 3.2.2, we have a homomorphism $\pi : N_{\mathrm{GL}(W)}(R) \to O_{2\ell}^-(2)$ with $\operatorname{Ker} \pi = RZ$ since $R$ is absolutely irreducible.

Thus for the proof of (i) and (ii), it will suffice to check that the images of $C_1'$, $E_1'$, $C_t$, $D_{1t}'$, $D_{st}$ generate $O_{2\ell}^-(2)$.

Let $G$ be the group generated by $\pi(C_1'), \pi(E_1'), \pi(C_t), \pi(D_{1t}'), \pi(D_{st})$.

Denote the images of $A_i$ and $B_i$ in $\overline{R} := R/Z(R)$ by $e_i := \overline{A_i}$ and $f_i := \overline{B_i}$ for $2 \leq i \leq \ell$, and $e_1 := \overline{A_1'}$ and $f_1 := \overline{B_1'}$. Then with respect to the alternating bilinear form $\xi$ on $\overline{R}$, the vectors $e_1, f_1, \ldots, e_\ell, f_\ell$ form a standard symplectic basis of $R/Z(R)$, with $\xi(e_i, f_j) = \delta_{i,j}$. Moreover $\vartheta(e_1) = 1 = \vartheta(f_1)$, and $\vartheta(e_i) = 0 = \vartheta(f_i)$ for all $2 \leq i \leq \ell$.

If $\ell = 1$, we have

$$G = \langle \pi(C_1'), \pi(E_1') \rangle = \left\langle \begin{pmatrix} 0 & 1 \\ 1 & 0 \end{pmatrix}, \begin{pmatrix} 1 & 1 \\ 0 & 1 \end{pmatrix} \right\rangle = O_2^-(2),$$

as required.

Suppose then that $\ell > 1$. Denote by $H$ the subgroup of $G$ generated by $\pi(C_t)$ for $2 \leq t \leq \ell$ and $\pi(D_{st})$ for $2 \leq s < t \leq \ell$. Then by Theorem 3.2.2, the subgroup $H$ generates all of $O_{2\ell-2}^+(2)$ in its action on $\langle e_i, f_j : 2 \leq i, j \leq \ell \rangle$. Using this fact, arguing as in the proof of Theorem 3.2.2 we see that $G$ acts transitively on the set of pairs $(e, f)$ satisfying $\xi(e, f) = 1$ and $\vartheta(e) = 1 = \vartheta(f)$. From this a straightforward argument as in the proof Theorem 3.2.2 (penultimate paragraph) completes the proof that $G = O_{2\ell}^-(2)$. □

### 3.2.5 Unitary Representation of $2_-^{1+2\ell}$

Consider the construction of Sect. 3.2.4 in the case where $\mathbb{K} = \mathbb{F}_{q^2}$ and $q$ is odd. Let $\zeta \in \mathbb{K}^\times$ be a primitive element. For the definition of $A'$ we can choose $b = 0$, so

$$A' = \begin{pmatrix} a & 0 \\ 0 & -a \end{pmatrix},$$

where $a^2 = -1$.

With respect to the automorphism $\xi \mapsto \xi^q$ of $\mathbb{K}$, we will construct a non-degenerate $R$-invariant Hermitian form $\widehat{b} : W \times W \to \mathbb{K}$ as follows. Note that we have a non degenerate $R_0$-invariant Hermitian form $\widehat{b_0}$ on $V$ by taking $v_0, v_1$ as an orthonormal basis. Then with a non-degenerate $R_0'$-invariant Hermitian form $\widehat{b_0'}$ on $V$, we can define $\widehat{b} = \widehat{b_0'} \otimes \widehat{b_0} \otimes \cdots \otimes \widehat{b_0}$.

For $\widehat{b_0'}$, if $q \equiv 3 \mod 4$, we can choose $\widehat{b_0'} = \widehat{b_0}$. If $q \equiv 1 \mod 4$, then defining

$$\widehat{b_0''}(v_0, v_1) = 1 = -\widehat{b_0''}(v_1, v_0)$$

$$\widehat{b_0''}(v_0, v_0) = 0 = \widehat{b_0''}(v_1, v_1)$$

and extending sesquilinearly provides non-degenerate $R_0'$-invariant sesquilinear form. The form $\widehat{b_0''}$ is skew-Hermitian, so we can define $\widehat{b_0'} := \zeta^{\frac{q+1}{2}} \widehat{b_0''}$ to get a non-degenerate $R_0'$-invariant Hermitian form. (This is because $\xi^q = -\xi$ for $\xi = \zeta^{\frac{q+1}{2}}$.)

Thus on the basis elements, the form $\widehat{b}$ is defined as follows. First if $q \equiv 3 \mod 4$, the basis $\{v_{\xi_1,\ldots,\xi_\ell} : 0 \leq \xi_1, \ldots, \xi_\ell \leq 1\}$ is an orthonormal basis for $\widehat{b}$. For

$q \equiv 1 \mod 4$, we have

$$\widehat{b}(v_{\xi_1,\dots,\xi_\ell}, v_{\xi'_1,\dots,\xi'_\ell}) = \zeta^{\frac{q+1}{2}}(-1)^{\xi_1}\delta_{\xi_1,\xi'_1+1}\delta_{\xi_2,\xi'_2}\cdots\delta_{\xi_\ell,\xi'_\ell}$$

for all $0 \le \xi_1, \xi'_1, \dots, \xi_\ell, \xi'_\ell \le 1$. (cf. the basis of "hyperbolic pairs" mentioned in Sect. 2.7.)

We have $A_t, B_t, D_{st} \in \mathrm{GU}(W, \widehat{b})$ for all $1 \le t \le \ell$ and $2 \le s < t \le \ell$, and $D'_{1t} \in \mathrm{GU}(W, \widehat{b})$ for all $2 \le t \le \ell$. Moreover by choosing $c \in \mathbb{K}^\times$ such that $2c^{q+1} = 1$, we have $cE'_1 \in \mathrm{GU}(W, \widehat{b})$, $cC'_1 \in \mathrm{GU}(W, \widehat{b})$, and $cC_t \in \mathrm{GU}(W, \widehat{b})$ for all $2 \le t \le \ell$. Then we have the following result.

**Theorem 3.2.6** *Let $Z$ be the group of scalar matrices in* $\mathrm{GL}(W)$ *and denote* $Z' = Z \cap \mathrm{GU}(W, \widehat{b})$. *Then the following statements hold:*

*(i)* $N_{\Delta U(W,\widehat{b})}(R) = N_{\mathrm{GL}(W)}(R)$;
*(ii)* $N_{\mathrm{GU}(W,\widehat{b})}(R)$ *is generated by* $Z'$ *together with the linear maps* $A'_1$, $B'_1$, $cC'_1$, $cE'_1$, $A_t$, $B_t$, $cC_t$, $D'_{1t}$, $D_{st}$ *for* $2 \le t \le \ell$ *and* $2 \le s < t \le \ell$;
*(iii)* $N_{\mathrm{GU}(W,\widehat{b})}(R)/RZ' \cong O^-_{2\ell}(2)$.

***Proof*** Applying Theorem 3.2.5, the result follows by arguing similarly to the proof of Theorem 3.2.3. □

### 3.2.6 *Symplectic Representation of* $2^{1+2\ell}_-$

Consider the construction of Sect. 3.2.4, so $\mathbb{K} = \mathbb{F}_q$ with $q$ odd.

The group $R'_0 \cong 2^{1+2}_-$ has a non-degenerate invariant alternating bilinear form $b'_0$ on $V$ with $b(v_0, v_1) = 1$. Furthermore $R_0 \cong 2^{1+2}_+$ has a non-degenerate invariant symmetric bilinear form $b_0$ on $V$ by taking $\{v_0, v_1\}$ as an orthonormal basis. Thus $b = b'_0 \otimes b_0 \otimes \cdots \otimes b_0$ is a non-degenerate alternating $R$-invariant bilinear form on $W = V \otimes \cdots \otimes V$.

Then the basis $\{v_{\xi_1,\dots,\xi_\ell} : 0 \le \xi_1, \dots, \xi_\ell \le 1\}$ is a standard symplectic basis for $b$, with hyperbolic pairs $(v_{0,\xi_2,\dots,\xi_\ell}, v_{1,\xi_2,\dots,\xi_\ell})$ for $0 \le \xi_2, \dots, \xi_\ell \le 1$.

We have $A'_1, B'_1, A_t, B_t, D'_{1t} \in \mathrm{Sp}(W, b)$ for all $2 \le t \le \ell$, and $D_{st} \in \mathrm{Sp}(W, b)$ for all $2 \le s < t \le \ell$. Moreover $E'_1, C'_1 \in \mathrm{GSp}(W, b)$ and $C_t \in \mathrm{GSp}(W, b)$ for all $2 \le t \le \ell$, with $\tau(E'_1) = \tau(C'_1) = \tau(C_t) = 2$ for all $2 \le t \le \ell$.

**Theorem 3.2.7** *Let $Z$ be the group of scalar matrices in* $\mathrm{GL}(W)$, *and let* $\pi : N_{\mathrm{GL}(W)}(R) \to O^-_{2\ell}(2)$ *be the homomorphism corresponding to the action of* $N_{\mathrm{GL}(W)}(R)$ *on* $R/Z(R)$. *Then the following statements hold:*

*(i)* $N_{\mathrm{GSp}(W,b)}(R) = N_{\mathrm{GL}(W)}(R)$;
*(ii)* $N_{\mathrm{GSp}(W,b)}(R)$ *is generated by* $Z$ *together with the linear maps together with the linear maps* $A'_1$, $B'_1$, $C'_1$, $E'_1$, $A_t$, $B_t$, $C_t$, $D'_{1t}$, $D_{st}$ *for* $2 \le t \le \ell$ *and* $2 \le s < t \le \ell$;

*(iii)* $N_{\mathrm{GSp}(W,b)}(R)/RZ \cong O^-_{2\ell}(2)$;

*(iv)* $N_{\mathrm{Sp}(W,b)}(R)/R \cong \begin{cases} O^-_{2\ell}(2), & \text{if 2 is a square in } \mathbb{K}, \\ \Omega^-_{2\ell}(2), & \text{otherwise.} \end{cases}$

*(v)* *Suppose that 2 is not a square in* $\mathbb{K}$*, and let* $g \in N_{\mathrm{GSp}(W,Q)}(R)$*. Then* $\pi(g) \in \Omega^-_{2\ell}(2)$ *if and only if* $\tau(g)$ *is a square in* $\mathbb{K}$*.*

*(vi)* *Suppose that 2 is a square in* $\mathbb{K}$*. Then* $\tau(g)$ *is a square in* $\mathbb{K}$ *for all* $g \in N_{\mathrm{GSp}(W,Q)}(R)$*.*

***Proof*** Applying Theorem 3.2.5, the claims follow similarly to the proof of Theorem 3.2.4. □

# Chapter 4
# Metrically Primitive Maximal Irreducible Solvable Subgroups

A group $G$ is solvable if and only if it admits a series

$$1 = G_0 \trianglelefteq G_1 \trianglelefteq G_2 \trianglelefteq \cdots \trianglelefteq G_k = G$$

such that $G_i/G_{i-1}$ is abelian for all $1 \leq i \leq k$. For the construction of maximal solvable subgroups, determining the groups $G_1, G_2, \ldots$ successively is what Jordan calls the "essence of his method"; see for example [42, p. 963], [45, p. 121], [47, p. 397].

For a metrically primitive maximal irreducible solvable $G \leq \Delta(V, \kappa)$, previously in Sects. 2.4–2.9 we have completed the first step of Jordan's "method" by identifying and constructing an abelian normal subgroup $F_0$ explicitly. In this chapter, we move on to the next step and identify a certain subgroup $F_0 \leq A \trianglelefteq G$ such that $A/F_0$ is abelian. We will then be able to complete the recursive description of the general structure of maximal irreducible solvable subgroups of $\Delta(V, \kappa)$.

## 4.1 The Fitting Subgroup of $C_{G^\circ}(F_0)$

We begin with two general results on nilpotent groups of class two.

**Lemma 4.1.1** *Let $Z \leq T \leq E$ be finite groups such that $Z = Z(T) = Z(E)$, $Z$ is cyclic, and $E/Z$ is abelian. Then $E/Z = T/Z \times C_E(T)/Z$.*

***Proof*** This is [64, Corollary 1.7], alternatively adapt the proof of [40, Theorem 2]. □

Let $W$ be a vector space over a field $\mathbb{F}$. For a subset $S$ of End($W$), we will denote by $[S]_{\mathbb{F}}$ the $\mathbb{F}$-subalgebra generated by $S$.


M. Korhonen, *Maximal Solvable Subgroups of Finite Classical Groups*,
Lecture Notes in Mathematics 2346, https://doi.org/10.1007/978-3-031-62915-0_4

**Lemma 4.1.2** *Let $T \leq \mathrm{GL}(W)$, where $W$ is a vector space over a field $\mathbb{F}$. Suppose that $[T, T] \leq Z(T) \leq \mathbb{F}^\times$ and that $T$ is completely reducible. Then the following statements hold:*

*(i)* $\dim_{\mathbb{F}}([T]_{\mathbb{F}}) = [T : Z(T)]$;
*(ii)* $Z([T]_{\mathbb{F}}) = \mathbb{F}$;
*(iii)* *$W \downarrow T$ is homogeneous with absolutely irreducible composition factors;*
*(iv)* *$[T : Z(T)] = d^2$, where $d$ is the dimension of a composition factor of $W \downarrow T$.*

***Proof*** Since $Z(T) \subset \mathbb{F}^\times$, representatives for the cosets of $Z(T)$ in $T$ span $[T]_{\mathbb{F}}$. Now arguing as in [77, Lemma 19.4] or [87, Proposition 2.2 (1)], we see that if $x_1$, ..., $x_t \in T$ lie in distinct cosets of $Z(T)$, then $\{x_1, \ldots, x_t\}$ is linearly independent over $\mathbb{F}$. Thus the coset representatives of $Z(T)$ in $T$ form a basis of $[T]_{\mathbb{F}}$, which proves (i).

Claim (ii) follows with the same proof as [77, Lemma 20.2] or [87, Proposition 2.2 (2)], using $Z(T) \leq \mathbb{F}^\times$ and the fact that representatives for the cosets of $Z(T)$ in $T$ form a $\mathbb{F}$-basis of $[T]_{\mathbb{F}}$.

For (iii) and (iv), first note that by complete reducibility the algebra $[T]_{\mathbb{F}}$ is semisimple [77, Theorem 14.3], and thus a direct sum of matrix algebras over division rings. Then it follows from (ii) that $[T]_{\mathbb{F}}$ is simple and $[T]_{\mathbb{F}} \cong \mathrm{Mat}_d(\mathbb{F})$ for some integer $d > 0$. Therefore $W \downarrow T$ is homogeneous, and all composition factors of $W \downarrow T$ are absolutely irreducible of dimension $d$ [77, Theorem 14.4]. Since $\dim_{\mathbb{F}} \mathrm{Mat}_d(\mathbb{F}) = d^2$, it follows from (i) that $[T : Z(T)] = d^2$. □

For the rest of this section, let $G \leq \Delta(V, \kappa)$ be maximal irreducible solvable, and let $F_0 \leq G$ be a subgroup satisfying properties (F1)–(F3) as in Sect. 2.4. We consider subgroups $A \leq G$ with the following properties:

(A1) $F_0 \leq A \trianglelefteq G$ and $A \leq C_{G^\circ}(F_0)$;
(A2) $[A, A] \leq F_0$;
(A3) $A$ is maximal among the subgroups of $G$ satisfying (A1) and (A2).

Note that a subgroup $A$ satisfying properties (A1)–(A3) always exists, and in view of Lemma 2.9.3 and Theorem 2.9.4, we have $F_0 \lneqq A$ if and only if $\mu > 1$.

**Remark 4.1.3** The properties (A1)–(A3) generalize a definition given by Suprunenko in [77, §20.2, p. 139] for maximal irreducible solvable subgroups of $\mathrm{GL}_n(q)$. The subgroup $A$ turns out to be what Jordan calls the *noyau* of $G$, the definition given by Jordan [52, §29, p. 284] is similar to Proposition 4.1.9.

We will first prove the following lemma, which shows that a subgroup $A$ satisfying (A1)–(A3) is uniquely determined by $F_0$.

**Lemma 4.1.4** *Suppose that $A$ satisfies (A1)–(A3). Then* $A = \mathrm{Fit}(C_{G^\circ}(F_0))$.

***Proof*** We provide a proof which is essentially the same as [87, Proposition 2.13] and [17, Theorem 4.2]. First, it is clear that $A$ is a nilpotent normal subgroup of $C_{G^\circ}(F_0)$, so $A \leq \mathrm{Fit}(C_{G^\circ}(F_0))$.

Next consider a nilpotent normal subgroup $N$ of $C_{G^\circ}(F_0)$ such that $F_0 \le N \trianglelefteq G$. Consider the lower central series

$$N = \gamma_1(N) > \gamma_2(N) > \cdots > \gamma_{d-1}(N) > \gamma_d(N) = 1$$

of $N$. We have $[\gamma_i(N), \gamma_j(N)] \le \gamma_{i+j}(N)$, so $\gamma_i(N)$ is abelian for $i \ge d/2$. Since also $\gamma_i(N) \trianglelefteq G$, it follows from the maximality of $F_0$ that $\gamma_i(N) \le F_0$ for $i \ge d/2$. Because $F_0$ centralizes $N$, we conclude that $\gamma_{i+1}(N) = 1$ for $i \ge d/2$. Therefore $d \le 3$, so $[N, N] \le F_0$. Thus conditions (A1) and (A2) hold for $N$.

In particular, they hold for $N = \mathrm{Fit}(C_{G^\circ}(F_0))$ which contains $A$, so by the maximality of $A$ we conclude that $A = \mathrm{Fit}(C_{G^\circ}(F_0))$. □

**Lemma 4.1.5** *Suppose that $A$ satisfies (A1)–(A3). Then $Z(A) = F_0$, and the centralizer of $A$ in $C_{G^\circ}(F_0)$ is equal to $F_0$.*

***Proof*** We have $F_0 \le Z(A) \trianglelefteq G$, so $Z(A) = F_0$ by maximality of $F_0$. Since $C_{G^\circ}(F_0)$ is solvable, its Fitting subgroup contains its centralizer. Thus by Lemma 4.1.4 we have $C_{C_{G^\circ}(F_0)}(A) = Z(A) = F_0$, which completes the proof of the lemma. □

**Lemma 4.1.6** *Let $A \trianglelefteq G$ be a subgroup satisfying properties (A1) and (A2). Then the action of $G$ on $A/F_0$ is completely reducible in the following sense: if $X/F_0 \le A/F_0$ is such that $X \trianglelefteq G$, then $A/F_0 = X/F_0 \times C_A(X)/F_0$.*

***Proof*** We have $F_0 \le Z(A) \trianglelefteq G$ and $F_0 \le Z(X) \le G$, so $Z(A) = Z(X) = F_0$ by the maximality of $F_0$. Furthermore $A/F_0$ is abelian and $F_0$ is cyclic, so the result follows from Lemma 4.1.1. □

Next we consider subgroups $R \le G$ with the following properties.

(R1) $R \trianglelefteq G$ and $R \le C_{G^\circ}(F_0)$;
(R2) $[R, R] \le F_0$ and $R \not\le F_0$;
(R3) $R$ is minimal among the subgroups of $G$ satisfying (R1) and (R2).

By the uniqueness of $A$ (Lemma 4.1.4), we conclude that $R \le A$ for every subgroup satisfying properties (R1)–(R3).

**Proposition 4.1.7** *Let $R \le G$ be a subgroup satisfying properties (R1)–(R3). Then the following statements hold:*

*(i) $R$ is an extraspecial $r$-group for some prime $r$ that divides $|F_0|$;*
*(ii) If $r > 2$, then $R$ has exponent $r$;*
*(iii) $Z(R) = R \cap F_0$.*

***Proof*** The group $R$ is nilpotent of class 2 since $[R, R] \le F_0$. Let $P \le R$ be an $r$-Sylow subgroup of $R$ such that $P \not\le F_0$. Then $P \trianglelefteq G$, so by minimality of $R$ we have $R = P$ and $R$ is an $r$-group.

The quotient $R/R \cap F_0$ is abelian, so $R_0 = \{x \in R : x^r \in F_0\}$ is a subgroup of $R$. We have $R_0 \trianglelefteq G$ and $R_0 \not\le F_0$, so again by minimality of $R$ we have $R = R_0$. In other words, the quotient $R/R \cap F_0$ is an elementary abelian $r$-group.

Since $Z(R)F_0$ is an abelian normal subgroup of $G$, it follows from the maximality of $F_0$ that $Z(R) \leq F_0$. Thus $Z(R) = R \cap F_0$, so (iii) holds.

Because $Z(R) \leq F_0$, it follows from property (R2) that $[R, R] \neq 1$. Next we prove the following two claims.

**Claim 1: $[R, R]$ is cyclic of order $r$, and $r \mid |F_0|$**
For all $x, y \in R$ we have $[x, y]^r = [x^r, y] = 1$ since $[R, R]$ is central in $R$ and $x^r \in F_0$. Because $F_0$ is cyclic and $[R, R] \leq F_0$, we conclude that $[R, R]$ is cyclic of order $r$ with $r \mid |F_0|$.

**Claim 2: There exists $x \in R \setminus F_0$ such that $x^r = 1$ if $r > 2$, and $x^4 = 1$ if $r = 2$**
Since $R$ is nonabelian, we can find $x, y \in R \setminus F_0$ such that $[x, y] \neq 1$. Let $t$ be a generator for the Sylow $r$-subgroup of $F_0$. Then $x^r = t^\alpha$ and $y^r = t^\beta$ for some integers $\alpha, \beta$ coprime to $r$. Since $[R, R]$ is central in $R$, we have

$$(x^k y)^r = x^{kr} y^r [y, x^k]^{\frac{r(r-1)}{2}} = t^{k\alpha+\beta}[y, x]^{\frac{kr(r-1)}{2}}$$

for all integers $k$. By choosing $k$ such that $k \equiv -\beta/\alpha \mod |t|$, we have

$$(x^k y)^r = [y, x]^{\frac{kr(r-1)}{2}}.$$

Since $[R, R]$ is cyclic of order $r$, by choosing $x_0 = x^k y$ we have $x_0^r = 1$ if $r > 2$, and $x_0^4 = 1$ if $r = 2$. Furthermore $x_0 \in R \setminus F_0$ since $[x^k, y] = [x, y]^k \neq 1$.

Now define $S = \{x \in R : x^r = 1\}$ if $r > 2$, and $S = \{x \in R : x^4 = 1\}$ if $r = 2$. Then $S \trianglelefteq G$ and $S \not\leq F_0$, so by the minimality of $R$ we have $S = R$. In other words, $R$ is an $r$-group of exponent $\gcd(r, 2)r$, and in particular (ii) holds.

It remains to prove that $R$ is extraspecial. If $r > 2$, then $Z(R) = R \cap F_0$ must be cyclic of order $r$ since $R$ has exponent $r$. Since $[R, R] \leq Z(R)$, we conclude that $[R, R] = Z(R)$ is cyclic of order $r$ and so $R$ is extraspecial. The same argument also works when $r = 2$ and $F_0$ contains no element of order 4.

Thus we can assume that $r = 2$, and that $R \cap F_0 = \langle g \rangle$, where $g$ has order 4. We will see that this will lead to a contradiction.

In this case $[R, R] = \langle \theta \rangle$, where $\theta = g^2$. Considering $R/R \cap F_0$ as a vector space over $\mathbb{F}_2$, we define an alternating bilinear form $\xi$ on $R/R \cap F_0$ by

$$[x, y] = \theta^{\xi(\overline{x}, \overline{y})}$$

for all $x, y \in R$. Since $R \cap F_0 = Z(R)$, it is clear that $\xi$ is a non-degenerate alternating bilinear form. By the minimality of $R$, the action of $G$ on $R/R \cap F_0$ by conjugation is an irreducible $\mathbb{F}_2[G]$-module. Since $G$ is solvable, by a theorem of Willems (Theorem 2.1.14) there exists a $G$-invariant quadratic form $Q$ on $R/R \cap F_0$ such that $Q$ polarizes to $\xi$.

Now consider the subgroup $R_Q = \{x \in R : x^2 = \theta^{Q(\overline{x})}\}$. Then $R_Q$ is a subgroup of $R$ since

$$(xy)^2 = x^2y^2[y, x] = \theta^{Q(\overline{x})+Q(\overline{y})+\xi(\overline{y},\overline{x})}$$

for all $x, y \in R_Q$. Furthermore, we have $R_Q \trianglelefteq G$ since $Q$ is $G$-invariant. Also, for all $x \in R$ we have $x \in R_Q$ or $xg \in R_Q$, so $R_Q \not\leq F_0$. Therefore by the minimality of $R$, we must have $R = R_Q$. However $g \notin R_Q$ since $Q(\overline{g}) = 0$ and $g^2 = \theta$, so this is a contradiction. □

**Remark 4.1.8** A subgroup $R$ satisfying (R1)–(R3) is called a *second faisceau* by Jordan [51, §VII]. Jordan also (essentially) claims that a *second faisceau* is extraspecial, but his proof seems to have the following gap.

In [51, pp. 400–401] Jordan argues correctly that $R$ is an $r$-group for some prime $r$, and that $[R, R] = \langle\theta\rangle$ is cyclic of order $r$, where $\theta \in F_0$. He also shows that $R$ is of exponent $r$ if $r > 2$, and exponent 4 if $r = 2$. Then in [51, p. 401] his arguments show that

$$R = \langle x_1, y_1, \dots, x_\ell, y_\ell, R \cap F_0\rangle,$$

where $[x_i, y_i] = \theta$ for all $i$ and $[x_i, y_j] = [x_i, x_j] = [y_i, y_j] = 1$ for all $i \neq j$.

For $r > 2$ we know that $R \cap F_0 = \langle\theta\rangle$, and this proves that $R$ is extraspecial. For $r = 2$, one should still rule out the possibility that $R \cap F_0$ is cyclic of order 4, and for this Jordan provides no argument. Here one would need to show that for a solvable group $G$ acting on a symplectic-type group $C_4 \circ 2_\pm^{1+2\ell}$, there is a $G$-invariant subgroup of type $2_+^{1+2\ell}$ or $2_-^{1+2\ell}$ in $C_4 \circ 2_\pm^{1+2\ell}$. This is equivalent to the statement that the corresponding action of $G$ on $\mathbb{F}_2^{2\ell}$ is of quadratic type, which holds by Theorem 2.1.14.

**Proposition 4.1.9** *Let $A \trianglelefteq G$ be a subgroup satisfying properties (A1) and (A2). Then there exist subgroups $S_1, \dots, S_t \trianglelefteq G$ satisfying properties (R1)–(R3) such that all of the following hold:*

*(i)* $A = F_0S_1 \cdots S_t$;
*(ii)* $[S_i, S_j] = 1$ *for all* $i \neq j$;
*(iii)* $Z(S_i) = S_i \cap F_0$ *for all* $i$;
*(iv)* $A/F_0 = \overline{S_1} \times \cdots \times \overline{S_t}$, *where* $\overline{S_i} = S_iF_0/F_0 \cong S_i/Z(S_i)$.

***Proof*** *(Cf. [64, Corollary 1.8])* If $A = F_0$, then we are done with $t = 0$. Otherwise $A/F_0$ is nontrivial, so choose a minimal nontrivial $S_1/F_0 \trianglelefteq G/F_0$ such that $S_1 \leq A$. Then $S_1$ satisfies properties (R1)–(R3), so $S_1 \cap F_0 = Z(S_1)$ by Proposition 4.1.7 and $A/F_0 = \overline{S_1} \times C_A(S_1)/F_0$ by Lemma 4.1.6. The lemma follows by applying induction on $C_A(S_1)$. □

**Proposition 4.1.10** *Let $A \trianglelefteq G$ be a subgroup satisfying properties (A1) and (A2). Let $S_1, \ldots, S_t \trianglelefteq G$ be subgroups as in Proposition 4.1.9. Then the following statements hold.*

*(i)* $[A, A] = Z(S_1) \cdots Z(S_t)$.
*(ii) If $r$ is a prime divisor of $|A/F_0|$, then $r \mid |F_0|$.*

***Proof*** It follows from Proposition 4.1.7 that $S_i$ is an extraspecial $r_i$-group for some prime $r_i$, for all $1 \leq i \leq t$. Then $Z(S_i) = [S_i, S_i]$ for all $1 \leq i \leq t$, so claim (i) is immediate from Proposition 4.1.9. Claim (ii) follows from Proposition 4.1.9 (iv) and the fact that $[S_i, S_i] \leq F_0$. □

**Proposition 4.1.11** *Let $A \trianglelefteq G$ be a subgroup satisfying properties (A1) and (A2). Denote the prime divisors of $|A/F_0|$ by $r_1, \ldots, r_k$. Then there exist subgroups $R_1, \ldots, R_k \trianglelefteq G$ such that the following hold:*

*(i)* $A = F_0 R_1 \cdots R_k$;
*(ii)* $[R_i, R_j] = 1$ *for all* $i \neq j$;
*(iii)* $Z(R_i) = R_i \cap F_0$ *for all* $i$;
*(iv)* $A/F_0 = \overline{R_1} \times \cdots \times \overline{R_k}$, *where* $\overline{R_i} = R_i F_0/F_0 \cong R_i/Z(R_i)$;
*(v)* $R_i$ *is an extraspecial* $r_i$*-group of exponent* $r_i \gcd(r_i, 2)$ *for all* $i$;
*(vi)* $[A, A] = Z(R_1) \cdots Z(R_k)$.

***Proof*** Let $S_1, \ldots, S_t \trianglelefteq G$ be subgroups as in Proposition 4.1.9. For $1 \leq i \leq k$, define $R_i = \prod_j S_j$, where the product runs over $j$ such that $S_j$ is an extraspecial $r_i$-group. Then it follows from Proposition 4.1.9 that (i)–(iv) hold, and that $R_i$ is an extraspecial $r_i$-group of exponent $r_i \gcd(r_i, 2)$ for all $1 \leq i \leq k$. Property (vi) follows from Proposition 4.1.10. □

Let $f$ be a generator of $F_0$. As in the notation of Sect. 2.4, we let $\mathbb{K}$ be the splitting field for $f$ on $V$ and denote $V' := \mathbb{K} \otimes_{\mathbb{F}_q} V$. Then $V'$ splits into a sum of $f$-eigenspaces; we choose an $f$-eigenvalue $\lambda \in \mathbb{K}$ as in Sect. 2.4 and denote $\mu = \dim_{\mathbb{K}} W'_\lambda$. As in Sects. 2.5–2.9, we define $\nu = [\mathbb{K} : \mathbb{F}_q]/2$ if $G$ is of type $\mathcal{B}_2$, and $\nu = [\mathbb{K} : \mathbb{F}_q]$ otherwise.

Recall that the action of $g \in C_{G^\circ}(F_0)$ on $V' = \mathbb{K} \otimes_{\mathbb{F}_q} V$ is determined by its action on $W'_\lambda$, so we can consider $C_{G^\circ}(F_0)$ as a subgroup of $\mathrm{GL}(W'_\lambda) \cong \mathrm{GL}_\mu(\mathbb{K})$. (See Lemma 2.5.2 (ii), Lemma 2.6.4 (iv), Lemma 2.7.3 (iv), and Lemma 2.8.1.) Note that in case $\mathcal{B}_2$ we have $C_{G^\circ}(F_0)$ as a subgroup of $\mathrm{GU}_\mu(\mathbb{K})$ (Lemma 2.7.3 (iv)), while in case $\mathcal{B}_3$ we have $V = W'_\lambda$ and $C_{G^\circ}(F_0) = G^\circ \leq I(V, \kappa)$ (Lemma 2.8.1).

Our next goal will be to prove that a subgroup $A \leq G$ satisfying properties (A1)–(A3) is an absolutely irreducible subgroup of $\mathrm{GL}(W'_\lambda)$.

**Lemma 4.1.12** *Suppose that $A$ satisfies properties (A1)–(A3). Then the following statements hold:*

*(i)* $\dim_{\mathbb{K}}([A]_{\mathbb{K}}) = [A : F_0]$.
*(ii)* $Z([A]_{\mathbb{K}}) = \mathbb{K}$.
*(iii)* $W'_\lambda \downarrow A$ *is completely reducible, and homogeneous with absolutely irreducible composition factors.*

***Proof*** Since $A \trianglelefteq G$, it follows from Clifford's theorem that $V \downarrow A$ is completely reducible. Then by a standard result in representation theory [14, Corollary 69.9], the action of $A$ on $V' = \mathbb{K} \otimes_{\mathbb{F}_q} V$ is completely reducible, hence also on the direct summand $W'_\lambda$. We have $Z(A) = F_0$ (Lemma 4.1.5), so $[A, A] \leq Z(A) \leq \mathbb{K}^\times \leq \mathrm{GL}(W'_\lambda)$. Now the claims follow from Lemma 4.1.2. □

**Lemma 4.1.13** *Suppose that $A$ satisfies properties (A1) and (A2). Write $A = F_0 R_1 \cdots R_k$, where $R_1, \ldots, R_k \trianglelefteq G$ are as in Proposition 4.1.11. Let $W$ be a faithful absolutely irreducible $\mathbb{K}[A]$-module. Then the following statements hold:*

*(i) $W \cong W_1 \otimes \cdots \otimes W_k$, where $W_i$ is a faithful absolutely irreducible $\mathbb{K}[R_i]$-module;*
*(ii) If $G$ is of type $\mathcal{B}_2$, then $W$ admits a non-degenerate $A$-invariant Hermitian form, with respect to the automorphism $\xi \mapsto \xi^{q^\nu}$ of $\mathbb{K}$;*
*(iii) If $G$ is of type $\mathcal{B}_3$, then $W$ admits a non-degenerate $A$-invariant bilinear form, which is alternating or symmetric.*

***Proof*** We begin with the proof of (i). By Clifford's theorem the restriction $W \downarrow R_1$ is completely reducible. Because $F_0 R_2 \cdots R_k$ centralizes $R_1$, each homogeneous component of $W \downarrow R_1$ is $A$-invariant, so by irreducibility $W \downarrow R_1$ must be homogeneous. Moreover by Lemma 4.1.10 (ii) and Theorem 3.2.1 (i), the composition factors of $W \downarrow R_1$ must be absolutely irreducible. Hence we can write $W \downarrow R_1 = W_1 \otimes W'$, where $W_1$ is a faithful absolutely irreducible $\mathbb{K}[R_1]$-module and $R_1$ acts trivially on $W'$. Because $F_0 R_2 \cdots R_k$ centralizes $R_1$, by Lemma 2.1.16 we have $A \leq \mathrm{GL}(W_1) \otimes \mathrm{GL}(W')$ and $R_2 \cdots R_k \leq 1 \otimes \mathrm{GL}(W')$. Now repeat the same argument for $R_2 \cdots R_k$ for its action on $W'$.

For claim (ii), note that it follows from Proposition 4.1.7 that $R_i$ is an extraspecial $r_i$-group of exponent $r_i \gcd(r_i, 2)$ for all $1 \leq i \leq k$. In type $\mathcal{B}_2$ we have $r_i \mid q^\nu + 1$ for all $1 \leq i \leq k$ (Lemma 4.1.10 (ii) and Lemma 2.7.5). Therefore by the construction in Sects. 3.2.2 and 3.2.5, there exists a non-degenerate $R_i$-invariant Hermitian form $b_i$ on $W_i$ for all $1 \leq i \leq k$. Then $b_1 \otimes \cdots \otimes b_k$ is a non-degenerate $A$-invariant Hermitian form on $W_1 \otimes \cdots \otimes W_k$.

For (iii), in type $\mathcal{B}_3$ we have $q$ odd and $|F_0| = 2$ (Lemma 2.8.1). Thus it follows from Lemma 4.1.10 (ii) that for all $1 \leq i \leq k$ we have $R_i = 2_{\varepsilon_i}^{1+2\ell_i}$ for some $\varepsilon_i \in \{+, -\}$. By the constructions given in Sects. 3.2.3 and 3.2.6, for all $1 \leq i \leq k$ there exists a non-degenerate $R_i$-invariant reflexive bilinear form $b_i$ on $W_i$, with $\mathrm{sgn}(b_i) = \varepsilon_i$. Then $b_1 \otimes \cdots \otimes b_k$ is a non-degenerate $A$-invariant bilinear form on $W_1 \otimes \cdots \otimes W_k$, and clearly $b_1 \otimes \cdots \otimes b_k$ is alternating or symmetric. □

**Lemma 4.1.14** *Suppose that $A$ satisfies properties (A1)–(A3). Then the following statements hold:*

*(i) $W'_\lambda \downarrow A = W \otimes_{\mathbb{K}} U$, where $W$ is an absolutely irreducible $\mathbb{K}[A]$-module, and $A$ acts trivially on $U$. Furthermore $C_{\mathrm{GL}(W'_\lambda)}(A) = 1 \otimes \mathrm{GL}(U)$ and $N_{\mathrm{GL}(W'_\lambda)}(A) = N_{\mathrm{GL}(W)}(A) \otimes \mathrm{GL}(U)$.*

(ii) *Suppose that $F_0$ is of type $\mathcal{B}_2$, and let $\widehat{b}$ be the Hermitian form on $W'_\lambda$ as in Sect. 2.7. Then $\widehat{b} = b_1 \otimes b_2$, where $b_1$ and $b_2$ are non-degenerate $A$-invariant Hermitian forms on $W$ and $U$, respectively.*

(iii) *Suppose that $F_0$ is of type $\mathcal{B}_3$. Then $b = b_1 \otimes b_2$, where $b_1$ and $b_2$ are non-degenerate $A$-invariant bilinear forms on $W$ and $U$, respectively. Furthermore, the forms $b_i$ are reflexive, with* $\operatorname{sgn}(b) = \operatorname{sgn}(b_1)\operatorname{sgn}(b_2)$.

***Proof***

(i) By Lemma 4.1.12, the $\mathbb{K}[A]$-module $W'_\lambda \downarrow A$ is a direct sum of copies of some absolutely irreducible $\mathbb{K}[A]$-module $W$. Then clearly $W'_\lambda \downarrow A = W \otimes_{\mathbb{K}} U$ for a trivial $\mathbb{K}[A]$-module $U$ of suitable dimension. The remaining claims follow from Lemma 2.1.16.

(ii) First, if $\xi_1$ and $\xi_2$ are two non-degenerate $A$-invariant Hermitian forms on $W'_\lambda$, then we will see by the following standard argument that there exists $\alpha \in C_{\mathrm{GL}(W'_\lambda)}(A)$ such that $\xi_2(v, v') = \xi_1(\alpha(v), v')$ for all $v, v' \in W'_\lambda$.

Indeed, the $\xi_i$ induce semilinear and $A$-invariant bijections $\alpha_i : W'_\lambda \to (W'_\lambda)^*$, defined by

$$\alpha_i(v)(v') = \xi_i(v)(v')$$

for all $v, v' \in W'_\lambda$. Then $\alpha = \alpha_2^{-1}\alpha_1$ is a linear bijection such that $\alpha \in C_{\mathrm{GL}(W'_\lambda)}(A)$ and $\xi_2(v, v') = \xi_1(\alpha(v), v')$ for all $v, v' \in W'_\lambda$.

By Lemma 4.1.13, there exists a non-degenerate $A$-invariant Hermitian form $b_1$ on $W$. Now let $b_0$ be any non-degenerate Hermitian form on $U$, in which case $\xi_1 = b_1 \otimes b_0$ is a non-degenerate $A$-invariant form on $W'_\lambda$. Arguing as in the previous paragraph with $\xi_2 = \widehat{b}$, we conclude that there exists $\alpha \in C_{\mathrm{GL}(W'_\lambda)}(A)$ such that $\widehat{b}(v, v') = \xi_1(\alpha(v), v')$ for all $v, v' \in W'_\lambda$. By (i) we have $\alpha = 1 \otimes \beta$ for some $\beta \in \mathrm{GL}(U)$, so $\widehat{b} = b_1 \otimes b_2$, where $b_2(v, v') = b_0(\beta(v), v')$ for all $v, v' \in U$.

(iii) The fact that $b$ decomposes into a tensor product $b = b_1 \otimes b_2$ as claimed follows similarly to (ii). Since $W$ is absolutely irreducible, a non-degenerate $A$-invariant bilinear form on $W$ is unique up to a scalar, and must be alternating or symmetric. Hence $b_1$ is alternating or symmetric, and since $b$ is alternating or symmetric, it follows that $b_2$ must also be alternating or symmetric.

□

**Lemma 4.1.15** *Let $b_1$ and $b_2$ be non-degenerate bilinear forms on vector spaces $V_1$ and $V_2$, respectively. Let $V = V_1 \otimes V_2$ and denote $b = b_1 \otimes b_2$. Then*

$$(\mathrm{GL}(V_1) \otimes \mathrm{GL}(V_2)) \cap \Delta(V_1 \otimes V_2, b) = \Delta(V_1, b_1) \otimes \Delta(V_2, b_2).$$

***Proof*** It is clear that $\Delta(V_1, b_1) \otimes \Delta(V_2, b_2) \leq (\mathrm{GL}(V_1) \otimes \mathrm{GL}(V_2)) \cap \Delta(V_1 \otimes V_2, b)$. For the other inclusion, suppose that $g = x \otimes y$ is contained in $\Delta(V_1 \otimes V_2, b)$, where $x \in \mathrm{GL}(V_1)$ and $y \in \mathrm{GL}(V_2)$. We have $b(g(v \otimes w), g(v' \otimes w')) = b_1(xv, xv')b_2(yw, yw')$, so

$$b_1(xv, xv')b_2(yw, yw') = b_1(v, v')b_2(w, w')$$

for all $v, v' \in V_1$ and $w, w' \in V_2$.

Choose $w, w' \in V_2$ such that $b_2(yw, yw') \neq 0$. Then $b_1(xv, xv') = c \cdot b_1(v, v')$ for all $v, v' \in V$, where $c = b_2(w, w')/b_2(yw, yw')$. Therefore $x \in \Delta(V_1, b_1)$, and similarly it follows that $y \in \Delta(V_2, b_2)$. □

**Theorem 4.1.16** *Suppose that A satisfies properties (A1)–(A3). Then A is an absolutely irreducible subgroup of* $\mathrm{GL}(W'_\lambda)$.

***Proof*** By Lemma 4.1.14, we have $W'_\lambda \downarrow A = W \otimes_{\mathbb{K}} U$, where $W$ is an absolutely irreducible $\mathbb{K}[A]$-module and $A$ acts trivially on $U$. Using the fact that $G$ is maximal solvable, we will prove that $\dim U = 1$, so then $W'_\lambda \downarrow A = W$. We consider each of the different types of $G$ in turn.

**Case 1: $G$ is of type $\mathcal{B}_0$ or $\mathcal{B}_1$**

Since $C_{G^\circ}(F_0)$ normalizes $A$, by Lemma 4.1.14 we have $C_{G^\circ}(F_0) \leq N_{\mathrm{GL}(W)}(A) \otimes \mathrm{GL}(U)$. Let $X/\mathbb{K}^\times$ be the image of $C_{G^\circ}(F_0)$ in $\mathrm{GL}(U)/\mathbb{K}^\times$ under the map $\mathrm{GL}(W) \otimes \mathrm{GL}(U) \to \mathrm{GL}(U)/\mathbb{K}^\times$. Then $1 \otimes X$ is solvable, and normalized by $G$ since $G$ normalizes $C_{\mathrm{GL}(W'_\lambda)}(A) = 1 \otimes \mathrm{GL}(U)$ and $\mathrm{GL}(W) \otimes 1$. Because $G$ is maximal solvable in $\Delta(V, \kappa)$, it follows that $1 \otimes X$ is contained in $G$. Then $1 \otimes X$ is a subgroup of $C_{G^\circ}(F_0)$ that centralizes $A$, which means that $1 \otimes X \leq F_0$.

Hence $1 \otimes X = \mathbb{K}^\times$, so in fact $C_{G^\circ}(F_0) \leq N_{\mathrm{GL}(W)}(A) \otimes 1$. Then $1 \otimes \mathrm{GL}(U)$ centralizes $C_{G^\circ}(F_0)$, so $1 \otimes \mathrm{GL}(U) \leq G$ by Lemma 2.9.2. In this case $1 \otimes \mathrm{GL}(U) \leq C_{G^\circ}(F_0)$, so $\dim U = 1$ since the image of $C_{G^\circ}(F_0)$ in $\mathrm{GL}(U)/\mathbb{K}^\times$ is trivial.

**Case 2: $G$ is of type $\mathcal{B}_2$**

Let $\widehat{b}$ be the Hermitian form on $W'_\lambda$, as defined in Sect. 2.7. By Lemma 4.1.14 we have $\widehat{b} = b_1 \otimes b_2$, where $b_1$ and $b_2$ are non-degenerate $A$-invariant Hermitian forms on $W$ and $U$, respectively. Since $C_{G^\circ}(F_0)$ normalizes $A$, by Lemmas 4.1.14 and 4.1.15 we have $C_{G^\circ}(F_0) \leq \Delta(W, b_1) \otimes \Delta(U, b_2)$. As in the previous case, let $X/\mathbb{K}^\times$ be the image of $C_{G^\circ}(F_0)$ in $\Delta(U, b_2)/\mathbb{K}^\times$. Then $1 \otimes X$ is solvable and normalized by $G$, so $1 \otimes X \leq G$ since $G$ is maximal solvable.

Since $b_2$ is a Hermitian form, every $g \in \Delta(U, b_2)$ can be written in the form $g = xy$, where $x \in I(U, b_2)$ and $y \in \mathbb{K}^\times$. Hence $X = X^\circ\mathbb{K}^\times$. Then $1 \otimes X^\circ$ is contained in $C_{G^\circ}(F_0)$ and centralizes $A$, so it must be contained in $F_0$ (Lemma 4.1.5). In other words $X^\circ$ consists of scalars, so $1 \otimes X = \mathbb{K}^\times$. Consequently $C_{G^\circ}(F_0) \leq I(W, b_1) \otimes 1$.

Now $1 \otimes I(U, b_2)$ centralizes $C_{G^\circ}(F_0)$, so as in the previous case, we have $1 \otimes I(U, b_2) \leq C_{G^\circ}(F_0)$ by Lemma 2.9.2. Thus $\dim U = 1$, since the image of $C_{G^\circ}(F_0)$ in $\mathrm{GL}(U)/\mathbb{K}^\times$ is trivial.

**Case 3: $G$ is of type $\mathcal{B}_3$**
By Lemma 4.1.14 we have $b = b_1 \otimes b_2$, where $b_1$ and $b_2$ are non-degenerate $A$-invariant bilinear forms on $W$ and $U$, respectively. Since $G$ normalizes $A$, by Lemmas 4.1.14 and 4.1.15 we have $G \leq \Delta(W, b_1) \otimes \Delta(U, b_2)$. Because $G$ is maximal solvable, we must have $G = Y \otimes X$, where $Y/\mathbb{K}^\times$ is the image of $G$ in $\Delta(W, b_1)/\mathbb{K}^\times$ and $X/\mathbb{K}^\times$ is the image of $G$ in $\Delta(U, b_2)/\mathbb{K}^\times$.

The subgroup $1 \otimes X^\circ$ is contained in $G^\circ$ and centralizes $A$, so it must be contained in $F_0$ (Lemma 4.1.5). Hence $X^\circ = \{\pm I_U\}$. Now because $G = Y \otimes X$ is maximal solvable, clearly $X$ must be maximal solvable in $\Delta(U, b_2)$ as well. It follows then from Lemma 2.1.6 that $\dim U = 1$. □

**Proposition 4.1.17** *Suppose that $A$ satisfies properties (A1)–(A3). Then we have* $[A : F_0] = \mu^2$.

***Proof*** Since $A$ is an absolutely irreducible subgroup of $\mathrm{GL}(W'_\lambda)$ (Theorem 4.1.16), it follows from Lemma 4.1.12 (i) that $[A : F_0] = \mu^2$. □

**Proposition 4.1.18** *If $r$ is prime divisor of $\mu$, then $r \mid |F_0|$.*

***Proof*** Follows from Propositions 4.1.17 and 4.1.10 (ii). □

As a consequence of Proposition 4.1.17, we can deduce the following.

**Lemma 4.1.19** *Suppose that $G$ is of type $\mathcal{B}_3$. Then $n = 2^\ell$ for some integer $\ell \geq 0$.*

***Proof*** For groups of type $\mathcal{B}_3$ we have $|F_0| = 2$ (Lemma 2.8.1). Thus it follows from Proposition 4.1.18 that $n = \mu$ is a power of two. □

**Proposition 4.1.20** *Suppose that $\kappa \neq 0$ is a quadratic form and that $n > 1$ is odd. If $H \leq \Delta(V, \kappa)$ is irreducible and solvable, then $H$ is metrically imprimitive.*

***Proof*** It will suffice to prove the result in the case where $H$ is maximal irreducible solvable. For metrically primitive groups of type $\mathcal{B}_1$ and $\mathcal{B}_2$ we have $n = 2\mu\nu$ even, and for groups of type $\mathcal{B}_3$ we have $n = 2^\ell$ for some $\ell \geq 0$ (Lemma 4.1.19). From this it follows that $H$ cannot be metrically primitive. □

## 4.2 Structure of Metrically Primitive Maximal Solvable Subgroups

In this section, we will complete the recursive description of the general structure of maximal irreducible solvable subgroups of $\Delta(V, \kappa)$. Continuing with the notation from the previous section, let $G$ be metrically primitive maximal irreducible solvable in $\Delta(V, \kappa)$. Furthermore, let $F_0$ and $A$ be subgroups of $G$ satisfying properties (F1)–(F3) and (A1)–(A3), respectively. We also use notation such as $\mu = \dim_{\mathbb{K}}(W'_\lambda)$ and $\nu$, as in Sects. 2.5–2.8.

Write $A = F_0 R_1 \cdots R_k$, with $R_i \trianglelefteq G$ extraspecial groups as in Proposition 4.1.11. For all $i$ we have $R_i = (r_i)_{\varepsilon_i}^{1+2\ell_i}$ for some prime $r_i$, integer $\ell_i > 0$, and $\varepsilon_i \in \{+, -\}$. By Proposition 4.1.7, we have $\varepsilon_i = +$ if $r_i > 2$.

Denote $\overline{R_i} := R_i/Z(R_i)$ for all $1 \leq i \leq k$. We have $A/F_0 \cong \overline{R_1} \times \cdots \times \overline{R_k}$ by Proposition 4.1.11 (iv), so $[A : F_0] = r_1^{2\ell_1} \cdots r_k^{2\ell_k}$. On the other hand $[A : F_0] = \mu^2$ by Proposition 4.1.17, so

$$\mu = r_1^{\ell_1} \cdots r_k^{\ell_k}$$

is the prime factorization of $\mu$.

Recall that we can consider $\overline{R_i}$ as a $2\ell_i$-dimensional vector space over $\mathbb{F}_r$, equipped with a non-degenerate alternating bilinear form $\xi$ (Sect. 3.1). In the case where $r_i = 2$, the form $\xi$ is the polarization of a quadratic form $\vartheta$ defined by $x^2 = \theta^{\vartheta(\overline{x})}$ for all $x \in R_i$. For any $g \in N_{\Delta(V,\kappa)}(F_0)$ that normalizes $R_i$, the action of $g$ on $\overline{R_i}$ is a similarity with respect to $\xi$ (and also $\vartheta$ if $r_i = 2$). Furthermore, the action of $g$ on $\overline{R_i}$ is an isometry if and only if $g$ centralizes $Z(R_i)$.

We have a homomorphism

$$\pi : N_{\Delta(V,\kappa)}(F_0, R_1, \ldots, R_k) \to \prod_{i=1}^{k} \mathrm{GSp}_{2\ell_i}(r_i)$$

defined by $\pi(g) = (g_1, \ldots, g_k)$, where $g_i$ is the action of $g$ on $\overline{R_i}$. Note that $g_i \in O_{2\ell_i}^{\varepsilon_i}(2)$ if $r_i = 2$.

**Lemma 4.2.1** $\operatorname{Ker}\pi \cap C_{I(V,\kappa)}(F_0) = A$.

***Proof*** Since $[A, A] \leq F_0$, it is clear that $A \leq \operatorname{Ker}\pi$. Conversely, suppose that $g \in \operatorname{Ker}\pi \cap C_{I(V,\kappa)}(F_0)$. Then $g$ acts trivially on $R_i/Z(R_i)$ and $Z(R_i)$ for all $1 \leq i \leq k$, so by Lemma 3.1.1 there exist $x_i \in R_i$ such that $g x_1^{-1} \cdots x_k^{-1}$ centralizes $A$.

Now $A$ is absolutely irreducible as a subgroup of $C_{I(V,\kappa)}(F_0) \leq \mathrm{GL}(W'_\lambda)$ (Theorem 4.1.16), so $C_{I(V,\kappa)}(A) = I(V, \kappa) \cap \mathbb{K}^\times = F_0$. Therefore $g x_1^{-1} \cdots x_k^{-1} \in F_0$, and $g \in A$. □

**Lemma 4.2.2** $\operatorname{Ker}\pi$ *is solvable.*

***Proof*** First note that the quotient $\operatorname{Ker}\pi/\operatorname{Ker}\pi \cap C_{I(V,\kappa)}(F_0)$ embeds into $N_{\Delta(V,\kappa)}(F_0)/C_{I(V,\kappa)}(F_0)$, which is an abelian group (Lemma 2.9.1). Combining this with Lemma 4.2.1, we conclude that $\operatorname{Ker}\pi$ is solvable. □

**Lemma 4.2.3** *The action of $G$ on $\overline{R_i}$ is metrically completely reducible for all $1 \leq i \leq k$.*

***Proof*** Any totally isotropic $G$-invariant subspace of $\overline{R_i} = R_i/Z(R_i)$ is equal to $T/Z(R_i)$, where $Z(R_i) \leq T \leq R_i$ and $T$ is an abelian normal subgroup of $G$. Then $TF_0$ is abelian and $TF_0 \trianglelefteq G$, so by maximality of $F_0$ we have $TF_0 = F_0$. Therefore $T \leq F_0$. On the other hand by Proposition 4.1.11 (iii) we have $F_0 \cap R_i = Z(R_i)$, so $T = Z(R_i)$ and $T/Z(R_i)$ is trivial. □

**Proposition 4.2.4** $G = \pi^{-1}(X_1 \times \cdots \times X_k)$, *for some* $X_i \leq \mathrm{GSp}_{2\ell_i}(r_i)$ *such that all of the following hold:*

*(i) If* $r_i > 2$, *then* $X_i$ *is metrically completely reducible maximal solvable in* $\mathrm{GSp}_{2\ell_i}(r_i)$.
*(ii) If* $r_i = 2$, *then* $X_i$ *is metrically completely reducible maximal solvable in* $O^{\varepsilon_i}_{2\ell_i}(2)$.

***Proof*** It follows from Lemma 4.2.3 that $\pi(G) \leq X_1 \times \cdots \times X_k$ for some metrically completely reducible maximal solvable $X_i \leq \mathrm{GSp}_{2\ell_i}(r_i)$, where $X_i \leq O^{\varepsilon_i}_{2\ell_i}(2)$ if $r_i = 2$.

Hence $G \leq \pi^{-1}(X_1 \times \cdots \times X_k)$. By Lemma 4.2.2 the group $\pi^{-1}(X_1 \times \cdots \times X_k)$ is solvable, so by maximality of $G$ we have $G = \pi^{-1}(X_1 \times \cdots \times X_k)$. □

**Remark 4.2.5** With Proposition 4.2.4, we have a description of $G$ in terms of metrically completely reducible maximal solvable subgroups in smaller dimensions; any such group will be of the form $\pi^{-1}(X_1 \times \cdots \times X_k)$, where $X_i$ is metrically completely reducible maximal solvable in $\mathrm{GSp}_{2\ell_i}(r_i)$ (or in $O^{\varepsilon_i}_{2\ell_i}(2)$, if $r_i = 2$). To be precise, for $n = 2$ or $n = 4$ the degree is not necessarily smaller; but in these cases $X_i \leq O^-_2(2)$ or $X_i \leq O^\pm_4(2)$, which can easily be handled separately.

Furthermore, the groups $X_i$ can be described in terms of maximal irreducible solvable subgroups, as seen in Sect. 2.3. Indeed, by Lemma 2.3.9, for $r_i > 2$ we have

$$X_i = \left(Y^{(i)}_1 \times \cdots \times Y^{(i)}_{t_i}\right) \cap \mathrm{GSp}_{2\ell_i}(r_i)$$

for some maximal irreducible solvable $Y^{(i)}_j \leq \mathrm{GSp}_{2n^{(i)}_j}(r_i)$, where $\sum_{j=1}^{t_i} n^{(i)}_j = \ell_i$. Similarly if $r_i = 2$, by Lemma 2.3.7 we have

$$X_i = Y^{(i)}_1 \times \cdots \times Y^{(i)}_{t_i}$$

for some maximal irreducible solvable $Y^{(i)}_j \leq O^{\sigma_j}_{2n^{(i)}_j}(2)$, where $\sum_{j=1}^{t_i} n^{(i)}_j = \ell_i$ and the signs $\sigma_j \in \{+, -\}$ satisfy $\varepsilon_i = \sigma_1 \cdots \sigma_{t_i}$.

In the next section, we will construct and describe $\pi^{-1}(X_1 \times \cdots \times X_k)$ more precisely. This provides us with a list of candidates for metrically primitive maximal irreducible subgroups of $\Delta(V, \kappa)$. As the next step, there are various questions about the properties of groups of the form $G = \pi^{-1}(X_1 \times \cdots \times X_k)$ that we should answer. For example:

- When is $G$ irreducible and metrically primitive?
- When is $G$ of multiplier 1? (This is needed e.g. for the construction of imprimitive maximal solvable subgroups of semiprimary type, see Theorem 2.2.14 (ii).)
- When is $G$ a maximal irreducible solvable subgroup of $\Delta(V, \kappa)$?

Roughly speaking, the answer to all of the questions above is "almost always", with a few exceptions which we shall determine explicitly in later sections. For example, we will show that $G$ is of multiplier 2 if and only if $G$ is of type $\mathcal{B}_3$ and $X_1 \leq \Omega^{\varepsilon_1}_{2\ell_1}(2)$ (Lemma 5.1.1).

## 4.3 Description of $G^{\mathcal{B}}_{\mu,\nu}(X_1, \ldots, X_k)$

At the end of the previous section, we have described the generic structure of maximal metrically primitive irreducible solvable subgroups of $\Delta(V, \kappa)$, in terms of groups in smaller degrees. Essentially, all of them are either as in the case $\mu = 1$ described in Sect. 2.9, or they are of the form $\pi^{-1}(X_1 \times \cdots \times X_k)$ as in the previous section. We will denote groups of the form $\pi^{-1}(X_1 \times \cdots \times X_k)$ by

$$G^{\mathcal{B}}_{\mu,\nu}(X_1, \ldots, X_k) := \pi^{-1}(X_1 \times \cdots \times X_k),$$

where $\mathcal{B} = \mathcal{B}_0, \mathcal{B}_1, \mathcal{B}_2$, or $\mathcal{B}_3$. (For $k = 0$, we have $\mu = 1$ and this defines the group $G^{\mathcal{B}}_{1,\nu}$ of Definition 2.9.5.) The purpose of this section is to define and construct $G^{\mathcal{B}}_{\mu,\nu}(X_1, \ldots, X_k)$, and to establish some of their basic properties. Conditions required for the definition and irreducibility of $G^{\mathcal{B}}_{\mu,\nu}(X_1, \ldots, X_k)$ are given in Table 4.1, see also Sect. 5.2.

We consider the different types $\mathcal{B}_0$–$\mathcal{B}_3$ case by case in the subsections that follow. The basic outline is as follows. In each subsection, we first recall the structure of $F_0$ from Sects. 2.5–2.8. (One can construct $F_0 \leq I(V, \kappa)$ explicitly, for example as described in Remarks 2.5.4, 2.6.7, 2.7.6, and Sect. 2.9.)

Then the subgroup $A$ is constructed as a subgroup of $\mathrm{GL}(W'_\lambda)$ by taking the tensor product of suitable extraspecial groups $R_1, \ldots, R_k$. We use notation as in the previous section, so $R_i = (r_i)^{1+2\ell_i}_{\varepsilon_i}$. Furthermore, we define

$$u_i := \begin{cases} 1, & \text{if } (r_i, \varepsilon_i) = (2, -) \\ 0, & \text{otherwise.} \end{cases}$$

for all $1 \leq i \leq k$.

We have $A = F_0 R_1 \cdots R_k$, after which we can define

$$G^{\mathcal{B}}_{\mu,\nu}(X_1, \ldots, X_k) := \pi^{-1}(X_1, \ldots, X_k)$$

with $\pi$ the homomorphism

$$\pi : N_{\Delta(V,\kappa)}(F_0, R_1, \ldots, R_k) \to \prod_{i=1}^{k} \mathrm{GSp}_{2\ell_i}(r_i)$$

as in Sect. 4.2.

**Table 4.1** Conditions required for $G^{\mathcal{B}}_{\mu,\nu}(X_1, \ldots, X_k) \leq \Delta(V, \kappa)$ to be irreducible, where $n = \dim V > 1$

| | |
|---|---|
| All types | $(r_i, \ell_i, \varepsilon_i) \neq (2, 1, +)$ |
| Type $\mathcal{B}_0$ | $\kappa = 0, n = \nu\mu$ |
| | $\mathbb{K} = \mathbb{F}_{q^\nu}$ |
| | $q^\nu > 2$ |
| | $\mu = r_1^{\ell_1} \cdots r_k^{\ell_k}$, where $r_i$ prime and $r_i \mid q^\nu - 1$ |
| Type $\mathcal{B}_1$ | $\kappa \neq 0, n = 2\nu\mu$ |
| | $\mathbb{K} = \mathbb{F}_{q^\nu}$ |
| | $q^\nu > 4$ or $(q, \nu) = (4, 1)$ |
| | $\operatorname{sgn}(\kappa) = +$ if $\kappa$ is a quadratic form |
| | $\mu = r_1^{\ell_1} \cdots r_k^{\ell_k}$, where $r_i$ primes and $r_i \mid q^\nu - 1$ |
| | One of the following conditions holds: (a) $\left(\frac{-1}{r_i}\right) = +$ for all $X_i$ of multiplier 2 (b) $\left(\frac{-q}{r_i}\right) = +$ for all $X_i$ of multiplier 2 |
| Type $\mathcal{B}_2$ | $\kappa \neq 0, n = 2\nu\mu$ |
| | $\mathbb{K} = \mathbb{F}_{q^{2\nu}}$ |
| | $\operatorname{sgn}(\kappa) = (-1)^\mu$ if $\kappa$ is a quadratic form |
| | $\mu = r_1^{\ell_1} \cdots r_k^{\ell_k}$, where $r_i$ prime and $r_i \mid q^\nu + 1$ |
| Type $\mathcal{B}_3$ | $\kappa \neq 0, n = \mu$ |
| | $\mathbb{K} = \mathbb{F}_q$, and $q$ is odd |
| | $\operatorname{sgn}(b) = \varepsilon$ |
| | $\operatorname{sgn}(\kappa) = (-1)^{\frac{n(q-1)}{4}}$ if $\kappa$ is a quadratic form |
| | $\mu = 2^\ell$ |

### 4.3.1 Groups of Type $\mathcal{B}_0$

In this case $\kappa = 0$, so $\Delta(V, \kappa) = \mathrm{GL}(V) = \mathrm{GL}_n(q)$. Let $n = \mu\nu$ for integers $\mu, \nu > 0$. Denote $\mathbb{K} = \mathbb{F}_{q^\nu}$, and let $\lambda \in \mathbb{K}^\times$ be a primitive element.

In view of Lemma 2.5.1 and Proposition 4.1.18, we make the following assumptions.

- $q^\nu > 2$,
- For every prime $r \mid \mu$, we have $r \mid q^\nu - 1$.

Denote $V' = \mathbb{K} \otimes_{\mathbb{F}_q} V$, and let $F_0 = \langle f \rangle \in \mathrm{GL}(V) \leq \mathrm{GL}(V')$ be as in Sect. 2.5, so

$$V' = W'_\lambda \oplus \cdots \oplus W'_{\lambda^{q^{\nu-1}}}$$

where $W'_{\lambda^{q^i}}$ is the $f$-eigenspace for eigenvalue $\lambda^{q^i}$, and $\dim_{\mathbb{K}} W'_{\lambda^{q^i}} = \mu$ for all $0 \le i < \nu$.

Let

$$\mu = r_1^{\ell_1} \cdots r_k^{\ell_k}$$

be the prime factorization of $\mu$, where $r_1 < \cdots < r_k$ are primes. Choose metrically completely reducible maximal solvable subgroups $X_i \le \mathrm{GSp}_{2\ell_i}(r_i)$ for all $1 \le i \le k$. If $r_1 = 2$, then by Theorem 2.1.14 and Lemma 2.3.7 there exists $\varepsilon_1 \in \{+, -\}$ such that $X_1 \le O^{\varepsilon_1}_{2\ell_1}(2)$. For $1 \le i \le k$ with $r_i > 2$, we denote $\varepsilon_i = +$.

Define $R_i = (r_i)^{1+2\ell_i}_{\varepsilon_i}$ for all $1 \le i \le k$. Since $r_i \mid q^\nu - 1$, there exists a primitive $r_i$th root of unity $\theta_i \in \mathbb{K}^\times$, and we choose

$$\theta_i = \lambda^{\frac{q^\nu - 1}{r_i}}.$$

Then by Theorem 3.2.1, there exists a faithful absolutely irreducible $\mathbb{K}[R_i]$-module $W_i$ of dimension $\dim_{\mathbb{K}}(W_i) = r_i^{\ell_i}$.

If $R_i = (r_i)^{1+2\ell_i}_+$, then by the construction from Sect. 3.2.1, we have a basis $v^{(i)}_{\xi_1,\ldots,\xi_{\ell_i}} := v_{\xi_1,\ldots,\xi_{\ell_i}}$ $(0 \le \xi_1, \ldots, \xi_{\ell_i} < r_i)$ of $W_i$ and generators

$$R_i = \left\langle A^{(i)}_1, B^{(i)}_1, \ldots, A^{(i)}_{\ell_i}, B^{(i)}_{\ell_i} \right\rangle$$

such that

$$\begin{aligned} A^{(i)}_j v_{\xi_1,\ldots,\xi_{\ell_i}} &= \theta_i^{\xi_j} v_{\xi_1,\ldots,\xi_{\ell_i}} \\ B^{(i)}_j v_{\xi_1,\ldots,\xi_{\ell_i}} &= v_{\xi_1,\ldots,\xi_{j-1},\xi_j+1,\xi_{j+1},\ldots,\xi_{\ell_i}} \end{aligned} \tag{4.1}$$

for all $0 \le \xi_1, \ldots, \xi_{\ell_i} < r_i$.

If $R_i = 2^{1+2\ell_i}_-$, then choose $a, b \in \mathbb{F}_q$ such that $a^2 + b^2 = -1$. By the construction from Sect. 3.2.4, we have a basis $v^{(i)}_{\xi_1,\ldots,\xi_{\ell_i}} := v_{\xi_1,\ldots,\xi_{\ell_i}}$ $(0 \le \xi_1, \ldots, \xi_{\ell_i} \le 1)$ of $W_i$ and generators

$$R_i = \left\langle A^{(i)}_1, B^{(i)}_1, \ldots, A^{(i)}_{\ell_i}, B^{(i)}_{\ell_i} \right\rangle$$

where $A^{(i)}_j$ and $B^{(i)}_j$ for $2 \le j \le \ell_i$ act as in (4.1), and

$$\begin{aligned} A^{(i)}_1 v_{\xi_1,\ldots,\xi_{\ell_i}} &= (-1)^{\xi_1} a v_{\xi_1,\ldots,\xi_{\ell_i}} + b v_{\xi_1+1,\xi_2,\ldots,\xi_{\ell_i}} \\ B^{(i)}_1 v_{\xi_1,\ldots,\xi_{\ell_i}} &= (-1)^{\xi_1} v_{\xi_1+1,\xi_2,\ldots,\xi_{\ell_i}} \end{aligned}$$

for all $0 \le \xi_1, \ldots, \xi_{\ell_i} \le 1$.

We identify $W'_\lambda = W_1 \otimes \cdots \otimes W_k$ as a $\mathbb{K}$-vector space, taking as a basis the tensor products $w_1 \otimes \cdots \otimes w_k$, where $w_i = v^{(i)}_{\xi_1,\ldots,\xi_{\ell_i}}$. We define $A \leq C_{\mathrm{GL}(V)}(F_0)$ to be the subgroup corresponding to $R_1 \otimes \cdots \otimes R_k \leq \mathrm{GL}(W'_\lambda) = \mathrm{GL}_\mu(\mathbb{K})$. Then $A = F_0 R_1 \cdots R_k$ and $R_i \cap F_0 = Z(R_i) = \langle \theta_i I_V \rangle$ for all $1 \leq i \leq k$. Note that as a tensor product of absolutely irreducible groups, the action of $A$ on $W'_\lambda$ is absolutely irreducible (Lemma 2.1.16 (i)).

We have a homomorphism

$$\pi : N_{\mathrm{GL}(V)}(F_0, R_1, \ldots, R_k) \to \prod_{i=1}^{k} \mathrm{GSp}_{2\ell_i}(r_i)$$

defined by $\pi(g) = (g_1, \ldots, g_k)$, where $g_i$ is the action of $g$ on $\overline{R_i} := R_i/Z(R_i)$. Here $g_1 \in O^{\varepsilon_1}_{2\ell_1}(2)$ if $r_1 = 2$.

Having fixed a basis of $W'_\lambda$, we let $\psi : V' \to V'$ be the $\mathbb{K}$-linear map defined as in Sect. 2.5, so $\psi \in N_{\mathrm{GL}(V)}(F_0)$ (Lemma 2.5.2 (i)). Note that the generators $B_j := B^{(i)}_j$ are matrices with entries in $\mathbb{F}_q$, so they commute with $\psi$. The generators $A_j := A^{(i)}_j$ are diagonal matrices, except possibly $A_1 := A^{(i)}_1$ for $(r_i, \varepsilon_i) = (2, -)$ which has its entries in $\mathbb{F}_q$. Therefore

$$\begin{aligned}
\psi^{-1} A_1 \psi &= \begin{cases} A_1^q, & \text{if } u_i = 0, \\ A_1, & \text{if } u_i = 1. \end{cases} \\
\psi^{-1} A_j \psi &= A_j^q \qquad \text{for all } 2 \leq j \leq \ell_i. \\
\psi^{-1} B_j \psi &= B_j \qquad \text{for all } 1 \leq j \leq \ell_i.
\end{aligned} \tag{4.2}$$

Therefore $\psi$ normalizes $R_i$ for all $1 \leq i \leq k$, and the action of $\psi$ on $\overline{R_i}$ is a diagonal matrix with respect to the basis $\overline{A_1}, \ldots, \overline{A_{\ell_i}}, \overline{B_1}, \ldots, \overline{B_{\ell_i}}$.

**Lemma 4.3.1** *We have*

$$\pi\left(N_{C_{\mathrm{GL}(V)}(F_0)}(R_1, \ldots, R_k)\right) = \begin{cases} \prod_{i=1}^k \mathrm{Sp}_{2\ell_i}(r_i), & \text{if } r_1 > 2. \\ O^{\varepsilon_1}_{2\ell_1}(2) \times \prod_{i=2}^k \mathrm{Sp}_{2\ell_i}(r_i), & \text{if } r_1 = 2. \end{cases}$$

***Proof*** It follows from Lemma 2.1.16 (ii) that as a subgroup of $\mathrm{GL}(W'_\lambda)$, we have

$$N_{C_{\mathrm{GL}(V)}(F_0)}(R_1, \ldots, R_k) = N_{\mathrm{GL}(W_1)}(R_1) \otimes_{\mathbb{K}} \cdots \otimes_{\mathbb{K}} N_{\mathrm{GL}(W_k)}(R_k).$$

Now the result follows by combining this with the fact that for $r_1 = 2$, the homomorphism $N_{\mathrm{GL}(W_1)}(R_1) \to O^{\varepsilon_1}_{2\ell_1}(2)$ is surjective (Theorem 3.2.2 (iii)), and for $r_i > 2$ the homomorphism $N_{\mathrm{GL}(W_i)}(R_i) \to \mathrm{Sp}_{2\ell_i}(r_i)$ is surjective (Theorem 3.2.2 (ii)). □

**Lemma 4.3.2** $\operatorname{Ker}\pi \cap C_{\mathrm{GL}(V)}(F_0) = A$.

***Proof*** Since $A$ is absolutely irreducible on $W'_\lambda$, the result follows with the same proof as Lemma 4.2.1. □

**Lemma 4.3.3** $\operatorname{Ker}\pi = A \rtimes \langle \psi^\alpha \rangle$, *where* $0 < \alpha \le \nu$ *is the smallest integer such that* $r_i \mid q^\alpha - 1$ *for all* $1 \le i \le k$.

***Proof*** Let $g \in \operatorname{Ker}\pi$. By Lemma 2.5.2 (i), we have $g = x\psi^\delta$ for some $x \in C_{\mathrm{GL}(V)}(F_0)$ and integer $\delta$. By (4.2), the action of $\psi$ on $\overline{R_i}$ is a diagonal matrix with respect to the basis $\overline{A_1}, \ldots, \overline{A_{\ell_i}}, \overline{B_1}, \ldots, \overline{B_{\ell_i}}$. Therefore since $g$ acts trivially on $\overline{R_i}$, the action of $x$ on $\overline{R_i}$ must also be a diagonal matrix, of the form $\operatorname{diag}(c_1, \ldots, c_{\ell_i}, c_1^{-1}, \ldots, c_{\ell_i}^{-1})$ since the bilinear form on $\overline{R_i}$ is $x$-invariant. On the other hand $\psi$ and $g$ both act trivially $\overline{A_1}, \ldots, \overline{A_{\ell_i}}$, so the same must be true for $x$. Therefore $c_1 = \cdots = c_{\ell_i} = 1$, and we conclude that $\pi(x) = 1$.

Therefore $x \in A$ by Lemma 4.3.2. We have proved that $\operatorname{Ker}\pi = A(\operatorname{Ker}\pi \cap \langle\psi\rangle)$, from which the lemma follows. □

Since $\operatorname{Ker}\pi$ is solvable by Lemma 4.3.3, the group $\pi^{-1}(X_1 \times \cdots \times X_k)$ is a solvable subgroup of $\mathrm{GL}(V)$. We now define the group of type $\mathcal{B}_0$ corresponding to $\mu$, $\nu$, and $X_1, \ldots, X_k$ by

$$G^{\mathcal{B}_0}_{\mu,\nu}(X_1, \ldots, X_k) := \pi^{-1}(X_1 \times \cdots \times X_k).$$

Note that in the case $\mu = 1$ we have $k = 0$, $A = F_0$, and $\pi$ is trivial, so this defines the group $G^{\mathcal{B}_0}_{1,\nu}$ of Definition 2.9.5.

The following result is a special case of Proposition 4.2.4.

**Proposition 4.3.4** *Let $G$ be a primitive maximal irreducible solvable subgroup of* $\mathrm{GL}(V)$. *Then $G$ is conjugate to a group of the form* $G^{\mathcal{B}_0}_{\mu,\nu}(X_1, \ldots, X_k)$.

Although by definition $\pi(G) \le X_1 \times \cdots \times X_k$ for $G = G^{\mathcal{B}_0}_{\mu,\nu}(X_1, \ldots, X_k)$, it is not usually the case that $\pi(G) = X_1 \times \cdots \times X_k$. The following lemma describes the elements of $G$ and $\pi(G)$ more precisely.

**Lemma 4.3.5** *Let* $G = G^{\mathcal{B}_0}_{\mu,\nu}(X_1, \ldots, X_k)$ *be of type* $\mathcal{B}_0$. *Then the following statements hold:*

*(i)* $\pi(C_G(F_0)) = X_1^\circ \times \cdots \times X_k^\circ$.

*(ii) $G$ contains an element of the form $g\psi^\beta$, $g \in C_{\mathrm{GL}(V)}(F_0)$ if and only if* $\left(\dfrac{q^\beta}{r_i}\right) = +1$ *for all $1 \le i \le k$ such that $X_i$ is of multiplier* 2.

***Proof*** Claim (i) is immediate from Lemma 4.3.1.

For (ii), suppose first that there exists $g \in C_{\mathrm{GL}(V)}(F_0)$ such that $g\psi^\beta \in G$, so $\pi(g\psi^\beta) = (x_1, \ldots, x_k)$ for some $x_i \in X_i$. The action of $g$ on $\overline{R_i}$ is an isometry, while by (4.2) the action of $\psi^\beta$ multiplies the alternating bilinear form on $\overline{R_i}$ by $q^\beta$. Therefore under the map $\tau_i : \mathrm{GSp}_{2\ell_i}(r_i) \to \mathbb{F}_{r_i}^\times$, we have $\tau_i(x_i) = q^\beta$. In the case where $X_i$ is of multiplier 2, we have $\tau_i(X_i) = (\mathbb{F}_{r_i}^\times)^2$, so $q^\beta$ must be a square modulo $r_i$.

Conversely, suppose that $\left(\frac{q^\beta}{r_i}\right) = +1$ for all $1 \leq i \leq k$ such that $X_i$ is of multiplier 2. In this case, for all $1 \leq i \leq k$ there exists $x_i \in X_i$ such that $\tau_i(x_i) = q^\beta$. Then $(x_1, \ldots, x_k)\pi(\psi^{-\beta}) = (g_1, \ldots, g_k)$, where $g_i \in \mathrm{Sp}_{2\ell_i}(r_i)$ for all $1 \leq i \leq k$. By Lemma 4.3.1 there exists $g \in C_{\mathrm{GL}(V)}(F_0)$ such that $\pi(g) = (g_1, \ldots, g_k)$. Then $\pi(g\psi^\beta) = (x_1, \ldots, x_k)$, so $g\psi^\beta \in G$. □

**Remark 4.3.6** It follows from Lemma 4.3.5 that there exists $g \in C_{\mathrm{GL}(V)}(F_0)$ such that $G = \langle C_G(F_0), g\psi^\beta \rangle$, where

$$\beta = \begin{cases} 1, & \text{if } \left(\frac{q}{r_i}\right) = +1 \text{ for all } i \text{ such that } X_i \text{ is of multiplier 2.} \\ 2, & \text{otherwise.} \end{cases}$$

Therefore $G$ is an extension of the form $G = C_G(F_0).H$, where $H$ is cyclic of order $\nu$ or $\nu/2$.

The survey by Dieudonné [16] on Jordan's work suggests that $G$ is always a semidirect product $G = C_G(F_0) \rtimes H$, with $H$ cyclic of order $\nu$ [16, p. xxxvii]. However, as seen from the above, it is possible that $G = C_G(F_0).H$ with $H$ cyclic of order $\nu/2$.

For example, consider $n = 14$, $\mu = 7$, $\nu = 2$, and assume that $q \equiv -1 \mod 7$. In $\mathrm{GSp}_2(7)$, the normalizer of an irreducible extraspecial group $2_-^{1+2}$ is a maximal irreducible solvable subgroup $X$. (Here $X = G_{1,1}^{\mathcal{B}_3}(O_2^-(2))$ as defined later in Sect. 4.3.4.) Furthermore, since $\left(\frac{2}{7}\right) = +$, in this case $X$ is of multiplier 2. Consider then $G = G_{7,2}^{\mathcal{B}_0}(X)$, which is a primitive maximal irreducible solvable subgroup of $\mathrm{GL}_{14}(q)$, by results proven later (Theorem 7.1.12). The claim in [16, p. xxxvii] suggests that $G = C_G(F_0).H$, with $H$ cyclic of order 2. But because $\left(\frac{q}{7}\right) = -$, it follows from Lemma 4.3.5 that $G = C_G(F_0)$. (This example appears for $q = 13$ in [74, Example 2.5.42].)

We also note that in general it is not true that $G$ is a semidirect product $G = C_G(F_0) \rtimes H$. For example, consider $n = 20$, $\mu = 5$, $\nu = 4$, and assume that $q \equiv \pm 2 \mod 5$. Then $X = \Gamma\mathrm{L}_1(5^2)$ is maximal irreducible solvable in $\mathrm{GSp}_2(5)$ of multiplier 1, and it follows from results proven later (Theorem 7.1.12) that $G = G_{5,4}^{\mathcal{B}_0}(X)$ is primitive maximal irreducible solvable in $\mathrm{GL}_{20}(q)$.

It follows from Lemma 4.3.5 that $G = C_G(F_0).H$, with $H$ cyclic of order 4. We claim that this extension is nonsplit. To this end, suppose for the sake of contradiction $G = C_G(F_0) \rtimes H$. Since $\mathrm{Ker}\,\pi = A$ (Lemma 4.3.3) and $\pi(C_G(F_0)) = X^\circ$ (Lemma 4.3.5), it follows that $\pi(G) = X^\circ \rtimes \pi(H) = X$. But a calculation shows that $X^\circ = X \cap \mathrm{Sp}_2(5)$ does not have a complement in $X = \Gamma\mathrm{L}_1(5^2)$, so this is a contradiction.

**Lemma 4.3.7** *Let $G = G^{\mathcal{B}_0}_{\mu,\nu}(X_1,\ldots,X_k)$ be of type $\mathcal{B}_0$. Then the following statements hold:*

*(i)* $|C_G(F_0)| = |F_0| \cdot \mu^2 \cdot |X_1^\circ| \cdots |X_k^\circ|$.

*(ii)* $|G/C_G(F_0)| = \begin{cases} \nu, & \text{if } \left(\frac{q}{r_i}\right) = +1 \text{ for all } i \text{ such that } X_i \text{ is of multiplier } 2. \\ \nu/2, & \text{otherwise.} \end{cases}$

***Proof*** For (i), first note that $C_G(F_0) \cap \operatorname{Ker}\pi = A$ by Lemma 4.3.2. Then (i) follows from $|A| = |F_0|\mu^2$ together with Lemma 4.3.5 (i). Claim (ii) follows from Lemma 4.3.5 (ii). □

### 4.3.2 Groups of Type $\mathcal{B}_1$

In this case $n$ is even, and $n = 2\mu\nu$ for integers $\mu, \nu > 0$. Denote $\mathbb{K} = \mathbb{F}_{q^\nu}$, and let $\lambda \in \mathbb{K}^\times$ be a primitive element.

In view of Lemma 2.6.1 and Proposition 4.1.18, we make the following assumptions:

- $q^\nu > 4$ or $(q, \nu) = (4, 1)$,
- For every prime $r \mid \mu$, we have $r \mid q^\nu - 1$.

Denote $V' = \mathbb{K} \otimes_{\mathbb{F}_q} V$, and let the action of $f \in \mathrm{GL}(V) \leq \mathrm{GL}(V')$ on $V'$ be as in Sect. 2.6, so

$$V' = Q'_\lambda \oplus Q'_{\lambda^{-1}} = W'_\lambda \oplus \cdots \oplus W'_{\lambda^{q^{\nu-1}}} \oplus W'_{\lambda^{-1}} \oplus \cdots \oplus W'_{\lambda^{-q^{\nu-1}}}$$

where $W'_{\lambda^{q^i}}$ is the $f$-eigenspace for eigenvalue $\lambda^{q^i}$, and $\dim_{\mathbb{K}} W'_{\lambda^{\pm q^i}} = \mu$ for all $0 \leq i < \nu$. Note that $V = Q_\lambda \oplus Q_{\lambda^{-1}}$, where $Q'_\lambda = \mathbb{K} \otimes_{\mathbb{F}_q} Q_\lambda$.

Denote $F_0 = \langle f \rangle$. We will first describe $A$ and the form $\kappa$ such that $F_0 \leq A \leq I(V, \kappa)$.

We can identify $W'_{\lambda^{-1}} = W^*_\lambda$ as $\mathbb{K}[F_0]$-modules. Let $s : V' \to V'$ be the semilinear map as introduced in Sect. 2.4. Each $v, w \in V$ can be written as $v = \sum_{0 \leq i < \nu}(s^i(x) + s^i(x^*))$ and $w = \sum_{0 \leq i < \nu}(s^i(y) + s^i(y^*))$ for unique $x, y \in W'_\lambda$ and $x^*, y^* \in W^*_\lambda$. Then we can define a bilinear form $b : V \times V \to \mathbb{F}_q$ by

$$b(v, w) = \mathrm{Tr}_{\mathbb{K}/\mathbb{F}_q}(y^*(x) + (-1)^e x^*(y))$$

for all $v, w \in V$. When $e = 0$, we can also define a quadratic form $Q : V \to \mathbb{F}_q$ by

$$Q(v) = \mathrm{Tr}_{\mathbb{K}/\mathbb{F}_q}(x^*(x))$$

for all $v \in V$, in which case $Q$ polarizes to $b$.

In the case $e = 0$ we define $\kappa = Q$, and in the case $e = 1$ we define $\kappa = b$. Extending $\mathbb{K}$-linearly we get a bilinear form $b : V' \times V' \to \mathbb{K}$, and in case $e = 0$ a quadratic form $Q : V' \to \mathbb{K}$ that polarizes to $b$. Then $\kappa$ is precisely the form described in Sect. 2.6, and furthermore $F_0 \leq I(V, \kappa) \leq I(V', \kappa)$ and $\Delta(V, \kappa) \leq \Delta(V', \kappa)$.

For the construction of groups of type $\mathcal{B}_1$, let

$$\mu = r_1^{\ell_1} \cdots r_k^{\ell_k}$$

be the prime factorization of $\mu$, with $r_1 < \cdots < r_k$ primes and $\ell_i > 0$ for all $1 \leq i \leq k$. Choose metrically completely reducible maximal solvable subgroups $X_1$, ..., $X_k$ as in Sect. 4.3.1. Let $A \leq \mathrm{GL}(W'_\lambda)$ be the absolutely irreducible subgroup corresponding to $R_1 \otimes \cdots \otimes R_k \leq \mathrm{GL}(W'_\lambda)$, constructed as in Sect. 4.3.1. We can then consider $A$ as a subgroup of $C_{\mathrm{GL}(V)}(F_0) = \mathrm{GL}(W'_\lambda)$, so $F_0 \leq A \leq I(V, \kappa)$.

We have a homomorphism

$$\pi : N_{\Delta(V,\kappa)}(F_0, R_1, \ldots, R_k) \to \prod_{i=1}^{k} \mathrm{GSp}_{2\ell_i}(r_i)$$

defined by $\pi(g) = (g_1, \ldots, g_k)$, where $g_i$ is the action of $g$ on $\overline{R_i} := R_i/Z(R_i)$. Here $g_1 \in O_{2\ell_1}^{\varepsilon_1}(2)$ if $r_1 = 2$.

Having fixed a basis of $W'_\lambda$ in the construction of $A$, we can consider $\mathbb{K}$-linear maps $\eta, \psi, \varphi : V' \to V'$ as defined in Sect. 2.6. Then $\eta, \psi, \varphi \in N_{\Delta(V,\kappa)}(F_0)$. Furthermore $\eta$ centralizes $C_{\mathrm{GL}(V)}(F_0)$, so $\eta \in \mathrm{Ker}\,\pi$. As seen in (2.3), the action of $\psi$ on $\mathrm{GL}(W'_\lambda)$ is as in type $\mathcal{B}_0$. Thus the action of $\psi$ on the generators of $R_1$, ..., $R_k$ is as described in (4.2), and in particular $\psi \in N_{\Delta(V,\kappa)}(F_0, R_1, \ldots, R_k)$.

The action of $\varphi$ on the generators of $R_1$, ..., $R_k$ corresponds to taking the inverse transpose, see (2.3). The generators $A_j := A_j^{(i)}$ are symmetric matrices for all $1 \leq j \leq \ell_i$. For $B_j := B_j^{(i)}$ the corresponding matrices are permutation matrices, or monomial matrices with nonzero entries in $\{1, -1\}$. Thus $\varphi \in N_{\Delta(V,\kappa)}(F_0, R_1, \ldots, R_k)$, and

$$\begin{aligned} \varphi A_j \varphi^{-1} &= A_j^{-1} \\ \varphi B_j \varphi^{-1} &= B_j \end{aligned} \tag{4.3}$$

for all $1 \leq j \leq \ell_i$.

**Lemma 4.3.8** *We have*

$$\pi\left(N_{C_{I(V,\kappa)}(F_0)}(R_1, \ldots, R_k)\right) = \begin{cases} \prod_{i=1}^{k} \mathrm{Sp}_{2\ell_i}(r_i), & \textit{if } r_1 > 2. \\ O_{2\ell_1}^{\varepsilon_1}(2) \times \prod_{i=2}^{k} \mathrm{Sp}_{2\ell_i}(r_i), & \textit{if } r_1 = 2. \end{cases}$$

***Proof*** Follows with the same proof as Lemma 4.3.1. □

**Lemma 4.3.9** $\operatorname{Ker}\pi \cap C_{I(V,\kappa)}(F_0) = A$.

***Proof*** As in Lemma 4.3.2, we can repeat the argument from Lemma 4.2.1. □

**Lemma 4.3.10** $\operatorname{Ker}\pi = AT \rtimes \langle\eta\rangle$, *where T is the subgroup*

$$T = \{\varphi^\alpha \psi^\beta : (-1)^\alpha q^\beta \equiv 1 \mod r_i \textit{ for all } 1 \leq i \leq k\}.$$

***Proof*** Since $\eta \in \operatorname{Ker}\pi$ and $\tau(\eta)$ is a primitive element of $\mathbb{F}_q$, it is clear that $\operatorname{Ker}\pi = (\operatorname{Ker}\pi)^\circ \rtimes \langle\eta\rangle$. By (4.3) and (4.2), the action of $\varphi^\alpha\psi^\beta$ on $\overline{R_i}$ is a diagonal matrix of the form

$$\operatorname{diag}(1, \ldots, 1, (-1)^\alpha q^\beta, \ldots, (-1)^\alpha q^\beta).$$

Thus $T$ consists of the $\varphi^\alpha\psi^\beta$ that act trivially on $\overline{R_i}$ for all $1 \leq i \leq k$, and $T$ is a subgroup since $\varphi$ and $\psi$ commute. Hence $AT \leq (\operatorname{Ker}\pi)^\circ$.

It remains to prove that $(\operatorname{Ker}\pi)^\circ \leq AT$. To this end, every element $g \in (\operatorname{Ker}\pi)^\circ$ can be written in the form $g = x\varphi^\alpha\psi^\beta$, where $x \in C_{I(V,\kappa)}(F_0)$ (Lemma 2.6.4 (ii)). Because $\varphi^\alpha\psi^\beta$ acts as a diagonal matrix on $\overline{R_i}$, arguing as in the proof of Lemma 4.3.3 shows that $x \in \operatorname{Ker}\pi$ and $\varphi^\alpha\psi^\beta \in \operatorname{Ker}\pi$. Then $\varphi^\alpha\psi^\beta \in T$ and $x \in A$ by Lemma 4.3.9, so we conclude that $(\operatorname{Ker}\pi)^\circ = AT$. □

Since $\operatorname{Ker}\pi$ is solvable by Lemma 4.3.10, the group $\pi^{-1}(X_1 \times \cdots \times X_k)$ is a solvable. We define the group of type $\mathcal{B}_1$ corresponding to $\mu$, $\nu$, and $X_1, \ldots, X_k$ as

$$G^{\mathcal{B}_1}_{\mu,\nu}(X_1, \ldots, X_k) := \pi^{-1}(X_1 \times \cdots \times X_k).$$

As in the previous section, we have the following special case of Proposition 4.2.4.

**Proposition 4.3.11** *Let G be a metrically primitive maximal irreducible solvable subgroup of type $\mathcal{B}_1$ in $\Delta(V, \kappa)$. Then G is conjugate to a group of the form $G^{\mathcal{B}_1}_{\mu,\nu}(X_1, \ldots, X_k)$.*

The following lemma gives a description of the elements of $N_{\Omega(V,\kappa)}(F_0)$, in the case where $n$ and $q$ are even and $\kappa$ is a quadratic form.

**Lemma 4.3.12** *Suppose that n and q are even, and that $\kappa$ is a quadratic form. Let $G = G^{\mathcal{B}_1}_{\mu,\nu}(X_1, \ldots, X_k)$ be of type $\mathcal{B}_1$. Then the following hold:*

*(i)* $\dim(V^g) \equiv 0 \mod 2$ *for all* $g \in C_{I(V,\kappa)}(F_0)$;
*(ii)* $\dim(V^\psi) \equiv 0 \mod 2$;
*(iii)* $\dim(V^\varphi) \equiv n/2 \mod 2$.

***Proof*** Note first that extending the field does not change the dimension of a fixed point space, so for any $g \in \mathrm{GL}(V)$ we have $\dim_{\mathbb{F}_q} V^g = \dim_{\mathbb{K}}(V')^g$. Then for claim (i), for $g = g_A \in C_{I(V,\kappa)}(F_0)$ we have

$$\dim V^g = \sum_{0 \leq k < \nu} \dim_{\mathbb{K}}(W'_{\lambda^{q^k}})^g + \sum_{0 \leq k < \nu} \dim_{\mathbb{K}}(W'_{\lambda^{-q^k}})^g.$$

The action of $g$ on $W'_{\lambda^{\pm q^k}}$ is by the matrix $B^{\pm 1}$, where $B_{ij} = A_{ij}^{q^k}$. Note that applying a field automorphism on a matrix does not change the dimension of the fixed point space, and neither does taking the inverse. Therefore we conclude that $\dim V^g = 2\nu \dim_{\mathbb{K}}(W'_\lambda)^g \equiv 0 \mod 2$.

The action of $\psi$ on $V'$ permutes the basis vectors, so $\dim_{\mathbb{K}} V^\psi$ is equal to the number of orbits of $\psi$ on the basis vectors. In notation of Sect. 2.6, the orbits are $\{x_i^{(k)} : 0 \le k < \nu\}$ and $\{y_i^{(k)} : 0 \le k < \nu\}$ for $1 \le i \le \mu$, so we conclude that $\dim V^\psi = 2\mu \equiv 0 \mod 2$.

Similarly $\varphi$ permutes basis vectors with orbits $\{x_i^{(k)}, y_i^{(k)}\}$, so $\dim V^\varphi = \mu\nu = n/2$. □

**Lemma 4.3.13** *Let $G = G^{\mathcal{B}_1}_{\mu,\nu}(X_1, \ldots, X_k)$ be of type $\mathcal{B}_1$. Then the following statements hold:*

*(i) $G$ is of multiplier* 1.
*(ii)* $\pi(C_{G^\circ}(F_0)) = X_1^\circ \times \cdots \times X_k^\circ$.
*(iii) $G$ contains an element of the form $g\varphi^\alpha\psi^\beta$ with $g \in C_{I(V,\kappa)}(F_0)$ if and only if* $\left(\dfrac{(-1)^\alpha q^\beta}{r_i}\right) = +1$ *for all $1 \le i \le k$ such that $X_i$ is of multiplier* 2.

***Proof*** We have $\eta \in G$ and $\tau(\eta)$ is a primitive element of $\mathbb{F}_q$, so (i) follows. Claim (ii) is immediate from Lemma 4.3.8.

For (iii), note that $\varphi^\alpha\psi^\beta$ multiplies the alternating bilinear form on $\overline{R_i}$ by $(-1)^\alpha q^\beta$. Thus arguing as in the proof of Lemma 4.3.5 shows that there exists $g\varphi^\alpha\psi^\beta \in G$ with $g \in C_{I(V,\kappa)}(F_0)$ if and only if there exist $x_i \in X_i$ such that $\tau_i(x_i) = (-1)^\alpha q^\beta$ for all $1 \le i \le k$. From this (iii) follows. □

In order to guarantee that the group is irreducible, we need it to contain an element of the form $g_A\varphi\psi^\beta$ for some $\beta$, as otherwise $Q_\lambda$ and $Q_{\lambda^{-1}}$ would be $G$-invariant. Therefore by Lemma 4.3.13, we have the following.

**Lemma 4.3.14** *Let $G = G^{\mathcal{B}_1}_{\mu,\nu}(X_1, \ldots, X_k)$ be of type $\mathcal{B}_1$. If $G$ is irreducible, then one of the following statements holds:*

*(i)* $\left(\dfrac{-1}{r_i}\right) = +1$ *for all $1 \le i \le k$ such that $X_i$ is of multiplier 2. In this case $G$ contains an element of the form $g_A\varphi$.*
*(ii)* $\left(\dfrac{-q}{r_i}\right) = +1$ *for all $1 \le i \le k$ such that $X_i$ is of multiplier 2. In this case $G$ contains an element of the form $g_A\varphi\psi$.*

**Lemma 4.3.15** *Let $G = G^{\mathcal{B}_1}_{\mu,\nu}(X_1, \ldots, X_k)$ be of type $\mathcal{B}_1$. Let $I$ be the set of indices $1 \leq i \leq k$ such that $X_i$ is of multiplier 2. Then the following statements hold:*

*(i)* $|G| = (q-1)|G^\circ|$.
*(ii)* $|C_{G^\circ}(F_0)| = |F_0| \cdot \mu^2 \cdot |X_1^\circ| \cdots |X_k^\circ|$.

$$(iii)\ |G^\circ/C_{G^\circ}(F_0)| = \begin{cases} 2\nu, & \text{if } \left(\frac{-1}{r_i}\right) = \left(\frac{q}{r_i}\right) = +1 \text{ for all } i \in I. \\ \nu, & \text{if } \left(\frac{-1}{r_i}\right) = +1 \text{ for all } i \in I \text{ and} \\ & \left(\frac{q}{r_j}\right) = -1 \text{ for some } j \in I. \\ \nu, & \text{if } \left(\frac{-q}{r_i}\right) = +1 \text{ for all } i \in I \text{ and} \\ & \left(\frac{-1}{r_j}\right) = -1 \text{ for some } j \in I. \\ \nu, & \text{if } \left(\frac{q}{r_i}\right) = +1 \text{ for all } i \in I \text{ and} \\ & \left(\frac{-1}{r_j}\right) = -1 \text{ for some } j \in I. \\ \nu/2, & \text{if } \left(\frac{q}{r_j}\right) = \left(\frac{-1}{r_{j'}}\right) = \left(\frac{q}{r_{j''}}\right) \\ & = -1 \text{ for some } j, j', j'' \in I. \end{cases}$$

***Proof*** Claim (i) follows from Lemma 4.3.13 (i). Claim (ii) follows similarly to Lemma 4.3.7 (i), applying Lemma 4.3.9 and Lemma 4.3.13 (ii).

Next we consider (iii). Denote the images of $\varphi$ and $\psi$ in $N_{\Delta(V,\kappa)}(F_0)/C_{\Delta(V,\kappa)}(F_0)$ by $\overline{\varphi}$ and $\overline{\psi}$. By Lemma 4.3.13 (iii), in the cases listed in (iii) we have

$$G^\circ/C_{G^\circ}(F_0) \cong \langle \overline{\varphi}, \overline{\psi} \rangle,$$

$$G^\circ/C_{G^\circ}(F_0) \cong \langle \overline{\varphi}, \overline{\psi^2} \rangle \ (\nu \text{ even}),$$

$$G^\circ/C_{G^\circ}(F_0) \cong \langle \overline{\varphi\psi} \rangle \ (\nu \text{ even}),$$

$$G^\circ/C_{G^\circ}(F_0) \cong \langle \overline{\psi} \rangle,$$

$$G^\circ/C_{G^\circ}(F_0) \cong \langle \overline{\psi^2} \rangle \ (\nu \text{ even}),$$

respectively. From this the claim follows. □

**Remark 4.3.16** Note that by Lemma 4.3.14, the first three conditions listed in (iii) cover the possibilities when $G$ is irreducible.

### 4.3.3 Groups of Type $\mathcal{B}_2$

In this case $n$ is even, and $n = 2\mu\nu$ for integers $\mu, \nu > 0$. Denote $\mathbb{K} = \mathbb{F}_{q^{2\nu}}$, and let $\zeta \in \mathbb{K}^\times$ be a primitive element. Set $\lambda = \zeta^{q^\nu - 1}$, so $\lambda$ has order $q^\nu + 1$.

In view of Proposition 4.1.18, we assume the following.

- For every prime $r \mid \mu$, we have $r \mid q^\nu + 1$.

Denote $V' = \mathbb{K} \otimes_{\mathbb{F}_q} V$, and let $f \in \mathrm{GL}(V) \leq \mathrm{GL}(V')$ act on $V'$ as in Sect. 2.7, so

$$V' = W'_\lambda \oplus \cdots \oplus W'_{\lambda^{q^{2\nu-1}}}$$

where $W'_{\lambda^{q^i}}$ is the $f$-eigenspace for eigenvalue $\lambda^{q^i}$, and $\dim_{\mathbb{K}} W'_{\lambda^{q^i}} = \mu$ for all $0 \leq i < 2\nu$.

Denote $F_0 = \langle f \rangle$. We begin with the construction of $A$ and the form $\kappa$ such that $F_0 \leq A \leq I(V, \kappa)$. As in the previous sections, let

$$\mu = r_1^{\ell_1} \cdots r_k^{\ell_k}$$

be the prime factorization of $\mu$, where $r_1 < \cdots < r_k$ are primes and $\ell_i > 0$ for all $1 \leq i \leq k$. Choose metrically completely reducible maximal solvable subgroups $X_1, \ldots, X_k$ as in Sect. 4.3.1. Let $A \leq C_{\mathrm{GL}(V)}(F_0)$ be the absolutely irreducible subgroup corresponding to $R_1 \otimes \cdots \otimes R_k \leq \mathrm{GL}(W'_\lambda)$, constructed as in Sect. 4.3.1.

As noted in Sects. 3.2.2 and 3.2.5 there is a non-degenerate $R_i$-invariant Hermitian form $\widehat{b_i}$ on $W_i$, with respect to the automorphism $\xi \mapsto \xi^{q^\nu}$ of $\mathbb{K}$. Recall that the form $\widehat{b_i}$ is defined simply by taking the basis used in Sect. 3.2.1 or 3.2.4 as an orthonormal basis, except when $r_1 = 2$, $\varepsilon_1 = -$, and $q^\nu \equiv 1 \mod 4$. In the exceptional case, for the basis vectors $v_{\xi_1,\ldots,\xi_{\ell_1}} := v_{\xi_1^{(1)},\ldots,\xi_{\ell_1}^{(1)}}$ we have

$$\widehat{b_1}(v_{\xi_1,\ldots,\xi_{\ell_1}}, v_{\xi'_1,\ldots,\xi'_{\ell_1}}) = \zeta^{\frac{q^\nu+1}{2}}(-1)^{\xi_1}\delta_{\xi_1,\xi'_1+1}\delta_{\xi_2,\xi'_2}\cdots\delta_{\xi_{\ell_1},\xi'_{\ell_1}}$$

for all $0 \leq \xi_1, \xi'_1, \ldots, \xi_{\ell_1}, \xi'_{\ell_1} \leq 1$.

The tensor product $\widehat{b} = \widehat{b_1} \otimes \cdots \otimes \widehat{b_k}$ is a non-degenerate $A$-invariant Hermitian form on $W'_\lambda = W_1 \otimes \cdots \otimes W_k$, so we have $A \leq I(W'_\lambda, \widehat{b}) = \mathrm{GU}_\mu(\mathbb{K})$.

Let $s : V' \to V'$ be the semilinear map as introduced in Sect. 2.4. Each $v, w \in V$ can be written as $v = \sum_{0 \leq i < 2\nu} s^i(x)$ and $w = \sum_{0 \leq i < 2\nu} s^i(y)$ for unique $x, y \in W'_\lambda$. Then we define a bilinear form $b : V \times V \to \mathbb{F}_q$ by

$$b(v, w) = \begin{cases} \mathrm{Tr}_{\mathbb{K}/\mathbb{F}_q}(\widehat{b}(x, y)), & \text{if } e = 0. \\ \mathrm{Tr}_{\mathbb{K}/\mathbb{F}_q}(\zeta^{\frac{q^\nu+1}{2}}\widehat{b}(x, y)), & \text{if } e = 1. \end{cases}$$

For $e = 1$ we define $\kappa = b$.

If $e = 0$, we define a quadratic form $Q : V \to \mathbb{F}_q$ by

$$Q(v) = \mathrm{Tr}_{\mathbb{F}_{q^\nu}/\mathbb{F}_q}(\widehat{b}(x,x)),$$

which has polarization equal to $b$. In this case we define $\kappa = Q$.

Then as a subgroup of $\mathrm{GL}(V)$, we have $F_0 \leq A \leq \Delta(V,\kappa)$ since $\widehat{b}$ is $A$-invariant. We extend $\mathbb{K}$-linearly to get a bilinear form $b : V' \times V' \to \mathbb{K}$, and then

$$b(W'_{\lambda^{q^i}}, W'_{\lambda^{q^j}}) = 0, \text{ if } |i-j| \neq \nu.$$

$$b(s^k(v), s^{k+\nu}(w)) = \begin{cases} \widehat{b}(v,w)^{q^k}, & \text{if } e = 0. \\ \zeta^{\frac{q^k(q^\nu+1)}{2}} \widehat{b}(v,w)^{q^k}, & \text{if } e = 1. \end{cases}$$

$$b(s^{k+\nu}(v), s^k(w)) = (-1)^e b(s^k(w), s^{k+\nu}(v))$$

for all $v, w \in W'_\lambda$ and $0 \leq k < \nu$. Similarly $Q$ extends $\mathbb{K}$-linearly to $V'$, and in $V'$ the polarization of $Q$ is $b$ and $Q(W'_{\lambda^{q^i}}) = 0$ for all $0 \leq i < 2\nu$.

Thus $F_0$ and $\kappa$ are exactly as in the setup of Sect. 2.7. Having fixed a basis of $(W'_\lambda, \widehat{b})$ in the construction of $A$, we can define maps $\eta, \psi : V' \to V'$ as in Sect. 2.7. Then $\eta$ centralizes $C_{\mathrm{GL}(V)}(F_0)$, so $\eta \in N_{\Delta(V,\kappa)}(F_0, R_1, \dots, R_k)$.

Furthermore, we have $\psi \in N_{I(V,\kappa)}(F_0)$. Recall that here the definition of $\psi$ involves a scalar which depends on $e \in \{0,1\}$ and the type of basis on $W'_\lambda$. The basis of $W'_\lambda$ is orthonormal unless $q^\nu \equiv 1 \mod 4$ and $(r_1, \varepsilon_1) = (2, -)$. In the case where $q^\nu \equiv 1 \mod 4$ and $(r_1, \varepsilon_1) = (2, -)$, we have a hyperbolic basis with products as in (2.5). Here the basis vector

$$v = v_{\xi_1^{(1)},\dots,\xi_{\ell_1}^{(1)}} \otimes v_{\xi_1^{(2)},\dots,\xi_{\ell_2}^{(2)}} \otimes \cdots \otimes v_{\xi_1^{(k)},\dots,\xi_{\ell_k}^{(k)}}$$

is paired with

$$w = v_{\xi_1^{(1)}+1,\xi_2^{(1)},\dots,\xi_{\ell_1}^{(1)}} \otimes v_{\xi_1^{(2)},\dots,\xi_{\ell_2}^{(2)}} \otimes \cdots \otimes v_{\xi_1^{(k)},\dots,\xi_{\ell_k}^{(k)}}$$

and

$$\widehat{b}(v,w) = (-1)^{\xi_1^{(1)}} \zeta^{\frac{q^\nu+1}{2}}.$$

Thus in this case, in the definition (2.6) of $\psi$ we choose

$$\alpha = \zeta^{-\left(\frac{q-1}{2}\right)(e+1)}.$$

The action of $\psi$ on the generators of $R_1$, …, $R_k$ is determined similarly to (4.2). The matrices $B_j^{(i)}$ have their entries in $\mathbb{F}_q$, so they commute with $\psi$. In the construction of unitary representations of $R_i$ (Sects. 3.2.1 and 3.2.4), all the generators $A_j^{(i)}$ are diagonal matrices. Therefore for the generators $A_j := A_j^{(i)}$ and $B_j := B_j^{(i)}$, we have relations

$$\begin{aligned} \psi^{-1} A_j \psi &= A_j^q \\ \psi^{-1} B_j \psi &= B_j \end{aligned} \tag{4.4}$$

for all $1 \le j \le \ell_i$. In particular $\psi \in N_{I(V,\kappa)}(F_0, R_1, \ldots, R_k)$.

We have a homomorphism

$$\pi : N_{\Delta(V,\kappa)}(F_0, R_1, \ldots, R_k) \to \prod_{i=1}^{k} \mathrm{GSp}_{2\ell_i}(r_i)$$

defined by $\pi(g) = (g_1, \ldots, g_k)$, where $g_i$ is the action of $g$ on $\overline{R_i} := R_i/Z(R_i)$. Note that $g_1 \in O_{2\ell_1}^{\varepsilon_1}(2)$ if $r_1 = 2$.

**Lemma 4.3.17** *We have*

$$\pi\left(N_{C_{I(V,\kappa)}(F_0)}(R_1, \ldots, R_k)\right) = \begin{cases} \prod_{i=1}^{k} \mathrm{Sp}_{2\ell_i}(r_i), & \textit{if } r_1 > 2. \\ O_{2\ell_1}^{\varepsilon_1}(2) \times \prod_{i=2}^{k} \mathrm{Sp}_{2\ell_i}(r_i), & \textit{if } r_1 = 2. \end{cases}$$

***Proof*** Using Theorem 3.2.3 (iii)–(iv) and Theorem 3.2.6 (iii), the result follows with the same proof as Lemma 4.3.1. □

**Lemma 4.3.18** $\mathrm{Ker}\,\pi \cap C_{I(V,\kappa)}(F_0) = A$.

***Proof*** As in Lemma 4.3.2, we can repeat the argument from Lemma 4.2.1. □

**Lemma 4.3.19** $\mathrm{Ker}\,\pi = \langle A, \psi^\alpha, \eta\rangle$, *where* $0 < \alpha \le 2v$ *is the smallest integer such that* $r_i \mid q^\alpha - 1$ *for all* $1 \le i \le k$.

***Proof*** The $R_i$ are centralized by $\eta$, so $\eta \in \mathrm{Ker}\,\pi$. Since $\tau(\eta)$ is a primitive element of $\mathbb{F}_q$, we have $\mathrm{Ker}\,\pi = \langle(\mathrm{Ker}\,\pi)^\circ, \eta\rangle$. Arguing as in the proof of Lemma 4.3.3 shows that $(\mathrm{Ker}\,\pi)^\circ = \langle A, \psi^\alpha\rangle$. □

By Lemma 4.3.19 the group $\pi^{-1}(X_1 \times \cdots \times X_k)$ is a solvable. Similarly to the previous sections, we define the group of type $\mathcal{B}_2$ corresponding to $X_1, \ldots, X_k$ as

$$G_{\mu,\nu}^{\mathcal{B}_2}(X_1, \ldots, X_k) := \pi^{-1}(X_1 \times \cdots \times X_k).$$

We have the following special case of Proposition 4.2.4.

**Proposition 4.3.20** *Let $G$ be a metrically primitive maximal irreducible solvable subgroup of type $\mathcal{B}_2$ in $\Delta(V,\kappa)$. Then $G$ is conjugate to a group of the form $G^{\mathcal{B}_2}_{\mu,\nu}(X_1,\ldots,X_k)$.*

**Lemma 4.3.21** *Suppose that $n$ and $q$ are even, and that $\kappa$ is a quadratic form. Let $G = G^{\mathcal{B}_2}_{\mu,\nu}(X_1,\ldots,X_k)$ be of type $\mathcal{B}_2$. Then the following hold:*

*(i)* $\dim(V^g) \equiv 0 \mod 2$ *for all* $g \in C_{I(V,\kappa)}(F_0)$*;*
*(ii)* $\dim(V^\psi) \equiv 1 \mod 2$.

***Proof*** As in the proof of Lemma 4.3.12, for $g = g_A \in C_{I(V,\kappa)}(F_0)$ we see that $\dim V^g = 2\nu \dim_{\mathbb{K}}(W'_\lambda)^g \equiv 0 \mod 2$.

Moreover $\psi$ permutes the basis vectors in $V'$, with orbits $\{x_i^{(k)} : 0 \le k < 2\nu\}$ (in the notation of Sect. 2.7). As in the proof of Lemma 4.3.12, it follows that $\dim(V^\psi) = \mu$. On the other hand, every prime divisor of $\mu$ divides $q^\nu + 1$, so $\mu$ is odd and we conclude that $\dim(V^\psi) \equiv 1 \mod 2$. □

**Lemma 4.3.22** *Let $G = G^{\mathcal{B}_2}_{\mu,\nu}(X_1,\ldots,X_k)$ be of type $\mathcal{B}_2$. Then the following statements hold:*

*(i)* *$G$ is of multiplier* 1.
*(ii)* $\pi(C_{G^\circ}(F_0)) = X_1^\circ \times \cdots \times X_k^\circ$.
*(iii)* *$G$ contains an element of the form $g\psi^\beta$, $g \in C_{I(V,\kappa)}(F_0)$ if and only if* $\left(\dfrac{q^\beta}{r_i}\right) = +1$ *for all $1 \le i \le k$ such that $X_i$ is of multiplier* 2.

***Proof*** We have $\eta \in G$ and $\tau(\eta)$ is a primitive element of $\mathbb{F}_q$, so claim (i) holds. Claim (ii) is immediate from Lemma 4.3.17. For (iii), argue as in the proof of Lemma 4.3.5. □

**Lemma 4.3.23** *Let $G = G^{\mathcal{B}_2}_{\mu,\nu}(X_1,\ldots,X_k)$ be of type $\mathcal{B}_2$. Then the following statements hold:*

*(i)* $|G| = (q-1)|G^\circ|$.
*(ii)* $|C_{G^\circ}(F_0)| = |F_0| \cdot \mu^2 \cdot |X_1^\circ| \cdots |X_k^\circ|$.
*(iii)* $$|G^\circ/C_{G^\circ}(F_0)| = \begin{cases} 2\nu, & \text{if } \left(\dfrac{q}{r_i}\right) = +1 \text{ for all } i \text{ such that } X_i \text{ is of multiplier } 2. \\ \nu, & \text{otherwise.} \end{cases}$$

***Proof*** Claim (i) follows from Lemma 4.3.22 (i). Claims (ii)–(iii) follow similarly to Lemma 4.3.7 (i)–(ii), applying Lemmas 4.3.18 and 4.3.22. □

### 4.3.4 Groups of Type $\mathcal{B}_3$

In this case $\mathbb{K} = \mathbb{F}_q$ for $q$ odd, $\nu = 1$, and $F_0 = \langle -I_V \rangle$. In view of Lemma 4.1.19, we assume that

$$n = \mu = 2^\ell$$

for some $\ell \geq 0$. For the construction of a group of type $\mathcal{B}_3$, we require a metrically completely reducible maximal solvable subgroup $X \leq O^\varepsilon_{2\ell}(2)$, where $\varepsilon \in \{+, -\}$. Denote $R = 2^{1+2\ell}_\varepsilon$.

Let $A \leq \mathrm{GL}(V)$ be the absolutely irreducible subgroup corresponding to $R \leq \mathrm{GL}(V)$, constructed as in Sect. 4.3.1. As seen in Sects. 3.2.6 and 3.2.3, there is a non-degenerate $R$-invariant reflexive bilinear form $b$ on $V$, where $\mathrm{sgn}(b) = \varepsilon$. Recall that here the basis used for $V$ is an orthonormal basis if $\varepsilon = +$, and a standard symplectic basis if $\varepsilon = -$.

If $\mathrm{sgn}(b) = -1$, we let $\kappa = b$. If $\mathrm{sgn}(b) = +1$, we let $\kappa = Q$ to be the quadratic form corresponding to $b$, defined by $Q(v) = \frac{1}{2}b(v, v)$ for all $v \in V$. If $b$ is symmetric, then by [55, Lemma 4.4.2] the corresponding quadratic form $Q$ has $\mathrm{sgn}(Q) = +$ if $\ell > 1$. As noted in Sect. 3.2.3, in the case where $\ell = 1$, we have $\mathrm{sgn}(Q) = -$ if and only if $q \equiv 3 \mod 4$.

As in the previous sections, we have a homomorphism

$$\pi : N_{\Delta(V,\kappa)}(R) \to O^\varepsilon_{2\ell}(2),$$

where $\pi(g)$ is the action of $g$ on $R/Z(R)$.

**Lemma 4.3.24** *The homomorphism $\pi$ is surjective. Moreover, the following statements hold:*

*(i) If 2 is a square in $\mathbb{F}_q$, then the restriction of $\pi$ to $N_{I(V,\kappa)}(R)$ is also surjective.*
*(ii) If 2 is not a square in $\mathbb{F}_q$, then the image of $N_{I(V,\kappa)}(R)$ under $\pi$ is equal to $\Omega^\varepsilon_{2\ell}(2)$.*

***Proof*** The result follows from Theorems 3.2.4 and 3.2.7. □

**Lemma 4.3.25** $\mathrm{Ker}\,\pi = AZ$, *where $Z$ is the group of scalar matrices in* $\mathrm{GL}(V)$.

***Proof*** Since $A$ is absolutely irreducible on $V$, the result follows similarly to the proof of Lemma 4.2.1. □

As in the previous sections, by Lemma 4.3.25 the group $\pi^{-1}(X)$ is a solvable, and we define the group of type $\mathcal{B}_3$ corresponding to $X$ as

$$G^{\mathcal{B}_3}_{\mu,\nu}(X) := \pi^{-1}(X).$$

We have the following special case of Proposition 4.2.4.

**Proposition 4.3.26** *Let $G$ be a maximal irreducible metrically primitive solvable subgroup of type $\mathcal{B}_3$ in $\Delta(V,\kappa)$. Then $G$ is conjugate to a group of the form $G^{\mathcal{B}_3}_{\mu,\nu}(X)$.*

**Lemma 4.3.27** *Let $G = G^{\mathcal{B}_3}_{\mu,\nu}(X)$ be of type $\mathcal{B}_3$. Then the following statements hold:*

(i) $\pi(G) = X$.
(ii) *If 2 is a square in $\mathbb{F}_q$, then $\pi(G^\circ) = X$.*
(iii) *If 2 is not a square in $\mathbb{F}_q$, then $\pi(G^\circ) = X^\Omega$.*

***Proof*** All of the claims are immediate from Lemma 4.3.24. □

**Lemma 4.3.28** *Let $G = G^{\mathcal{B}_3}_{\mu,\nu}(X)$ be of type $\mathcal{B}_3$. Then the following statements hold:*

(i) $|G| = (q-1)n^2|X|$.
(ii) *If 2 is a square in $\mathbb{F}_q$, then $|G^\circ| = 2n^2|X|$.*
(iii) *If 2 is not a square in $\mathbb{F}_q$, then $|G^\circ| = 2n^2|X^\Omega|$.*

***Proof*** Follows from Lemmas 4.3.25 and 4.3.27. □

**Lemma 4.3.29** *Let $G = G^{\mathcal{B}_3}_{\mu,\nu}(X)$ be of type $\mathcal{B}_3$. Write $X = X_1 \times \cdots \times X_t$, where $X_i \le O^{\varepsilon_i}_{2\ell_i}(2)$ is maximal irreducible solvable for all $1 \le i \le t$, with $\ell = \ell_1 + \cdots + \ell_t$ and $\varepsilon = \varepsilon_1 \cdots \varepsilon_t$ (Lemma 2.3.7). Then*

$$G \cong G^{\mathcal{B}_3}_{\mu_1,\nu_1}(X_1) \otimes \cdots \otimes G^{\mathcal{B}_3}_{\mu_t,\nu_t}(X_t),$$

*where $\mu_i = 2^{\ell_i}$ and $\nu_i = 1$ for all $1 \le i \le t$.*

***Proof*** Since $\pi(G) = X$ (Lemma 4.3.27), we have an orthogonal decomposition

$$R/Z(R) = R_1/Z(R) \perp \cdots \perp R_t/Z(R),$$

where the action of $G$ on $R_i/Z(R)$ is equal to $X_i \le O^{\varepsilon_i}_{2\ell_i}(2)$. Then $R$ is a central product $R = R_1 \cdots R_t$, where $R_i = 2^{1+2\ell_i}_{\varepsilon_i}$. Arguing as in the proof of Lemma 4.1.13, we can identify $V = W_1 \otimes \cdots \otimes W_t$ and $R = R_1 \otimes \cdots \otimes R_t$, where $W_i$ is an absolutely irreducible $\mathbb{F}_q[R_i]$-module of dimension $2^{\ell_i}$.

Because $V$ is absolutely irreducible, an $R$-invariant bilinear form on $V$ is unique up to a scalar. Thus we have $b = b_1 \otimes \cdots \otimes b_t$, where $b_i$ is a non-degenerate reflexive bilinear form on $W_i$ with $\mathrm{sgn}(b_i) = \varepsilon_i$. By Lemma 4.1.15 and Lemma 2.1.16 (ii) we have

$$N_{\Delta(V,b)}(R) = N_{\Delta(W_1,b_1)}(R_1) \otimes \cdots \otimes N_{\Delta(W_t,b_t)}(R_t).$$

Since the action of $G$ on $R_i/Z(R) = R_i/Z(R_i)$ is equal to $X_i$ for all $1 \le i \le t$, the result follows. □

### 4.3.5 General Properties of $G^{\mathcal{B}}_{\mu,\nu}(X_1, \ldots, X_k)$

We end this section by summarizing some of the general properties of groups of the form $G^{\mathcal{B}}_{\mu,\nu}(X_1, \ldots, X_k)$. In later sections we will examine various other properties of $G = G^{\mathcal{B}}_{\mu,\nu}(X_1, \ldots, X_k)$ and determine for example when $G$ is irreducible, and when $G$ is irreducible and metrically primitive. Our ultimate aim is to classify completely the cases where $G$ is maximal irreducible solvable.

**Lemma 4.3.30** *Let $G \leq \Delta(V, \kappa)$ be metrically primitive maximal irreducible solvable of type $\mathcal{B} = \mathcal{B}_i$ for some $0 \leq i \leq 3$. Then $G$ is conjugate to a group of the form $G^{\mathcal{B}}_{\mu,\nu}(X_1, \ldots, X_k)$ in $\Delta(V, \kappa)$.*

***Proof*** As noted in Sects. 4.3.1–4.3.4, this follows from Proposition 4.2.4. □

**Remark 4.3.31** The special case where $G \leq \mathrm{GL}_n(q)$ and $\mu$ is a prime power was considered in [74, Theorem 2.5.35, Theorem 2.5.37]. For example, the proof given by Short in [74, Theorem 2.5.35, Theorem 2.5.37] (partially due to L. G. Kovács) shows that in this special case Lemma 4.3.30 holds and $G$ is irreducible and primitive. He also describes $\mathrm{Ker}\,\pi$ in this case [74, Theorem 2.5.29], similarly to Lemma 4.3.3.

**Lemma 4.3.32** *Let $G = G^{\mathcal{B}}_{\mu,\nu}(X_1, \ldots, X_k)$. Then $\mathrm{Ker}\,\pi \cap C_{I(V,\kappa)}(F_0) = A$.*

***Proof*** As noted in Sects. 4.3.1–4.3.4, the result follows similarly to Lemma 4.2.1. □

**Lemma 4.3.33** *Let $G = G^{\mathcal{B}}_{\mu,\nu}(X_1, \ldots, X_k)$. Then $N_{\Delta(V,\kappa)}(F_0)/C_{I(V,\kappa)}(F_0)$ and $G/C_{G^\circ}(F_0)$ are abelian.*

***Proof*** Since $F_0$ is as described in Sects. 2.4–2.8, the lemma follows exactly as in the proofs of Lemmas 2.9.1 and 2.9.2. □

**Lemma 4.3.34** *Let $G = G^{\mathcal{B}}_{\mu,\nu}(X_1, \ldots, X_k)$. We have*

$$\pi\left(N_{C_{I(V,\kappa)}(F_0)}(R_1, \ldots, R_k)\right) = \begin{cases} \prod_{i=1}^{k} \mathrm{Sp}_{2\ell_i}(r_i), & \text{if } r_1 > 2. \\ O^{\varepsilon_1}_{2\ell_1}(2) \times \prod_{i=2}^{k} \mathrm{Sp}_{2\ell_i}(r_i), & \text{if } r_1 = 2. \end{cases}$$

***Proof*** The result follows from Lemmas 4.3.1, 4.3.8, 4.3.17, and 4.3.24. □

**Lemma 4.3.35** *Let $G = G^{\mathcal{B}}_{\mu,\nu}(X_1, \ldots, X_k)$. Then the following statements hold:*

*(i)* $X_1^\circ \times \cdots \times X_k^\circ \leq \pi(G) \leq X_1 \times \cdots \times X_k$.
*(ii) If $G$ is not of type $\mathcal{B}_3$, then $\pi(C_{G^\circ}(F_0)) = X_1^\circ \times \cdots \times X_k^\circ$.*
*(iii) Assume that $q$ is even and $\kappa$ is a quadratic form. Then $\pi(C_{G^\Omega}(F_0)) = X_1^\circ \times \cdots \times X_k^\circ$.*

***Proof*** Claims (i) and (ii) follow for the different types from Lemmas 4.3.5, 4.3.13, 4.3.22, and 4.3.27. In claim (iii), we have $C_{G^\Omega}(F_0) = C_{G^\circ}(F_0)$ by Lemmas 4.3.12 and 4.3.21. Thus $\pi(C_{G^\Omega}(F_0)) = X_1^\circ \times \cdots \times X_k^\circ$ follows from (ii). □

Since $G/G^\circ$ is cyclic, in view of Lemma 4.3.35 (i) it is not usually the case that $\pi(G) = X_1 \times \cdots \times X_k$. The results from the previous sections will allow us to give a more precise description of $\pi(G)$ (Lemmas 4.3.5, 4.3.13, 4.3.22, and 4.3.27).

Denote by $\pi_i$ the homomorphism $\pi_i : N_{\Delta(V,\kappa)}(F_0, R_1, \ldots, R_k) \to \mathrm{GSp}_{2\ell_i}(r_i)$, where $\pi_i(g)$ is the action of $g$ on $R_i/Z(R_i)$. Let $Z_i$ be the group of scalar matrices in $\mathrm{GL}_{2\ell_i}(r_i)$.

In the next lemma, we will classify the cases where $\pi_i(G) \leq X_i^\circ Z_i$. One motivation for this is the fact that there are some cases where $X_i$ is maximal solvable in $\mathrm{GSp}_{2\ell_i}(r_i)$, but $X_i^\circ$ is not maximal solvable in $\mathrm{Sp}_{2\ell_i}(r_i)$. (One such example will be given in Lemma 5.5.12.) In these cases if $\pi_i(G) \leq X_i^\circ Z_i$, one can prove that $G$ is not maximal solvable; we will see this later in the proof of Theorem 7.1.12.

**Lemma 4.3.36** *Let $G = G^{\mathcal{B}}_{\mu,\nu}(X_1, \ldots, X_k)$ be irreducible. Let $I$ be the set of indices $1 \leq i \leq k$ such that $X_i$ is of multiplier 2.*

*Then $\pi_i(G) \leq X_i^\circ Z_i$ if and only if one of the following conditions holds:*

*(i) $G$ is of type $\mathcal{B}_0$ or $\mathcal{B}_2$, and one of the following holds:*

*(a)* $\left(\frac{q}{r_i}\right) = +1$.

*(b)* $\left(\frac{q}{r_j}\right) = -1$ *for some* $j \in I$.

*(ii) $G$ is of type $\mathcal{B}_1$, and one of the following holds:*

*(a)* $\left(\frac{q}{r_i}\right) = \left(\frac{-1}{r_i}\right) = +1$, *and* $\left(\frac{q}{r_j}\right) = \left(\frac{-1}{r_j}\right) = +1$ *for all* $j \in I$.

*(b)* $\left(\frac{-1}{r_i}\right) = \left(\frac{-1}{r_j}\right) = +1$ *for all* $j \in I$, *and* $\left(\frac{q}{r_j}\right) = -1$ *for some* $j \in I$.

*(c)* $\left(\frac{-q}{r_i}\right) = \left(\frac{-q}{r_j}\right) = +1$ *for all* $j \in I$, *and* $\left(\frac{-1}{r_j}\right) = -1$ *for some* $j \in I$.

*(iii) $G$ is of type $\mathcal{B}_3$.*

***Proof*** Denote by $\tau_i$ the homomorphism $\tau_i : \mathrm{GSp}_{2\ell_i}(r_i) \to \mathbb{F}_{r_i}^\times$. Then $\pi_i(G) \leq X_i^\circ Z_i$ if and only if $\tau_i(\pi_i(g)) \in \left(\mathbb{F}_{r_i}^\times\right)^2$ for all $g \in G$. Note that $\pi_i(C_{I(V,\kappa)}(F_0)) \leq X_i^\circ$, since $Z(R_i) \leq F_0$.

We consider the three different possibilities case-by-case.

**Case 1: $G$ is of type $\mathcal{B}_0$ or $\mathcal{B}_2$**
In this case $\pi_i(G) = \pi_i(G^\circ)$.

Suppose first that $\left(\frac{q}{r_j}\right) = +$ for all $j \in I$. It follows from Lemmas 4.3.5 and 4.3.22 that $G^\circ = \langle C_{G^\circ}(F_0), g\psi\rangle$ for some $g \in C_{I(V,\kappa)}(F_0)$. Since $\tau_i(\pi_i(\psi)) = q$, we have $\pi_i(G) \leq X_i^\circ Z_i$ if and only if $\left(\frac{q}{r_i}\right) = +1$, as claimed by the lemma.

The other possibility is that $\left(\frac{q}{r_j}\right) = -$ for some $j \in I$. In this case it follows from Lemmas 4.3.5 and 4.3.22 that $G^\circ = \langle C_{G^\circ}(F_0), g\psi^2\rangle$ for some $g \in C_{I(V,\kappa)}(F_0)$. Therefore $\pi_i(G^\circ) \leq X_i^\circ Z_i$ in this case, as claimed.

**Case 2: $G$ is of type $\mathcal{B}_1$**
In this case $\pi_i(G) = \pi_i(G^\circ)$.

Suppose first that $\left(\frac{-1}{r_j}\right) = \left(\frac{q}{r_j}\right) = +$ for all $j \in I$. Then by Lemma 4.3.13, we have $G^\circ = \langle C_{G^\circ}(F_0), g\varphi, h\psi\rangle$ for some $g, h \in C_{I(V,\kappa)}(F_0)$. Thus $\pi_i(G^\circ) \leq X_i^\circ Z_i$ if and only if $\left(\frac{-1}{r_i}\right) = \left(\frac{q}{r_i}\right) = +$, as claimed by the lemma.

Suppose next that $\left(\frac{-1}{r_j}\right) = +$ for all $j \in I$, and $\left(\frac{q}{r_j}\right) = -$ for some $j \in I$. Then by Lemma 4.3.13, we have $G^\circ = \langle C_{G^\circ}(F_0), g\varphi, h\psi^2\rangle$ for some $g, h \in C_{I(V,\kappa)}(F_0)$. In this case $\pi_i(G^\circ) \leq X_i^\circ Z_i$ if and only if $\left(\frac{-1}{r_i}\right) = +$, as claimed.

It remains to consider the case where $\left(\frac{-1}{r_j}\right) = -1$ for some $j \in I$. Because $G$ is irreducible, it follows from Lemma 4.3.14 that $\left(\frac{-q}{r_j}\right) = +$ for all $j \in I$. In this case it follows from Lemma 4.3.13 that $G^\circ = \langle C_{G^\circ}(F_0), g\varphi\psi\rangle$ for some $g \in C_{I(V,\kappa)}(F_0)$. Thus $\pi_i(G^\circ) \leq X_i^\circ Z_i$ if and only if $\left(\frac{-q}{r_i}\right) = +$, as claimed.

**Case 3: $G$ is of type $\mathcal{B}_3$**
In this case $F_0 = Z(R_1) = \langle -I_V\rangle$, so $\pi_i(G) \leq X_i^\circ$. This completes the proof of the lemma.

□

## 4.4 Maximal Irreducible Solvable Subgroups of $\mathrm{GO}_n(q)$ for $n$ Odd

Suppose that $\kappa \neq 0$ is a quadratic form. As seen in Proposition 4.1.20, metrically primitive groups of types $\mathcal{B}_1, \mathcal{B}_2, \mathcal{B}_3$ exist in $\Delta(V, \kappa)$ only when $\dim V = n$ is even. As a consequence, we have the following result for maximal irreducible solvable subgroups of $\Delta(V, \kappa)$.

**Theorem 4.4.1** *Suppose that $\kappa \neq 0$ is a quadratic form and that $n$ is odd. Let $Z \leq \mathrm{GL}(V)$ be the group of scalar matrices. If $G \leq \Delta(V, \kappa)$ is maximal irreducible solvable, then all of the following statements hold:*

*(i)* $G = G^\circ Z$.
*(ii)* $G^\circ = O_1(q) \wr K$, *where $K \leq S_n$ is maximal transitive solvable.*

***Proof*** Suppose that $G \leq \Delta(V, \kappa)$ is maximal irreducible solvable. By maximality we have $Z \leq G$. Since $n$ is odd, by [55, (2.6.3), p. 34] we have $\tau(\Delta(V, \kappa)) = (\mathbb{F}_q^\times)^2$, so $\Delta(V, \kappa) = I(V, \kappa)Z$. Therefore $G = G^\circ Z$, so (i) holds.

When $n = 1$ we have $G = \mathrm{GO}_1(q)$ and $G^\circ = O_1(q)$, so (ii) holds. Suppose then that $n > 1$.

It follows from Proposition 4.1.20 that $G$ is metrically imprimitive. Since $n$ is odd, by Theorem 2.2.14 we have $G = (H \wr K) \cap \Delta(V, \kappa)$ as in Theorem 2.2.14 (i). Here $H$ is metrically primitive, so by Proposition 4.1.20 we have $H = \mathrm{GO}_1(q)$. It other words $G^\circ = O_1(q) \wr K$, where $K \leq S_n$ is maximal transitive solvable. This completes the proof of the theorem. □

**Remark 4.4.2** For irreducible solvable subgroups of $O(n, \mathbb{R})$, a result similar to Proposition 4.1.20 and Theorem 4.4.1 was proved by Gow in [29]. The main result of [29] implies that if $G \leq O(n, \mathbb{R})$ is irreducible and solvable, then $G$ is monomial. Consequently, every irreducible solvable subgroup of $O(n, \mathbb{R})$ is conjugate to a subgroup of $O_1(\mathbb{R}) \wr S_n$. This is generalized to all fields of odd characteristic in [56].

**Theorem 4.4.3** *Suppose that $\kappa \neq 0$ is a quadratic form and that $n$ is odd. Let $Z \leq \mathrm{GL}(V)$ be the group of scalar matrices. Suppose that $K \leq S_n$ maximal transitive solvable. Then all of the following statements hold:*

*(i) $O_1(q) \wr K$ is maximal irreducible solvable in $I(V, \kappa)$.*
*(ii) $(O_1(q) \wr K)Z$ is maximal irreducible solvable in $\Delta(V, \kappa)$.*

***Proof*** As noted in Theorem 4.4.1, by [55, (2.6.3), p. 34] we have $\Delta(V, \kappa) = I(V, \kappa)Z$. Thus $G \leq \Delta(V, \kappa)$ is maximal irreducible solvable if and only if $G = G^\circ Z$ and $G^\circ$ is maximal irreducible solvable in $I(V, \kappa)$. Therefore it will suffice to prove that (i) holds.

Let $X' \leq I(V, \kappa)$ be a maximal irreducible solvable subgroup such that $X \leq X'$. We will prove that $X = X'$, from which (i) follows. By Theorem 4.4.1 the subgroup $X'$ is of the form $X' = O_1(q) \wr K'$ for some maximal transitive solvable $K' \leq S_n$. Since $n$ is odd, it follows from [57, Theorem 1.1] that $X$ has a unique nonrefinable system of imprimitivity. Thus the system of imprimitivity defining $X$ is the same as the system of imprimitivity defining $X'$. This provides an embedding $K \leq K'$, so $K = K'$ by maximality of $K$. Therefore $X = X'$, as required. □

Suppose that $n$ and $q$ are odd. With Theorems 4.4.1 and 4.4.3, the classification of maximal irreducible solvable subgroups of $\mathrm{GO}_n(q)$ and $O_n(q)$ is completely reduced to the classification of maximal transitive solvable subgroups of $S_n$. This in turn is reduced to the classification of maximal irreducible solvable subgroups of $\mathrm{GL}_n(p)$ for $p$ prime, as we have seen earlier in Theorems 1.1.1 and 1.1.2. The classification of maximal irreducible solvable subgroups of $\mathrm{GL}_n(p)$ is a special case of the main results presented in this book.

# Chapter 5
# Basic Properties of $G^{\mathcal{B}}_{\mu,\nu}(X_1, \ldots, X_k)$

In the previous chapter, we have proved that every metrically completely reducible maximal irreducible solvable subgroup of $\Delta(V, \kappa)$ is conjugate to a subgroup of the form $G^{\mathcal{B}}_{\mu,\nu}(X_1, \ldots, X_k)$. Furthermore, we established some basic properties of groups of the form $G^{\mathcal{B}}_{\mu,\nu}(X_1, \ldots, X_k)$, such as their order.

In this chapter, we will establish some further properties of groups of the form $G^{\mathcal{B}}_{\mu,\nu}(X_1, \ldots, X_k)$. Specifically, in this chapter, we will:

- Determine when $G^{\mathcal{B}}_{\mu,\nu}(X_1, \ldots, X_k)$ is of multiplier 2 (Sect. 5.1).
- Determine when $G^{\mathcal{B}}_{\mu,\nu}(X_1, \ldots, X_k)$ is irreducible (Sect. 5.2).
- Determine when $G^{\mathcal{B}}_{\mu,\nu}(X_1, \ldots, X_k)$ is irreducible and metrically primitive (Sect. 5.6).
- Show that the subgroup $F_0$ used in the construction of $G^{\mathcal{B}}_{\mu,\nu}(X_1, \ldots, X_k)$ satisfies properties (F1)–(F3) (Sect. 5.3).
- Show that the subgroup $A$ used in the construction of $G^{\mathcal{B}}_{\mu,\nu}(X_1, \ldots, X_k)$ satisfies properties (A1)–(A3) (Sect. 5.3).
- Prove uniqueness of $F_0$ and $A$ in the case where $G^{\mathcal{B}}_{\mu,\nu}(X_1, \ldots, X_k)$ is metrically primitive and irreducible (Sect. 5.7).
- Describe when two metrically primitive irreducible subgroups of the form $G^{\mathcal{B}}_{\mu,\nu}(X_1, \ldots, X_k)$ are conjugate in $\Delta(V, \kappa)$ (Sect. 5.8).
- Provide some families of examples where $G^{\mathcal{B}}_{\mu,\nu}(X_1, \ldots, X_k)$ is not maximal solvable (Sect. 5.5).

Additionally, in Sect. 5.4 we will provide examples of the construction of maximal irreducible solvable subgroups in some special cases.



M. Korhonen, *Maximal Solvable Subgroups of Finite Classical Groups*, Lecture Notes in Mathematics 2346, https://doi.org/10.1007/978-3-031-62915-0_5

## 5.1 Maximal Irreducible Solvable Subgroups of Multiplier 2

In this section, we will complete the description of maximal irreducible solvable groups of multiplier 2. As seen earlier, this information is needed at various point in the recursive construction of maximal irreducible subgroups. For example, groups of multiplier 2 appear in the construction of semiprimary groups (Example 2.2.7), and in necessary conditions on $G^{\mathcal{B}}_{\mu,\nu}(X_1, \ldots, X_k)$ such as Lemma 4.3.14. We will also need to know the multiplier to determine the conjugacy classes of maximal irreducible solvable subgroups of $I(V, \kappa)$ (Lemma 2.1.10).

A semiprimary group is of multiplier 1 (Example 2.2.7). Therefore, an imprimitive maximal irreducible solvable $G \leq \Delta(V, \kappa)$ of multiplier 2 must be of the form $G = (H \wr K) \cap \Delta(V, \kappa)$, where $H$ is metrically primitive maximal irreducible solvable of multiplier 2.

Thus the question of multipliers reduces to the case of metrically primitive groups. By Lemmas 4.3.13 and 4.3.22, a group $G^{\mathcal{B}}_{\mu,\nu}(X_1, \ldots, X_k)$ of multiplier 2 must be of type $\mathcal{B}_3$. For these groups we will first prove the following.

**Lemma 5.1.1** *Let* $G = G^{\mathcal{B}_3}_{\mu,\nu}(X)$ *be of type* $\mathcal{B}_3$. *Then* $G$ *is of multiplier* 2 *if and only if one of the following conditions holds:*

*(i)* 2 *is a square in* $\mathbb{F}_q$.
*(ii)* $X \leq \Omega^{\varepsilon}_{2\ell}(2)$.

***Proof*** Let $Z$ be the group of scalar matrices in $\mathrm{GL}(V)$. Suppose first that (i) or (ii) holds. Then by Lemma 4.3.27 (i)–(iii) we have $\pi(G) = \pi(G^\circ) = \pi(G^\circ Z)$. Therefore $G = G^\circ Z$ by Lemma 4.3.25, so $G$ is of multiplier 2.

Conversely, suppose that $G$ is of multiplier 2, in which case $G = G^\circ Z$ and thus $\pi(G) = \pi(G^\circ)$. Suppose that 2 is not a square in $\mathbb{F}_q$. Then by Lemma 4.3.24 we have $\pi(G) = X$ and $\pi(G^\circ) = X^\Omega$. Therefore $X = X^\Omega$, in which case (ii) holds. □

In order to make Lemma 5.1.1 more precise, we will determine when a maximal irreducible solvable subgroup of $O^\pm_n(q)$ ($q$ even) is contained in $\Omega^\pm_n(q)$. This was originally determined by Jordan for $q = 2$ in [51, p. 39], [52, §34].

In the metrically imprimitive case we only need to consider isometric imprimitive subgroups, since semiprimary groups are only defined in odd characteristic. Then for subgroups of the form $G = H \wr K \leq O^\pm_n(q)$, we have $G = G^\Omega$ if and only if $H = H^\Omega$ (Lemma 2.2.20 (iii)). Thus it suffices to consider the metrically primitive case, which we will consider next.

**Lemma 5.1.2** *Suppose that* $n$ *and* $q$ *are even, and let* $G = G^{\mathcal{B}_1}_{\mu,\nu}(X_1, \ldots, X_k)$ *be a group of type* $\mathcal{B}_1$ *in* $\mathrm{GO}^+_n(q)$. *Assume that* $G$ *is irreducible. Then* $G^\circ \leq \Omega^+_n(q)$ *if and only if* $n/2$ *is even.*

***Proof*** Every element of $G^\circ$ is of the form $g_A \varphi^\alpha \psi^\beta$. If $n/2$ is even, then $g_A, \varphi, \psi \in \Omega^+_n(q)$ by Lemma 4.3.12 so $G^\circ \leq \Omega^+_n(q)$. If $n/2$ is odd, we have $g_A, \psi \in \Omega^+_n(q)$

and $\varphi \notin \Omega_n^+(q)$ by Lemma 4.3.12. By irreducibility $G^\circ$ must contain an element of the form $g_A\varphi\psi^\beta$, so we conclude that $G^\circ \not\leq \Omega_n^+(q)$. □

**Lemma 5.1.3** *Suppose that $n$ and $q$ are even, and let $G \leq \mathrm{GO}_n^\varepsilon(q)$ be a metrically primitive maximal irreducible solvable group of type $\mathcal{B}_2$. Then $\varepsilon = -$, and $G^\circ \not\leq \Omega_n^-(q)$.*

***Proof*** We use the notation and setup as described in Sect. 4.3.3, starting with $n = 2\nu\mu$. As noted in the proof of Lemma 4.3.21, we know that $\mu$ is odd and thus $\varepsilon = (-1)^\mu = -1$ (Lemma 2.7.2) as claimed.

Next note that every element of $G^\circ$ can be written in the form $g_A\psi^\beta$ for some $0 \leq \beta < 2\nu$. By Lemma 4.3.21 we have $g_A \in \Omega_n^-(q)$ and $\psi \notin \Omega_n^-(q)$, so by Lemma 4.3.22 we have $G^\circ \not\leq \Omega_n^-(q)$ if and only if $G^\circ$ contains an element of the form $g_A\psi$.

To prove that $G^\circ \not\leq \Omega_n^-(q)$, we proceed by induction on $n$. The claim is obvious for $n = 2$, since in this case $G = \mathrm{GO}_2^-(q)$. Suppose then that $n > 2$. If $\mu = 1$, we have $\psi \in G^\circ$, so $G^\circ \not\leq \Omega_n^-(q)$. Suppose then that $\mu > 1$ and write $G = G_{\mu,\nu}^{\mathcal{B}_2}(X_1, \ldots, X_k)$, where $X_i \leq \mathrm{GSp}_{2\ell_i}(r_i)$ for all $1 \leq i \leq k$ and $\mu = r_1^{\ell_1} \cdots r_k^{\ell_k}$.

If $\left(\dfrac{q}{r_i}\right) = +$ for all $X_i$ of multiplier 2, it follows from Lemma 4.3.22 that $G^\circ$ contains an element of the form $g_A\psi$, and thus $G^\circ \not\leq \Omega_n^-(q)$.

Consider then the possibility that $\left(\dfrac{q}{r_i}\right) = -$ for some $X_i$ of multiplier 2, in which case $\left(\dfrac{2}{r_i}\right) = -$ since $q$ is even. By Lemma 2.3.9, we have

$$X_i = (Y_1 \times \cdots \times Y_t) \cap \mathrm{GSp}_{2\ell_i}(r_i),$$

where $Y_j \leq \mathrm{GSp}_{2n_j}(r_i)$ is maximal irreducible solvable for all $1 \leq j \leq t$, and $\sum_{1\leq j\leq t} n_j = \ell_i$.

Because $X_i$ is of multiplier 2, there exists some $1 \leq j \leq t$ such that $Y_j$ is of multiplier 2. Then $Y_j = Y_j^\circ Z$, where $Z$ is the group of scalar matrices in $\mathrm{GSp}_{2n_j}(r_i)$. Furthermore, by Theorem 2.2.14 we can write

$$Y_j^\circ = Y^\circ \wr T,$$

where $Y \leq \mathrm{GSp}_{2\ell'}(r_i)$ is metrically primitive maximal irreducible solvable and $T \leq S_{\ell''}$, with possibly $\ell'' = 1$. Here $2n_j = 2\ell'\ell''$, and $Y$ is of multiplier 2 since $Y_j$ is.

Since $Y$ is of multiplier 2, it must be of type $\mathcal{B}_3$. Therefore $2\ell' = 2^\alpha$ and $Y = G_{2\ell',1}^{\mathcal{B}_3}(\Sigma)$, where $\Sigma$ is metrically completely reducible maximal solvable in $O_{2\alpha}^{\varepsilon'}(2)$. Because $Y$ is of multiplier 2, by Lemma 5.1.1 we must have $\Sigma \leq \Omega_{2\alpha}^{\varepsilon'}(2)$. By induction $\varepsilon' = +$, but this is in contradiction with the fact that $Y \leq \mathrm{GSp}_{2\ell'}(r_i)$. □

**Lemma 5.1.4** *Suppose that n and q are even, and let $G \leq \mathrm{GO}^{\varepsilon}_n(q)$ be metrically completely reducible maximal solvable. If $G^\circ \leq \Omega^{\varepsilon}_n(q)$, then $\varepsilon = +$.*

***Proof*** Let $V = \mathbb{F}_q^n$. Since $q$ is even, we have $G^\circ$ maximal solvable in $O^{\varepsilon}_n(q)$. It follows from Lemma 2.3.7 that $G^\circ = X_1 \times \cdots \times X_t$, where $X_i \leq O^{\varepsilon_i}_{2\ell_i}(q)$ is maximal irreducible solvable for all $1 \leq i \leq t$. Here $\ell_i = \ell_i'\ell_i''$ and $\ell_i'' \geq 1$.

We have $G^\circ = G^\Omega$ if and only if $X_i = X_i^\Omega$ for all $1 \leq i \leq t$, so it suffices to consider the case where $G$ is maximal irreducible solvable. In this case $G^\circ = H \wr K$, where $H \leq O^{\varepsilon'}_{n'}(q)$ is metrically primitive maximal irreducible solvable, and $K \leq S_\ell$ is maximal transitive solvable (Theorem 2.2.14). Here $n = n'\ell$ with $\ell \geq 1$, and $\varepsilon = (\varepsilon')^\ell$.

Assume that $G^\circ \leq \Omega^{\varepsilon}_n(q)$. Then $H \leq \Omega^{\varepsilon'}_{n'}(q)$ by Lemma 2.2.20 (iii), so $\varepsilon' = +$ by Lemma 5.1.3. Thus $\varepsilon = (\varepsilon')^\ell = +$, as claimed. □

**Lemma 5.1.5** *Let $G = G^{\mathcal{B}_3}_{\mu,\nu}(X)$ be of type $\mathcal{B}_3$ in $\Delta(V, \kappa)$. Then G is of multiplier 2 if and only if one of the following conditions holds:*

*(i) 2 is a square in $\mathbb{F}_q$.*
*(ii) $\kappa$ is a quadratic form and $X \leq \Omega^{+}_{2\ell}(2)$.*

*In particular, if $\kappa$ is an alternating bilinear form, then G is of multiplier 2 if and only if 2 is a square in $\mathbb{F}_q$.*

***Proof*** If (i) or (ii) holds, it follows from Lemma 5.1.1 that $G$ is of multiplier 2.

Conversely, suppose that $G$ is of multiplier 2. If 2 is not a square in $\mathbb{F}_q$, by Lemma 5.1.1 we have $X \leq \Omega^{\varepsilon}_{2\ell}(2)$. It follows from Lemma 5.1.4 that $\varepsilon = +$, in which case $\kappa$ must be a quadratic form. □

**Lemma 5.1.6** *Suppose that n is even. Let $G \leq \mathrm{GSp}_n(q)$ be metrically completely reducible maximal solvable. If G is of multiplier 2, then 2 is a square in $\mathbb{F}_q$.*

***Proof*** By Lemma 2.3.9, we have $G = (H_1 \times \cdots \times H_t) \cap \mathrm{GSp}_n(q)$, where $H_i \leq \mathrm{GSp}_{n_i}(q)$ is maximal irreducible solvable for all $1 \leq i \leq t$. Here $n_i$ is even for all $i$ and $n = n_1 + \cdots + n_t$.

If $G$ is of multiplier 2, then $H_i$ must be of multiplier 2 for some $1 \leq i \leq t$. Therefore it suffices to consider the case where $G$ is irreducible of multiplier 2.

Groups of semiprimary type are of multiplier 1 (Theorem 2.2.19). Thus by Theorems 2.2.14 and 2.2.18, it suffices to consider the case where $G$ is metrically primitive. In this case $G$ must be of type $\mathcal{B}_3$ by Lemmas 4.3.13 and 4.3.22. For metrically primitive $G$ of type $\mathcal{B}_3$, the result is given by Lemma 5.1.5. □

**Lemma 5.1.7** *Suppose that n and q are even, and that $\kappa$ is a quadratic form. Let $G = G^{\mathcal{B}_1}_{\mu,\nu}(X_1, \ldots, X_k)$ be irreducible of type $\mathcal{B}_1$. Then the following statements hold:*

*(i) $G^\circ = \langle C_{G^\circ}(F_0), g\varphi, h\psi\rangle$ for some $g, h \in C_{I(V,\kappa)}(F_0)$.*
*(ii) $G^\Omega = G^\circ$ if $n/2$ is even.*
*(iii) $G^\Omega = \langle C_{G^\circ}(F_0), h\psi\rangle$ if $n/2$ is odd.*

***Proof*** Since $q$ is even, by Lemma 5.1.6 we have $\left(\frac{q}{r_i}\right) = +1$ for all $1 \leq i \leq k$ such that $X_i$ is of multiplier 2. Furthermore we are assuming that $G$ is irreducible, so by Lemma 4.3.14 we have $\left(\frac{-1}{r_i}\right) = +1$ for all $1 \leq i \leq k$ such that $X_i$ is of multiplier 2. Thus it follows from Lemma 4.3.13 (iii) that $g\varphi, h\psi \in G^\circ$ for some $g, h \in C_{I(V,\kappa)}(F_0)$. Because $N_{I(V,\kappa)}(F_0)/C_{I(V,\kappa)}(F_0)$ is generated by the images of $\varphi$ and $\psi$, we conclude that (i) holds.

If $n/2$ is even, then $G^\Omega = G^\circ$ by Lemma 5.1.2, so (ii) holds. For (iii), suppose that $n/2$ is odd. By (i), we have

$$G^\circ/C_{G^\circ}(F_0) = \langle \overline{\varphi}, \overline{\psi} \rangle \cong C_2 \times C_\nu.$$

Since $\psi \in G^\Omega$ and $\varphi \notin G^\Omega$ (Lemma 4.3.12), we conclude that $G^\Omega = \langle C_{G^\circ}(F_0), h\psi \rangle$. □

**Lemma 5.1.8** *Suppose that $n$ and $q$ are even, and that $\kappa$ is a quadratic form. Let $G = G^{\mathcal{B}_2}_{\mu,\nu}(X_1, \ldots, X_k)$ be irreducible of type $\mathcal{B}_2$. Then the following statements hold:*

*(i) $G^\circ = \langle C_{G^\circ}(F_0), g\psi \rangle$ for some $g \in C_{I(V,\kappa)}(F_0)$.*
*(ii) $G^\Omega = \langle C_{G^\circ}(F_0), h\psi^2 \rangle$ for some $h \in C_{I(V,\kappa)}(F_0)$.*

***Proof*** Since $q$ is even, by Lemma 5.1.6 we have $\left(\frac{q}{r_i}\right) = +1$ for all $1 \leq i \leq k$ such that $X_i$ is of multiplier 2. Thus it follows from Lemma 4.3.22 that $g\psi \in G^\circ$ for some $g \in C_{I(V,\kappa)}(F_0)$.

Since $N_{I(V,\kappa)}(F_0)/C_{I(V,\kappa)}(F_0)$ is generated by the image of $\psi$, we conclude that $G^\circ = \langle C_{G^\circ}(F_0), g\psi \rangle$, as claimed by (i). Then we have $G^\circ/C_{G^\circ}(F_0) \cong C_{2\nu}$. Since $\psi \notin G^\Omega$ (Lemma 4.3.21 (ii)), we conclude that

$$G^\Omega = \langle C_{G^\circ}(F_0), (g\psi)^2 \rangle = \langle C_{G^\circ}(F_0), h\psi^2 \rangle$$

for some $h \in C_{I(V,\kappa)}(F_0)$. □

## 5.2 Irreducibility of $G^{\mathcal{B}}_{\mu,\nu}(X_1, \ldots, X_k)$

In this section, we will prove that a group of the form $G = G^{\mathcal{B}}_{\mu,\nu}(X_1, \ldots, X_k)$ is irreducible if and only if the assumptions of Table 4.1 are satisfied. We will also see that $G^\circ$ is irreducible when assumptions of Table 4.1 hold, and for orthogonal groups in characteristic two we shall determine when $G^\Omega$ is irreducible.

**Lemma 5.2.1** *Let $G = G^{\mathcal{B}}_{\mu,\nu}(X_1, \ldots, X_k) \leq \Delta(V, \kappa)$. If $G$ is irreducible, then assumptions of Table 4.1 hold.*

***Proof*** The assumption $(r_i, \ell_i, \varepsilon_i) \neq (2, 1, +)$ is necessary since $O_2^+(2)$ is not metrically completely reducible. The remaining assumptions are necessary by definition, and by Lemma 4.3.14 for groups of type $\mathcal{B}_1$. □

**Lemma 5.2.2** *Let* $G = G^{\mathcal{B}}_{\mu,\nu}(X_1, \ldots, X_k) \leq \Delta(V, \kappa)$. *Then the following hold:*

*(i) A is absolutely irreducible as a subgroup of* $\mathrm{GL}(W'_\lambda)$.
*(ii) The action of A is faithful and absolutely irreducible on any eigenspace of f on* $V' = \mathbb{K} \otimes_{\mathbb{F}_q} V$.
*(iii) Suppose that assumptions of Table 4.1 hold. Then* $G^\circ$ *is an irreducible subgroup of* $I(V, \kappa)$.

***Proof***

(i) $A \leq C_{G^\circ}(F_0)$ is absolutely irreducible on $\mathrm{GL}(W'_\lambda)$, since a tensor product of absolutely irreducible representations is absolutely irreducible (Lemma 2.1.16 (i)).
(ii) Note that $\psi$ normalizes $A$ and $\psi^k(W'_\lambda) = W'_{\lambda^{q^k}}$, it follows that as a $\mathbb{K}[A]$-module $W'_{\lambda^{q^k}} \downarrow A$ is isomorphic to $W'_\lambda \downarrow A$ twisted by the automorphism $x \mapsto \psi^{-k} x \psi^k$ of $A$. Hence $W'_{\lambda^{q^k}} \downarrow A$ is absolutely irreducible for all $0 \leq k < [\mathbb{K} : \mathbb{F}_q]$.
This proves the claim in all cases, except for case $\mathcal{B}_1$. In this case, note that $W'_{\lambda^{-q^k}} \downarrow A \cong (W'_{\lambda^{q^k}})^* \downarrow A$ for all $0 \leq k < [\mathbb{K} : \mathbb{F}_q]$. Hence $W'_{\lambda^{-q^k}} \downarrow A$ is absolutely irreducible for all $0 \leq k < [\mathbb{K} : \mathbb{F}_q]$.
(iii) Let $W \subseteq V$ be a nonzero $\mathbb{F}_q[G^\circ]$-submodule. Denote $W' = \mathbb{K} \otimes_{\mathbb{F}_q} W$. Since $W'$ is $f$-invariant and $f$ acts diagonalizably on $V'$, it follows that $W'$ contains an eigenvector $w \neq 0$ of $f$. By (ii), the action of $A$ on $w$ produces the entire eigenspace, so $W'$ contains an eigenspace of $f$, say $W' \supseteq W'_\alpha$.
Now $W'$ is defined over $\mathbb{F}_q$, so by Lemma 2.4.4 it contains $W'_{\alpha^{q^k}}$ for all $k \geq 0$. In cases $\mathcal{B}_0$, $\mathcal{B}_2$, and $\mathcal{B}_3$ this proves that $W' = V'$, and thus $W = V$. In case $\mathcal{B}_1$ it follows that $W'$ contains $Q_\lambda$ or $Q_{\lambda^{-1}}$. By the assumptions required for groups of type $\mathcal{B}_1$, we know that $G^\circ$ contains an element that swaps $Q_\lambda$ and $Q_{\lambda^{-1}}$ (Lemma 4.3.14). Thus we conclude that $W' = V'$ and $W = V$ in this case as well.

□

**Example 5.2.3** Since 2 is a square in $\mathbb{F}_7$, in $\mathrm{GSp}_2(7)$ the group

$$X = G^{\mathcal{B}_3}_{2,1}(O_2^-(2))$$

is metrically primitive irreducible of multiplier 2 (Lemma 5.1.5). It will follow from Theorem 7.4.6 proven later that $X$ is maximal solvable. (Alternatively, it is easy to see directly from the list of possible maximal irreducible solvable subgroups of $\mathrm{GSp}_2(7)$ that $X$ is maximal solvable. See also [77, Theorem 21.6].)

Let $q$ be odd and suppose that $q \equiv 1 \mod 7$ (for example $q = 29$ or $q = 43$). Then in $\mathrm{GSp}_{14}(q)$, we have a group of type $\mathcal{B}_1$ defined as $G = G^{\mathcal{B}_1}_{7,1}(X)$. Because $X$ is of multiplier 2 and $\left(\frac{-1}{7}\right) = \left(\frac{-q}{7}\right) = -$, it follows from Lemma 4.3.14 that $G$ is not irreducible.

**Lemma 5.2.4** *Assume that $n$ and $q$ are even, and that $\kappa$ is a quadratic form. Let $G = G^{\mathcal{B}}_{\mu,\nu}(X_1, \ldots, X_k)$ be irreducible. Then $G^{\Omega}$ is irreducible, except when the all of the following conditions hold:*

*(i) $\varepsilon = +$ and $n/2$ is odd;*
*(ii) $G$ is of type $\mathcal{B}_1$.*

*In the exceptional case, we have $V = Q_\lambda \oplus Q_{\lambda^{-1}}$ with $Q_\lambda$ and $Q_{\lambda^{-1}}$ nonisomorphic irreducible $\mathbb{F}_q[G^{\Omega}]$-modules.*

***Proof*** Let $W \subseteq V$ be a nonzero $\mathbb{F}_q[G^{\Omega}]$-module. Since $A \leq G^{\Omega}$ (Lemma 4.3.12 (i) and 4.3.21 (i)), arguing as in the proof of Lemma 5.2.2 (iii) shows that $V = W$ if $G$ is of type $\mathcal{B}_2$. When $G$ is of type $\mathcal{B}_1$, we see similarly that $W$ contains $Q_\lambda$ or $Q_{\lambda^{-1}}$.

Suppose then that $G$ is of type $\mathcal{B}_1$, in which case $\varepsilon = +$. Because $G$ is irreducible, it contains an element which swaps $Q_\lambda$ and $Q_{\lambda^{-1}}$ (Lemma 4.3.14), and any such element is of the form $g\chi\varphi\psi^{\beta}$. If $n/2$ is even, it follows from Lemma 4.3.12 that $g\chi\varphi\psi^{\beta} \in G^{\Omega}$, so $V = W$.

Suppose then that $n/2$ is odd. In this case it follows from Lemma 4.3.12 that elements of $G^{\Omega}$ leave $Q_\lambda$ and $Q_{\lambda^{-1}}$ invariant. We have shown that any nonzero $G^{\Omega}$-invariant subspace in $V$ is equal to $Q_\lambda$, $Q_{\lambda^{-1}}$, or $V$. Thus $V = Q_\lambda \oplus Q_{\lambda^{-1}}$ with $Q_\lambda$ and $Q_{\lambda^{-1}}$ irreducible $\mathbb{F}_q[G^{\Omega}]$-modules. Furthermore $Q_\lambda \not\cong Q_{\lambda^{-1}}$ as $\mathbb{F}_q[G^{\Omega}]$-modules, since they are not isomorphic as $\mathbb{F}_q[F_0]$-modules. □

**Remark 5.2.5** In the exceptional case of Lemma 5.2.4, the subgroup $G^{\Omega}$ is not maximal solvable if $n > 2$. Indeed, in this case we have a totally singular decomposition $V \downarrow G^{\Omega} = Q_\lambda \oplus Q_{\lambda^{-1}}$ with $Q_{\lambda^{-1}} \cong Q_\lambda^*$ irreducible $\mathbb{F}_q[G^{\Omega}]$-modules. Then with respect to a suitable basis, every element of $G^{\Omega}$ is of the form

$$\begin{pmatrix} B & 0 \\ 0 & B^{-T} \end{pmatrix}$$

for some $B \in \mathrm{GL}_{n/2}(q)$. These normalize

$$U = \left\{ \begin{pmatrix} I & X \\ 0 & I \end{pmatrix} : X^T = X \text{ and } X_{ii} = 0 \text{ for all } i \right\},$$

which can be shown to be a subgroup of $\Omega(V, \kappa)$. As a group of unitriangular matrices $U$ is nilpotent. Therefore $G^{\Omega}$ is not maximal solvable if $n > 2$, since it is properly contained in the solvable group $G^{\Omega}U \leq \Omega(V, \kappa)$.

## 5.3 Properties of $F_0$ and $A$ in $G^{\mathcal{B}}_{\mu,\nu}(X_1, \ldots, X_k)$

We continue to study the basic properties of groups of the form $G^{\mathcal{B}}_{\mu,\nu}(X_1, \ldots, X_k)$. In this section, we will show that the subgroups $F_0$ and $A$ satisfy various properties that we would expect in view of previous analysis, for example the fact that $A = \mathrm{Fit}(C_{G^\circ}(F_0))$ (Lemma 4.1.4).

**Lemma 5.3.1** *Let $G = G^{\mathcal{B}}_{\mu,\nu}(X_1, \ldots, X_k)$ and $1 \leq i \leq k$. Then the following statements hold:*

*(i) The action of $C_G(F_0)$ on $R_i/Z(R_i)$ is metrically completely reducible.*
*(ii) Suppose that $G$ is not of type $\mathcal{B}_3$. Then the action of $C_{G^\circ}(F_0)$ on $R_i/Z(R_i)$ is metrically completely reducible.*
*(iii) Suppose that $n$ and $q$ are even and that $\kappa$ is a quadratic form. Then the action of $C_{G^\Omega}(F_0)$ on $R_i/Z(R_i)$ is metrically completely reducible.*

***Proof*** We first prove (ii), so suppose that $G$ is not of type $\mathcal{B}_3$. It follows from Lemma 4.3.5 (i), Lemma 4.3.13 (ii) and Lemma 4.3.22 that the action of $C_{G^\circ}(F_0)$ on $R_i/Z(R_i)$ is equal to $X_i^\circ$. It follows from Lemma 2.3.9 and Lemma 2.3.10 (i) that $X_i^\circ$ is metrically completely reducible, so (ii) holds.

By (ii), it suffices to prove (i) in the case where $G$ is of type $\mathcal{B}_3$. In this case it follows from Lemma 4.3.27 that the action of $C_G(F_0) = G$ on $R_i/Z(R_i)$ is equal to $X_i$, which is metrically completely reducible.

For (iii), suppose that $n$ and $q$ are even and that $\kappa$ is a quadratic form. We first note that the action of $C_{G^\Omega}(F_0)$ on $R_i/Z(R_i)$ is equal to $X_i^\circ$. For type $\mathcal{B}_1$ this follows from Lemma 4.3.12 (i), Lemma 4.3.8, and for type $\mathcal{B}_2$ from Lemma 4.3.21 (i), Lemma 4.3.17. As noted in the beginning of the proof, the action of $X_i^\circ$ is metrically completely reducible, so we conclude that (iii) holds. □

**Lemma 5.3.2** *Let $G = G^{\mathcal{B}}_{\mu,\nu}(X_1, \ldots, X_k)$. Suppose that $N \trianglelefteq C_{G^\circ}(F_0)$ is nilpotent. Then $N \leq A$.*

***Proof*** Let $\mu = r_1^{\ell_1} \cdots r_k^{\ell_k}$ be the prime factorization of $\mu$, with $r_1 < \cdots < r_k$ primes. Then $A = F_0 R_1 \cdots R_k$ with $R_i \trianglelefteq G$ extraspecial of order $r_i^{1+2\ell_i}$.

For $1 \leq i \leq k$, we can write $N = O_{r_i}(N) \times O_{r_i'}(N)$ since $N$ is nilpotent. Since $O_{r_i'}(N)$ and $R_i$ are normal subgroups of $C_{G^\circ}(F_0)$ with coprime order, they commute. The action of $G$ on $R_i/Z(R_i)$ is completely reducible by Lemma 5.3.1. Furthermore $O_{r_i}(N) \trianglelefteq C_{G^\circ}(F_0) \trianglelefteq G$, so by Clifford's theorem the action of $O_{r_i}(N)$ is completely reducible on $R_i/Z(R_i)$. But $O_{r_i}(N)$ is an $r_i$-group, so we conclude that $O_{r_i}(N)$ acts trivially on $R_i/Z(R_i)$.

We have shown that $N$ acts trivially on $R_i/Z(R_i)$ for all $1 \leq i \leq k$. Since $N \leq C_{G^\circ}(F_0)$, it follows from Lemma 4.3.32 that $N \leq A$. □

**Lemma 5.3.3** *Let $G = G^{\mathcal{B}}_{\mu,\nu}(X_1, \ldots, X_k)$. Assume that $n$ and $q$ are even, and that $\kappa$ is a quadratic form. Suppose that $N \trianglelefteq C_{G^\Omega}(F_0)$ is nilpotent. Then $N \leq A$.*

***Proof*** As in the proof of Lemma 5.3.2, write $N = O_{r_i}(N) \times O_{r_i'}(N)$ for $1 \leq i \leq k$. We have $O_{r_i}(N) \trianglelefteq C_{G^\Omega}(F_0) \trianglelefteq G$, so arguing as in the proof of Lemma 5.3.2 shows that $N$ acts trivially on $R_i/Z(R_i)$ for all $1 \leq i \leq k$. Therefore $N \leq A$ by Lemma 4.3.32, as required. □

As a corollary of Lemma 5.3.2, we have the following results.

**Corollary 5.3.4** *Let $G = G^{\mathcal{B}}_{\mu,\nu}(X_1,\ldots,X_k)$. Then $F_0$ is maximal among the abelian subgroups of $G^\circ$ that are normalized by $C_G(F_0)$.*

***Proof*** Suppose that $F_0 \leq F \leq G^\circ$ is abelian and that $F \trianglelefteq C_G(F_0)$. By Lemma 5.3.2 we have $F \leq A$.

We have

$$A/F_0 = O_{r_1}(A/F_0) \times \cdots \times O_{r_k}(A/F_0),$$

where $O_{r_i}(A/F_0) = R_iF_0/F_0$ for all $1 \leq i \leq k$. Then

$$F/F_0 = O_{r_1}(F/F_0) \times \cdots \times O_{r_k}(F/F_0).$$

Because $F$ is abelian, each $O_{r_i}(F/F_0)$ is a totally isotropic $C_G(F_0)$-invariant subspace of $O_{r_i}(A/F_0)$. It follows from Lemma 5.3.1 (i) that the action of $C_G(F_0)$ on $O_{r_i}(F/F_0) = R_iF_0/F_0 \cong R_i/Z(R_i)$ is metrically completely reducible. Thus $O_{r_i}(F/F_0)$ is trivial for all $1 \leq i \leq k$, from which we conclude that $F = F_0$. □

**Corollary 5.3.5** *Let $G = G^{\mathcal{B}}_{\mu,\nu}(X_1,\ldots,X_k)$. Then the subgroup $F_0$ satisfies properties (F1)–(F3).*

***Proof*** Properties (F1)–(F2) are satisfied by definition, and (F3) is immediate from Corollary 5.3.4. □

**Corollary 5.3.6** *Let $G = G^{\mathcal{B}}_{\mu,\nu}(X_1,\ldots,X_k)$. Then the subgroup $A$ satisfies properties (A1)–(A3). Furthermore, we have $A = \mathrm{Fit}(C_{G^\circ}(F_0))$.*

***Proof*** Property (A1) is clear from the construction of $G$, and (A2) holds since $A$ is a central product of extraspecial groups. For (A3), suppose that $A \leq B \leq C_{G^\circ}(F_0)$ is such that $B \trianglelefteq G$ and $[B, B] \leq F_0$. Then $B$ is nilpotent, so $B = A$ by Lemma 5.3.2. Similarly $A = \mathrm{Fit}(C_{G^\circ}(F_0))$ by Lemma 5.3.2. □

**Lemma 5.3.7** *Let $G = G^{\mathcal{B}}_{\mu,\nu}(X_1,\ldots,X_k)$. The subgroup $F_0$ is maximal abelian normal in $C_{G^\circ}(F_0)$, except when all of the following conditions hold:*

*(i) $G = G^{\mathcal{B}_3}_{\mu,\nu}(X)$ is of type $\mathcal{B}_3$;*
*(ii) 2 is not a square in $\mathbb{F}_q$;*
*(iii) $X^\Omega \leq \Omega^\varepsilon_{2\ell}(2)$ is not metrically completely reducible.*

*Furthermore, if conditions (i)–(iii) hold, then $G^\circ$ is metrically imprimitive, $\varepsilon = +$, and $\kappa$ is a quadratic form.*

***Proof*** Suppose that $F_0 \leq F \leq G^\circ$ such that $F$ is abelian and $F \trianglelefteq C_{G^\circ}(F_0)$. By Lemma 5.3.2 we have $F \leq A$.

Suppose first that $G$ is not of type $\mathcal{B}_3$. Then by Lemma 5.3.1 (ii), the action of $C_{G^\circ}(F_0)$ on $O_{r_i}(A/F_0) = R_i F_0/F_0 \cong R_i/Z(R_i)$ is metrically completely reducible for all $1 \leq i \leq k$. Thus arguing as in the proof of Corollary 5.3.4 shows that $F = F_0$.

Suppose then that $G$ is of type $\mathcal{B}_3$, so $G = G^{\mathcal{B}_3}_{\mu,\nu}(X)$ and $C_{G^\circ}(F_0) = G^\circ$. Here $\mu = 2^\ell$, $\nu = 1$, and $X$ is metrically completely reducible maximal solvable in $O^{\varepsilon}_{2\ell}(2)$.

If 2 is a square in $\mathbb{F}_q$, by Lemma 4.3.27 the action of $G^\circ$ on $A/F_0 = R/Z(R)$ is equal to $X$, which is metrically completely reducible. Since $F/F_0$ is a totally isotropic $G^\circ$-invariant subspace of $A/F_0$, it follows that $F = F_0$.

Suppose then that 2 is not a square in $\mathbb{F}_q$. In this case, by Lemma 4.3.27 the action of $G^\circ$ on $R/Z(R)$ is equal to $X^\Omega$. Therefore if $X^\Omega$ is metrically completely reducible, we conclude that $F = F_0$. We have proved that $F_0$ is maximal abelian normal in $C_{G^\circ}(F_0)$, except possibly when all of (i)–(iii) hold.

Conversely, suppose that (i)–(iii) hold. Let $\vartheta$ be the $G$-invariant quadratic form on $A/F_0 = R/Z(R)$. By Lemma 2.3.7, we have $X = Y_1 \times \cdots \times Y_t$ with respect to a decomposition

$$A/F_0 = W_1 \perp \cdots \perp W_t$$

with $Y_i \leq O(W_i, \vartheta) = O^{\varepsilon_i}_{2\alpha_i}(2)$ maximal irreducible solvable for all $1 \leq i \leq t$. Here $\ell = \alpha_1 + \cdots + \alpha_t$ and $\varepsilon = \varepsilon_1 \cdots \varepsilon_t$.

Since $X^\Omega$ is not metrically completely reducible, by Lemma 2.3.12 there exists $i$ such that $Y_i^\Omega$ is not irreducible, and $Y_j = Y_j^\Omega$ for all $j \neq i$. By Lemmas 2.2.20 and 5.1.3 we have $\varepsilon_j = +$ for all $j \neq i$. Furthermore, it follows from Lemmas 2.2.21 and 5.2.4 that $Y_i$ is metrically primitive and $\varepsilon_i = +$. Thus $\varepsilon = \varepsilon_1 \cdots \varepsilon_t = +$, in which case $\kappa$ must be a quadratic form.

It remains to check that $G^\circ$ is metrically imprimitive. By Lemma 5.2.4, the group $Y_i$ must be metrically primitive of type $\mathcal{B}_1$, with $\alpha_i$ odd. Because $Y_i \leq O^{\varepsilon_i}_{2\alpha_i}(2)$ is assumed to be irreducible and $O^+_2(2)$ is not, we must have $\alpha_i \geq 3$. The action of $G^\circ$ on $A/F_0$ is equal to $X^\Omega$ (Lemma 4.3.27), so by Lemma 5.2.4 we have a decomposition $W_i = W_i' \perp W_i''$ with $W_i'$ and $W_i''$ totally singular irreducible $\mathbb{F}_2[G^\circ]$-modules.

Then $W_i' = Q/F_0$, where $Q$ is an abelian normal subgroup of $G^\circ$. Since $W_i'$ is totally singular, we have $Q \cong C_2^{d+1}$, where $d = \dim_{\mathbb{F}_2} W_i' = \alpha_i$. We have shown that $G^\circ$ has a noncyclic abelian normal subgroup, so by Lemma 2.4.3 it is metrically imprimitive. □

**Lemma 5.3.8** *Let $G = G^{\mathcal{B}}_{\mu,\nu}(X_1, \ldots, X_k)$. Suppose that $n$ and $q$ are even, and that $\kappa$ is a quadratic form. Then $F_0$ is maximal abelian normal in $C_{G^\Omega}(F_0)$.*

***Proof*** Suppose that $F$ is abelian such that $F_0 \leq F \trianglelefteq C_{G^\Omega}(F_0)$. It follows from Lemma 5.3.3 that $F \leq A$. By Lemma 5.3.1 (iii), the action of $C_{G^\Omega}(F_0)$ on

$O_{r_i}(A/F_0) = R_i F_0/F_0 \cong R_i/Z(R_i)$ is metrically completely reducible for all $1 \leq i \leq k$. Thus arguing as in the proof of Corollary 5.3.4 shows that $F = F_0$. □

As obvious corollaries of Lemmas 5.3.7 and 5.3.8, we have the following results.

**Corollary 5.3.9** *Let $G = G^{\mathcal{B}}_{\mu,\nu}(X_1, \ldots, X_k)$. The subgroup $F_0$ is maximal abelian normal in $G^\circ$, except when conditions (i)–(iii) of Lemma 5.3.7 hold.*

**Corollary 5.3.10** *Let $G = G^{\mathcal{B}}_{\mu,\nu}(X_1, \ldots, X_k)$. Suppose that $n$ and $q$ are even, and that $\kappa$ is a quadratic form. Then $F_0$ is maximal abelian normal in $G^\Omega$.*

## 5.4 Examples of the Construction in Some Special Cases

With the results of the previous sections (Sects. 2.2, 4.3), we have a completed the recursive construction which provides a list of candidates for the maximal irreducible solvable subgroups of $\Delta(V, \kappa)$. That is, any maximal irreducible solvable subgroup of $\Delta(V, \kappa)$ is conjugate to one of the groups generated by the construction. As the main result of this book, we will find that in general the groups constructed are maximal solvable, with a few exceptions that will be classified.

The purpose of this section is to illustrate the recursive construction in some special cases. Information about these examples will also be useful in later sections. We begin with the next two lemmas, which we will prove simultaneously.

**Lemma 5.4.1** *Let $G \leq O^+_{2^\alpha}(2)$ be maximal irreducible solvable, where $\alpha \geq 2$. Then one of the following holds:*

*(i) $G$ is metrically primitive and $G = G^{\mathcal{B}_1}_{1,2^{\alpha-1}}$, where $\alpha \geq 3$.*

*(ii) $\alpha = \alpha' + \alpha''$ such that $G = G^{\mathcal{B}_1}_{1,2^{\alpha'}-1} \wr K$ or $G = G^{\mathcal{B}_2}_{1,2^{\alpha'}-1} \wr K$, where $K \leq S_{2^{\alpha''}}$ is maximal transitive solvable.*

**Lemma 5.4.2** *Let $G \leq O^-_{2^\alpha}(2)$ be maximal irreducible solvable, where $\alpha \geq 1$. Then $G = G^{\mathcal{B}_2}_{1,2^{\alpha-1}}$.*

***Proof of Lemmas 5.4.1 and 5.4.2*** Let $G \leq O^\varepsilon_{2^\alpha}(2)$ be metrically primitive maximal irreducible solvable. Then $G$ must be of type $\mathcal{B}_1$ or $\mathcal{B}_2$, and $2^\alpha = 2\mu\nu$, where $|F_0| = 2^\nu \pm 1$. Since $|F_0|$ is odd and every prime divisor of $\mu$ divides $|F_0|$ (Lemma 4.1.18), we have $\mu = 1$.

Thus either $G = G^{\mathcal{B}_1}_{1,2^{\alpha-1}}$ and $\varepsilon = +$, or $G = G^{\mathcal{B}_2}_{1,2^{\alpha-1}}$ and $\varepsilon = -$. There are no metrically imprimitive maximal irreducible solvable subgroups in $O^-_{2^\alpha}(2)$, so we conclude that $G^{\mathcal{B}_2}_{1,2^{\alpha-1}}$ is the unique maximal irreducible solvable subgroup of $O^-_{2^\alpha}(2)$, as claimed by Lemma 5.4.2.

Note that for $G = G^{\mathcal{B}_1}_{1,2^{\alpha}-1}$ we must have $\alpha \geq 3$, since groups of type $\mathcal{B}_1$ are only defined when $q^{\nu} > 4$ or $(q, \nu) = (4, 1)$. Thus if $G \leq O^+_{2^\alpha}(2)$ is metrically primitive, Lemma 5.4.1 (i) holds.

Let $G \leq O^+_{2^\alpha}(2)$ is metrically imprimitive maximal irreducible solvable. Then by Theorem 2.2.14 we have $\alpha = \alpha' + \alpha''$ such that $G = H \wr K$, where $H \leq O^{\varepsilon'}_{2^{\alpha'}}(2)$ is metrically primitive maximal irreducible solvable, and $K \leq S_{2^{\alpha''}}$ is maximal transitive solvable. As seen earlier, we have $H = G^{\mathcal{B}_1}_{1,2^{\alpha'}-1}$ or $H = G^{\mathcal{B}_2}_{1,2^{\alpha'}-1}$ so Lemma 5.4.1 (ii) holds. □

**Example 5.4.3** In Table 5.1, we list the candidates for maximal irreducible solvable subgroups of $\mathrm{GL}_2(q)$. There is one possible imprimitive maximal irreducible solvable subgroup, which is $\mathrm{GL}_1(q) \wr S_2$ for $q > 2$. There are two possibilities for primitive maximal irreducible subgroups, which are $G^{\mathcal{B}_0}_{1,2} = \Gamma\mathrm{L}_1(q^2)$ and $G^{\mathcal{B}_0}_{2,1}(O^-_2(2))$ (for $q$ odd).

On the other hand $\mathrm{GL}_2(q) = \mathrm{GSp}_2(q)$, and in Table 5.2 we give the list of subgroups provided by the construction for $\mathrm{GSp}_2(q)$. In this case all the subgroups are metrically primitive, and there are three possibilities. In terms of the subgroups of $\mathrm{GL}_2(q)$, we have $G^{\mathcal{B}_1}_{1,1} = \mathrm{GL}_1(q) \wr S_2$, $G^{\mathcal{B}_2}_{1,1} = G^{\mathcal{B}_0}_{1,2}$, and $G^{\mathcal{B}_3}_{2,1}(O^-_2(2)) = G^{\mathcal{B}_0}_{2,1}(O^-_2(2))$.

Not all the subgroups listed in Tables 5.1 and 5.2 are maximal solvable. For example, it turns out that $\mathrm{GL}_1(5) \wr S_2 = G^{\mathcal{B}_1}_{1,1}$ is not maximal solvable in $\mathrm{GL}_2(5) = \mathrm{GSp}_2(5)$ (Theorem 7.2.22, Lemma 5.5.5).

**Example 5.4.4** In Tables 5.3 and 5.4, we will similarly list the candidates for maximal irreducible solvable subgroups of $\mathrm{GSp}_4(q)$ and $\mathrm{GO}^+_4(q)$. Here the groups $X$ of type $\mathcal{B}_1$ and $\mathcal{B}_2$ are of multiplier 1 (Lemma 4.3.13 (i), Lemma 4.3.22 (i)), and the values of $|X|$ are determined by Lemmas 4.3.15 and 4.3.23. For groups $X$ of type $\mathcal{B}_3$, the order $|X|$ is determined by Lemma 4.3.28, and the multiplier by Lemma 5.1.1.

**Table 5.1** Candidates for maximal irreducible solvable subgroups of $\mathrm{GL}_2(q)$

| $X$ | $\|X\|$ | Conditions |
|---|---|---|
| $\mathrm{GL}_1(q) \wr S_2$ | $2(q-1)^2$ | $q > 2$ |
| $G^{\mathcal{B}_0}_{1,2}$ | $2(q^2-1)$ | |
| $G^{\mathcal{B}_0}_{2,1}(O^-_2(2))$ | $24(q-1)$ | $q$ odd |

**Table 5.2** Candidates for maximal irreducible solvable subgroups of $\mathrm{GSp}_2(q)$. In the table, $(*)$ denotes the condition that 2 is a square in $\mathbb{F}_q$

| $X$ | $\|X\|$ | Multiplier | Conditions |
|---|---|---|---|
| $G^{\mathcal{B}_1}_{1,1}$ | $2(q-1)^2$ | 1 | $q > 3$ |
| $G^{\mathcal{B}_2}_{1,1}$ | $2(q^2-1)$ | 1 | |
| $G^{\mathcal{B}_3}_{2,1}(O^-_2(2))$ | $24(q-1)$ | 1 if not $(*)$<br>2 if $(*)$ | $q$ odd |

**Table 5.3** Candidates for maximal irreducible solvable subgroups of $\mathrm{GSp}_4(q)$. In the table, $(*)$ denotes the condition that 2 is a square in $\mathbb{F}_q$

| $X$ | $\lvert X\rvert$ | Multiplier | Conditions |
|---|---|---|---|
| $\mathrm{semiwr}\left(\left(G_{2,1}^{\mathcal{B}_3}\right)^{\circ}\right)$ | $2304(q-1)$ | 1 | $q$ odd and $(*)$ holds |
| $G_{1,1}^{\mathcal{B}_1} \wr S_2$ | $8(q-1)^3$ | 1 | $q > 3$ |
| $G_{1,1}^{\mathcal{B}_2} \wr S_2$ | $8(q+1)^2(q-1)$ | 1 | |
| $G_{2,1}^{\mathcal{B}_3} \wr S_2$ | $1152(q-1)$ if not $(*)$<br>$2304(q-1)$ if $(*)$ | 1 if not $(*)$<br>2 if $(*)$ | $q$ odd |
| $G_{1,2}^{\mathcal{B}_1}$ | $4(q^2-1)(q-1)$ | 1 | $q > 2$ |
| $G_{2,1}^{\mathcal{B}_1}(O_2^-(2))$ | $48(q-1)^2$ | 1 | $q$ odd |
| $G_{1,2}^{\mathcal{B}_2}$ | $4(q^2+1)(q-1)$ | 1 | |
| $G_{2,1}^{\mathcal{B}_2}(O_2^-(2))$ | $48(q+1)(q-1)$ | 1 | $q$ odd |
| $G_{4,1}^{\mathcal{B}_3}(G_{1,2}^{\mathcal{B}_2})$ | $320(q-1)$ | 1 if not $(*)$<br>2 if $(*)$ | $q$ odd |

**Table 5.4** Candidates for maximal irreducible solvable subgroups of $\mathrm{GO}_4^+(q)$. In the table, $(*)$ denotes the condition that 2 is a square in $\mathbb{F}_q$

| $X$ | $\lvert X\rvert$ | Multiplier | Conditions |
|---|---|---|---|
| $O_2^-(q) \wr S_2$ | $8(q+1)^2(q-1)$ | 1 | |
| $O_2^+(q) \wr S_2$ | $8(q-1)^3$ | 1 | $q > 3$ |
| $O_1(q) \wr S_4$ | $192(q-1)$ | 2 | $q$ odd |
| $G_{1,2}^{\mathcal{B}_1}$ | $4(q^2-1)(q-1)$ | 1 | $q > 2$ |
| $G_{2,1}^{\mathcal{B}_1}(O_2^-(2))$ | $48(q-1)^2$ | 1 | $q$ odd |
| $G_{2,1}^{\mathcal{B}_2}(O_2^-(2))$ | $48(q+1)(q-1)$ | 1 | $q$ odd |
| $G_{4,1}^{\mathcal{B}_3}(O_4^+(2))$ | $1152(q-1)$ | 1 if not $(*)$<br>2 if $(*)$ | $q$ odd |

**Example 5.4.5** In Table 5.5, we list the candidates for maximal irreducible solvable subgroups of $O_n^{\varepsilon}(2)$ provided by the construction, with $2 \leq n \leq 12$ even. The metrically imprimitive groups are provided by Theorem 2.2.14 (i). For metrically primitive groups we only consider groups of type $\mathcal{B}_1$ and $\mathcal{B}_2$, which are discussed in Sect. 4.3.2 and Sect. 4.3.3, respectively.

We also give the values of $|X|$ and $[X : X^{\Omega}]$ in Table 5.5. In general the orders $|X|$ can be determined with Lemmas 4.3.15 and 4.3.23, and $[X : X^{\Omega}] \in \{1, 2\}$ is determined by Lemmas 5.1.2, 5.1.3, and 2.2.20.

As in the other tables of this section, not all the subgroups in Table 5.5 are maximal solvable. For example, we will see later that $G_{1,3}^{\mathcal{B}_2}$ is metrically imprimitive (Lemma 5.6.6). Later in Sect. 8.1, we will give tables which include only maximal solvable subgroups; see for example Table 8.2 there.

**Table 5.5** Candidates for maximal irreducible solvable subgroups $X$ of $\Delta(V, \kappa) = O^{\varepsilon}_n(2)$ with $2 \leq n \leq 12$ even

| $\Delta(V, \kappa)$ | $X$ | $\lvert X \rvert$ | $[X : X^{\Omega}]$ |
|---|---|---|---|
| $O^+_2(2)$ | None | n/a | n/a |
| $O^-_2(2)$ | $O^-_2(2) = G^{\mathcal{B}_2}_{1,1}$ | 6 | 2 |
| $O^+_4(2)$ | $O^+_4(2) = O^-_2(2) \wr S_2$ | 72 | 2 |
| $O^-_4(2)$ | $G^{\mathcal{B}_2}_{1,2}$ | 20 | 2 |
| $O^+_6(2)$ | $G^{\mathcal{B}_1}_{1,3}$ | 42 | 2 |
| $O^-_6(2)$ | $O^-_2(2) \wr S_3$ | 1296 | 2 |
| | $G^{\mathcal{B}_2}_{1,3}$ | 54 | 2 |
| | $G^{\mathcal{B}_2}_{3,1}(\mathrm{GSp}_2(3))$ | 1296 | 2 |
| $O^+_8(2)$ | $O^-_2(2) \wr S_4$ | 31,104 | 2 |
| | $G^{\mathcal{B}_2}_{1,2} \wr S_2$ | 800 | 2 |
| | $G^{\mathcal{B}_1}_{1,4}$ | 120 | 1 |
| $O^-_8(2)$ | $G^{\mathcal{B}_2}_{1,4}$ | 136 | 2 |
| $O^+_{10}(2)$ | $G^{\mathcal{B}_1}_{1,5}$ | 310 | 2 |
| $O^-_{10}(2)$ | $O^-_2(2) \wr \mathrm{AGL}_1(5)$ | 155,520 | 2 |
| | $G^{\mathcal{B}_2}_{1,5}$ | 330 | 2 |
| $O^+_{12}(2)$ | $O^-_2(2) \wr S_3 \wr S_2$ | 3,359,232 | 2 |
| | $O^-_2(2) \wr S_2 \wr S_3$ | 2,239,488 | 2 |
| | $G^{\mathcal{B}_1}_{1,3} \wr S_2$ | 3528 | 2 |
| | $G^{\mathcal{B}_2}_{1,3} \wr S_2$ | 5832 | 2 |
| | $G^{\mathcal{B}_2}_{3,1}(\mathrm{GSp}_2(3)) \wr S_2$ | 3,359,232 | 2 |
| | $G^{\mathcal{B}_1}_{1,6}$ | 756 | 1 |
| $O^-_{12}(2)$ | $G^{\mathcal{B}_2}_{1,2} \wr S_3$ | 48,000 | 2 |
| | $G^{\mathcal{B}_2}_{1,6}$ | 780 | 2 |

## 5.5 Some Examples Where $G^{\mathcal{B}}_{\mu,\nu}(X_1, \ldots, X_k)$ Is Not Maximal Solvable

So far we have shown that every metrically primitive maximal irreducible solvable subgroup of $\Delta(V, \kappa)$ is conjugate to one of the groups $G = G^{\mathcal{B}}_{\mu,\nu}(X_1, \ldots, X_k)$ constructed in Sect. 4.3. In this section, we will give some examples where $G$ is not maximal solvable $\Delta(V, \kappa)$. We will also provide further examples where it is possible that $G$ is maximal solvable, but $G^{\circ}$ is not maximal solvable in $I(V, \kappa)$.

Most of the examples given in this section correspond to the *cas d'exclusion* listed by Jordan in [52, §41–47]. Jordan only considers the case where $q$ is a prime, so in our setup we find some additional examples (Lemma 5.5.14). Throughout this section, we will denote $G = G^{\mathcal{B}}_{\mu,\nu}(X_1, \ldots, X_k)$ and assume that $G$ is irreducible.

We begin with some observations in the case $\mu = 2^\ell$ that most of the examples are based on.

**Remark 5.5.1** Let $G$ be a group of type $\mathcal{B}_1$ such that $\mu = 2^\ell$ for some $\ell \geq 0$. If $\mu > 1$, then $G = G^{\mathcal{B}_1}_{\mu,\nu}(X_1)$, where $X_1 \leq O^{\varepsilon_1}_{2\ell}(2)$ is metrically completely reducible maximal irreducible solvable, and $A = F_0 R_1$ with $R_1 = 2^{1+2\ell}_{\varepsilon_1}$. For $\mu = 1$, we denote $R_1 = 1$, $\varepsilon_1 = +$, and $X_1 = 1$.

It follows from Lemma 4.3.10 that $\varphi$ acts trivially on $R_1/Z(R_1)$, so $\varphi \in G^\circ$. More specifically, we see from (4.3) that the action of $\varphi$ centralizes the generators of $R_1$, except $A_1^{(1)}$ for $\mu > 1$ and $u_1 = 1$, in which case

$$\varphi A_1^{(1)} \varphi^{-1} = \left(A_1^{(1)}\right)^{-1}.$$

Thus we define

$$\varphi_0 := \begin{cases} \varphi, & \text{if } \mu = 1; \\ \left(B_1^{(1)}\right)^{u_1} \varphi, & \text{if } \mu > 1; \end{cases}$$

so $\varphi_0$ centralizes $R_1$ and $\varphi_0 \in G^\circ$. (A similar observation is made by Jordan in [52, §32].)

Furthermore, as seen in Sect. 4.3.2, the action of $\psi$ on the generators of $R_1$ is as given in (4.2), so $\psi$ centralizes $R_1$. In particular $\psi$ acts trivially on $R_1/Z(R_1)$, so $\psi \in G^\circ$. Since the images of $\varphi$ and $\psi$ generate $N_{I(V,\kappa)}(F_0)/C_{I(V,\kappa)}(F_0)$ (Lemma 2.6.4 (ii)), it follows that $G^\circ = \langle C_{G^\circ}(F_0), \varphi_0, \psi\rangle$.

**Lemma 5.5.2** *Let $G$ be a group of type $\mathcal{B}_1$ with $\mu = 2^\ell$ for some $\ell \geq 0$, and $F_0 = \langle f\rangle$ with splitting field $\mathbb{K} = \mathbb{F}_{q^\nu}$. Define $\varphi_0$ as in Remark 5.5.1. If $\mu > 1$, let $\pi$ be the homomorphism $\pi : N_{\Delta(V,\kappa)}(F_0, R_1) \to O^{\varepsilon_1}_{2\ell}(2)$ used in the construction of G (Sect. 4.3.2).*

*Then all of the following hold:*

*(i)* $\varphi_0 \in G^\circ$ *and* $\varphi_0$ *centralizes* $R_1$.
*(ii)* $[\varphi_0, g] \in F_0$ *for all* $g \in G$.
*(iii)* *If* $\mu = 1$, *then* $[\varphi_0, g] \in \langle f^2\rangle$ *for all* $g \in G^\circ$.
*(iv)* *Suppose that* $\mu > 1$, *and let* $g \in G^\circ$. *Then* $[\varphi_0, g] \in \langle f^2\rangle$ *if and only if one of the following holds:*

*(a)* *2 is a square in* $\mathbb{K}$.
*(b)* $\pi(g) \in \Omega^{\varepsilon_1}_{2\ell}(2)$.

***Proof*** We have already noted in Remark 5.5.1 that $\varphi_0 \in G^\circ$ and $\varphi_0$ centralizes $R_1$, so (i) holds.

For (ii), note that since $G/C_{G^\circ}(F_0)$ is abelian (Lemma 4.3.33), for all $g \in G$ the commutator $[\varphi_0, g]$ centralizes $F_0$. Furthermore $[\varphi_0, g]$ centralizes $R_1$ since $\varphi_0$ does, so by Lemma 4.3.32 we conclude that $[\varphi_0, g] \in C_{G^\circ}(A) = F_0$ for all $g \in G$.

If $\mu = 1$, we have $G^\circ = N_{I(V,\kappa)}(F_0) = \langle f, \psi, \varphi \rangle$. By the relations (2.3), we have $[\varphi, f] = f^2$ and $[\varphi, \psi] = 1$, so (iii) holds.

We assume then $\mu > 1$ for the rest of the proof and consider claim (iv). As seen in Remark 5.5.1, we have $G^\circ = \langle C_{G^\circ}(F_0), \varphi_0, \psi \rangle$. Since $[\varphi_0, \varphi_0] = [\varphi_0, \psi] = 1$, it suffices to consider (iv) in the case where $g \in C_{G^\circ}(F_0)$.

Let $b_1$ be the non-degenerate $R_1$-invariant bilinear on $W'_\lambda$, as described in Sects. 3.2.3 and 3.2.6. Then $\operatorname{sgn}(b_1) = \varepsilon_1$, and the basis of $W'_\lambda$ used in the construction is an orthonormal basis if $\varepsilon_1 = +$, and a standard symplectic basis if $\varepsilon_1 = -$. Thus if $\Omega$ denotes the matrix of $b_1$ with respect to the basis of $W'_\lambda$, we have

$$\left(B_1^{(1)}\right)^{u_1} = (-1)^{u_1}\Omega = \Omega^{-1}.$$

Let $\tau_1 : \Delta(W'_\lambda, b_1) \to \mathbb{K}^\times$ be the homomorphism corresponding to $b_1$. By Theorems 3.2.4 and 3.2.7 we have $N_{\Delta(W'_\lambda, b_1)}(R_1) = N_{\mathrm{GL}(W'_\lambda)}(R_1)$, so the action of $C_{G^\circ}(F_0)$ on $W'_\lambda$ preserves $b_1$ up to a scalar. That is, for $g \in C_{G^\circ}(F_0)$ we have $g^T \Omega g = \tau_1(g)\Omega$ in $\mathrm{GL}(W'_\lambda)$.

Conjugation by $\varphi$ corresponds to taking the inverse transpose (2.3). Note that $\left(B_1^{(1)}\right)^{u_1}$ is equal to its inverse transpose. Therefore for $g \in C_{G^\circ}(F_0)$, as an element of $\mathrm{GL}(W'_\lambda)$ the commutator $[\varphi_0, g]$ is equal to

$$[\varphi_0, g] = \left(B_1^{(1)}\right)^{-u_1} g^T \left(B_1^{(1)}\right)^{u_1} g = \Omega g^T \Omega^{-1} g = \Omega^{-1} g^T \Omega g = \tau_1(g) I_{W'_\lambda}$$

In other words, we have $[\varphi_0, g] = f^r$, where $\lambda^r = \tau_1(g)$.

Since $\lambda \in \mathbb{K}^\times$ is a primitive element, we conclude that $[\varphi_0, g] \in \langle f^2 \rangle$ if and only if $\tau_1(g)$ is a square in $\mathbb{K}$. With this observation we can complete the proof of (iv). If 2 is a square in $\mathbb{K}$, it follows from Theorem 3.2.4 (vi) and Theorem 3.2.7 (vi) that $\tau_1(x)$ is a square for all $x \in N_{\Delta(W'_\lambda, b_1)}(R_1) = N_{\mathrm{GL}(W'_\lambda)}(R_1)$, so $[\varphi_0, g] \in \langle f^2 \rangle$ for all $g \in C_{G^\circ}(F_0)$.

Suppose then that 2 is not a square in $\mathbb{K}$. In this case, it follows from Theorem 3.2.4 (v) and Theorem 3.2.7 (v) that $\tau_1(x)$ is a square for $x \in N_{\Delta(W'_\lambda, b_1)}(R_1)$ if and only if the action of $x$ on $R_1$ is contained in $\Omega^{\varepsilon_1}_{2\ell}(2)$. Therefore for $g \in C_{G^\circ}(F_0)$, we have $[\varphi_0, g] \in \langle f^2 \rangle$ if and only if $\pi(g) \in \Omega^{\varepsilon_1}_{2\ell}(2)$, as claimed. □

**Remark 5.5.3** Suppose that $\mu = 2^\ell$ and let $G = G^{\mathcal{B}_2}_{\mu,\nu}(X)$ be a group of type $\mathcal{B}_2$. Furthermore, assume that $q \equiv 3 \mod 4$.

Let $G$ be a group of type $\mathcal{B}_2$ such that $\mu = 2^\ell$ for some $\ell \geq 0$. If $\mu > 1$, then $G = G^{\mathcal{B}_2}_{\mu,\nu}(X_1)$, where $X_1 \leq O^{\varepsilon_1}_{2\ell}(2)$ is metrically completely reducible maximal irreducible solvable, and $A = F_0 R_1$ with $R_1 = 2^{1+2\ell}_{\varepsilon_1}$. For $\mu = 1$, we denote $R_1 = 1$, $\varepsilon_1 = +$, and $X_1 = 1$.

It follows from Lemma 4.3.19 that $\psi$ acts trivially on $R_1/Z(R_1)$, so $\psi \in G^\circ$. Since $q \equiv 3 \mod 4$, it follows from (4.4) that $\psi$ centralizes the generators of $R_1$, except $A_1^{(1)}$ for $\mu > 1$ and $u_1 = 1$, in which case

$$\psi A_1^{(1)} \psi^{-1} = \left(A_1^{(1)}\right)^{-1}.$$

Thus we define

$$\psi_0 := \begin{cases} \psi, & \text{if } \mu = 1; \\ \left(B_1^{(1)}\right)^{u_1} \psi, & \text{if } \mu > 1; \end{cases}$$

so $\psi_0$ centralizes $R_1$ and $\psi_0 \in G^\circ$. (As in Remark 5.5.1, this observation is similar to [52, §32].)

Since the image of $\psi$ in $N_{I(V,\kappa)}(F_0)/C_{I(V,\kappa)}(F_0)$ (Lemma 2.7.3 (ii)), it follows that in this case $G^\circ = \langle C_{G^\circ}(F_0), \psi_0\rangle$.

**Lemma 5.5.4** *Let $G$ be a group of type $\mathcal{B}_2$ with $\mu = 2^\ell$ for some $\ell \geq 0$, and $F_0 = \langle f \rangle$ with splitting field $\mathbb{K} = \mathbb{F}_{q^{2\nu}}$. Assume that $q \equiv 3 \mod 4$.*

*Define $\psi_0$ as in Remark 5.5.3. If $\mu > 1$, let $\pi$ be the homomorphism $\pi : N_{\Delta(V,\kappa)}(F_0, R_1) \to O^{\varepsilon_1}_{2\ell}(2)$ used in the construction of G (Sect. 4.3.3).*

*Then all of the following hold:*

*(i) $\psi_0 \in G^\circ$ and $\psi_0$ centralizes $R_1$.*
*(ii) $[\psi_0, g] \in F_0$ for all $g \in G$.*
*(iii) If $\mu = 1$, then $[\psi_0, g] \in \langle f^2 \rangle$ for all $g \in G^\circ$.*
*(iv) Suppose that $\mu > 1$, and let $g \in G^\circ$. Then $[\varphi_0, g] \in \langle f^2 \rangle$ if and only if one of the following holds:*

*(a) 2 is a square in $\mathbb{F}_{q^\nu}$.*
*(b) $\pi(g) \in \Omega^{\varepsilon_1}_{2\ell}(2)$.*

***Proof*** The fact that $\psi_0 \in G^\circ$ and $\psi_0$ centralizes $R_1$ was observed in Remark 5.5.3, so (i) holds. Then (ii) follows similarly to the proof of Lemma 5.5.2, using the fact that $G/C_{G^\circ}(F_0)$ is abelian (Lemma 4.3.33).

If $\mu = 1$, then $G^\circ = N_{I(V,\kappa)}(F_0) = \langle f, \psi\rangle$. We have $[\psi_0, f] = f^{-q+1} \in \langle f^2 \rangle$ since $q$ is odd, so (iii) holds.

We assume then $\mu > 1$ for the rest of the proof and consider (iv). We have $G^\circ = \langle C_{G^\circ}(F_0), \psi_0\rangle$ and $\psi_0 \in \operatorname{Ker} \pi$ (Remark 5.5.3), so as in the proof of Lemma 5.5.2, it suffices to consider (iv) in the case $g \in C_{G^\circ}(F_0)$.

Let $b_1$ be the non-degenerate $R_1$-invariant bilinear on $W'_\lambda$, as described in Sects. 3.2.3 and 3.2.6, so $\operatorname{sgn}(b_1) = \varepsilon_1$. Let $\tau_1 : \Delta(W'_\lambda, b_1) \to \mathbb{K}^\times$ be the homomorphism corresponding to $b_1$.

We can consider $g \in C_{G^\circ}(F_0)$ as an element of $I(W'_\lambda, \widehat{b})$, where $\widehat{b}$ is the non-degenerate Hermitian form $\widehat{b}$ on $W'_\lambda$ used in the construction of $G$ (Sect. 4.3.3). In this setup, conjugation by $\psi$ corresponds to taking the conjugate of the matrix with respect to the automorphism $\xi \mapsto \xi^{q^\nu}$ of $\mathbb{K}$, as see in (2.7). That is, as an element of $\mathrm{GL}(W'_\lambda) = \mathrm{GL}_\mu(\mathbb{K})$, we have $\psi^{-1} g \psi = \overline{g}$, where $\overline{g}_{ij} = g_{ij}^{q^\nu}$ for all $1 \leq i, j \leq \mu$.

Since $q \equiv 3 \mod 4$, the basis of $W'_\lambda$ used in the construction is an orthonormal basis (see Sects. 3.2.2, 3.2.5, 4.3.3) for $\widehat{b}$. Therefore $g^T \overline{g} = I$ for all $g \in C_{G^\circ}(F_0)$, so

$$\psi^{-1} g \psi = g^{-T}.$$

In other words, conjugation by $\psi$ corresponds to taking the inverse transpose. It follows then as in the proof of Lemma 5.5.2 that $[\psi_0, g] = \tau_1(g) I_{W'_\lambda}$ for all $g \in C_{G^\circ}(F_0)$.

In other words, for all $g \in C_{G^\circ}(F_0)$, we have $[\psi_0, g] = f^r$, where $\lambda^r = \tau_1(g)$. Therefore we conclude that for $g \in C_{G^\circ}(F_0)$, we have

$$[\psi_0, g] \in \langle f^2 \rangle \Leftrightarrow \tau_1(g) \in \langle \lambda^2 \rangle \Leftrightarrow \tau_1(g)^{(q^\nu+1)/2} = 1. \tag{5.1}$$

(Here the last equivalence follows from the fact that $\lambda$ has order $q^\nu + 1$ in $\mathbb{K}$.)

With this we will be able to complete the proof of (iv). Consider first the case $\varepsilon_1 = +$. Note that $f$ generates the scalar matrices in $I(W'_\lambda, \widehat{b})$. Thus by Theorem 3.2.3, the group $N_{I(W'_\lambda, \widehat{b})}(R_1)$ is generated by $f$ and the matrices $A_t$, $B_t$, $cC_t$, $D_{st}$ described in Sects. 3.2.1 and 3.2.2. Here $2c^{q^\nu+1} = 1$.

The generators of $N_{I(W'_\lambda, \widehat{b})}(R_1)$ preserve the bilinear form $b_1$ up to a scalar, as seen in Sect. 3.2.3 and Theorem 3.2.4. In other words, we have $N_{I(W'_\lambda, \widehat{b})}(R_1) \leq N_{\mathrm{GO}(W'_\lambda, b_1)}(R_1)$.

We have $\tau_1(A_t) = \tau_1(B_t) = 1$ for all $1 \leq t \leq \ell$, and $\tau_1(D_{st}) = 1$ for all $1 \leq s < t \leq \ell$, as noted in Sect. 3.2.3. Furthermore $\tau_1(cC_t) = 2c^2$, so by (5.1) we get

$$\begin{aligned}
\tau_1(cC_t) \in \langle \lambda^2 \rangle &\Leftrightarrow (2c^2)^{\frac{q^\nu+1}{2}} = 1 \\
&\Leftrightarrow 2^{\frac{q^\nu+1}{2}} c^{q^\nu+1} = 1 \\
&\Leftrightarrow 2^{\frac{q^\nu-1}{2}} = 1 \\
&\Leftrightarrow 2 \text{ is a square in } \mathbb{F}_{q^\nu}.
\end{aligned}$$

Therefore if 2 is a square in $\mathbb{F}_{q^\nu}$, we conclude that $\tau_1(g) \in \langle \lambda^2 \rangle$ for all $g \in N_{I(W'_\lambda, \widehat{b})}(R_1)$, and in particular $[\psi_0, g] \in \langle f^2 \rangle$ for all $g \in C_{G^\circ}(F_0)$ by (5.1).

Suppose then that 2 is not a square in $\mathbb{F}_{q^\nu}$. In this case we have

$$\begin{aligned}
\tau_1(cC_t) &\notin \langle \lambda^2 \rangle & \qquad \pi(cC_t) &\notin \Omega^+_{2\ell}(2) \\
\tau_1(D_{st}) &\in \langle \lambda^2 \rangle & \qquad \pi(D_{st}) &\in \Omega^+_{2\ell}(2)
\end{aligned}$$

for all $1 \leq t \leq \ell$ and $1 \leq s < t \leq \ell$. Furthermore, for the generators of $g \in N_{I(W'_\lambda,\widehat{b})}(R_1)$ in $\operatorname{Ker}\pi = R_1F_0$, we have $\tau_1(g) \in \langle\lambda^2\rangle$, since $\tau_1(f) = \lambda^2$ and $\tau_1(A_t) = \tau_1(B_t) = 1$ for all $1 \leq t \leq \ell$.

Thus we conclude that for $g \in N_{I(W'_\lambda,\widehat{b})}(R_1)$, we have $\tau_1(g) \in \langle\lambda^2\rangle$ if and only if $\pi(g) \in \Omega^+_{2\ell}(2)$. By (5.1), for $g \in C_{G^\circ}(F_0)$ we get $[\psi_0, g] \in \langle f^2\rangle$ if and only if $\pi(g) \in \Omega^+_{2\ell}(2)$, as claimed. This completes the proof in the case $\varepsilon_1 = +$.

For $\varepsilon_1 = -$, the result follows similarly, using the generators of $N_{I(W'_\lambda,\widehat{b})}(R_1) \leq N_{\mathrm{GSp}(W'_\lambda,b_1)}(R_1)$ described in Sects. 3.2.4, 3.2.5, and Theorem 3.2.6. □

**Lemma 5.5.5** *Suppose that $G$ is of type $\mathcal{B}_1$, with $q = 5$ and $\nu = 1$. If $e+\sum_{i=1}^k u_i \equiv 1 \mod 2$, then $G$ is not maximal solvable in $\Delta(V,\kappa)$.*

***Proof*** *([52, §43])* Suppose that $e+\sum_{i=1}^k u_i \equiv 1 \mod 2$. We have $|F_0| = q^\nu - 1 = 4$, so $n = 2\mu$ and $\mu = 2^\ell$ for some $\ell \geq 0$. If $\mu > 1$, then $G = G^{\mathcal{B}_1}_{\mu,1}(X_1)$, where $X_1 \leq O^{\varepsilon_1}_{2\ell}(2)$ is metrically completely reducible maximal irreducible solvable, and $A = F_0R_1$ with $R_1 = 2^{1+2\ell}_{\varepsilon_1}$. For $\mu = 1$, we denote $R_1 = 1$ and $X_1 = 1$. Define $\varphi_0$ as in Remark 5.5.1. Then by Lemma 5.5.2, we have $\varphi_0 \in G^\circ$ and $[\varphi_0, g] \in F_0$ for all $g \in G$.

Therefore $R := \langle\varphi_0, f\rangle$ is a normal subgroup of $G$. We have

$$\begin{aligned} f^2 &= -1, \\ \varphi_0^2 &= (-1)^{e+\sum_{i=1}^k u_i} = -1, \\ [\varphi_0, f] &= -1, \end{aligned}$$

so $R \cong 2^{1+2}_-$. Thus $G$ normalizes $\overline{A} = R_1R$ and is contained in a group

$$\overline{G} = G^{\mathcal{B}_3}_{n,1}(Y)$$

of type $\mathcal{B}_3$, where $X_1 \times O^-_2(2) \leq Y \leq O^{-\varepsilon_1}_{2\ell+2}(2)$ with $Y$ metrically completely reducible maximal solvable.

The action of $G$ on $R/Z(R)$ fixes the image of $f$ and thus does not generate all of $O^-_2(2)$, so $G$ is properly contained in $\overline{G}$. Hence $G$ is not maximal solvable in $\Delta(V,\kappa)$. □

**Example 5.5.6** It follows from Lemma 5.5.5 that in $\Delta(V,\kappa) = \mathrm{GSp}_2(5) = \mathrm{GL}_2(5)$, the subgroup $G = G^{\mathcal{B}_1}_{1,1} = \mathrm{GL}_1(5) \wr S_2$ is not maximal solvable. (In this case $e = 1$ and $k = 0$.)

More specifically, the proof of Lemma 5.5.5 shows that $G \lneqq G^{\mathcal{B}_3}_{2,1}(O^-_2(2))$. In this example

$$G^{\mathcal{B}_3}_{2,1}(O^-_2(2)) = N_{\Delta(V,\kappa)}(R) = (\mathbb{F}_5^\times \circ 2^{1+2}_-).O^-_2(2),$$

where $R = 2^{1+2}_-$ is an absolutely irreducible subgroup of $\mathrm{GL}_2(5)$.

For $q = 5$, $\nu = 1$, $e = 0$, and $k = 0$ the hypotheses of Lemma 5.5.5 are not satisfied. In this case it turns out that $G = G^{\mathcal{B}_1}_{1,1}$ is maximal solvable, since $G = \mathrm{GO}_2^+(5) = \Delta(V, \kappa)$.

**Lemma 5.5.7** *Suppose that G is of type $\mathcal{B}_2$, with $q = 3$ and $\nu = 1$. If $e+\sum_{i=1}^k u_i \equiv 1 \mod 2$, then G is not maximal solvable in $\Delta(V, \kappa)$.*

***Proof*** *([52, §44])* The result follows similarly to Lemma 5.5.5. In this case $|F_0| = q^\nu + 1 = 4$, so $n = 2\mu$ and $\mu = 2^\ell$ for some $\ell \geq 0$. If $\mu > 1$, then $G = G^{\mathcal{B}_2}_{\mu,1}(X_1)$, where $X_1 \leq O^{\varepsilon_1}_{2\ell}(2)$ is metrically primitive maximal irreducible solvable, and $A = F_0 R_1$ with $R_1 = 2^{1+2\ell}_{\varepsilon_1}$. For $\mu = 1$, we denote $R_1 = 1$ and $X_1 = 1$.

Assume that $e + \sum_{i=1}^k u_i \equiv 1 \mod 2$. Define $\psi_0$ as in Remark 5.5.3, so $\psi_0 \in G^\circ$ and $[\psi_0, g] \in F_0$ for all $g \in G$ by Lemma 5.5.4. Then as in the proof of Lemma 5.5.5, we see that $R := \langle \psi_0, f \rangle$ is a normal subgroup of $G$ with $R \cong 2^{1+2}_-$; so $G$ normalizes $\overline{A} = R_1 R$, and is properly contained in a group of type $\mathcal{B}_3$. □

**Example 5.5.8 (Cf. Example 5.5.6)** For $\mu = 1$, Lemma 5.5.7 is the statement that in $\Delta(V, \kappa) = \mathrm{GSp}_2(3) = \mathrm{GL}_2(3)$, the subgroup $G = G^{\mathcal{B}_2}_{1,1} = \Gamma\mathrm{L}_1(3^2)$ is not maximal solvable. (In this case $e = 1$ and $k = 0$.) This is clear since $\mathrm{GL}_2(3)$ is itself solvable. Furthermore, we have $\mathrm{GSp}_2(3) = G^{\mathcal{B}_3}_{2,1}(O_2^-(2))$.

For $q = 3$, $\nu = 1$, $e = 0$, and $k = 0$ the hypotheses of Lemma 5.5.7 are not satisfied for $G = G^{\mathcal{B}_2}_{1,1}$. In this case $G$ is maximal solvable, since $G = \mathrm{GO}_2^-(3) = \Delta(V, \kappa)$.

**Lemma 5.5.9** *Suppose that G is of type $\mathcal{B}_0$, with $q = 3$ and $\nu = 2$. Then G is not maximal solvable in $\Delta(V, \kappa)$.*

***Proof*** *([52, §41])* In this case $\mathbb{K} = \mathbb{F}_9$ and $|F_0| = 3^2 - 1 = 8$, so $n = 2\mu$ with $\mu = 2^\ell$ for some $\ell \geq 0$. If $\mu > 1$, then $G = G^{\mathcal{B}_2}_{\mu,1}(X_1)$, where $X_1 \leq O^{\varepsilon}_{2\ell}(2)$ is metrically primitive maximal irreducible solvable, and $A = F_0 R_1$ with $R_1 = 2^{1+2\ell}_{\varepsilon}$. For $\mu = 1$, we denote $R_1 = 1$ and $X_1 = 1$. As seen from the relations (4.2), the map $\psi$ centralizes $R_1$. In particular $\psi \in \operatorname{Ker} \pi$, so $\psi \in G$.

Since $G/C_G(F_0)$ is abelian (Lemma 4.3.33), it follows as in the proof of Lemma 5.5.2 (ii) that $[g, \psi] \in F_0$ for all $g \in G$. Hence all $G$-conjugates of $\psi$ are of the form $\psi f^d$ for some $d \in \mathbb{Z}$. We have $\psi^2 = 1$, so for any such $G$-conjugate

$$1 = (\psi f^d)^2 = \psi^2 f^{4d} = f^{4d},$$

which implies that $d$ is even.

Therefore $R = \langle f^2, f\psi \rangle$ is a normal subgroup of $G$. Furthermore

$$f^4 = (f\psi)^2 = [f^2, f\psi] = -1,$$

so $R \cong 2_-^{1+2}$. Thus $G$ normalizes $\overline{A} = F_0 R_1 R$ and is contained in a group

$$\overline{G} = G^{\mathcal{B}_0}_{n,1}(X_1 \times O_2^-(2))$$

of type $\mathcal{B}_0$. The action of $G$ on $R/Z(R)$ is not irreducible, since it leaves invariant the subspace spanned by the image of $f^2$. In particular the action of $G$ on $R/Z(R)$ does not generate all of $O_2^-(2)$, so $G$ is properly contained in $\overline{G}$. Hence $G$ is not maximal solvable in $\Delta(V, \kappa) = \mathrm{GL}(V)$. □

**Example 5.5.10** For $\mu = 1$, Lemma 5.5.9 corresponds to the fact that $G^{\mathcal{B}_0}_{1,2} = \Gamma\mathrm{L}_1(3^2)$ is not maximal solvable in $\mathrm{GL}_2(3) = G^{\mathcal{B}_0}_{2,1}(O_2^-(2))$.

**Lemma 5.5.11** *Suppose that $G$ is of type $\mathcal{B}_1$, with $q = 5$ and $\nu = 2$. Then $G$ is not maximal solvable in $\Delta(V, \kappa)$.*

***Proof*** *([52, §42])* In this case $|F_0| = 5^2 - 1 = 24$, so $n = 4\mu$ with $\mu = 2^{\ell_1}3^{\ell_2}$ for some $\ell_1, \ell_2 \geq 0$. We have $A = F_0 R_1 R_2$, where

$$R_1 = \begin{cases} 1, & \text{if } \ell_1 = 0. \\ 2_{\varepsilon_1}^{1+2\ell_1}, & \text{if } \ell_1 > 0. \end{cases} \qquad R_2 = \begin{cases} 1, & \text{if } \ell_2 = 0. \\ 3_+^{1+2\ell_2}, & \text{if } \ell_2 > 0. \end{cases}$$

There is a homomorphism

$$\pi : N_{\Delta(V,\kappa)}(F_0, R_1, R_2) \to \mathrm{GSp}_{2\ell_1}(2) \times \mathrm{GSp}_{2\ell_2}(3)$$

where we interpret $\mathrm{GSp}_{2\ell_i}(r) = 1$ for $\ell_i = 0$. By definition, we have $G = \pi^{-1}(X_1, X_2)$ for some $X_1$ and $X_2$. Here $X_1 = 1$ if $\ell_1 = 0$, and otherwise $X_1 \leq O^{\varepsilon_1}_{2\ell_1}(2)$ is metrically completely reducible maximal solvable. Similarly $X_2 = 1$ if $\ell_2 = 0$, and otherwise $X_2 \leq \mathrm{GSp}_{2\ell_2}(3)$ is metrically completely reducible maximal solvable.

As seen in Sect. 4.3.2, the action of $\psi$ and $\varphi$ on the generators of $R_1$ and $R_2$ is as given in (4.2) and (4.3). In particular, we find that $\psi\varphi$ centralizes all of the generators of $R_1$ and $R_2$, except for $A_1^{(1)}$ for $\ell_1 > 0$, $u_1 = 1$, in which case

$$\psi\varphi A_1^{(1)}(\psi\varphi)^{-1} = \left(A_1^{(1)}\right)^{-1}.$$

Thus $\phi_0$ defined by

$$\phi_0 := \begin{cases} \psi\varphi, & \text{if } \ell_1 = 0. \\ \left(B_1^{(1)}\right)^{u_1} \psi\varphi, & \text{if } \ell_1 > 0. \end{cases}$$

centralizes $R_1$ and $R_2$. In particular $\phi_0 \in \operatorname{Ker}\pi$, so $\phi_0 \in G^\circ$.

As in the proof of Lemma 5.5.9, it follows that $[g, \phi_0] \in Z(A) = F_0$ for all $g \in G$, so all $G$-conjugates of $\phi_0$ are of the form $\phi_0 f^d$ for some $d \in \mathbb{Z}$. Now $(\psi\varphi)^2 = (-1)^e$, so

$$\phi_0^2 = (-1)^{e+\sum_{i=1}^k u_i}.$$

Therefore for any $G$-conjugate $\phi_0 f^d$ of $\phi_0$, we have

$$(-1)^{e+\sum_{i=1}^k u_i} = \phi_0^2 = (\phi_0 f^d)^2 = \phi_0^2 f^{-4d}.$$

Thus $f^{-4d} = 1$, so $d \equiv 0 \mod 6$.

Suppose first that $e + \sum_{i=1}^k u_i \equiv 1 \mod 2$. Because every $G$-conjugate of $\phi_0$ is of the form $\phi_0 f^{6d}$, the subgroup $R = \langle \phi_0, f^6 \rangle$ is a normal subgroup of $G$ that centralizes $R_1$ and $R_2$. Furthermore

$$\phi_0^2 = f^{12} = [\phi_0, f^6] = f^{12} = -I_V,$$

so $R \cong 2_-^{1+2}$. Now $\overline{F_0} := \langle f^4 \rangle$ centralizes $R$. Thus $G$ normalizes $\overline{A} = \overline{F_0} R_1 R_2 R$ and is contained in a group $\overline{G}$ of type $\mathcal{B}_2$ corresponding to $X_1 \times O_2^-(2)$ and $X_2$. The action of $G$ on $R/Z(R)$ is not irreducible, since it leaves invariant the subspace spanned by the image of $f^6$. In particular the action of $G$ on $R/Z(R)$ does not generate all of $O_2^-(2)$, so $G$ is properly contained in $\overline{G}$. In particular $G$ is not maximal solvable in $\Delta(V, \kappa)$.

Suppose then that $e + \sum_{i=1}^k u_i \equiv 0 \mod 2$. In this case we use $R = \langle \phi_0 f^3, f^6 \rangle$ instead, which is similarly seen to be a normal subgroup of $G$ and $R \cong 2_-^{1+2}$. Then $\overline{F_0} = \langle f^4 \rangle$ is centralized by $R$, and $G$ normalizes $\overline{A} := \overline{F_0} R_1 R_2 R$. It follows that $G$ is contained in a group $\overline{G}$ of type $\mathcal{B}_2$ corresponding to $X_1 \times O_2^-(2)$ and $X_2$. As in the previous paragraph, we see that the action of $G$ on $R/Z(R)$ does not generate all of $O_2^-(2)$, so $G$ is properly contained in $\overline{G}$. □

**Lemma 5.5.12** *Let $n$ be a power of 2. Suppose that $G$ is of type $\mathcal{B}_1$ with $q = 7$, $\nu = 1$, and $e + \sum_{i=1}^k u_i \equiv 1 \mod 2$. Then there exists $\overline{G}$ of type $\mathcal{B}_3$ such that $G^\circ \lneqq \overline{G^\circ}$. For any such $\overline{G}$, we have $G \not\leq \overline{G}$.*

***Proof*** Because 2 is a square modulo 7, by Lemma 5.1.5 every group of type $\mathcal{B}_3$ in $\Delta(V, \kappa)$ is of multiplier 2. On the other hand $G$ is of type $\mathcal{B}_1$, so it is of multiplier 1 (Lemma 4.3.13 (i)). Consequently $G \not\leq \overline{G}$ for every group $\overline{G} \leq \Delta(V, \kappa)$ of type $\mathcal{B}_3$.

Assume that $e + \sum_{i=1}^k u_i \equiv 1 \mod 2$. We will show that $G^\circ \leq \overline{G^\circ}$ for some $\overline{G}$ of type $\mathcal{B}_3$. Recall that the construction of $G$ starts with a totally singular decomposition

$$V = W_\lambda \oplus W_{\lambda^{-1}},$$

where $\lambda \in \mathbb{F}_7^\times$ is a primitive element and $\dim W_\lambda = \dim W_{\lambda^{-1}} = \mu$. Without loss of generality we may assume that $\lambda = 3$, so $\lambda^{-1} = 5$.

Because $n$ is a power of 2, we have $n = 2\mu$ with $\mu = 2^\ell$ for some $\ell \geq 0$. If $\mu > 1$, then $G = G^{\mathcal{B}_2}_{\mu,1}(X_1)$, where $X_1 \leq O^{\varepsilon}_{2\ell}(2)$ is metrically completely reducible maximal solvable, and $A = F_0 R_1$ with $R_1 = 2^{1+2\ell}_{\varepsilon_1}$. In the case where $\mu = 1$, we denote $R_1 = 1$ and $X_1 = 1$.

From the construction of $G$, as an $\mathbb{F}_7[R_1]$-module $W_\lambda$ is absolutely irreducible. Furthermore, it follows from Theorem 3.2.1 (iv) that $W_\lambda \cong W_{\lambda^{-1}}$ as $\mathbb{F}_7[R_1]$-modules. Thus as $\mathbb{F}_7[R_1]$-modules we can identify $V = W \otimes U$, where $U$ is a trivial $\mathbb{F}_7[R_1]$-module with $\dim U = 2$, and $W$ is a faithful absolutely irreducible $\mathbb{F}_7[R_1]$-module. Then we can choose a basis $\{u, v\}$ of $U$ such that $W \otimes \langle u \rangle = W_\lambda$ and $W \otimes \langle v \rangle = W_{\lambda^{-1}}$.

As seen in Sects. 3.2.6 and 3.2.3, there exists a non-degenerate $R_1$-invariant reflexive bilinear form $b_W$ on $W$. Furthermore, the bilinear form $b_W$ is unique up to a scalar, and

$$\operatorname{sgn}(b_W) = (-1)^{\sum_{i=1}^{k} u_i} = (-1)^{e+1}.$$

Because $W$ is an absolutely irreducible $\mathbb{F}_7[R_1]$-module, it follows as in the proof of Lemma 4.1.14 (iii) that the bilinear form $b$ corresponding to $\kappa$ is of the form

$$b = b_W \otimes b_U,$$

where $b_U$ is a non-degenerate bilinear form on $U$. We have $\operatorname{sgn}(b) = (-1)^e$ and $\operatorname{sgn}(b) = \operatorname{sgn}(b_W)\operatorname{sgn}(b_U) = (-1)^{e+1}\operatorname{sgn}(b_U)$, so $b_U$ must be alternating.

Let $x_1, \ldots, x_\mu$ the basis used in the construction of $W$ and $b_W$ (see Sects. 3.2.6, 3.2.3, 4.3.4). We have a unique dual basis $y_1, \ldots, y_\mu$ of $W$ with $b_W(x_i, y_j) = \delta_{i,j}$. By replacing $u$ with a scalar if necessary, we can assume $b_U(u, v) = 1$. Then by the definition of $\varphi$ and $g_X$ for $X \in \mathrm{GL}_\mu(\mathbb{K})$, we have

$$\varphi(x_i \otimes u) = y_i \otimes v, \qquad \varphi(y_i \otimes v) = (-1)^e x_i \otimes u$$

$$g_X(x_i \otimes u) = \sum_{j=1}^{\mu} X_{ji} x_j \otimes u \qquad g_X(y_i \otimes v) = \sum_{j=1}^{\mu} (X^{-T})_{ji} y_j \otimes v$$

for all $1 \leq i \leq \mu$.

Denote by $\Omega$ be the matrix of $b_W$ with respect to the basis $x_1, \ldots, x_\mu$. (Recall that the basis is orthonormal if $\operatorname{sgn}(b_W) = +$, and a standard symplectic basis if $\operatorname{sgn}(b_W) = -$.) Then $y_i = \Omega^{-1} x_i$ for all $1 \leq i \leq \mu$. Note that if $\mu > 1$, then $\Omega^{-1} = \Omega^T$ is precisely the matrix of the action of $\left(B_1^{(1)}\right)^{u_1}$ on $W$. Furthermore, we have $\Omega^2 = (-1)^{e+1} I$, $\Omega^T = (-1)^{e+1}\Omega$, and $\Omega^{-T} = \Omega$.

As seen in Remark 5.5.1, the element

$$\varphi_0 := g_{\Omega^T}\varphi = \begin{cases} \varphi, & \text{if } \mu = 1. \\ \left(B_1^{(1)}\right)^{u_1}\varphi, & \text{if } \mu > 1. \end{cases}$$

is contained in $G^\circ$ and centralizes $R_1$. A calculation shows that

$$\varphi_0(x_i \otimes u) = (-1)^{e+1} x_i \otimes v, \qquad \varphi_0(x_i \otimes v) = (-1)^e x_i \otimes u$$

for all $1 \leq i \leq \mu$.

Thus by defining $\varphi_1 := (-1)^{e+1}\varphi_0$, we have

$$\varphi_1 = I_W \otimes \begin{pmatrix} 0 & 1 \\ -1 & 0 \end{pmatrix},$$

$$f = I_W \otimes \begin{pmatrix} 3 & 0 \\ 0 & 5 \end{pmatrix}.$$

Since $W$ is an absolutely irreducible $\mathbb{F}_7[R_1]$-module, we have

$$N_{\mathrm{GL}(V)}(R_1) = N_{\mathrm{GL}(W)}(R_1) \otimes \mathrm{GL}(U)$$
$$C_{\mathrm{GL}(V)}(R_1) = I_W \otimes \mathrm{GL}(U)$$

by Lemma 2.1.16. In particular $G^\circ$ is contained in $N_{\mathrm{GL}(W)}(R_1) \otimes \mathrm{GL}(U)$.

The image of $\varphi$ generates $G^\circ / C_{G^\circ}(F_0)$, so we have $G^\circ = \langle C_{G^\circ}(F_0), \varphi_1 \rangle$. Because $f$ centralizes $\mathrm{GL}(W) \otimes I_U$, for every $g \in C_{G^\circ}(F_0)$ we have $g = g' \otimes g''$ with $g' \in \mathrm{GL}(W)$ and $g'' \in C_{\mathrm{GL}(U)}(f)$. The centralizer of $\begin{pmatrix} 3 & 0 \\ 0 & 5 \end{pmatrix}$ in $\mathrm{GL}_2(7)$ is generated by $\begin{pmatrix} 3 & 0 \\ 0 & 5 \end{pmatrix}$, so we have $g = (g' \otimes I_U) f^d$ for some $d \geq 0$. Thus

$$G^\circ = \langle f, \varphi_1, (\mathrm{GL}(W) \otimes I_U) \cap C_{G^\circ}(F_0) \rangle . \tag{5.2}$$

We will now construct an extraspecial subgroup $2^{1+2}_-$ in $I(V, \kappa)$ that is normalized by $G^\circ$. This will allow us construct the group of type $\mathcal{B}_3$ containing $G^\circ$. Define

$$x := I_W \otimes \begin{pmatrix} 4 & 4 \\ 1 & 3 \end{pmatrix}$$

and

$$y := \varphi_1 x \varphi_1^{-1} = I_W \otimes \begin{pmatrix} 3 & 6 \\ 3 & 4 \end{pmatrix}.$$

We have $x^2 = y^2 = (xy)^2 = -I$, so $R = \langle x, y \rangle \cong 2_-^{1+2}$.

Another calculation shows that

$$\varphi_1 x \varphi_1^{-1} = y \qquad f x f^{-1} = y^{-1}$$

$$\varphi_1 y \varphi_1^{-1} = x \qquad f y f^{-1} = xy$$

so $R$ is normalized by $f$ and $\varphi_1$. It follows then from (5.2) that $R$ is normalized by $G^\circ$. In particular $G^\circ$ normalizes $\overline{A} = R_1 R$, so $G^\circ$ is contained in $\overline{G}^\circ$, where

$$\overline{G} = G^{\mathcal{B}_3}_{n,1}(Y)$$

with $X_1 \times O_2^-(2) \leq Y \leq O_{2\ell+2}^{-\varepsilon_1}(2)$ metrically completely reducible maximal solvable.

Here we have $G^\circ \lneq \overline{G}^\circ$, since $G^\circ$ acts on $\{W_\lambda, W_{\lambda^{-1}}\}$, but $x \in \overline{G}^\circ$ does not. □

**Lemma 5.5.13** *Suppose that $G$ is of type $\mathcal{B}_2$ with $q = 7$, $\nu = 1$, and $e + \sum_{i=1}^k u_i \equiv 1 \mod 2$. Then there exists $\overline{G}$ of type $\mathcal{B}_3$ such that $G^\circ \lneq \overline{G}^\circ$. For any such $\overline{G}$, we have $G \not\leq \overline{G}$.*

***Proof*** As in the proof of Lemma 5.5.12, for every $\overline{G} \leq \Delta(V, \kappa)$ of type $\mathcal{B}_3$ we have $G \not\leq \overline{G}$, since $G$ is of multiplier 1 (Lemma 4.3.22 (i)) and $\overline{G}$ is of multiplier 2 (Lemma 5.1.5).

Suppose that $e + \sum_{i=1}^k u_i \equiv 1 \mod 2$. We will show that $G^\circ \lneq \overline{G}^\circ$ for some $\overline{G}$ of type $\mathcal{B}_3$. We have $|F_0| = 7 + 1 = 8$, so $n = 2\mu$ with $\mu = 2^\ell$ for some $\ell \geq 0$. If $\mu > 1$, then $G = G^{\mathcal{B}_2}_{\mu,1}(X_1)$ with $X_1 \leq O_{2\ell}^{\varepsilon_1}(2)$ metrically completely reducible maximal irreducible solvable, and $A = F_0 R_1$ with $R_1 = 2_{\varepsilon_1}^{1+2\ell}$. For $\mu = 1$, we denote $R_1 = 1$ and $X_1 = 1$.

Define $\psi_0$ as in Remark 5.5.3, so $\psi_0 \in G^\circ$ and $[\psi_0, g] \in F_0$ for all $g \in G$ by Lemma 5.5.4. Since 2 is a square modulo 7, it follows also from Lemma 5.5.4 that $[\psi_0, g] \in \langle f^2 \rangle$ for all $g \in C_{G^\circ}(F_0)$. Furthermore, the image of $\psi_0$ generates $N_{I(V,\kappa)}(F_0)/C_{I(V,\kappa)}(F_0)$ (Lemma 2.7.3 (iii)), so $G^\circ = \langle C_{G^\circ}(F_0), \psi_0 \rangle$.

Thus it follows that $[\psi_0, g] \in \langle f^2 \rangle$ for all $g \in G^\circ$, so $R = \langle \psi_0, f^2 \rangle$ is a normal subgroup of $G^\circ$. We have

$$\psi_0^2 = (-1)^{e + \sum_{i=1}^k u_i} = -1,$$

$$(f^2)^2 = (\psi_0 f^2)^2 = -1,$$

so $R \cong 2_-^{1+2}$.

Consequently $G^\circ$ normalizes $\overline{A} = R_1 R$ and is contained in a group

$$\overline{G} := G^{\mathcal{B}_3}_{n,1}(Y)$$

of type $\mathcal{B}_3$, where $X_1 \times O_2^-(2) \leq Y \leq O^{-\varepsilon_1}_{2\ell+2}(2)$ is metrically completely reducible maximal solvable. We have a proper inclusion $G^\circ \lneqq \overline{G^\circ}$, since $G^\circ$ fixes the image of $f^2$ in $R/Z(R)$, and thus the image of $G^\circ$ does not generate all of $O_2^-(2)$. □

**Lemma 5.5.14** *Suppose that G is of type $\mathcal{B}_1$ with $q = 9$, $\nu = 1$, and $e + \sum_{i=1}^{k} u_i \equiv 1 \mod 2$. Then there exists $\overline{G}$ of type $\mathcal{B}_3$ such that $G^\circ \lneqq \overline{G^\circ}$. For any such $\overline{G}$, we have $G \not\leq \overline{G}$.*

***Proof*** The fact that $G \not\leq \overline{G}$ follows as in the beginning of the proof of Lemma 5.5.12. Assume that $e + \sum_{i=1}^{k} u_i \equiv 1 \mod 2$. We will show that $G^\circ \lneqq \overline{G^\circ}$ for some group $\overline{G}$ of type $\mathcal{B}_3$.

Note that $|F_0| = 9 - 1 = 8$, so $n = 2\mu$ with $\mu = 2^\ell$ for some $\ell \geq 0$. Thus if $\mu > 1$, then $G = G^{\mathcal{B}_1}_{\mu,1}(X_1)$ with $X_1 \leq O^{\varepsilon_1}_{2\ell}(2)$ and $A = F_0 R_1$ with $R_1 = 2^{1+2\ell}_{\varepsilon_1}$. If $\mu = 1$, we denote $R_1 = 1$ and $X_1 = 1$.

Let $\varphi_0$ be as in Remark 5.5.1, so $\varphi_0 \in G^\circ$ by Lemma 5.5.2. Since 2 is a square in $\mathbb{K} = \mathbb{F}_9$, it follows from Lemma 5.5.2 that $[\varphi_0, g] \in \langle f^2 \rangle$ for all $g \in G^\circ$. Thus $G^\circ$ normalizes $R = \langle \varphi_0, f^2 \rangle$, and $R \cong 2^{1+2}_-$ since

$$\varphi_0^2 = (f^2)^2 = [\varphi_0, f^2] = -1.$$

Now $G^\circ$ normalizes the central product $\overline{A} = R_1 R$, so as in the proof of Lemma 5.5.13, we see that $G^\circ \lneqq \overline{G^\circ}$ for a group $\overline{G}$ of type $\mathcal{B}_3$. □

**Example 5.5.15** Consider $\Delta(V, \kappa) = \mathrm{GSp}_2(q) = \mathrm{GL}_2(q)$ with $q$ odd. It is known that $G = G^{\mathcal{B}_1}_{1,1} = \mathrm{GL}_1(q) \wr S_2$ is maximal irreducible solvable in $\mathrm{GSp}_2(q)$ if $q \neq 3, 5$ (for example [77, Theorem 21.6]), and this fact will also follow from results proven later (Theorem 7.4.6).

In particular $G$ is maximal solvable for $q \in \{7, 9\}$, but in this case $G^\circ$ is not maximal solvable in $\mathrm{Sp}_2(q) = \mathrm{SL}_2(q)$ by Lemmas 5.5.13 and 5.5.14. In this case $G$ is also a maximal subgroup of $\mathrm{GSp}_2(q)$.

In the case where $q = 11$, the following hold:

- $G$ maximal in $\mathrm{GSp}_2(11)$;
- $G^\circ$ is maximal solvable in $\mathrm{Sp}_2(11)$.

But $G^\circ \lneqq \mathrm{Sp}_2(5) \lneqq \mathrm{Sp}_2(11)$ [7, Theorem 6.3.10 (i)], so $G^\circ$ is not a maximal subgroup of $\mathrm{Sp}_2(11)$.

**Example 5.5.16** It is claimed in [23, p. 34, Class (5)] that if $G$ is a maximal irreducible solvable subgroup of $\mathrm{GL}_n(q)$ that is contained in $\Gamma\mathrm{L}_{n/m}(q^m)$ (Aschbacher class $\mathcal{C}_3$), then $G \cap \mathrm{GL}_{n/m}(q^m)$ is maximal irreducible solvable in $\mathrm{GL}_{n/m}(q^m)$. This turns out to be false in general, and counterexamples can be constructed as follows.

As seen in Example 5.5.15, the subgroup $X = G^{\mathcal{B}_1}_{1,1} = \mathrm{GL}_1(7) \wr S_2$ is maximal irreducible solvable in $\mathrm{GSp}_2(7)$, but $X^\circ$ is not maximal solvable in $\mathrm{Sp}_2(7)$. Assume that $q \equiv -1 \mod 7$, for example $q = 13$. Consider the subgroup $G = G^{\mathcal{B}_0}_{7,2}(X)$ of $\mathrm{GL}_{14}(q)$. It will follow from results proven later that $G$ is primitive maximal irreducible solvable in $\mathrm{GL}_{14}(q)$, see Theorem 7.1.12.

Let $F_0$ be the maximal abelian normal subgroup used in the construction of $G$. Then $G$ is maximal solvable in $N_{\mathrm{GL}_{14}(q)}(F_0) = \Gamma\mathrm{L}_7(q^2)$. However, the subgroup $C_G(F_0)$ is not maximal solvable in $C_{\mathrm{GL}_{14}(q)}(F_0) = \mathrm{GL}_7(q^2)$, which follows from the fact that $X^\circ$ is not maximal solvable in $\mathrm{Sp}_2(7)$. Indeed, we have $C_G(F_0) \lneqq C_{\overline{G}}(F_0)$ for $\overline{G} = G^{\mathcal{B}_0}_{7,2}(Y)$, where $X^\circ \lneqq Y^\circ$.

Thus in certain degrees there are issues with the algorithm presented in [23] for the construction of irreducible solvable subgroups of $\mathrm{GL}_n(q)$. However, we note that for the values of $(q, n)$ implemented in the GAP library IRREDSOL based on [23], this particular problem does not arise.

## 5.6 Primitivity of $G^{\mathcal{B}}_{\mu,\nu}(X_1, \ldots, X_k)$

In this section, we will determine when $G^{\mathcal{B}}_{\mu,\nu}(X_1, \ldots, X_k)$ satisfying the assumptions of Table 4.1 is metrically primitive (Theorem 5.6.9). More specifically, we will see that $G^{\mathcal{B}}_{\mu,\nu}(X_1, \ldots, X_k)$ is solvable, irreducible, and metrically primitive, except when $(q, \nu, \mathcal{B}) = (3, 2, \mathcal{B}_1)$ or $(2, 3, \mathcal{B}_2)$.

Furthermore, for $G = G^{\mathcal{B}}_{\mu,\nu}(X_1, \ldots, X_k)$ satisfying the conditions in Table 4.1, we will prove (Theorem 5.6.9) that either $G^\circ$ is metrically primitive, or belongs to a small number of exceptional cases. In the exceptional cases, we will show that either $G^\circ$ is metrically primitive, or $G$ is not maximal solvable. With some additional effort one could describe completely the cases where $G^\circ$ is metrically primitive, but this is not necessary for the purposes of this book.

**Proposition 5.6.1** *Let $G = G^{\mathcal{B}}_{\mu,\nu}(X_1, \ldots, X_k)$. Let $A \leq H \leq G$ be such that $F_0$ is maximal among the abelian subgroups of $H^\circ$ normalized by $C_H(F_0)$. Then the action of $C_H(F_0)$ on $W'_\lambda$ is absolutely irreducible and primitive.*

***Proof*** Since $A$ the action of $A$ on $W'_\lambda$ is absolutely irreducible (Lemma 5.2.2 (i)), it follows that the action of $C_H(F_0)$ on $W'_\lambda$ is absolutely irreducible. For primitivity, suppose that

$$W'_\lambda = Z_1 \oplus \cdots \oplus Z_t$$

such that $C_H(F_0)$ acts on $\{Z_1, \ldots, Z_t\}$. We will prove that $t = 1$.

First note that if $x \in A$ stabilizes $Z_1$, then the same is true for any $A$-conjugate of $x$ since $[A, A] \leq F_0$ and $F_0$ stabilizes $Z_1$. In other words, the subgroup

$T = N_A(Z_1)$ is a normal subgroup of $A$. Furthermore, we note that the action of $A$ on $\{Z_1, \ldots, Z_t\}$ is transitive since $A$ is irreducible on $W'_\lambda$. Therefore

$$T = \bigcap_{i=1}^{t} N_A(Z_i)$$

meaning that if $x \in A$ stabilizes any summand $Z_i$, it stabilizes all of them.

Hence $T$ is normalized by $C_H(F_0)$, since $C_H(F_0)$ acts on $\{Z_1, \ldots, Z_t\}$. Because $F_0$ is maximal among the abelian subgroups of $H^\circ$ normalized by $C_H(F_0)$, we have $F_0 = Z(T)$. Therefore it follows from Lemma 4.1.1 that $A/F_0 = T/F_0 \times C_A(T)/F_0$.

Because $T \trianglelefteq A$, by Clifford's theorem the action of $T$ is completely reducible on $W'_\lambda$. Hence we can apply Lemma 4.1.2 to conclude that $W'_\lambda \downarrow T$ is homogeneous, with composition factors absolutely irreducible of dimension $d$, where $[T : F_0] = d^2$. Now since the subspace $Z_1$ is $T$-invariant, we have $\dim Z_1 \geq d$, and thus $t \leq \mu/d$.

On the other hand, by transitivity of $A$ on $\{Z_1, \ldots, Z_t\}$ we have

$$t = [A : T] = \frac{[A : F_0]}{[T : F_0]} = \frac{\mu^2}{d^2}.$$

Therefore $\mu^2/d^2 = t \leq \mu/d$, which forces $\mu = d$ and $t = 1$. □

**Remark 5.6.2** When $H \leq G^\circ$, the assumption of Proposition 5.6.1 simply says that $F_0$ is maximal abelian normal in $C_H(F_0)$.

**Corollary 5.6.3** *Let $G = G^{\mathcal{B}}_{\mu,\nu}(X_1, \ldots, X_k)$. Then the following statements hold:*

*(i) $W'_\lambda \downarrow C_G(F_0)$ is absolutely irreducible and primitive.*
*(ii) $W'_\lambda \downarrow C_{G^\circ}(F_0)$ is absolutely irreducible and primitive, unless $G$ is of type $\mathcal{B}_3$ as in Lemma 5.3.7 (i)–(iii).*
*(iii) If $\kappa$ is a quadratic form and $q$ is even, then $W'_\lambda \downarrow C_{G^\Omega}(F_0)$ is absolutely irreducible and primitive.*

***Proof*** Claim (i) follows from Corollary 5.3.4 and Proposition 5.6.1 with $H = G$. Similarly (ii) follows from Lemma 5.3.7 and Proposition 5.6.1 with $H = G^\circ$, and (iii) follows from Lemma 5.3.8 and Proposition 5.6.1 with $H = G^\Omega$. □

**Lemma 5.6.4** *Let $G = G^{\mathcal{B}}_{\mu,\nu}(X_1, \ldots, X_k) \leq \Delta(V, \kappa)$. Let $V' = Z'_1 \oplus \cdots \oplus Z'_t$ be a system of imprimitivity for $F_0$ on $V' = \mathbb{K} \otimes_{\mathbb{F}_q} V$. Then each $Z'_i$ contains an $f$-eigenvector, except possibly in the following cases:*

*(i) $G$ is of type $\mathcal{B}_1$ and $(q, \nu) = (3, 2)$ or $(5, 1)$.*
*(ii) $G$ is of type $\mathcal{B}_2$ and $(q, \nu) = (2, 3)$ or $(3, 1)$.*

***Proof*** *([52, §51–§54])* Each $v \in V'$ can be written in the form $v = w_1 + \cdots + w_m$, where $w_i \neq 0$ are eigenvectors for $f$ with distinct eigenvalues. Choose $v$ from $Z'_i$ such that $m > 0$ is minimal. We will show that aside from the exceptional cases listed, we must have $m = 1$, in which case $v$ is an eigenvector.

We first note that if $f^k(v) \in Z'_i$, then $f^k(v)$ must be a scalar multiple of $v$. Indeed, let $c$ be the eigenvalue corresponding to $w_1$. If $f^k(v) \in Z'_i$, then $f^k(v) - c^k v \in Z'_i$ and by minimality of $m$ we must have $f^k(v) - c^k v = 0$, so $f^k(v) = c^k v$.

Therefore in the orbit of $F_0$ on $v$, there are vectors in $|F_0|/|T|$ different summands $Z'_j$, where $T$ is the stabilizer of $\langle v \rangle$ in $F_0$. In particular, the $F_0$-orbit of $v$ has at least $|F_0|/|T|$ linearly independent vectors; since all of these lie in the $m$-dimensional subspace $\langle w_1, \ldots, w_m \rangle$, we conclude that

$$\frac{|F_0|}{|T|} \leq m. \tag{5.3}$$

We will consider each of the types in turn and show that aside from the exceptional cases (i)–(ii), this inequality forces $m = 1$. Type $\mathcal{B}_3$ is trivial, so it will suffice to consider types $\mathcal{B}_0$, $\mathcal{B}_1$, and $\mathcal{B}_2$.

**Case 1: $G$ is of type $\mathcal{B}_0$**
We have $\mathbb{K} = \mathbb{F}_{q^\nu}$. Let

$$\lambda^{q^{\alpha_1}}, \ldots, \lambda^{q^{\alpha_m}}$$

be the eigenvalues corresponding to $w_1, \ldots, w_m$. Here $\lambda \in \mathbb{K}^\times$ is a primitive element and $0 \leq \alpha_i < \nu$ for all $1 \leq i \leq m$. Then $f^k$ stabilizes $\langle v \rangle$ if and only if $\lambda^{q^{\alpha_i} k} = \lambda^{q^{\alpha_j} k}$ for all $i \neq j$. Since $\lambda$ has order $q^\nu - 1$, this is equivalent to $q^\nu - 1 \mid k(q^{\alpha_i} - q^{\alpha_j})$ for all $i \neq j$.

Let $\delta = \gcd(\nu, \alpha_i - \alpha_j : i \neq j)$. It follows from Lemma 2.1.19 (i) that $q^\nu - 1 \mid k(q^{\alpha_i} - q^{\alpha_j})$ for all $i \neq j$ if and only if $q^\nu - 1 \mid k(q^\delta - 1)$. Therefore we conclude that $f^k$ stabilizes $\langle v \rangle$ if and only if

$$\frac{q^\nu - 1}{q^\delta - 1} \mid k,$$

so $|T| = q^\delta - 1$. Applying this to (5.3) we get $(q^\nu - 1)/(q^\delta - 1) \leq m$. On the other hand, we have $\alpha_i \equiv \alpha_j \mod \delta$ for all $i$ and $j$, so $m \leq \nu/\delta$. Therefore

$$\frac{q^\nu - 1}{q^\delta - 1} \leq \frac{\nu}{\delta},$$

which is only possible for $\nu = \delta$, forcing $m = 1$.

**Case 2: $G$ is of type $\mathcal{B}_1$**
We have $\mathbb{K} = \mathbb{F}_{q^\nu}$ for some $\nu \geq 1$. Let

$$\lambda^{q^{\alpha_1}}, \ldots, \lambda^{q^{\alpha_{m'}}}, \lambda^{-q^{\beta_1}}, \ldots, \lambda^{-q^{\beta_{m''}}}$$

be the eigenvalues corresponding to $w_1, \ldots, w_m$, where $\lambda \in \mathbb{K}^\times$ is a primitive element, $m = m' + m''$, and $0 \leq \alpha_i, \beta_j < \nu$ for all $1 \leq i \leq m'$ and $1 \leq j \leq m''$.

Let $\delta = \gcd(\nu, \alpha_i - \alpha_j, \beta_i - \beta_j : i \neq j)$. If $m' = 0$ or $m'' = 0$, then as in the $\mathcal{B}_0$ case, we get the inequality $(q^\nu - 1)/(q^\delta - 1) \leq m \leq \nu/\delta$, which forces $\nu = \delta$ and $m = 1$.

Suppose then that $m', m'' > 0$. As in the $\mathcal{B}_0$ case, we see that $f^k$ stabilizes $\langle v \rangle$ if and only if $q^\nu - 1 \mid k(q^\delta - 1)$ and $q^\nu - 1 \mid k(q^{\alpha_1} + q^{\beta_1})$. Therefore $f^k$ stabilizes $\langle v \rangle$ if and only if

$$\frac{q^\nu - 1}{\gcd(q^\delta - 1, q^{\alpha_1} + q^{\beta_1})} \mid k,$$

so $|T| = \gcd(q^\delta - 1, q^{\alpha_1} + q^{\beta_1})$.

Let $\delta' = \gcd(\delta, \alpha_1 - \beta_1)$. By Lemma 2.1.19 we have

$$\gcd(q^\delta - 1, q^{\alpha_1} + q^{\beta_1}) = \begin{cases} q^{\delta'} + 1, & \text{if } \nu_2(\delta) > \nu_2(\alpha_1 - \beta_1). \\ \gcd(q - 1, 2), & \text{if } \nu_2(\delta) \leq \nu_2(\alpha_1 - \beta_1). \end{cases}$$

Thus if $\nu_2(\delta) > \nu_2(\alpha_1 - \beta_1)$, then $|T| = q^{\delta'} + 1$ and so $(q^\nu - 1)/(q^{\delta'} + 1) \leq m$ by (5.3). On the other hand $m', m'' \leq \nu/\delta$ since the $\alpha_i$ and $\beta_j$ are congruent modulo $\delta$, so $m \leq 2\nu/\delta$. Therefore

$$\frac{q^\nu - 1}{q^{\delta'} + 1} \leq \frac{2\nu}{\delta}$$

Using the fact that $\delta' < \nu$, it is readily seen that the only solutions to this inequality occur for $\delta = 2$ and $\delta' = 1$, with $(q, \nu)$ equal to $(2, 2)$ or $(3, 2)$. The case $(q, \nu) = (2, 2)$ is excluded from the definition of groups of type $\mathcal{B}_1$, while $(q, \nu) = (3, 2)$ is one of the exceptions in (i).

Suppose then that $\nu_2(\delta) \leq \nu_2(\alpha_1 - \beta_1)$. Then $|T| = \gcd(q - 1, 2)$, and we get

$$\frac{q^\nu - 1}{\gcd(q - 1, 2)} \leq m \leq \frac{2\nu}{\delta}.$$

The only solutions to this inequality occur for $\delta = 1$, with $(q, \nu)$ equal to $(2, 1)$, $(2, 2)$, $(3, 1)$, $(3, 2)$, or $(5, 1)$. The cases $(2, 1)$, $(2, 2)$ or $(3, 1)$ are excluded from the definition of type $\mathcal{B}_1$, while the remaining cases are the exceptions given in (i).

**Case 3: $G$ is of type $\mathcal{B}_2$**
In this case $\mathbb{K} = \mathbb{F}_{q^{2\nu}}$ for some $\nu \geq 1$. Let

$$\lambda^{q^{\alpha_1}}, \ldots, \lambda^{q^{\alpha_m}}$$

be the eigenvalues corresponding to $w_1, \ldots, w_m$. Here $\lambda \in \mathbb{K}^\times$ has order $q^\nu + 1$, and $0 \leq \alpha_i < 2\nu$ for all $1 \leq i \leq m$.

Let $\delta = \gcd(\alpha_i - \alpha_j : i \neq j)$. As in previous cases, we see that $f^k$ stabilizes $\langle v \rangle$ if and only if $q^\nu + 1 \mid k(q^\delta - 1)$, so $|T| = \gcd(q^\nu + 1, q^\delta - 1)$.

We have $m \leq 2\nu/\delta$ since $\alpha_i \equiv \alpha_j \mod \delta$ for all $i$ and $j$, so (5.3) gives

$$\frac{q^\nu - 1}{\gcd(q^\nu + 1, q^\delta - 1)} \leq m \leq \frac{2\nu}{\delta}.$$

Denote $\delta' = \gcd(\nu, \delta)$.

If $\nu_2(\nu) < \nu_2(\delta)$, then $\gcd(q^\nu + 1, q^\delta - 1) = q^{\delta'} + 1$ by Lemma 2.1.19, which gives $(q^\nu + 1)/(q^{\delta'} + 1) \leq 2\nu/\delta$. First if $\delta' = \nu$, then $\nu \mid \delta$. Thus $\delta = 2\nu$ since $\nu_2(\nu) < \nu_2(\delta)$, so $m = 1$. If $\delta' < \nu$, the only solution to the inequality is $q = 2$, $\nu = 3$, $\delta = \delta' = 1$, which is one of the exceptions in (ii).

If $\nu_2(\nu) \geq \nu_2(\delta)$, then $\gcd(q^\nu + 1, q^\delta - 1) = \gcd(q - 1, 2)$ by Lemma 2.1.19, so $(q^\nu + 1)/\gcd(q - 1, 2) \leq 2\nu/\delta$. The only solution to this inequality is $q = 3$, $\nu = 1$, which is one of the exceptions in (ii).

This completes the proof of the lemma. □

We will now deal with the exceptional cases of Lemma 5.6.4.

**Lemma 5.6.5** *Let $G = G^{\mathcal{B}}_{\mu,\nu}(X_1, \ldots, X_k)$. Suppose that $G$ is of type $\mathcal{B}_1$, with $q = 3$ and $\nu = 2$. Then $G$ is metrically imprimitive.*

***Proof*** *([52, §38])* We have $|F_0| = 3^2 - 1 = 8$, so $n = 4\mu$ with $\mu = 2^\ell$ for some $\ell \geq 0$. If $\mu > 1$, then $G = G^{\mathcal{B}_1}_{\mu,2}(X_1)$, where $X_1 \leq O^\varepsilon_{2\ell}(2)$ is metrically completely reducible maximal solvable, and $A = F_0 R_1$ with $R_1 = 2^{1+2\ell}_\varepsilon$. If $\mu = 1$, we define $R_1 = 1$.

Note that $G$ is irreducible by Lemma 5.2.2, since $X_1$ is of multiplier 1 if $\mu > 1$. We will show that $G$ has a noncyclic normal abelian subgroup $N$ such that $N \leq G^\circ$, which by Lemma 2.4.3 implies that $G$ is metrically imprimitive.

We have $\operatorname{Ker} \pi^\circ = A \rtimes \langle \varphi, \psi \rangle$ (Lemma 4.3.10), so $\varphi, \psi \in G$. In particular $\varphi\psi \in G$. Now $(\varphi\psi)^{-1} f^2 (\varphi\psi) = f^{-6} = f^2$, so $N = \langle \varphi\psi, f^2 \rangle$ is an abelian subgroup isomorphic to $C_4 \times C_2$. We will show that $N$ is a normal subgroup of $G$.

To this end, note that $\varphi\psi$ centralizes $R_1$. Then for all $g \in G$, the commutator $[g, \varphi\psi]$ is contained in $C_{G^\circ}(F_0)$ and centralizes $A = F_0 R_1$, so $[g, \varphi\psi] \in Z(A) = F_0$ (Lemma 4.3.32). Therefore the $G$-conjugates of $\varphi\psi$ are of the form $\varphi\psi f^d$. All of these conjugates must square to $(-1)^e$ since $(\varphi\psi)^2 = (-1)^e$, so

$$(-1)^e = (\varphi\psi f^d)^2 = (\varphi\psi)^2 f^{2d} [f^d, \varphi\psi] = (-1)^e f^{-2d}.$$

Thus $2d \equiv 0 \mod 8$, which forces $d \equiv 0 \mod 4$ and proves that $N \trianglelefteq G$. □

**Lemma 5.6.6** *Let $G = G^{\mathcal{B}}_{\mu,\nu}(X_1,\ldots,X_k)$. Suppose that $G$ is of type $\mathcal{B}_2$, with $q = 2$ and $\nu = 3$. Then $G$ is metrically imprimitive.*

***Proof*** *([52, §39])* Note that $G$ is irreducible by Lemma 5.2.2. We will show that $G$ has a noncyclic normal abelian subgroup $N$ such that $N \leq G^\circ$, which by Lemma 2.4.3 implies that $G$ is metrically imprimitive.

We have $|F_0| = 2^3 + 1 = 9$, so $\mu = 3^\ell$ for some $\ell \geq 0$. If $\mu > 1$, then $G = G^{\mathcal{B}_2}_{\mu,3}(X_1)$, where $X_1 \leq \mathrm{GSp}_{2\ell}(3)$ is metrically completely reducible maximal solvable, and $A = F_0 R_1$ with $R_1 = 3^{1+2\ell}_+$. For $\mu = 1$, we define $R_1 = 1$.

We have $\operatorname{Ker}\pi^\circ = A \rtimes \langle \psi^2 \rangle$ (Lemma 4.3.19), so $\psi^2 \in G$. The commutator $[f, \psi^2] = f^3$ is centralized by $\psi^2$, so $N = \langle \psi^2, f^3 \rangle$ is an abelian subgroup of $G$ isomorphic to $C_3 \times C_3$. We will show that $N$ is a normal subgroup of $G$.

To this end, note that $\psi^2$ centralizes $R_1$, so as in the proof of Lemma 5.6.5, the $G$-conjugates of $\psi^2$ are of the form $\psi^2 f^d$. All of these conjugates must have order 3 since $\psi$ does, so

$$1 = (\psi^2 f^d)^3 = \psi^6 f^{3d} [f^d, \psi^2]^2 = f^{12d}.$$

Thus $12d \equiv 0 \mod 9$, which forces $d \equiv 0 \mod 3$ and proves that $N \trianglelefteq G$. □

**Lemma 5.6.7** *Let $G = G^{\mathcal{B}}_{\mu,\nu}(X_1,\ldots,X_k)$. Suppose that $G$ is of type $\mathcal{B}_1$, with $q = 5$ and $\nu = 1$. Then the following hold:*

*(i) $G$ is metrically primitive. Moreover $V = Q_\lambda \oplus Q_{\lambda^{-1}}$ is the unique system of imprimitivity for $G$ on $V$;*
*(ii) Every system of imprimitivity for $G^\circ$ on $V$ has exactly 2 summands;*
*(iii) If $\mu > 1$ and $X_1 \not\leq \Omega^{\varepsilon_1}_{2\ell_1}(2)$, then $G^\circ$ is metrically primitive. Moreover $V = Q_\lambda \oplus Q_{\lambda^{-1}}$ is the unique system of imprimitivity for $G^\circ$ on $V$.*
*(iv) If $\mu = 1$ or $X_1 \leq \Omega^{\varepsilon_1}_{2\ell_1}(2)$, then the following hold:*

*(a) If $e = 0$, then $G^\circ$ is metrically imprimitive.*
*(b) If $e = 1$, then $G$ is not maximal solvable.*

***Proof*** Note that $\mathbb{K} = \mathbb{F}_5$, so we have $V' = V$ and $V = W'_\lambda \oplus W'_{\lambda^{-1}}$, where $\lambda \in \mathbb{F}_5^\times$ is a primitive element and $n = 2\mu$. Here $\mu = \dim W'_\lambda = \dim W'_{\lambda^{-1}}$, and $Q_{\lambda^{\pm 1}} = W'_{\lambda^{\pm 1}}$.

Suppose that $V = Z_1 \oplus \cdots \oplus Z_t$ is a system of imprimitivity for $G^\circ$ on $V$, where $t \geq 2$. It follows from Lemma 5.3.7 and Proposition 5.6.1 that the action of $C_{G^\circ}(F_0)$ on $W'_\lambda$ and $W'_{\lambda^{-1}}$ is absolutely irreducible and primitive. Therefore if some $Z_i$ contains an eigenvector of $f$, it must contain the entire eigenspace, and thus $t = 2$ and $\{Z_1, Z_2\} = \{Q_\lambda, Q_{\lambda^{-1}}\}$.

We can assume then for the rest of the proof that no $Z_i$ contains an $f$-eigenvector. In this case, we claim that the action of $A$ must be transitive on $\{Z_1,\ldots,Z_t\}$. Indeed, suppose that $\{Z_{i_1},\ldots,Z_{i_s}\}$ is $A$-invariant for some $0 < s < t$. Then since $V = W'_\lambda \oplus W'_{\lambda^{-1}}$ with $W'_\lambda$, $W'_{\lambda^{-1}}$ nonisomorphic irreducible $\mathbb{F}_q[A]$-modules (Lemma 5.2.2), by Lemma 2.1.15 the $\mathbb{F}_q[A]$-submodule $Z_{i_1} \oplus \cdots \oplus Z_{i_s}$ must be equal to $W'_\lambda$ or $W'_{\lambda^{-1}}$. But then $Z_{i_1}$ consists of eigenvectors, contrary to our assumption.

We have $|F_0| = q^\nu - 1 = 4$, so $n = 2\mu$ with $\mu = 2^{\ell_1}$ for some $\ell_1 \geq 0$. If $\mu > 1$, then $G = G^{\mathcal{B}_1}_{\mu,1}(X_1)$, where $X_1 \leq O^{\varepsilon_1}_{2\ell_1}(2)$ is metrically completely reducible maximal solvable, and $A = F_0 R_1$ with $R_1 = 2^{1+2\ell_1}_{\varepsilon_1}$. In the case where $\mu = 1$, we define $X_1 = 1$ and $R_1 = 1$. Then $[A, A] = Z(R_1) = \langle f^2 \rangle = \langle -I_V \rangle$ if $\mu > 1$, and $[A, A] = 1$ if $\mu = 1$.

Let $T = N_A(Z_1)$. We have $[A, A] \leq T$, so $T \trianglelefteq A$. Since $A$ acts transitively on $\{Z_1, \ldots, Z_t\}$, we conclude that $T = \cap_{i=1}^t N_A(Z_i)$ and $[A : T] = t$. Because none of the $Z_i$ contain $f$-eigenvectors, we have $T \cap F_0 = \langle f^2 \rangle$.

Because $T = \cap_{i=1}^t N_A(Z_i)$ and $G^\circ$ acts on $\{Z_1, \ldots, Z_t\}$, we have $T \trianglelefteq G^\circ$. By Lemma 5.3.7 the subgroup $F_0$ is maximal abelian normal in $G^\circ$, so $Z(TF_0) = F_0$. Then $Z(T) = T \cap F_0$, so it follows from Lemma 4.1.2 that $V \downarrow T$ is homogeneous with absolutely irreducible composition factors. Furthermore

$$[T : Z(T)] = [T : T \cap F_0] = d^2,$$

where $d$ is the dimension of a composition factor in $V \downarrow T$.

Since $T$ acts on $Z_1$, we have $\dim Z_1 \geq d$, and so $t \leq 2\mu/d$. On the other hand

$$t = [A : T] = [A : TF_0][TF_0 : T] = \frac{[A : F_0]}{[TF_0 : F_0]}[F_0 : T \cap F_0] = 2\frac{\mu^2}{d^2},$$

so we conclude that $\mu = d$ and $t = 2$. This proves claim (ii) of the lemma.

For claim (i), suppose that $G$ acts on $\{Z_1, Z_2\}$. Note that $\varphi \in G^\circ$ and $\eta \in G$ by Lemma 4.3.10. From the relations (2.3) we have $[\varphi, \eta] = f$. But any commutator should act trivially on $\{Z_1, Z_2\}$, so this is a contradiction since $f \notin T$. Therefore $G$ does not act on $\{Z_1, Z_2\}$ and claim (i) follows.

For the remaining claims, consider $\varphi_0$ as defined in Remark 5.5.1. Then $\varphi_0 \in G^\circ$ and $[\varphi_0, g] \in F_0$ for all $g \in G$ by Lemma 5.5.2. Furthermore, since 2 is not a square modulo 5, it follows from Lemma 5.5.2 that for $g \in G^\circ$ we have $[\varphi_0, g] \in \langle f^2 \rangle$ if and only if $\pi(g) \in O^{\varepsilon_1}_{2\ell_1}(2)$. Here $\pi$ is the homomorphism $\pi : N_{\Lambda(V,\kappa)}(F_0, R_1) \to O^{\varepsilon_1}_{2\ell_1}(2)$ used in the definition of $G$.

Therefore if $\mu > 1$ and $X_1 \not\leq \Omega^{\varepsilon_1}_{2\ell_1}(2)$, there exists $g \in G^\circ$ such that $[\varphi_0, g] = f^{\pm 1}$. But commutators in $G^\circ$ must act trivially on $\{Z_1, Z_2\}$, contradicting the fact that $f \notin T$. This completes the proof of claim (iii).

It remains to consider claim (iv). Suppose that $\mu = 1$ or $X_1 \leq \Omega^{\varepsilon_1}_{2\ell_1}(2)$. In this case, it follows from Lemma 5.5.2 that $[\varphi_0, g] \in \langle f^2 \rangle$ for all $g \in G^\circ$. Therefore $N = \langle \varphi_0, f^2 \rangle$ is an abelian normal subgroup of $G^\circ$. If $e = 0$, then $\varphi_0^2 = 1$, so $N \cong C_2 \times C_2$ and $G^\circ$ is metrically imprimitive by Lemma 2.4.3. If $e = 1$, then it follows from Lemma 5.5.5 that $G$ is not maximal solvable. □

**Lemma 5.6.8** *Let* $G = G^{\mathcal{B}}_{\mu,\nu}(X_1, \ldots, X_k)$. *Suppose that* $G$ *is of type* $\mathcal{B}_2$, *with* $q = 3$ *and* $\nu = 1$. *Then the following hold:*

(i) *$G$ is primitive;*
(ii) *If $G^\circ$ has a system of imprimitivity on $V$, the number of summands is 2;*
(iii) *If $\mu > 1$ and $X_1 \not\leq \Omega^{\varepsilon_1}_{2\ell_1}(2)$, then $G^\circ$ is primitive.*
(iv) *If $\mu = 1$ or $X_1 \leq \Omega^{\varepsilon_1}_{2\ell_1}(2)$, then the following hold:*

(a) *If $e = 0$, then $G^\circ$ is metrically imprimitive.*
(b) *If $e = 1$, then $G$ is not maximal solvable.*

***Proof*** The result will follow similarly to the proof of Lemma 5.6.7. In this case $\mathbb{K} = \mathbb{F}_9$. We have $F_0 = \langle f \rangle$, with $f$-eigenspace decomposition $V' = W'_\lambda \oplus W'_{\lambda^3}$, where $\lambda = \zeta^2$ and $\zeta \in \mathbb{K}^\times$ is a primitive element. Furthermore $n = 2\mu$, where $\mu = \dim_{\mathbb{K}} W'_\lambda = \dim_{\mathbb{K}} W'_{\lambda^3}$.

Suppose that $V = Z_1 \oplus \cdots \oplus Z_t$ is a system of imprimitivity for $G^\circ$ on $V$, where $t \geq 2$. Then with $Z'_i := \mathbb{K} \otimes_{\mathbb{F}_3} Z_i$, we have system of imprimitivity $V' = Z'_1 \oplus \cdots \oplus Z'_t$ for $G^\circ$ on $V'$.

As in Lemma 5.6.7, we see that if some $Z'_i$ contains an $f$-eigenvector, then $Z'_i$ contains $W'_\lambda$ or $W'_{\lambda^3}$. Because $Z'_i$ is defined over $\mathbb{F}_q$, by Lemma 2.4.4 this forces $Z'_i = V'$ and thus $Z_i = V$, contrary to $t \geq 2$. Therefore none of the summands $Z'_i$ contains an $f$-eigenvector.

We have $|F_0| = q^\nu + 1 = 4$, so $\mu = 2^{\ell_1}$ for some $\ell_1 \geq 0$. If $\mu > 1$, then $G = G^{\mathcal{B}_2}_{\mu,1}(X_1)$, where $X_1 \leq O^{\varepsilon_1}_{2\ell_1}(2)$ is metrically completely reducible maximal solvable, and $A = F_0 R_1$ with $R_1 = 2^{1+2\ell_1}_{\varepsilon_1}$. In the case where $\mu = 1$, we define $X_1 = 1$ and $R_1 = 1$.

Note that if $\mu > 1$, then $[A, A] = Z(R_1) = \langle f^2 \rangle = \langle -I_V \rangle$. Now the same arguments as in the proof of Lemma 5.6.7 show that $t = 2$, and that $f$ acts nontrivially on $\{Z_1, Z_2\}$. Hence claim (ii) holds.

For claim (i), note that in type $\mathcal{B}_2$ we have $\eta \in G$ with $\eta^2 = f$. Therefore $\eta$ does not act on $\{Z_1, Z_2\}$, since $f$ acts nontrivially on $\{Z_1, Z_2\}$. Hence $G$ does not act on $\{Z_1, Z_2\}$, and we conclude that $G$ is primitive.

For the remaining claims, we consider $\psi_0$ as defined in Remark 5.5.3, so $\psi_0 \in G^\circ$ and $[\psi_0, g] \in F_0$ for all $g \in G$ by Lemma 5.5.4. Since 2 is not a square modulo 3, it follows from Lemma 5.5.4 that for $g \in G^\circ$ we have $[\psi_0, g] \in \langle f^2 \rangle$ if and only if $\pi(g) \in \Omega^{\varepsilon_1}_{2\ell}(2)$.

Therefore if $\mu > 1$ and $X_1 \not\leq \Omega^{\varepsilon_1}_{2\ell_1}(2)$, there exists $g \in G^\circ$ with $[\psi_0, g] = f^{\pm 1}$, contradicting the fact that $f$ acts nontrivially on $\{Z_1, Z_2\}$. This completes the proof of (iii).

For claim (iv), suppose that $\mu = 1$ or $X_1 \leq \Omega^{\varepsilon_1}_{2\ell_1}(2)$. Then $\varepsilon_1 = +$ by Lemma 5.1.3, and $[\psi_0, g] \in \langle f^2 \rangle$ for all $g \in G^\circ$ by Lemma 5.5.4. If $e = 0$, then as in the proof of Lemma 5.6.7 we see that $\langle \psi_0, f^2 \rangle \cong C_2 \times C_2$ is normal in $G^\circ$, so $G^\circ$ is metrically imprimitive (Lemma 2.4.3). If $e = 1$, it follows from Lemma 5.5.7 that $G$ is not maximal solvable. □

**Theorem 5.6.9** *Let $G = G^{\mathcal{B}}_{\mu,\nu}(X_1, \ldots, X_k)$ be irreducible. Then one of the following holds:*

*(i) $G$ is metrically primitive. Furthermore:*

*(a) $G$ is primitive, if it is not of type $\mathcal{B}_1$;*
*(b) If $G$ is of type $\mathcal{B}_1$, then $V = Q_\lambda \oplus Q_{\lambda^{-1}}$ is the unique system of imprimitivity for $G$ on $V$.*

*(ii) $G$ is of type $\mathcal{B}_1$ and $q = 3$, $\nu = 2$.*
*(iii) $G$ is of type $\mathcal{B}_2$ and $q = 2$, $\nu = 3$.*

*In cases (ii)–(iii), the group $G$ is metrically imprimitive.*

***Proof*** Suppose that $V = Z_1 \oplus \cdots \oplus Z_t$ such that $G$ acts on $\{Z_1, \ldots, Z_t\}$. Denoting $Z_i' := \mathbb{K} \otimes_{\mathbb{F}_q} Z_i$, we have $V' = Z_1' \oplus \cdots \oplus Z_t'$ and $G$ acts on $\{Z_1', \ldots, Z_t'\}$. The exceptional cases of Lemma 5.6.4 are handled by Lemmas 5.6.5, 5.6.6, 5.6.7, and 5.6.8.

Therefore by Lemma 5.6.4, we can assume that each $Z_i'$ contains an $f$-eigenvector. Without loss of generality we can assume that $Z_1' \cap W_\lambda' \neq 0$. Because $A$ acts irreducibly on $W_\lambda'$ (Lemma 5.2.2), we have

$$W_\lambda' = \bigoplus_{1 \leq i \leq t} Z_i' \cap W_\lambda'.$$

It follows from Corollary 5.6.3 (i) that the action of $C_G(F_0)$ on $W_\lambda'$ is primitive. Hence $Z_1' \cap W_\lambda' = W_\lambda'$, so $Z_1'$ contains $W_\lambda'$.

On the other hand $Z_1' = \mathbb{K} \otimes_{\mathbb{F}_q} Z_1$ is defined over $\mathbb{F}_q$, so by Lemma 2.4.4 it contains $W'_{\lambda^{q^i}}$ for all $0 \leq i < [\mathbb{K} : \mathbb{F}_q]$. If $G$ is of type $\mathcal{B}_0$, $\mathcal{B}_2$, or $\mathcal{B}_3$, this proves that $Z_1' = V'$ and hence $t = 1$. In the case where $G$ is of type $\mathcal{B}_1$, it follows that $Z_1'$ contains $Q_\lambda'$. Since $G^\circ$ contains an element that swaps $Q_\lambda'$ and $Q_{\lambda^{-1}}'$ (Lemma 4.3.14), we conclude that either $Z_1' = V'$, or $Z_1' = Q_\lambda'$, $Z_2' = Q_{\lambda^{-1}}'$. In other words, either $t = 1$, or $t = 2$ and $\{Z_1, Z_2\} = \{Q_\lambda, Q_{\lambda^{-1}}\}$. □

**Theorem 5.6.10** *Let $G = G^{\mathcal{B}}_{\mu,\nu}(X_1, \ldots, X_k)$ be irreducible. Then one of the following holds:*

*(i) $G^\circ$ is metrically primitive. Furthermore:*

*(a) $G^\circ$ is primitive, if it is not of type $\mathcal{B}_1$.*
*(b) If $G$ is of type $\mathcal{B}_1$, then $V = Q_\lambda \oplus Q_{\lambda^{-1}}$ is the unique system of imprimitivity for $G^\circ$ on $V$.*

*(ii) $G$ is of type $\mathcal{B}_1$ with $q = 3$, $\nu = 2$.*
*(iii) $G$ is of type $\mathcal{B}_1$ with $q = 5$, $\nu = 1$, and $X_1 \leq \Omega^+_{2\ell_1}(2)$ if $\mu > 1$.*
*(iv) $G$ is of type $\mathcal{B}_2$ with $q = 2$, $\nu = 3$.*
*(v) $G$ is of type $\mathcal{B}_2$ with $q = 3$, $\nu = 1$, and $X_1 \leq \Omega^+_{2\ell_1}(2)$ if $\mu > 1$.*
*(vi) $G$ is of type $\mathcal{B}_3$ as in Lemma 5.3.7 (i)–(iii).*

*In cases (ii)–(vi), either $G^\circ$ is metrically imprimitive, or $G$ is not maximal solvable in $\Delta(V, \kappa)$.*

***Proof*** We can assume that (vi) does not hold, so $F_0$ is maximal abelian normal in $G^\circ$ by Lemma 5.3.7. The exceptional cases of Lemma 5.6.4 are handled by Lemmas 5.6.5, 5.6.6, 5.6.7, and 5.6.8. Hence we can also assume that (ii)–(v) do not hold. Now with Lemma 5.6.4 and Corollary 5.6.3 (ii), arguing as in the proof of Theorem 5.6.9 shows that (i) holds. □

**Theorem 5.6.11** *Let $G = G^{\mathcal{B}}_{\mu,\nu}(X_1, \ldots, X_k)$ be irreducible. Suppose that $n$ and $q$ are even, and that $\kappa$ is a quadratic form. Then one of the following holds:*

*(i) $G^\Omega$ is metrically primitive. Furthermore:*

  *(a) $G^\Omega$ is primitive, if it is not of type $\mathcal{B}_1$.*
  *(b) If $G$ is of type $\mathcal{B}_1$, then $V = Q_\lambda \oplus Q_{\lambda^{-1}}$ is the unique system of imprimitivity for $G^\Omega$ on $V$.*

*(ii) $G$ is of type $\mathcal{B}_2$ with $q = 2$, $\nu = 3$.*

*In case (ii) $G$ is metrically imprimitive.*

***Proof*** By Lemma 5.6.6, we know that $G$ is metrically imprimitive in case (ii). Suppose then that (ii) does not hold. Applying Lemma 5.6.4 and Corollary 5.6.3 (iii), it follows as in the proof of Theorem 5.6.9 that (i) holds. □

**Remark 5.6.12** We have seen earlier in Lemma 5.2.4 that $G^\Omega$ is irreducible, except when $G$ is of type $\mathcal{B}_1$ with $n/2 = \mu\nu$ odd. However, in this exceptional case it is still true by Theorem 5.6.11 that $G^\Omega$ is metrically primitive, and $\{Q_\lambda, Q_{\lambda^{-1}}\}$ is the unique system of imprimitivity for $G^\Omega$ on $V$. Furthermore, the action of $G^\Omega$ on $Q_\lambda$ and $Q_{\lambda^{-1}}$ is irreducible and primitive.

As seen from Theorems 5.6.9 and 5.6.10, it is usually the case that if $G = G^{\mathcal{B}}_{\mu,\nu}(X_1, \ldots, X_k)$ is metrically primitive, the same is true for $G^\circ$ as well.

**Corollary 5.6.13** *Let $G = G^{\mathcal{B}}_{\mu,\nu}(X_1, \ldots, X_k)$ be metrically primitive, and suppose that $G^\circ$ is metrically imprimitive. Then one of the following holds:*

*(i) $G$ is of type $\mathcal{B}_1$ with $q = 5$, $\nu = 1$, and $X_1 \le \Omega^+_{2\ell_1}(2)$ if $\mu > 1$.*
*(ii) $G$ is of type $\mathcal{B}_2$ with $q = 3$, $\nu = 1$, and $X_1 \le \Omega^+_{2\ell_1}(2)$ if $\mu > 1$.*
*(iii) $G$ is of type $\mathcal{B}_3$ as in Lemma 5.3.7 (i)–(iii).*

In the case where $\kappa$ is a quadratic form, we have a similar result for $G^\Omega$.

**Corollary 5.6.14** *Suppose that $n$ and $q$ are even, and that $\kappa$ is a quadratic form. Let $G = G^{\mathcal{B}}_{\mu,\nu}(X_1, \ldots, X_k)$ be metrically primitive. Then $G^\Omega$ is metrically primitive.*

We will also need the following observation in the case where $\Delta(V, \kappa) = \mathrm{GSp}_n(q)$.

**Corollary 5.6.15** *Suppose that $n$ is even and $G \leq \mathrm{GSp}_n(q)$ is metrically primitive maximal irreducible solvable. Then $G^\circ$ is metrically primitive and irreducible.*

***Proof*** We know that $G^\circ$ is irreducible by Lemma 5.2.2. We will prove that $G^\circ$ is metrically primitive. If $q$ is even, this is clear since $G = G^\circ Z$ where $Z \leq \mathrm{GL}_n(q)$ is the group of scalar matrices.

Suppose then that $q$ is odd. Because $G$ is maximal solvable, it follows from Lemma 5.6.7 (iv) and Lemma 5.6.8 (iv) that Corollary 5.6.13 (i)–(ii) do not apply. Furthermore, the conditions of Lemma 5.3.7 (i)–(iii) are not satisfied because $G \leq \mathrm{GSp}_n(q)$, so Corollary 5.6.13 (iii) does not apply. Thus it follows from Corollary 5.6.13 that $G^\circ$ is metrically primitive. □

## 5.7 Uniqueness of $F_0$ and $A$

We have determined when groups of the form $G = G^{\mathcal{B}}_{\mu,\nu}(X_1, \ldots, X_k) \leq \Delta(V, \kappa)$ are metrically primitive and irreducible (Theorem 5.6.9, Lemmas 5.2.1 and 5.2.2). Furthermore, we have also seen that the subgroup $F_0$ satisfies properties (F1)–(F3) (Corollary 5.3.5). In this section, we will show that $F_0$ is uniquely determined, and as a corollary, so is the subgroup $A$ (Corollary 5.3.6).

We will also see that with a few exceptions $F_0$ is the unique maximal cyclic normal subgroup of $G^\circ$. This in turn implies that with a few exceptions, we have $G = N_{\Delta(V,\kappa)}(G^\circ)$, which in view of Lemma 2.1.5 is the expected behaviour.

**Lemma 5.7.1** *Suppose that $G = G^{\mathcal{B}}_{\mu,\nu}(X_1, \ldots, X_k) \leq \Delta(V, \kappa)$ and let $1 \leq i \leq k$. Then the action of $G^\circ$ on $R_i/Z(R_i)$ is completely reducible and contains no trivial submodules.*

***Proof*** If $G$ is not of type $\mathcal{B}_3$, the result follows from Lemma 5.3.1 (ii). Suppose then that $G = G^{\mathcal{B}_3}_{\mu,\nu}(X_1)$ is of type $\mathcal{B}_3$, so $n = \mu = 2^{\ell_1}$ and $X \leq O^{\varepsilon_1}_{2\ell_1}(2)$ is metrically completely reducible maximal solvable. We have $X_1 = H_1 \times \cdots \times H_t$ with respect to a decomposition $R_1/Z(R_1) = W_1 \perp \cdots \perp W_t$, where $H_j \leq O(W_j, \vartheta)$ is maximal irreducible solvable for all $1 \leq j \leq t$ (Lemma 2.3.7).

Since the action of $G$ on $R_1/Z(R_1)$ contains $X_1^\Omega$ (Lemma 4.3.27), it will suffice to check that $H_j^\Omega$ has no trivial submodules on $W_j$. Suppose for the sake of contradiction that $W_j^{H_j^\Omega} \neq 0$. Because $H_j^\Omega \trianglelefteq H_j$, the subspace $W_j^{H_j^\Omega}$ is $H_j$-invariant, so by irreducibility $W_j = W_j^{H_j^\Omega}$. Then $H_j^\Omega = 1$, in which case $H_j$ must be cyclic of order $\leq 2$. But this is clearly impossible when $H_j$ is irreducible. □

**Lemma 5.7.2** *Let $G = G^{\mathcal{B}}_{\mu,\nu}(X_1, \ldots, X_k)$ and $A = F_0 R_1 \cdots R_k$, where $R_i \trianglelefteq G$ are extraspecial groups as in the construction of $G$. Then $R_1 \cdots R_k \leq [G^\circ, G^\circ]$.*

***Proof*** We will prove that $R_i = [R_i, G^\circ]$ for all $1 \le i \le k$. To this end, it follows from Lemma 5.7.1 that the action of $G^\circ$ on $R_i/Z(R_i)$ is completely reducible and has no trivial submodules. By Lemma 2.1.17, the images of $[x, g]$ in $R_i/Z(R_i)$ with $x \in R_i$ and $g \in G^\circ$ generate all of $R_i/Z(R_i)$. Furthermore $[R_i, R_i] = Z(R_i)$, so we have $R_i = [R_i, G^\circ]$. Because $R_i \le G^\circ$, we conclude that $R_1 \cdots R_k \le [G^\circ, G^\circ]$. □

**Lemma 5.7.3** *Let* $G = G^{\mathcal{B}}_{\mu,\nu}(X_1, \ldots, X_k)$*. Suppose that* $n$ *and* $q$ *are even, and that* $\kappa$ *is a quadratic form. Let* $A = F_0R_1 \cdots R_k$*, where* $R_i \trianglelefteq G$ *are extraspecial groups as in the construction of* $G$*. Then* $R_1 \cdots R_k \le [G^\Omega, G^\Omega]$.

***Proof*** By Lemma 5.3.1 (iii), the action of $G^\Omega$ on $R_i/Z(R_i)$ is completely reducible and contains no trivial submodules. Thus arguing as in the proof of Lemma 5.7.2 shows that $R_i = [R_i, G^\Omega]$ for all $1 \le i \le k$. Since $R_i \le G^\Omega$, the result follows. □

**Theorem 5.7.4** *Let* $G = G^{\mathcal{B}}_{\mu,\nu}(X_1, \ldots, X_k)$*, and assume that* $G$ *is metrically primitive. Then one of the following holds.*

(i) $F_0$ *is the unique maximal cyclic normal subgroup of* $G^\circ$.
(ii) $G$ *is of type* $\mathcal{B}_1$*, with* $q = 5$, $\nu = 1$, $e = 1$*, and* $X_1 \le \Omega^+_{2\ell_1}(2)$ *if* $\mu > 1$.
(iii) $G$ *is of type* $\mathcal{B}_2$*, with* $q = 3$, $\nu = 1$, $e = 1$*, and* $X_1 \le \Omega^+_{2\ell_1}(2)$ *if* $\mu > 1$.

*In cases (ii) and (iii) there are exactly three maximal cyclic normal subgroups in* $G^\circ$*, all of which are isomorphic to* $F_0$*. Furthermore, in cases (ii) and (iii) the subgroup* $F_0$ *is the only maximal cyclic normal subgroup of* $G^\circ$ *that is normalized by* $G$*.*

***Proof*** Suppose that $F_0' \trianglelefteq G^\circ$ is maximal cyclic normal. We will first prove the result in three special cases.

**Case 1: $G$ is of type $\mathcal{B}_3$**
Since $F_0'$ is cyclic, the automorphism group $\mathrm{Aut}(F_0')$ is abelian, and thus the quotient $G^\circ/C_{G^\circ}(F_0')$ is abelian. It follows then from Lemma 5.7.2 that $F_0'$ centralizes $A = R_1$, so $F_0' \le A$ by Lemma 4.3.25. Then $F_0' \le Z(A) = F_0$, and $F_0' = F_0$ by maximality of $F_0'$. This proves that (i) holds in this case.

**Case 2: $G$ is of type $\mathcal{B}_1$ *with* $q = 5$ and $\nu = 1$**
In this case $|F_0| = 5 - 1 = 4$, so $\mu = 2^{\ell_1}$ for some $\ell_1 \ge 0$. If $\mu > 1$, we have $G = G^{\mathcal{B}_1}_{\mu,1}(X_1)$, where $X_1 \le O^{\varepsilon_1}_{2\ell_1}(2)$ is metrically completely reducible maximal solvable, and $A = F_0R_1$ with $R_1 = 2^{1+2\ell_1}_{\varepsilon_1}$. If $\mu = 1$, we define $X_1 = 1$ and $R_1 = 1$.

Let $\varphi_0$ be defined as in Remark 5.5.1, so $\varphi_0 \in G^\circ$ and $\varphi_0$ centralizes $R_1$ by Lemma 5.5.2. Denote $R := C_{G^\circ}(R_1)$. Then $R \le \mathrm{Ker}\,\pi$, and $\mathrm{Ker}\,\pi = \langle A, \varphi_0\rangle$ by Lemma 4.3.10. Because the centralizer of $R_1$ in $A$ is equal to $F_0$ (Lemma 4.3.32), we deduce that $R = \langle f, \varphi_0\rangle$.

As in the previous case, it follows from Lemma 5.7.2 that $F_0'$ centralizes $R_1$, so $F_0' \leq R = \langle f, \varphi_0 \rangle$. We have

$$\begin{aligned} f^2 &= -1, \\ \varphi_0^2 &= (-1)^{e+\sum_{i=1}^k u_i}, \\ [\varphi_0, f] &= -1. \end{aligned}$$

Therefore if $e + \sum_{i=1}^k u_i \equiv 0 \mod 2$, we have $R \cong 2_+^{1+2}$. In this case every cyclic normal subgroup of $R$ is contained in $\langle f \rangle$, so $F_0' \leq \langle f \rangle$ and thus $F_0' = \langle f \rangle = F_0$ by maximality of $F_0'$. This proves that (i) holds in this case.

Suppose then that $e + \sum_{i=1}^k u_i \equiv 1 \mod 2$, in which case $R \cong 2_-^{1+2}$. In this case every cyclic normal subgroup of $R$ is contained in $\langle f \rangle$, $\langle \varphi_0 \rangle$, or $\langle \varphi_0 f \rangle$. Thus $F_0' \leq R$ must be equal to one of these three subgroups.

If $\mu > 1$ and $X_1 \not\leq \Omega_{2\ell_1}^{\varepsilon_1}(2)$, it follows from Lemma 5.5.2 that there exists $g \in G^\circ$ such that $[\varphi_0, g] = f^{\pm 1}$. Thus in this case $\langle \varphi_0 \rangle$ and $\langle \varphi_0 f \rangle$ are not normal in $G^\circ$, so we must have $F_0' = \langle f \rangle = F_0$.

We consider then the case where $\mu = 1$ or $X_1 \leq \Omega_{2\ell_1}^{\varepsilon_1}(2)$. In this case $\varepsilon_1 = +$ by Lemma 5.1.3, so statement (ii) of the lemma holds. It remains to verify the claims of the lemma in this case. First note that since 2 is not a square modulo 5, by Lemma 5.5.2 we have $[\varphi_0, g] \in \langle f^2 \rangle$ for all $g \in G^\circ$. Therefore $\langle f \rangle$, $\langle \varphi_0 \rangle$, and $\langle \varphi_0 f \rangle$ are all maximal cyclic normal subgroups $G^\circ$.

Furthermore, by the relations (2.3), for the element $\eta \in G$ we have $[\eta, \varphi_0] = f$, so $\langle \varphi_0 \rangle$ and $\langle \varphi_0 f \rangle$ are not normal in $G$. This completes the proof of the lemma in this case.

**Case 3: $G$ is of type $\mathcal{B}_2$ *with* $q = 3$ *and* $\nu = 1$**
In this case $|F_0| = 3 + 1 = 4$, so $\mu = 2^{\ell_1}$ for some $\ell_1 \geq 0$. As in the previous case, if $\mu > 1$, we have $G = G_{\mu,1}^{\mathcal{B}_1}(X_1)$, where $X_1 \leq O_{2\ell_1}^{\varepsilon_1}(2)$ is metrically completely reducible maximal solvable, and $A = F_0 R_1$ with $R_1 = 2_{\varepsilon_1}^{1+2\ell_1}$. If $\mu = 1$, we define $X_1 = 1$ and $R_1 = 1$.

Let $\psi_0$ be defined as in Remark 5.5.3, so $\psi_0 \in G^\circ$ and $\psi_0$ centralizes $R_1$ by Lemma 5.5.4. Arguing as in Case 2 shows that for $R := C_{G^\circ}(R_1)$, we have $R = \langle \psi_0, f \rangle$.

It follows from Lemma 5.7.2 that $F_0'$ centralizes $R_1$, so $F_0 \leq R = \langle \psi_0, f \rangle$. If $e + \sum_{i=1}^k u_i \equiv 0 \mod 2$, as in Case 2 we see that $R \cong 2_+^{1+2}$ and thus $F_0' = F_0$.

In the case where $e + \sum_{i=1}^k u_i \equiv 1 \mod 2$, we see similarly to Case 2 that $R \cong 2_-^{1+2}$, and thus $F_0'$ must be equal to $\langle f \rangle$, $\langle \psi_0 \rangle$, or $\langle f \psi_0 \rangle$. If $\mu > 1$ and $X_1 \not\leq \Omega_{2\ell_1}^{\varepsilon_1}(2)$, it follows from Lemma 5.5.4 that there exists $g \in G^\circ$ such that $[\psi_0, g] = f^{\pm 1}$, so $F_0' = F_0$ as in Case 2.

Suppose then that $\mu = 1$ or $X_1 \leq \Omega_{2\ell_1}^{\varepsilon_1}(2)$, so $\varepsilon_1 = +$ by Lemma 5.1.3. We are then precisely in case (iii) of the lemma. For the remaining claims, note that $[\psi_0, g] \in \langle f^2 \rangle$ for all $g \in G^\circ$ by Lemma 5.5.4. Therefore as in Case 2, the subgroups $\langle f \rangle$, $\langle \psi_0 \rangle$, and $\langle \psi_0 f \rangle$ are all normal in $G^\circ$. Again similarly to Case 2, from (2.7) we have $[\eta, \psi_0] = \eta^2 = f$, so $\langle \psi_0 \rangle$ and $\langle \psi_0 f \rangle$ are not normal in $G$.

We can assume then for the remainder of the proof that Case 1, Case 2, and Case 3 do not hold. Then $G^\circ$ is metrically primitive by Corollary 5.6.13. Furthermore, by Lemma 5.3.7 the subgroup $F_0$ is maximal abelian normal in $G^\circ$.

If $[F_0, F_0'] = 1$, then $F_0 F_0'$ is an abelian normal subgroup of $G^\circ$, and thus $F_0' \leq F_0$ by the maximality of $F_0$. Suppose then that $[F_0, F_0'] \neq 1$. We will first use an argument from [87, Proposition 2.11] to prove the following.

**Claim:** $[F_0, F_0']$ **has order** 2
Since $F_0$ and $F_0'$ are abelian normal subgroups and $[F_0, F_0'] \neq 1$, there exist Sylow $r$-subgroups $R$ and $R'$ of $F_0$ and $F_0'$ respectively, such that $[R, R'] \neq 1$.

Let $D = R \cap R'$ and denote $|D| = r^\alpha$. We have $1 \lneqq D \lneqq R, R'$ since $[R, R'] \neq 1$, so $\alpha > 0$ and there exist $r$-subgroups $D < H < R$ and $D < H' < R'$ such that $|H| = |H'| = r^{\alpha+1}$.

Consider the subgroup $S = HH'$, which is a normal subgroup of $G^\circ$ with $|S| = r^{\alpha+2}$. Note that $[H, H'] \neq 1$. Indeed, if $[H, H'] = 1$, then $S$ would have exponent $r^{\alpha+1}$, and thus would be a noncyclic abelian normal subgroup of $G^\circ$, contradicting the fact that $G^\circ$ is metrically primitive (Lemma 2.4.3). Moreover $D$ is central in $S$, so it follows that $S/D \cong C_r \times C_r$. Since $H$ and $H'$ are cyclic, we can find generators $z$ and $w$ for $H$ and $H'$, respectively, such that $z^r = w^r$.

The commutator subgroup $[S, S] \leq D$ is central in $S$, so the identity

$$(ab)^k = a^k b^k [b, a]^{k(k-1)/2} \tag{5.4}$$

holds for all $a, b \in S$ and integers $k > 0$. Therefore if $r > 2$ or $\alpha > 1$, the set

$$T = \{x \in S : x^{r^\alpha} = 1\}$$

is a subgroup of $S$, and clearly $D \leq T \trianglelefteq G^\circ$. Furthermore, by (5.4) we have $zw^{-1} \in T$. Then $D \lneqq T \lneqq S$, so $|T| = r^{\alpha+1}$. Thus $T \cong C_{r^\alpha} \times C_r$ is a noncyclic abelian normal subgroup of $G^\circ$, which contradicts Lemma 2.4.3.

We conclude then that the only possibility is that $r = 2$ and $\alpha = 1$. In particular, for all odd primes the Sylow subgroups of $F_0$ and $F_0'$ commute, so $[F_0, F_0'] \leq [R, R'] \leq D$. Thus $[F_0, F_0'] = D$ has order 2, as claimed.

In particular, we have shown that $F_0$ has a subgroup of order 2, which for all types $\mathcal{B}_0$–$\mathcal{B}_2$ implies that $q$ must be odd. Then both $F_0$ and $F_0'$ must contain $-I_V$, so we conclude that $[F_0, F_0'] = \langle -I_V \rangle$. Let $f$ be a generator of $F_0$ and let $g$ be a generator of $F_0'$. Then $[f, g] = -I_V$, so $g^{-1} f g = -f$.

Let $\lambda \in \mathbb{K}$ be a eigenvalue of $f$ on $V' = \mathbb{K} \otimes_{\mathbb{F}_q} V$. Since $g^{-1} f g = -f$, it follows that $-\lambda$ is also an eigenvalue of $f$ on $V'$. We will use this fact to reach a contradiction in all cases.

**Case: $G$ is of type $\mathcal{B}_0$**
In this case $f$ has eigenvalues $\lambda, \lambda^q, \ldots, \lambda^{q^{\nu-1}}$ on $V'$, where $\mathbb{K} = \mathbb{F}_{q^\nu}$ and $\lambda \in \mathbb{K}^\times$ is a primitive element. Since one of the eigenvalues must be equal to $-\lambda$, we have $-\lambda = \lambda^{q^\delta}$ for some $0 \leq \delta < \nu$. Then $\lambda^{2(q^\delta-1)} = 1$, which implies $q^\nu - 1 \mid 2(q^\delta - 1)$. For $0 < \delta < \nu$ we have $q^\nu - 1 > 2(q^\delta - 1)$, so $\delta = 0$. But then $\lambda = -\lambda$, which is impossible since $q$ is odd.

**Case: $G$ is of type $\mathcal{B}_1$**
In this case $f$ has eigenvalues $\lambda, \lambda^q, \ldots, \lambda^{q^{\nu-1}}, \lambda^{-1}, \lambda^{-q}, \ldots, \lambda^{-q^{\nu-1}}$ on $V' = \mathbb{K} \otimes_{\mathbb{F}_q} V$, where $\mathbb{K} = \mathbb{F}_{q^\nu}$ and $\lambda \in \mathbb{K}^\times$ is a primitive element.

We know that one of the eigenvalues must be equal to $-\lambda$. If $-\lambda = \lambda^{q^\delta}$ for some $0 \leq \delta < \nu$, then we have a contradiction as in the previous case. Therefore $-\lambda = \lambda^{-q^\delta}$ for some $0 \leq \delta < \nu$. Then $\lambda^{2(q^\delta+1)} = 1$, so $q^\nu - 1 \mid 2(q^\delta + 1)$. It is straightforward to check that for $q$ odd, this can only happen for $(q, \nu)$ equal to $(5, 1)$, $(3, 2)$, or $(3, 1)$. Here the case $(q, \nu) = (5, 1)$ has already been excluded earlier in the proof. Furthermore, the case $(q, \nu) = (3, 2)$ is not applicable since $G$ would be metrically imprimitive (Lemma 5.6.5), and $(q, \nu) = (3, 1)$ is excluded from the definition of groups of type $\mathcal{B}_1$. Thus we have a contradiction.

**Case: $G$ is of type $\mathcal{B}_2$**
In this case $f$ has eigenvalues $\lambda, \lambda^q, \ldots, \lambda^{q^{2\nu-1}}$ on $V' = \mathbb{K} \otimes_{\mathbb{F}_q} V$, where $\mathbb{K} = \mathbb{F}_{q^{2\nu}}$ and $\lambda \in \mathbb{K}^\times$ has order $q^\nu + 1$.

One of the eigenvalues must be equal to $-\lambda$, so $-\lambda = \lambda^{q^\delta}$ for some $0 \leq \delta < 2\nu$. We have $\lambda^{2(q^\delta-1)} = 1$, so $q^\nu + 1 \mid 2(q^\delta - 1)$. It is straightforward to check that this can only happen when $q = 3$ and $\nu = 1$ (see Lemma 2.1.19), a case that we have already excluded.

We have reached a contradiction in all cases, so the theorem follows. □

As a corollary, we have the following result.

**Theorem 5.7.5** *Let $G = G^{\mathcal{B}}_{\mu,\nu}(X_1, \ldots, X_k)$, and assume that $G$ is metrically primitive and irreducible. Then $F_0$ is the unique subgroup of $G$ that satisfies properties (F1)–(F3).*

***Proof*** As seen in Corollary 5.3.5, the subgroup $F_0$ satisfies properties (F1)–(F3). Suppose then that $F_0' \leq G$ is another subgroup of $G$ with properties (F1)–(F3). Because $G$ is metrically primitive and $F_0'$ is abelian normal, it follows from Lemma 2.4.3 that $F_0'$ is cyclic. Thus $F_0'$ is a cyclic normal subgroup of $G^\circ$, so by Theorem 5.7.4 we have $F_0' \leq F_0$. By maximality of $F_0'$ we have $F_0' = F_0$, which completes the proof of the theorem. □

Since we know that every metrically primitive maximal irreducible solvable subgroup of $\Delta(V, \kappa)$ is of the form $G^{\mathcal{B}}_{\mu,\nu}(X_1, \ldots, X_k)$, we have the following corollary.

**Corollary 5.7.6** *Let $G \leq \Delta(V, \kappa)$ be metrically primitive maximal irreducible solvable. Then $G$ has a unique subgroup $F_0$ that satisfies properties (F1)–(F3).*

**Remark 5.7.7** In [77, Theorem 20.9], Suprunenko proves that a maximal irreducible primitive solvable subgroup of $\mathrm{GL}_n(\mathbb{K})$ ($\mathbb{K}$ any field) has a unique maximal abelian normal subgroup. This is also true for imprimitive maximal irreducible solvable subgroups of $\mathrm{GL}_n(\mathbb{K})$ ($\mathbb{K}$ any field), by a result of Kozel and Tyškevič [58].

For the subgroup $A$ we have previously seen that $A = \mathrm{Fit}(C_{G^\circ}(F_0))$ (Corollary 5.3.6), so from the uniqueness of $F_0$ we also get the uniqueness of $A$.

**Corollary 5.7.8** *Let $G = G^{\mathcal{B}}_{\mu,\nu}(X_1, \ldots, X_k)$, and assume that $G$ is metrically primitive and irreducible. Then $F_0$ is the unique subgroup of $G$ that satisfies properties (F1)–(F3), and $A$ is the unique subgroup of $G$ satisfying properties (A1)–(A3).*

**Theorem 5.7.9** *Let $G = G^{\mathcal{B}}_{\mu,\nu}(X_1, \ldots, X_k)$ with $A = F_0 R_1 \cdots R_k$, and assume that $G$ is metrically primitive and irreducible. Suppose that $R \leq A$ is such that $R \trianglelefteq G^\circ$ and $R \cong (r_i)^{1+2\ell_i}_{\varepsilon_i}$ for some $1 \leq i \leq k$. Then $R = R_i$.*

***Proof*** If $r_i > 2$, then we have $R_i = \{x \in A : x^{r_i} = 1\}$. Therefore $R \leq R_i$, and $R = R_i$ because $|R| = |R_i|$.

Suppose then that $r_i = 2$. We have $R_i \cap F_0 = Z(R_i) = \langle \theta \rangle$, where $\theta \in F_0$ has order 2. We have $O_2(A/F_0) = R_1 F_0/F_0 \cong C_2^{2\ell_1}$, and we have a non-degenerate $G$-invariant bilinear form $\xi$ on $O_2(A/F_0)$ defined by

$$[x, y] = \theta^{\xi(\overline{x},\overline{y})}$$

for all $x, y \in R_1 F_0$. Furthermore, there is a $G$-invariant quadratic form $\vartheta : O_2(A/F_0) \to \mathbb{F}_2$ with polarization $\xi$, defined by

$$x^2 = \theta^{\vartheta(\overline{xy})}$$

for all $x \in R_1$ and $y \in F_0$.

Because $R$ centralizes $F_0$, we have $R \cap F_0 = Z(R) = \langle \theta \rangle$. Consequently $RF_0/F_0 \cong C_2^{2\ell_1}$, so $O_2(A/F_0) = RF_0/F_0$. Because $R \trianglelefteq G^\circ$, we have a $G^\circ$-invariant quadratic form $\vartheta' : O_2(A/F_0) \to \mathbb{F}_2$ with polarization $\xi$, defined by

$$x^2 = \theta^{\vartheta'(\overline{xy})}$$

for all $x \in R$ and $y \in F_0$.

We have $O_2(A/F_0) = R_1 F_0/F_0 \cong R_1/Z(R_1)$ as $\mathbb{F}_2[G^\circ]$-modules, so it follows from Lemmas 5.7.1 and 2.3.4 that $\vartheta = \vartheta'$. On the other hand

$$R_1 = \{x \in R_1 F_0 : x^2 = \theta^{\vartheta(\overline{x})}\}$$

and

$$R = \{x \in RF_0 : x^2 = \theta^{\vartheta'(\overline{x})}\},$$

so we conclude that $R = R_1$. □

**Theorem 5.7.10** *Let $G = G^{\mathcal{B}}_{\mu,\nu}(X_1, \ldots, X_k)$, and assume that $G$ is metrically primitive and irreducible. Then one of the following holds.*

*(i)* $G = N_{\Delta(V,\kappa)}(G^\circ)$.
*(ii)* *$G$ is of type $\mathcal{B}_1$, with $q = 5$, $\nu = 1$, $e = 1$, and $X_1 \leq \Omega^+_{2\ell_1}(2)$ if $\mu > 1$.*
*(iii)* *$G$ is of type $\mathcal{B}_2$, with $q = 3$, $\nu = 1$, $e = 1$, and $X_1 \leq \Omega^+_{2\ell_1}(2)$ if $\mu > 1$.*

*Furthermore, in cases (ii) and (iii) the group $G$ is not maximal solvable in $\Delta(V, \kappa)$.*

***Proof*** The fact that $G$ is not maximal solvable in cases (ii) and (iii) follows from Lemma 5.5.5 and Lemma 5.5.7, respectively.

Thus we can assume that (ii) and (iii) do not hold. We have $G^\circ \trianglelefteq G$ so $G \leq N_{\Delta(V,\kappa)}(G^\circ)$, and we will show that equality holds. To this end, let $g \in N_{\Delta(V,\kappa)}(G^\circ)$.

Because (ii) and (iii) do not hold, it follows from Lemma 5.7.4 that $F_0$ is the unique maximal cyclic normal subgroup of $G^\circ$. Therefore $g$ must normalize $F_0$. If $\mu = 1$, we conclude that $g \in G$ (Definition 2.9.5) as required.

Suppose then for the rest of the proof that $\mu > 1$. Because $g$ normalizes $F_0$, it also normalizes $C_{G^\circ}(F_0)$. Consequently $g$ normalizes $A$, because $A = \mathrm{Fit}(C_{G^\circ}(F_0))$ (Corollary 5.3.6).

Since $g$ normalizes $A$ and $G^\circ$, it follows from Theorem 5.7.9 that $g$ normalizes the subgroups $R_1, \ldots, R_k$ with $A = F_0 R_1 \cdots R_k$. Thus $g \in N_{\Delta(V,\kappa)}(F_0, R_1, \ldots, R_k)$. Let

$$\pi : N_{\Delta(V,\kappa)}(F_0, R_1, \ldots, R_k) \to \prod_{i=1}^{k} \mathrm{GSp}_{2\ell_i}(r_i)$$

be the homomorphism defined in the construction of $G$.

For $1 \leq i \leq k$, let $G_i$ be the action of $G^\circ$ on $R_i/Z(R_i)$. We have $\pi(g) = (g_1, \ldots, g_k)$, where $g_i$ normalizes $G_i$. If $G$ is not of type $\mathcal{B}_3$, then $X_i^\circ \leq G_i \leq X_i$, so $g_i$ normalizes $G_i^\circ = X_i^\circ$. Because $X_i$ is maximal solvable in $\mathrm{GSp}_{2\ell_i}(r_i)$, it follows from Lemma 2.1.5 that $g_i \in X_i$ for all $1 \leq i \leq k$. Thus $\pi(g) \in X_1 \times \cdots \times X_k$, so $g \in G$.

Suppose then that $G$ is of type $\mathcal{B}_3$, in which case $k = 1$ and $X_1^\Omega \leq G_1 \leq X_1$, so $\pi(g) = g_1$ normalizes $G_1^\Omega = X_1^\Omega$. Because $X_1$ is maximal solvable in $O^{\varepsilon_1}_{2\ell_1}(2)$, we have $g_1 \in X_1$ by Lemma 2.1.5, and so $g \in G$. □

**Corollary 5.7.11** *Let $G = G^{\mathcal{B}}_{\mu,\nu}(X_1, \ldots, X_k)$, and assume that $G$ is metrically primitive. If $G^\circ$ is maximal solvable in $I(V, \kappa)$, then $G$ is maximal solvable in $\Delta(V, \kappa)$.*

***Proof*** Suppose that $G^\circ$ is maximal solvable in $I(V, \kappa)$. To prove that $G$ is maximal solvable, suppose that $G \leq X \leq \Delta(V, \kappa)$ with $X$ solvable. Then $G^\circ = X^\circ$ since $G^\circ$ is maximal solvable. If $G$ is not of type $\mathcal{B}_3$, then $G$ is of multiplier 1 (Lemma 4.3.13 (i), Lemma 4.3.22 (i)). Thus it follows from $[X : X^\circ] \leq q - 1$ that $X = G$. If $G$ is of type $\mathcal{B}_3$, then $X \leq N_{\Delta(V,\kappa)}(G^\circ) = G$ by Theorem 5.7.10, so $X = G$. Therefore we conclude that $G$ is maximal solvable. □

## 5.8 Conjugacy of $G^{\mathcal{B}}_{\mu,\nu}(X_1, \ldots, X_k)$

In this section, we will determine when metrically primitive irreducible groups of the form $G^{\mathcal{B}}_{\mu,\nu}(X_1, \ldots, X_k)$ are conjugate in $\Delta(V, \kappa)$. We begin with the following lemma, which shows that subgroups of different types are not conjugate.

**Lemma 5.8.1** *Let $G = G^{\mathcal{B}}_{\mu,\nu}(X_1, \ldots, X_k)$ and $\overline{G} = G^{\overline{\mathcal{B}}}_{\overline{\mu},\overline{\nu}}(Y_1, \ldots, Y_\ell)$ be metrically primitive irreducible subgroups of $\Delta(V, \kappa)$. Let $F_0$ be a subgroup satisfying properties (F1)–(F3) in $G$, and let $A$ be a subgroup satisfying properties (A1)–(A3) in $G$. Let $\overline{F_0}$ and $\overline{A}$ be the corresponding subgroups in $\overline{G}$.*

*Suppose that $G$ and $\overline{G}$ are conjugate in $\Delta(V, \kappa)$, with $gGg^{-1} = \overline{G}$ for some $g \in \Delta(V, \kappa)$. Then all of the following hold:*

*(i) $gF_0g^{-1} = \overline{F_0}$ and $gAg^{-1} = \overline{A}$.*
*(ii) $\mathcal{B} = \overline{\mathcal{B}}$, $\mu = \overline{\mu}$, and $\nu = \overline{\nu}$.*

***Proof*** Claim (i) follows from the uniqueness of $F_0$ and $A$ in $G$ (Corollary 5.7.8).

In particular, we have $|F_0| = |\overline{F_0}|$. Suppose first that $\kappa = 0$, in which case $\mathcal{B} = \overline{\mathcal{B}} = \mathcal{B}_0$. Then $|F_0| = q^\nu - 1$ and $|\overline{F_0}| = q^{\overline{\nu}} - 1$, so $q^\nu - 1 = q^{\overline{\nu}} - 1$ which implies $\nu = \overline{\nu}$. Since $\mu = n/\nu$, we also have $\mu = \overline{\mu}$.

We can assume then that $\kappa \neq 0$. If $G$ is of type $\mathcal{B}_3$, then $|F_0| = 2$. Since $|\overline{F_0}| > 2$ if $\overline{G}$ is of type $\mathcal{B}_1$ or $\mathcal{B}_2$, it follows that $\overline{G}$ is also of type $\mathcal{B}_3$. Then $\mathcal{B} = \overline{\mathcal{B}} = \mathcal{B}_3$, $\mu = \overline{\mu} = n$, and $\nu = \overline{\nu} = 1$. Similarly if $\overline{G}$ is of type $\mathcal{B}_3$, we see that (ii) holds.

Thus we assume that $G$ and $\overline{G}$ belong to type $\mathcal{B}_1$ or $\mathcal{B}_2$. If $G$ and $\overline{G}$ are of the same type, it follows immediately from $|F_0| = |\overline{F_0}|$ that $\nu = \overline{\nu}$. Then $n = 2\mu\nu = 2\overline{\mu}\,\overline{\nu}$ implies that (ii) holds.

It remains to consider the case where $G$ and $\overline{G}$ are of different types. We can assume without loss of generality that $G$ is of type $\mathcal{B}_1$, and $\overline{G}$ is of type $\mathcal{B}_2$. Then $q^\nu - 1 = q^{\overline{\nu}} + 1$. It is elementary to see that the only possible solution is $q = 2$, $\nu = 2$, $\overline{\nu} = 1$. But $(q, \nu) = (2, 2)$ is excluded from the definition of groups of type $\mathcal{B}_1$, so we have a contradiction. □

It follows from Lemma 5.8.1 that in order to describe subgroups of the form $G^{\mathcal{B}}_{\mu,\nu}(X_1, \ldots, X_k)$ up to conjugacy, it will suffice to consider those constructed using the same subgroups $F_0$ and $A = F_0 R_1 \cdots R_k$.

**Lemma 5.8.2** *Let $G = G^{\mathcal{B}}_{\mu,\nu}(X_1, \ldots, X_k)$ and $\overline{G} = G^{\mathcal{B}}_{\mu,\nu}(Y_1, \ldots, Y_k)$ be metrically primitive irreducible subgroups, constructed with the same $F_0$ and $A = F_0 R_1 \cdots R_k$. Then the following statements hold:*

*(i) If $G$ and $\overline{G}$ are conjugate in $\Delta(V, \kappa)$, then $X_i$ and $Y_i$ are conjugate in $\mathrm{GSp}_{2\ell_i}(r_i)$ for all $1 \leq i \leq k$.*
*(ii) If $X_i$ and $Y_i$ are $\mathrm{Sp}_{2\ell_i}(r_i)$-conjugate for all $1 \leq i \leq k$, then $G$ and $\overline{G}$ are conjugate in $\Delta(V, \kappa)$.*

***Proof*** For claim (i), suppose that $G$ and $\overline{G}$ are conjugate and let $g \in \Delta(V, \kappa)$ be such that $gGg^{-1} = \overline{G}$. It follows from the uniqueness of $F_0$ and $R_1, \ldots, R_k$ (Corollary 5.7.8, Theorem 5.7.9) that $g \in N_{\Delta(V,\kappa)}(F_0, R_1, \ldots, R_k)$.

Then $\pi(g) = (g_1, \ldots, g_k)$, where $g_i$ is the action of $g$ on $R_i/Z(R_i)$ for all $1 \leq i \leq k$. We have $\pi(g)\pi(G)\pi(g)^{-1} = \pi(\overline{G})$. Since $X_1^\circ \times \cdots \times X_k^\circ \leq \pi(G)$ and $Y_1^\circ \times \cdots \times Y_k^\circ \leq \pi(\overline{G})$ (Lemma 4.3.35), it follows that $g_i X_i^\circ g_i^{-1} = Y_i^\circ$ for all $1 \leq i \leq k$.

Because $X_i$ and $Y_i$ are maximal solvable in $\mathrm{GSp}_{2\ell_i}(r_i)$, by Lemma 2.1.5 we have

$$X_i = N_{\mathrm{GSp}_{2\ell_i}(r_i)}(X_i^\circ), \qquad Y_i = N_{\mathrm{GSp}_{2\ell_i}(r_i)}(Y_i^\circ).$$

Therefore we conclude that $g_i X_i g_i^{-1} = Y_i$ for all $1 \leq i \leq k$, so (i) holds.

For (ii), suppose that $X_i$ and $Y_i$ are $\mathrm{Sp}_{2\ell_i}(r_i)$-conjugate for all $1 \leq i \leq k$, and let $g_i \in \mathrm{Sp}_{2\ell_i}(r_i)$ be such that $g_i X_i g_i^{-1} = Y_i$. If $r_i = 2$, it follows from Lemma 2.3.4 that $g_i \in O^{\varepsilon_i}_{2\ell_i}(2)$. Therefore by Lemma 4.3.34, there exists $g \in N_{\Delta(V,\kappa)}(F_0, R_1, \ldots, R_k)$ such that $\pi(g) = (g_1, \ldots, g_k)$. Then

$$g\pi^{-1}(X_1 \times \cdots \times X_k)g^{-1} = \pi^{-1}(g_1 X_1 g_1^{-1} \times \cdots \times g_k X_k g_k^{-1}) = \pi^{-1}(Y_1 \times \cdots \times Y_k),$$

so $gGg^{-1} = \overline{G}$. □

In view of Lemma 5.8.2, for the conjugacy of $G = G^{\mathcal{B}}_{\mu,\nu}(X_1, \ldots, X_k)$ and $\overline{G} = G^{\mathcal{B}}_{\mu,\nu}(Y_1, \ldots, Y_k)$, it suffices to consider the case where $Y_i$ is $\mathrm{GSp}_{2\ell_i}(r_i)$-conjugate to $X_i$ for all $1 \leq i \leq k$. We proceed to consider this case.

Suppose that $X_i$ and $Y_i$ are conjugate in $\mathrm{GSp}_{2\ell_i}(r_i)$ for all $1 \leq i \leq k$. If $X_i$ is of multiplier 1, then it follows from Lemma 2.1.10 that the $\mathrm{GSp}_{2\ell_i}(r_i)$-conjugacy class of $X_i$ forms a single $\mathrm{Sp}_{2\ell_i}(r_i)$-conjugacy class. In the case where $X_i$ is of multiplier 2, the conjugacy class of $X_i$ splits into two $\mathrm{Sp}_{2\ell_i}(r_i)$-conjugacy classes. For $1 \leq i \leq k$ with $X_i$ of multiplier 2, let $X_i^+$ and $X_i^-$ be representatives for the two $\mathrm{Sp}_{2\ell_i}(r_i)$-classes. In the case where $X_i$ is of multiplier 1, we define $X_i^+ := X_i$.

Let $I$ be the set of indices $1 \le i \le k$ such that $X_i$ is of multiplier 2. It follows now from Lemma 5.8.2 (ii) that $\overline{G}$ is conjugate to a group of the form

$$G^{\mathcal{B}}_{\mu,\nu}(X_1^{\alpha_1}, \dots, X_k^{\alpha_k}),$$

where $\alpha_i \in \{+, -\}$ for all $1 \le i \le t$, and $\alpha_i = +$ if $i \notin I$.

Therefore to settle the question of conjugacy, it remains to determine when metrically primitive irreducible groups of the form

$$G = G^{\mathcal{B}}_{\mu,\nu}(X_1^{\alpha_1}, \dots, X_k^{\alpha_k}) \qquad \overline{G} = G^{\mathcal{B}}_{\mu,\nu}(X_1^{\beta_1}, \dots, X_k^{\beta_k}) \tag{5.5}$$

are conjugate. Here $\alpha_i, \beta_i \in \{+, -\}$ for all $1 \le i \le t$, and $\alpha_i = \beta_i = +$ for all $i \notin I$.

Suppose that $G$ and $\overline{G}$ as in (5.5) are conjugate in $\Delta(V, \kappa)$, and let $g \in \Delta(V, \kappa)$ be such that $gGg^{-1} = \overline{G}$. By uniqueness of $F_0$ and $R_1, \dots, R_k$ (Corollary 5.7.8, Theorem 5.7.9) we have $g \in N_{\Delta(V,\kappa)}(F_0, R_1, \dots, R_k)$. Let $\pi(g) = (g_1, \dots, g_k)$, where $g_i$ is the action of $g$ on $R_i/Z(R_i)$ for all $1 \le i \le k$. Then

$$gGg^{-1} = G^{\mathcal{B}}_{\mu,\nu}(Y_1, \dots, Y_k),$$

where $Y_i = g_i X_i g_i^{-1}$.

In the case where $X_i$ is of multiplier 1, we already know that $Y_i$ is $\mathrm{Sp}_{2\ell_i}(r_i)$-conjugate to $X_i$. If $X_i$ is of multiplier 2, then $Y_i$ is $\mathrm{Sp}_{2\ell_i}(r_i)$-conjugate to $X_i$ if and only if $\tau_i(g_i)$ is a square in $\mathbb{F}_{r_i}^\times$. Here $\tau_i$ is the homomorphism $\tau_i : \mathrm{GSp}_{2\ell_i}(r_i) \to \mathbb{F}_{r_i}^\times$.

Now by applying Lemma 5.8.2 (ii), we conclude that $gGg^{-1} = \overline{G}$ is conjugate in $\Delta(V, \kappa)$ to

$$G^{\mathcal{B}}_{\mu,\nu}(X_1^{\alpha'_1}, \dots, X_k^{\alpha'_k}),$$

where

$$\alpha'_i = \begin{cases} \alpha_i, & \text{if } i \notin I \text{ or } \tau_i(g_i) \in \left(\mathbb{F}_{r_i}^\times\right)^2. \\ -\alpha_i, & \text{if } \tau_i(g_i) \notin \left(\mathbb{F}_{r_i}^\times\right)^2. \end{cases} \tag{5.6}$$

We can now define an action of $N_{\Delta(V,\kappa)}(F_0, R_1, \dots, R_k)$ on the set of tuples

$$\Lambda = \{(\alpha_1, \dots, \alpha_k) : \alpha_i \in \{+, -\} \text{ for all } i, \text{ and } \alpha_i = + \text{ if } X_i \text{ is of multiplier } 1\}$$

by $g \cdot (\alpha_1, \dots, \alpha_k) = (\alpha'_1, \dots, \alpha'_k)$, where $\alpha'_i$ is defined by (5.6).

We conclude from the discussion so far that $G$ and $\overline{G}$ as in (5.5) are conjugate in $\Delta(V, \kappa)$ if and only if $(\alpha_1, \dots, \alpha_k)$ and $(\beta_1, \dots, \beta_k)$ are in the same $N_{\Delta(V,\kappa)}(F_0, R_1, \dots, R_k)$-orbit.

**Theorem 5.8.3** *Let $I$ be the indices $1 \le i \le k$ such that $X_i$ is of multiplier 2. Let $J \subset I$ be the set of $i \in I$ such that $\left(\frac{q}{r_i}\right) = +$.*

*For $i \in I$, let $X_i^+$ and $X_i^-$ be representatives for the $\mathrm{Sp}_{2\ell_i}(r_i)$-conjugacy classes of $X_i$. For $1 \le i \le k$ with $i \notin I$, define $X_i^+ := X_i$.*

*Let*

$$G = G^{\mathcal{B}}_{\mu,\nu}(X_1^{\alpha_1}, \dots, X_k^{\alpha_k}),$$
$$\overline{G} = G^{\mathcal{B}}_{\mu,\nu}(X_1^{\beta_1}, \dots, X_k^{\beta_k}),$$

*be metrically primitive irreducible subgroups of $\Delta(V, \kappa)$, constructed with the same subgroups $F_0$ and $A = F_0 R_1 \cdots R_k$. Assume that $\alpha_i, \beta_i \in \{+, -\}$ for all $1 \le i \le k$, and $\alpha_i = \beta_i = +$ for all $i \notin I$.*

*Then $G$ and $\overline{G}$ are conjugate in $\Delta(V, \kappa)$ if and only if one of the following holds:*

*(i) $\alpha_i = \beta_i$ for all $i \in I$.*
*(ii) $\alpha_i = \beta_i$ for all $i \in J$, and $\alpha_i = -\beta_i$ for all $i \in I \setminus J$.*

***Proof*** If $G$ is of type $\mathcal{B}_3$, then $I = \emptyset$ since every subgroup of $O^{\pm}_{2\ell}(2)$ is of multiplier 1. Thus the result holds in this case.

We assume then for the rest of the proof that $G$ is not of type $\mathcal{B}_3$. Denote $N_1 := N_{\Delta(V,\kappa)}(F_0, R_1, \dots, R_k)$ and $N_2 := N_1 \cap C_{I(V,\kappa)}(F_0)$. It is clear that $N_2$ acts trivially on $\Lambda$ (since it centralizes $F_0$), and so does the map $\psi^2$ since it is a square.

Consider first the case where $G$ is not of type $\mathcal{B}_1$. Then the quotient

$$N_1/\langle N_2, \psi^2\rangle = \langle\overline{\psi}\rangle$$

is trivial or isomorphic to $C_2$. Thus $(\alpha_1, \dots, \alpha_k)$ and $(\beta_1, \dots, \beta_k)$ are on the same $N_1$-orbit if and only if $(\beta_1, \dots, \beta_k) = \psi^c \cdot (\alpha_1, \dots, \alpha_k)$ for some $c \in \{0, 1\}$. Equivalently, the tuples $(\alpha_1, \dots, \alpha_k)$ and $(\beta_1, \dots, \beta_k)$ are on the same $N_1$-orbit if and only if (i) or (ii) holds. Since $G$ and $\overline{G}$ are $\Delta(V, \kappa)$-conjugate if and only if the tuples are on the same $N_1$-orbit, the result follows.

Suppose then that $G$ is of type $\mathcal{B}_1$. In this case

$$N_1/\langle N_2, \psi^2\rangle = \langle\overline{\varphi}, \overline{\psi}\rangle,$$

which is isomorphic to $C_2$ or $C_2 \times C_2$. Because $G$ is irreducible, it follows from Lemma 4.3.14 that either $\left(\frac{-1}{r_i}\right) = +$ for all $i \in I$, or $\left(\frac{-q}{r_i}\right) = +$ for all $i \in I$. Thus either $\varphi$ or $\varphi\psi$ acts trivially on $\Lambda$, and the action of $N_1$ on $\Lambda$ is determined by $\psi$. As in the previous paragraph, we find that $(\alpha_1, \dots, \alpha_k)$ and $(\beta_1, \dots, \beta_k)$ are on the same $N_1$-orbit if and only if (i) or (ii) holds, and the result follows. □

**Example 5.8.4** Since 2 is a square in $\mathbb{F}_7$, in $\mathrm{GSp}_2(7)$ the group

$$X := G^{\mathcal{B}_3}_{2,1}(O_2^-(2)) = (\mathbb{F}_7^\times \circ 2_-^{1+2}).(O_2^-(2))$$

is metrically primitive irreducible of multiplier 2 (Lemma 5.1.5). Then we know that $X$ is primitive maximal solvable in $\mathrm{GSp}_2(7)$, which will also follow from Theorem 7.1.12 proven later. Let $X^+$ and $X^-$ be representatives for the two $\mathrm{Sp}_2(7)$-conjugacy classes of $X$.

Consider then for example $\kappa = 0$ with $q = 2$, $\mu = 7$, $\nu = 3$, and $n = 21$. Since $7 \mid 2^3 - 1$, in $\mathrm{GL}_{21}(2)$ we can construct subgroups

$$G^{\mathcal{B}_0}_{7,3}(X^+) \qquad\qquad G^{\mathcal{B}_0}_{7,3}(X^-)$$

which are primitive, irreducible, and solvable. We have $\left(\frac{2}{7}\right) = +$, so by Theorem 5.8.3 the subgroups $G^{\mathcal{B}_0}_{7,3}(X^+)$ and $G^{\mathcal{B}_0}_{7,3}(X^-)$ are not conjugate in $\mathrm{GL}_{21}(2)$.

For another example, consider $\kappa = 0$ with $q = 13$, $\mu = 7$, $\nu = 2$, and $n = 14$. Since $7 \mid 13^2 - 1$, in $\mathrm{GL}_{14}(13)$ we have subgroups

$$G^{\mathcal{B}_0}_{7,2}(X^+) \qquad\qquad G^{\mathcal{B}_0}_{7,2}(X^-)$$

which are primitive, irreducible, and solvable. We have $\left(\frac{13}{7}\right) = -$, so by Theorem 5.8.3 the subgroups $G^{\mathcal{B}_0}_{7,2}(X^+)$ and $G^{\mathcal{B}_0}_{7,2}(X^-)$ are conjugate in $\mathrm{GL}_{14}(13)$.

# Chapter 6
# Fixed Point Spaces and Abelian Subgroups

In this chapter, we will find upper bounds for the dimension of the fixed point spaces of elements in metrically primitive irreducible solvable subgroups of $\Delta(V, \kappa)$. We will do this by finding such upper bounds for elements of prime order in $\mathrm{GL}_n(q)$ that normalize an absolutely irreducible extraspecial $r$-group of exponent $r \gcd(r, 2)$. Specifically, we will prove Theorem 1.1.5 stated in the introduction, and as a corollary we get the same upper bounds for metrically primitive irreducible solvable subgroups of $\Delta(V, \kappa)$ (Theorem 6.2.3).

In particular, it follows from Theorem 6.2.3 proven in this chapter that if $G \leq \Delta(V, \kappa)$ is metrically primitive irreducible and $g \in G$ has prime order $> 3$, then $\dim V^g \leq \frac{1}{2} \dim V$. In general, it is true that if $G \leq \Delta(V, \kappa)$ is metrically primitive irreducible and $g \in G \setminus \{1\}$, then $\dim V^g \leq \frac{3}{4} \dim V$ (Theorem 6.2.4). In Sect. 6.3, we apply these upper bounds in the proof of Theorem 1.1.4 from the introduction, which states that if $G \leq \mathrm{GL}(V)$ is irreducible solvable, then every abelian subgroup of the affine group $V \rtimes G$ has order $\leq |V|$.

The results of this chapter will be applied in Chap. 7, where we will complete the classification of maximal irreducible solvable subgroups. For example in the imprimitive case, it follows from the upper bound $\dim V^g \leq \frac{3}{4} \dim V$ that if $H \wr K$ is a subgroup of $\mathrm{GL}(V)$ as in Theorem 2.2.14 (i), then $H \wr K$ can be contained in a primitive irreducible solvable subgroup of $\mathrm{GL}(V)$ only if $K = S_2$, $K = S_3$, or $K = S_4$.

## 6.1 Fixed Point Spaces in Symplectic-Type Normalizers

In this section, we will consider the fixed point spaces of elements in $G$, where $G \leq \mathrm{GL}_n(q)$ belongs to the family of $\mathcal{C}_6$-subgroups (normalizers of symplectic-type groups) in Aschbacher's theorem [2] on maximal subgroups of classical groups.



M. Korhonen, *Maximal Solvable Subgroups of Finite Classical Groups*,
Lecture Notes in Mathematics 2346, https://doi.org/10.1007/978-3-031-62915-0_6

We recall that the family of $\mathcal{C}_6$-subgroups in $\mathrm{GL}_n(q)$ can be defined as follows. Let $r$ be a prime, and let $R$ be an extraspecial group of order $r^{1+2\ell}$ and exponent $r\gcd(r,2)$. In other words, either $R = r_+^{1+2\ell}$ or $R = 2_-^{1+2\ell}$. Suppose that $q$ is a prime such that $q \equiv 1 \mod r$. Then we have an embedding $R \leq \mathrm{GL}_n(q)$ for $n = r^\ell$. Furthermore, in this case $R$ is absolutely irreducible and unique up to conjugacy in $\mathrm{GL}_n(q)$, see Sect. 3.2 and Theorem 3.2.1.

In this setup, a $\mathcal{C}_6$-subgroup of $\mathrm{GL}_n(q)$ is a group of the form $G = N_{\mathrm{GL}_n(q)}(RZ)$, where $Z \leq \mathrm{GL}_n(q)$ is the group of scalar matrices in $\mathrm{GL}_n(q)$. In Sect. 3.2, we have given an explicit description of the embedding of $R$ and the generators of $G$.

Throughout this section, we assume that $G = N_{\mathrm{GL}_n(q)}(RZ)$ is a $\mathcal{C}_6$-subgroup, and denote by $V = \mathbb{F}_q^n$ the natural module. For $x \in G$ and $\lambda \in \mathbb{F}_q^\times$, we denote by $V_{\lambda,x}$ the $x$-eigenspace on $V$ corresponding to the eigenvalue $\lambda$.

We begin with the following lemma, which describes the fixed point spaces of elements of $R$.

**Lemma 6.1.1** *Let $x \in R$. Then the following statements hold.*

*(i) If $x \in Z(R) \setminus \{1\}$, then $V^x = 0$.*
*(ii) If $r = 2$ and $|x| = 4$, then $V^x = 0$.*
*(iii) If $x \in R \setminus Z(R)$ and $|x| = r$, then* $\dim V^x = n/r$.

***Proof*** Claim (i) follows from the fact that $Z(R)$ acts on $V$ as scalar matrices. For (ii), if $r = 2$ and $|x| = 4$, then $x^2 = -I_V$, so $V^x = 0$ since $V^{x^2} = 0$.

For claim (iii), we prove first that if $x, y \in R \setminus Z(R)$ are of order $r$, then they are conjugate in $N_{\mathrm{GL}_n(q)}(R)$. To see this, note that their images in $R/Z(R)$ are nonzero vectors and thus conjugate under the action of $N_{\mathrm{GL}_n(q)}(R)/R \cong \mathrm{Sp}_{2l}(r)$ if $r > 2$ (Theorem 3.2.2 (ii)). If $R$ is of type $2_\pm^{1+2\ell}$, their images are singular vectors in $R/Z(R)$ and thus conjugate under the action of $N_{\mathrm{GL}_n(q)}(R)/R \cong O_{2\ell}^\pm(2)$ (Theorem 3.2.2 (iii)). Thus we can assume that $y = xz$ for some $z \in Z(R)$. The $R$-conjugates of $x$ are elements of the form $x\theta^d$, where $0 \leq d < r$ and $Z(R) = \langle\theta\rangle$. Hence we can conclude that $x$ and $y$ are conjugate in $N_{\mathrm{GL}_n(q)}(R)$.

Therefore to prove (iii), we can assume that $x$ is one of the generators $x = A_i$ as described in Sect. 3.2.1. In this case it is clear from the action of $A_i$ (which is a diagonal matrix) that $\dim V^x = n/r$, as required. □

We will now prove the following result, which was stated as Theorem 1.1.5 in the introduction and gives an upper bound on the dimension of eigenspaces of elements of $G$. The proof will be based on arguments due to Jordan [52, §68–§74]. Some similar upper bounds were given by Guralnick and Maróti in [31, Section 2], with a different proof more in the lines of a result of Hall and Higman [32, Theorem 2.5.1]; see also [28, Theorem 2.1 (ii), p. 364]. Furthermore, the eigenvalues of elements of $G$ were considered for example in [85], [86, Section 3], and [5].

**Theorem 6.1.2** *Let $g \in G = N_{\mathrm{GL}_n(q)}(RZ)$ be an element of prime order $\varpi$ and suppose that $g$ is non-scalar. Then for all $\lambda \in \mathbb{F}_q^\times$, we have*

$$\dim V_{g,\lambda} \leq \begin{cases} 3n/4, & \text{if } \varpi = 2, \\ 2n/3, & \text{if } \varpi = 3, \\ n/2, & \text{if } \varpi > 3. \end{cases}$$

*Moreover, if $\varpi = 3$ and $n$ is not a multiple of 3, then we also have* $\dim V_{g,\lambda} \leq n/2$.

***Proof*** Since $\lambda$ is an eigenvalue of $g$, we have $\lambda^{\varpi} = 1$. Thus $g\lambda^{-1}$ has order $\varpi$. Since $V_{g,\lambda} = V_{g\lambda^{-1},1}$, by replacing $g$ with $g\lambda^{-1}$ we reduce to the case where $\lambda = 1$. In other words, it will suffice to prove the claimed bounds for the dimension $\dim V^g$ of the fixed point space of $g$.

Suppose that $\dim V^g > n/2$. First we note the following.

**Claim 1: We can assume that $g$ normalizes $R$**
If $r > 2$ we have $R = \{x \in RZ : x^r = 1\}$, so $g$ normalizes $R$. If $r = 2$ and $q \equiv 3 \mod 4$, similarly $R = \{x \in RZ : x^4 = 1\}$, so $g$ normalizes $R$.

Thus we only need to consider the case where $r = 2$ and $q \equiv 1 \mod 4$. In this case $N_{\mathrm{GL}_n(q)}(RZ)/RZ \cong \mathrm{Sp}_{2\ell}(2)$, while $N_{\mathrm{GL}_n(q)}(R)/RZ$ is isomorphic to $O_{2\ell}^+(2)$ or $O_{2\ell}^-(2)$, according to whether $R$ is of type $2_+^{1+2\ell}$ or $2_-^{1+2\ell}$. It is a standard fact that every element of $\mathrm{Sp}_{2\ell}(2)$ is conjugate to an element of $O_{2\ell}^+(2)$ or $O_{2\ell}^-(2)$, either apply Theorem 2.1.14 or see for example [69, Lemma 4.1], [21, Theorem 2]. Thus by replacing $g$ with a conjugate in $N_{\mathrm{GL}_n(q)}(RZ)$, we may assume that the image of $g$ lies in $O_{2\ell}^+(2)$ or $O_{2\ell}^-(2)$. In other words, we may assume that $g$ normalizes an extraspecial group of type $2_+^{1+2\ell}$ or $2_-^{1+2\ell}$; which is explicitly given as the group $R_Q$ as in the end of the proof of Proposition 4.1.7.

We will assume then for the rest of the proof that $g$ normalizes $R$.

**Claim 2: $g$ normalizes some abelian subgroup $H < R$ of order $r^{1+\ell}$**
We proceed with an inductive construction as in [52, §69]. First note that $g$ centralizes $Z(R) = \langle \theta \rangle$. Next we will show that if $g$ normalizes an abelian subgroup $H < R$ of order $< r^{1+\ell}$, we can always find a larger one which is normalized by $g$. Since every maximal abelian subgroup of $R$ has order $r^{1+\ell}$ (Lemma 3.1.2), it follows by induction that $g$ normalizes an abelian subgroup of $R$ with order $r^{1+\ell}$.

Suppose then that $G$ normalizes some abelian subgroup $H_s < R$ of order $< r^{1+\ell}$. Since $H_s$ is not maximal abelian, there exists $x \in R \setminus H_s$ such that $[H_s, x] = 1$. If $x' := [x, g] \in H_s$, then $g$ normalizes the abelian subgroup $\langle x, H_s \rangle \gneq H_s$.

Thus we can assume that $x' \notin H_s$. Then

$$V^{x'} \supseteq V^{x^{-1}g^{-1}x} \cap V^g$$

and furthermore $V^{x^{-1}g^{-1}x} \cap V^g \neq 0$ since $\dim V^g > n/2$, so there exists a nonzero vector $v_0 \in V$ that is fixed by both $x'$ and $g$. Then $v_0$ is fixed by all conjugates of the form

$$g^k x' g^{-k}, k \in \mathbb{Z}.$$

These conjugates commute pairwise, since their commutators are scalar matrices which have $v_0$ as a nonzero fixed point. Furthermore, all of these conjugates commute with $H_s$, since $g$ normalizes $H_s$ and $[H_s, x'] = 1$. Thus

$$\langle H_s, g^k x' g^{-k} : k \in \mathbb{Z}\rangle \gneqq H_s$$

is an abelian subgroup of $R$ normalized by $g$.

**Claim 3: If $R$ is of type $2_+^{1+2\ell}$, we can arrange $H$ to be elementary abelian** This claim is essentially equivalent to [52, §70]. If the subgroup $H$ normalized by $g$ is not elementary abelian, then $H \cong C_4 \times C_2^{\ell-1}$ (Lemma 3.1.2). In this case, we can find generators $H = \langle \theta, A_1, A_2, \ldots, A_\ell\rangle$ such that $A_1^2 = \theta$ and $A_i^2 = 1$ for all $1 < i \leq \ell$. The image of $H$ in $R/Z(R)$ is a maximal totally isotropic subspace, so by lifting a subspace dual to it we can find another maximal abelian subgroup $K < R$ such that $H \cap K = Z(R)$. Then $K$ has generators $K = \langle \theta, B_1, \ldots, B_\ell\rangle$ such that $[A_i, B_i] = \theta$ for all $1 \leq i \leq \ell$ and $[A_i, B_j] = 1$ for all $i \neq j$.

In this case we must have $B_1^2 = 1$. Indeed, otherwise $B_1^2 = \theta$, and then replacing $B_i$ with $B_1 B_i$ if necessary, we can assume that $B_i^2 = 1$ for all $1 < i \leq \ell$. But then the relations among the generators $A_1, \ldots, A_\ell, B_1, \ldots, B_\ell$ would make $R$ isomorphic to an extraspecial group of type $2_-^{1+2\ell}$.

Now $g$ normalizes $H$, so it also normalizes the subgroup $H' = \langle \theta, A_2, \ldots, A_\ell\rangle$ generated by the elements of order 2 in $H$. Moreover $H'$ is centralized by $B_1$, so $g$ must normalize the maximal abelian subgroup $\langle \theta, B_1, A_2, \ldots, A_\ell\rangle$, which is elementary abelian.

By Claim 3, we can choose generators $H = \langle \theta, A_1, \ldots, A_\ell\rangle$, where $A_i^r = 1$ for all $1 \leq i \leq \ell$ if $R$ is not of type $2_-^{1+2\ell}$. If $R$ is of type $2_-^{1+2\ell}$, we have $A_1^2 = \theta$ and $A_i^2 = 1$ for all $1 < i \leq \ell$. By lifting a subspace dual to $H$ in $R/Z(R)$, we can find elements $B_1, \ldots, B_\ell$ such that $[A_i, B_i] = \theta$, $[A_i, B_j] = 1$ for $i \neq j$, and $A_i^r = B_i^r$ for all $1 \leq i \leq \ell$.

Note that if $\mathbb{K}$ is an extension field of $\mathbb{F}_q$, then $\dim_{\mathbb{K}}(V \otimes_{\mathbb{F}_q} \mathbb{K})^g = \dim_{\mathbb{F}_q} V^g$. Therefore by extension of scalars, we may replace $\mathbb{F}_q$ with $\mathbb{F}_{q^d}$ if necessary. In particular when $r = 2$, by replacing $\mathbb{F}_q$ with $\mathbb{F}_{q^2}$ we can assume that $q \equiv 1 \mod 4$.

If $R$ is not type $2_-^{1+2\ell}$, let $\zeta \in \mathbb{F}_q^\times$ be a primitive $r$th root of unity such that $\theta = \zeta I$. By the construction given in Sect. 3.2.1, we can find a basis $\{v_{j_1,\ldots,j_k} : 0 \leq j_1, \ldots, j_k < r\}$ of $V$ such that

$$A_k v_{j_1,\ldots,j_\ell} = \zeta^{j_k} v_{j_1,\ldots,j_\ell} \text{ for } 1 \leq k \leq \ell,$$

$$B_k v_{j_1,\ldots,j_\ell} = v_{j_1,\ldots,j_{k-1},j_k+1,j_{k+1},\ldots,j_\ell} \text{ for } 1 \leq k \leq \ell,$$

where the indices $j_1, \ldots, j_\ell$ are interpreted modulo $r$.

When $R$ is of type $2_-^{1+2\ell}$, we let $\zeta \in \mathbb{F}_q^\times$ be a primitive 4th root of unity. Then by the construction given in Sect. 3.2.4, we have a basis $\{v_{j_1,\dots,j_k} : 0 \leq j_1, \dots, j_k < r\}$ of $V$ such that

$$\begin{aligned} A_1 v_{j_1,\dots,j_\ell} &= \zeta^{1+2j_1} v_{j_1,\dots,j_\ell}, \\ A_k v_{j_1,\dots,j_\ell} &= \zeta^{2j_k} v_{j_1,\dots,j_\ell} \text{ for } 2 \leq k \leq \ell, \\ B_1 v_{j_1,\dots,j_\ell} &= (-1)^{j_1} v_{j_1+1,j_2,\dots,j_\ell}, \\ B_k v_{j_1,\dots,j_\ell} &= v_{j_1,\dots,j_{k-1},j_k+1,j_{k+1},\dots,j_\ell} \text{ for } 2 \leq k \leq \ell, \end{aligned}$$

where again the indices are interpreted modulo $r$.

We now deal with two special cases.

**Claim 4: For $x \in R \setminus Z(R)$ of prime order, we have $\dim V^x = n/r$**
This is Lemma 6.1.1 (iii).

**Claim 5: The theorem holds when $\varpi = 2$**
We will use the argument from [52, §68]. Since $g$ is non-scalar and since $R$ is absolutely irreducible, there exists some $x \in R$ such that $[x, g] \neq 1$. Since

$$V^{x^{-1}g^{-1}x} \cap V^g \subseteq V^{[g,x]}$$

and $\dim V^{x^{-1}g^{-1}x} = \dim V^g$, we have $\dim V^{[g,x]} \geq 2\dim V^g - n$. On the other hand $\dim V^{[g,x]} \leq n/r$ by Claim 4, so

$$\dim V^g \leq \frac{n}{2} + \frac{n}{2r}$$

and therefore $\dim V^g < \frac{3n}{4}$.

Thus for the remainder of the proof, we shall assume that $\varpi \geq 3$. We will argue similarly to Jordan [52, §71–74].

Denote $V_{j_1,\dots,j_\ell} = \langle v_{j_1,\dots,j_\ell} \rangle$. Then

$$V = \bigoplus_{0 \leq j_1,\dots,j_\ell < r} V_{j_1,\dots,j_\ell},$$

and the summands $V_{j_1,\dots,j_\ell}$ are pairwise nonisomorphic $\mathbb{F}_q[H]$-modules, which is clear from the action of the generators of $H$ given above.

Thus since $g$ normalizes $H$, it permutes the summands $V_{j_1,\dots,j_\ell}$. Moreover the order of $g$ is a prime $\varpi$, so the action of $g$ on these summands consists of subspaces left invariant by $g$, and of summands permuted in a $\varpi$-cycle.

If $g$ acts as a cycle on 1-dimensional subspaces $W_1, \dots, W_\varpi$, then we have $W_1 = \langle w \rangle$, $W_2 = \langle gw \rangle, \dots, W_\varpi = \langle g^{\varpi-1} w \rangle$. Then in the subspace $W_1 \oplus \cdots \oplus W_\varpi$ the vector $w+gw+\cdots+g^{\varpi-1}w$ is the unique fixed point of $g$, up to a scalar. Therefore if $m_0$ is the number subspaces $V_{j_1,\ldots,j_\ell}$ left invariant by $g$, we have

$$\dim V^g \leq \frac{n-m_0}{\varpi} + m_0. \tag{6.1}$$

Thus $m_0 > 0$, as otherwise $\dim V^g \leq n/\varpi \leq n/2$.

For $1 \leq i \leq \ell$, we have

$$g^{-1}A_i g = \left(\prod_{t=1}^{\ell} A_t^{X_{ti}}\right)\theta^{\alpha_i}$$

for some integers $0 \leq X_{ti} < r$ and $0 \leq \alpha_i < r$. Note that if $R$ is of type $2_-^{1+2\ell}$, then $X_{11} = 1$ and $X_{1i} = 0$ for all $1 < i \leq \ell$. Let $X$ be the matrix $X = (X_{ij})$ with entries in $\mathbb{F}_r$. Then we have

$$g \cdot V_{j_1,\ldots,j_\ell} = V_{j_1',\ldots,j_\ell'},$$

where

$$\begin{pmatrix} j_1' \\ \vdots \\ j_\ell' \end{pmatrix} = X^T \begin{pmatrix} j_1 \\ \vdots \\ j_\ell \end{pmatrix} + \begin{pmatrix} \alpha_1 \\ \vdots \\ \alpha_\ell \end{pmatrix} \tag{6.2}$$

with indices interpreted modulo $r$. Hence $m_0 = n/r^{d_0}$, where $d_0$ is the rank of $X^T - I$ (as a matrix over $\mathbb{F}_r$), so

$$\dim V^g \leq \left(\frac{1}{\varpi} - \frac{1}{\varpi r^{d_0}} + \frac{1}{r^{d_0}}\right) n$$

by (6.1). Thus if $d_0 \geq 2$, it follows from $\varpi \geq 3$ and $r \geq 2$ that $\dim V^g \leq n/2$.

Next we consider the case where $d_0 = 1$. Then $X^T - I$ has rank 1, so after a change of basis $X$ becomes a matrix of the form

$$\begin{pmatrix} 1 & 0 & \cdots & 0 & \beta_1 \\ & 1 & \cdots & 0 & \beta_2 \\ & & \ddots & \vdots & \vdots \\ & & & 1 & \beta_{\ell-1} \\ & & & & \beta \end{pmatrix}.$$

In other words, we can find generators $H = \langle \theta, A'_1, \dots, A'_\ell \rangle$ such that $g^{-1}A'_i g = A'_i \theta^{\alpha'_i}$ for all $1 \le i < \ell$ and

$$g^{-1}A'_\ell g = \left( \prod_{t=1}^{\ell-1} (A'_t)^{\beta_t} \right) \cdot (A'_\ell)^\beta \cdot \theta^{\alpha_\ell}.$$

Since $g$ has order $\varpi$, we have $\beta^\varpi \equiv 1 \mod r$. If $\beta \not\equiv 1 \mod r$, then $\beta$ has order $\varpi$ in $\mathbb{F}_r^\times$ and thus $\varpi \mid r-1$. Since $\varpi \ge 3$, this implies $r \ge 7$ and thus

$$\dim V^g \le \left( \frac{1}{\varpi} - \frac{1}{\varpi r} + \frac{1}{r} \right) n < n/2$$

by (6.1). In the case where $\beta \equiv 1 \mod r$, the matrix $X$ is unipotent, so $g$ has order $\varpi = r$. Then

$$\dim V^g \le \left( \frac{2}{\varpi} - \frac{1}{\varpi^2} \right) n.$$

If $\varpi > 3$, this implies $\dim V^g \le n/2$. If $\varpi = 3$, we get $\dim V^g \le \frac{5}{9}n < \frac{2}{3}n$, as claimed by the theorem.

It remains to consider the case where $d_0 = 0$. In other words, in this case every 1-dimensional subspace $V_{j_1,\dots,j_\ell}$ is left invariant by $g$. It follows that $g$ centralizes $H$, so $gA_i g^{-1} = A_i$ for all $1 \le i \le \ell$. On the other hand, the alternating bilinear form on $R/Z(R)$ is $g$-invariant and the images $\overline{A_1}, \dots, \overline{A_\ell}, \overline{B_1}, \dots, \overline{B_\ell}$ form a standard symplectic basis. Thus the action of $g$ on $R/Z(R)$ must be a matrix of the form

$$\begin{pmatrix} I & Y \\ 0 & I \end{pmatrix}, \tag{6.3}$$

where $Y = (Y_{ti})$ is a matrix with entries $\mathbb{F}_r$ and $Y^T = Y$.

It follows that $g$ must have order $\varpi = r$. Indeed, if $g$ acts trivially on $R/Z(R)$ this is clear, since then $g \in R$. Otherwise $g$ has the same order as the unipotent matrix in (6.3), and every unipotent matrix in $\mathrm{Sp}_{2\ell}(r)$ has order a power of $r$, so $\varpi = r$. In particular, by Claim 5 we have $r > 2$.

Recall (Sect. 3.2.1) that $G$ contains linear maps $D_{st} : V \to V$ $(1 \le s < t \le \ell)$ and $E_t : V \to V$ $(1 \le t \le \ell)$ defined by

$$D_{st} v_{j_1,\dots,j_\ell} = \zeta^{j_s j_t} v_{j_1,\dots,j_\ell}$$

$$E_t v_{j_1,\dots,j_\ell} = \zeta^{\frac{j_t(j_t-1)}{2}} v_{j_1,\dots,j_\ell}$$

for all $0 \le j_1, \dots, j_\ell < r$.

The linear maps $D_{st}$ and $E_t$ centralize the generators of $R$, except for the ones given in (3.1). Hence the image of $D_{st}$ in $\mathrm{Sp}_{2\ell}(r)$ is of the form (6.3) with $Y =$

$E_{s,t} + E_{t,s}$, while the image of $E_t$ in $\mathrm{Sp}_{2\ell}(r)$ is of the form (6.3) with $Y = E_{t,t}$. Thus we can find a $g_0 \in G$ which is a product of $D_{st}$'s and $E_t$'s such that $u := g_0^{-1}g$ acts trivially on $R/Z(R)$. Then $u$ must act on $R$ as an inner automorphism (Lemma 3.1.1). Moreover $u$ centralizes $H$, so we can find a product $g_1 \in G$ of $A_i$'s such that $g_1^{-1}u$ centralizes $R$. Then $g_1^{-1}u$ must be a scalar, since $R$ is absolutely irreducible.

Hence $g$ is a product of $D_{st}$'s, $E_t$'s, $A_i$'s, and a scalar. By looking at the action of these elements on the basis vectors $v_{j_1,\dots,j_\ell}$, it follows that

$$gv_{j_1,\dots,j_\ell} = \zeta^{\xi(j_1,\dots,j_\ell)}v_{j_1,\dots,j_\ell},$$

where $\xi(j_1,\dots,j_\ell)$ is a polynomial of degree $\le 2$. Then $\dim V^g$ is precisely the number of solutions to

$$\xi(j_1,\dots,j_\ell) \equiv 0 \mod r.$$

By a linear change of variables, we can write

$$\xi(j_1,\dots,j_\ell) = \varphi(j_1',\dots,j_k') + \psi(j_{k+1}',\dots,j_\ell') + c,$$

where $\varphi(j_1',\dots,j_k')$ is a non-degenerate quadratic form, $\psi(j_{k+1}',\dots,j_\ell')$ is linear, and $c$ is a constant.

If $\psi \ne 0$, then

$$\varphi(j_1',\dots,j_k') + \psi(j_{k+1}',\dots,j_\ell') + c \equiv 0 \mod r$$

determines one variable in $\psi$ uniquely, and the rest of the $\ell - 1$ variables can be chosen arbitrarily. Thus the number of solutions to $\xi(j_1,\dots,j_\ell) \equiv 0 \mod r$ is $r^{\ell-1}$, in which case $\dim V^g = r^{\ell-1} = n/r < n/2$.

Suppose next that $\psi = 0$, and let $\delta$ be the determinant of the quadratic form $\varphi$. The number $N_k$ of solutions

$$\varphi(j_1',\dots,j_k') + c \equiv 0 \mod r \text{ with } 0 \le j_1',\dots,j_k' < r$$

is known explicitly, and goes back to Jordan [47, §199–200, pp. 159–161]—see for example [59, Theorem 6.26, Theorem 6.27]. In terms of the Legendre symbol, we have

$$N_k = \begin{cases} r^{2m} + \left(\dfrac{(-1)^m\delta c}{r}\right) r^m, & \text{if } k = 2m+1. \\ r^{2m+1} - \left(\dfrac{(-1)^m\delta}{r}\right) r^m, & \text{if } k = 2m+2 \text{ and } c \not\equiv 0 \mod r. \\ r^{2m+1} + \left(\dfrac{(-1)^m\delta}{r}\right) r^m(r-1), & \text{if } k = 2m+2 \text{ and } c \equiv 0 \mod r. \end{cases}$$

Then the number of solutions to $\xi(j_1, \ldots, j_\ell) \equiv 0 \mod r$ is equal to

$$\dim V^g = N_k r^{\ell-k} = \frac{N_k}{r^k} n.$$

It follows from the formula for $N_k$ that

$$\frac{N_k}{r^k} < 1/2,$$

except in the following cases:

- $k = 2, r = 3, \left(\frac{\delta}{r}\right) = 1$, in which case $N_k/r^k = 5/9 < 2/3$.
- $k = 1, r = 3, \left(\frac{\delta c}{r}\right) = 1$, in which case $N_k/r^k = 2/3$.

This completes the proof of the theorem. □

## 6.2 Fixed Point Spaces in Primitive Solvable Linear Groups

In this section, we will bound the fixed point spaces of elements in metrically primitive solvable subgroups of $\Delta(V, \kappa)$. These bounds can be used to show that in most cases, the imprimitive groups of Theorem 2.2.14 cannot be contained in a metrically primitive solvable subgroup of $\Delta(V, \kappa)$.

We need the following two lemmas, see for example [61, Lemma 3.7].

**Lemma 6.2.1** *Let $V_1$ and $V_2$ be $\mathbb{K}[G]$-modules, where $\mathbb{K}$ is a field. Suppose that $g \in G$ acts on $V_1$ and $V_2$ as a unipotent linear map. Then* $\dim(V_1 \otimes V_2)^g \leq \dim V_1^g \dim V_2$.

**Lemma 6.2.2** *Let $V_1$ and $V_2$ be $\mathbb{K}[G]$-modules, where $\mathbb{K}$ is a field. Suppose that $g \in G$ acts on $V_1$ and $V_2$ as a diagonalizable linear map. Suppose that every $g$-eigenspace of $V_1$ has dimension at most $c$. Then* $\dim(V_1 \otimes V_2)^g \leq c \cdot \dim V_2$.

We will now apply Theorem 6.1.2 to prove the following result [52, Théorème 74].

**Theorem 6.2.3** *Let $G$ be a metrically primitive irreducible solvable subgroup of $\Delta(V, \kappa)$, where $n = \dim V$. Suppose that $g \in G$ has prime order $\varpi$. Then*

$$\dim V^g \leq \begin{cases} 3n/4, & \text{if } \varpi = 2, \\ 2n/3, & \text{if } \varpi = 3, \\ n/2, & \text{if } \varpi > 3. \end{cases}$$

*Moreover, if $\varpi = 3$ and $n$ is not a multiple of 3, then we also have* $\dim V^g \leq n/2$.

***Proof*** If $n = 1$ the theorem is obvious, so we suppose that $n > 1$ and proceed by induction on $n$. Clearly we can assume that $G$ is maximal solvable in $\Delta(V, \kappa)$. Let $F_0 \leq G$ be an abelian normal subgroup of $G$ satisfying properties (F1)–(F3). Then $F_0 = \langle f \rangle$ is cyclic (Lemma 2.4.3).

Suppose that there exists some $g \in G \setminus \{1\}$ such that $N = \dim V^g > n/2$. For every $h \in G$, we have $\dim V^{h^{-1}gh} = N$. Therefore $\dim(V^g \cap V^{h^{-1}gh}) \geq 2N - n$, which implies that

$$\dim V^{[g,h]} \geq 2N - n.$$

In particular with $h = f$, we have $\dim V^{[g,f]} > 0$. Since $[g, f] \in F_0$, we have $[g, f] = 1$ because by Lemma 2.4.3 nontrivial elements of $F_0$ have no nonzero fixed points on $V$. Therefore $g \in C_G(F_0)$.

As in the usual setup (Sect. 2.4), let $\mathbb{K}$ be the splitting field of $f$, denote $V' = \mathbb{K} \otimes_{\mathbb{F}_q} V$, and let $\lambda \in \mathbb{K}$ be an eigenvalue of $f$ on $V'$. Moreover $W'_\lambda$ is the $f$-eigenspace corresponding to $\lambda$ and we denote $\mu = \dim_{\mathbb{K}}(W'_\lambda)$. Then it is readily seen as in the proof of Lemma 4.3.12 that

$$\dim_{\mathbb{F}_q} V^g = \dim_{\mathbb{K}}(V')^g = \begin{cases} 2[\mathbb{K} : \mathbb{F}_q] \dim_{\mathbb{K}}(W'_\lambda)^g, & \text{if } G \text{ is of type } \mathcal{B}_1 \\ [\mathbb{K} : \mathbb{F}_q] \dim_{\mathbb{K}}(W'_\lambda)^g, & \text{otherwise.} \end{cases} \tag{6.4}$$

It follows from Proposition 5.6.1 that the action of $C_G(F_0)$ on $W'_\lambda$ is absolutely irreducible and primitive. Thus if $\mu < n$, then by induction the theorem holds for $C_G(F_0)$, so $\dim_{\mathbb{K}}(W'_\lambda)^g \leq \alpha\mu$, where

$$\alpha = \begin{cases} 3/4, & \text{if } \varpi = 2 \\ 2/3, & \text{if } \varpi = 3 \text{ and } 3 \mid \mu \\ 1/2, & \text{otherwise.} \end{cases}$$

Then combining this with (6.4) gives $\dim_{\mathbb{F}_q} V^g \leq \alpha n$, as required by the theorem.

Therefore we can assume that $\mu = n$, in which case $\mathbb{K} = \mathbb{F}_q$ and $G = C_G(F_0)$ is absolutely irreducible and primitive on $V$. By primitivity, we can also reduce to the case where $\kappa = 0$, so $F_0 = \mathbb{F}_q^\times$ is the set of scalar matrices in $\mathrm{GL}_n(q)$.

Let $A = \mathrm{Fit}(G)$, which is the unique subgroup satisfying properties (A1)–(A3) in $G$ (Lemma 4.1.4). Then $A$ is absolutely irreducible on $V$ (Theorem 4.1.16). Furthermore, by Lemma 4.1.13 we have a tensor decomposition $V = W_1 \otimes \cdots \otimes W_k$ such that $A = R_1 \otimes \cdots \otimes R_k$, where $R_i \trianglelefteq G$ is an extraspecial $r_i$-group of exponent $r_i \gcd(r_i, 2)$ and $W_i$ is an absolutely irreducible $\mathbb{F}_q[R_i]$-module.

Note that $g$ cannot commute with $R_i$ for all $1 \leq i \leq t$. Indeed, otherwise $g \in C_G(A) = F_0$ because $A$ is absolutely irreducible. But this is a contradiction, as nontrivial elements of $F_0$ do not have nonzero fixed points on $V$. We can assume then without loss of generality that $[g, R_1] \neq 1$. The image of $g$ in $N_{\mathrm{GL}(W_1)}(R_1)$ must be non-scalar, so we will be able to apply Theorem 6.1.2.

If order of $g$ is equal to $p = \operatorname{char} \mathbb{F}_q$, then by Lemma 6.2.1 and Theorem 6.1.2 we get

$$\dim V^g \leq \dim(W_1^g) \cdot \frac{n}{\dim(W_1)} \leq \alpha n,$$

as claimed by the theorem.

It remains to consider the case where $\gcd(\varpi, p) = 1$. Let $\mathbb{K}'$ be a field extension of $\mathbb{F}_q$ which contains a primitive $\varpi$th root of unity. Then the action of $g$ on $\mathbb{K}' \otimes_{\mathbb{K}} V$ is diagonalizable, and by Theorem 6.1.2 every $g$-eigenspace of $\mathbb{K}' \otimes_{\mathbb{F}_q} W_1$ has dimension at most $\alpha \dim(W_1)$. Therefore

$$\dim_{\mathbb{F}_q}(V^g) = \dim_{\mathbb{K}'}(\mathbb{K}' \otimes_{\mathbb{F}_q} V)^g \leq \alpha \dim(W_1) \cdot \frac{n}{\dim(W_1)} = \alpha n$$

by Lemma 6.2.2. This completes the proof of the theorem. □

As an immediate corollary, we get the following result [52, Théorème 68], which was stated as Theorem 1.1.3 in the introduction.

**Theorem 6.2.4** *Let $G \leq \mathrm{GL}_n(q)$ be primitive irreducible solvable and $V = \mathbb{F}_q^n$. Then* $\dim V^g \leq 3n/4$ *for all $g \in G \setminus \{1\}$.*

## 6.3 Abelian Subgroups of Solvable Affine Groups

Let $G \leq \mathrm{GL}_n(q)$ be irreducible and solvable. In this section, we will prove Theorem 1.1.4 from the introduction, which states that every abelian subgroup of the affine group $\mathbb{F}_q^n \rtimes G$ has order $\leq q^n$.

As a consequence, we will find that if $G \leq \Delta(V, \kappa)$ is an irreducible solvable subgroup of the form $G = G^{\mathcal{B}}_{\mu,\nu}(X_1, \ldots, X_k)$, then every abelian subgroup of $G$ has order $\leq |F_0|\mu^2$ (Corollary 6.3.9). Here $F_0$ denotes the subgroup satisfying properties (F1)–(F3) in $G$. We will also see that such a $G$ contains an abelian subgroup of order $\geq |F_0|\mu$ (Lemma 6.3.10). Following the approach of Jordan, in later sections (for example Sects. 7.1–7.2) we will use these inequalities to rule out most inclusions among the candidates for maximal irreducible solvable subgroups of $\Delta(V, \kappa)$.

We begin with the following lemma.

**Lemma 6.3.1** *Let $q = p^f$, where $p$ is a prime. Suppose that $H$ is an abelian subgroup of $\mathrm{Sp}_{2\ell}(q)$ such that $p \nmid |H|$. Then $H$ is generated by $\leq \ell$ elements, and $|H| \leq (q+1)^\ell$.*

***Proof*** Let $V = (\mathbb{F}_q)^{2\ell}$ be the natural module of $\mathrm{Sp}_{2\ell}(q)$, equipped with a non-degenerate alternating bilinear form. Since $p \nmid |H|$, by Maschke's theorem the action of $H$ on $V$ is completely reducible. It is well known (see for example [84, Proposition 3]) that in this case $V$ decomposes into an orthogonal direct sum

$$V = Z_1 \perp \cdots \perp Z_t \perp (W_1 \oplus W_1^*) \perp \cdots \perp (W_s \oplus W_s^*),$$

where $Z_i$ and $W_j$ are irreducible $\mathbb{F}_q[H]$-modules for all $1 \leq i \leq t$ and $1 \leq j \leq s$. Moreover $Z_i$ is a non-degenerate subspace for all $1 \leq i \leq t$ and $\{W_j, W_j^*\}$ is dual pair of a totally isotropic subspaces for all $1 \leq j \leq s$. We have $\dim Z_i = 2\ell_i$ for some $\ell_i > 0$ for the non-degenerate summands $Z_i$, and $\dim W_j = \ell'_j$ for some $\ell'_j > 0$.

Since $H$ is abelian, by Schur's lemma its image in $\mathrm{Sp}(Z_i)$ is irreducible cyclic, and thus of order dividing $q^{\ell_i} + 1$ by [38, II, Satz 9.23]. Similarly the image of $H$ in $\mathrm{GL}(W_i)$ is irreducible cyclic, and thus of order dividing $q^{\ell'_j} - 1$ by [38, II, Satz 3.10]. This implies that $H$ is generated by $\leq t + s \leq \ell$ elements, and moreover

$$\begin{aligned}|H| &\leq (q^{\ell_1} + 1) \cdots (q^{\ell_t} + 1)(q^{\ell'_1} - 1) \cdots (q^{\ell'_s} - 1) \\ &\leq (q + 1)^{\ell_1 + \cdots + \ell_t + \ell'_1 + \cdots + \ell'_s} \\ &= (q + 1)^{\ell}\end{aligned}$$

as claimed. □

**Remark 6.3.2** As an alternative proof using the theory of groups of Lie type, one could prove Lemma 6.3.1 by noting that $H$ is contained in a maximal torus of $\mathrm{Sp}_{2\ell}(q)$, in which case $|H| \leq (q + 1)^{\ell}$ by [71, Lemma 2.4 (iii)].

**Theorem 6.3.3** *Let $V$ be a finite-dimensional vector space over $\mathbb{F}_q$ with $n = \dim V$, and suppose that $G \leq \mathrm{GL}(V)$ is irreducible solvable. If $D$ is an abelian subgroup of the affine group $V \rtimes G$, then $|D| \leq q^n$.*

***Proof*** Let $p$ be the characteristic of $\mathbb{F}_q$. Recall that $V \rtimes G$ is identified as the group of affine transformations on $V$:

$$V \rtimes G = \{\varphi_{A,v} : V \to V : A \in G, v \in V\}$$

where $\varphi_{A,v}(w) = Aw + v$ for all $w \in V$.

If $n = 0$ there is nothing to prove. For $n = 1$, suppose that $D \leq V \rtimes G$ is abelian. If $p \nmid |D|$, then $D \cap V$ is trivial, so $D$ embeds into $\mathrm{GL}(V) \cong \mathrm{GL}_1(q)$ and thus $|D| \leq q - 1$. If $p \mid |D|$, then $D \cap V$ is nontrivial. Since $D$ is abelian, this implies that the image of $D$ in $\mathrm{GL}_1(q)$ is trivial. In other words $D \leq V$, and in particular $|D| \leq q$.

For the rest of the proof, we will suppose that $n > 1$ and proceed by induction on $n$. We will take an approach similar to that of Jordan [52, §81–§90], with a few adjustments.

We begin with an observation from [52, §83]. Suppose that $D \leq V \rtimes G$ is abelian of maximal order. Let $H$ be the image of $D$ in $G$. Since $D$ is abelian, the action of $H$ fixes every vector in $V_0 := D \cap V$. Thus $V_0 \times H$ is an abelian subgroup of $V \rtimes G$, of order $|D|$ since $D/V_0 \cong H$. Therefore we may assume that $D$ is of the form $V_0 \times H$, where $H \leq \mathrm{GL}(V)$ is abelian and fixes every vector in $V_0$. On the other hand $V^H \times H$ is also abelian and contains $D$, so we may assume that $D = V^H \times H$, where $H \leq G$ is abelian.

It is also clear that we can assume that $G$ is maximal irreducible solvable. We first deal with the case where $G$ is imprimitive.

**Claim 1: The theorem holds when $G$ is imprimitive**

We argue similarly to [52, §81]. In this case, there exists a decomposition $V = W_1 \oplus \cdots \oplus W_k$ such that $k > 1$ and $G$ acts on the summands $\{W_1, \ldots, W_k\}$. Since $G$ is maximal solvable, by Theorem 2.2.14 we have $G = \Gamma \wr \Delta$, where $\Gamma \leq \mathrm{GL}_d(q)$ is irreducible primitive, $\Delta \leq S_k$ is transitive solvable, and $n = dk$. Here $G = (\Gamma_1 \times \cdots \times \Gamma_k) \rtimes \Delta$, where $\Gamma_i$ acts irreducibly and primitively on $W_i$ for all $1 \leq i \leq k$.

Consider first the case where $H$ acts intransitively on $\{W_1, \ldots, W_k\}$, say with $t > 1$ orbits $\{W_1^{(1)}, \ldots, W_{d_1}^{(1)}\}, \ldots, \{W_1^{(t)}, \ldots, W_{d_t}^{(t)}\}$. Then

$$V = Z_1 \oplus \cdots \oplus Z_t,$$

where $Z_i = W_1^{(i)} \oplus \cdots \oplus W_{d_i}^{(i)}$ for all $1 \leq i \leq t$. Thus $H \leq (\Gamma \wr \Delta_1) \times \cdots \times (\Gamma \wr \Delta_t)$, where $\Delta_i \leq S_{d_i}$ is solvable and transitive for all $1 \leq i \leq t$. We have

$$V^H = Z_1^H \oplus \cdots \oplus Z_t^H$$

since $H$ acts on the $Z_i$, so $D \leq (Z_1^H \times H_1) \times \cdots \times (Z_t^H \times H_t)$, where $H_i$ is the image of $H$ in $\Gamma \wr \Delta_i$. The wreath products $\Gamma \wr \Delta_i$ are irreducible (Lemma 2.2.6), so by induction

$$|D| \leq q^{n_1} \cdots q^{n_t} = q^n,$$

where $n_i = \dim Z_i$ for all $1 \leq i \leq t$.

Therefore we can assume that $H$ acts transitively on $\{W_1, \ldots, W_k\}$. The image of $H$ in $\Delta$ is a transitive abelian subgroup of $S_k$, and thus has order $k$. Hence $|H| = k|H_0|$, where $H_0 := H \cap (\Gamma_1 \times \cdots \times \Gamma_k)$. Note that $H$ acts transitively on $\{\Gamma_1, \ldots, \Gamma_k\}$ by conjugation. Combining this with the fact that $H$ is abelian, it follows that the projection map $H_0 \to \Gamma_1$ must be injective. Therefore $|H| = k|H_1|$, where $H_1$ is the image of $H_0$ in $\Gamma_1$. Similarly since $H$ acts transitively on the $W_i$'s, the projection map $V^H \to W_1^{H_1}$ is injective, so $|V^H| = |W_1^{H_1}|$. Thus by applying induction on $W_1^{H_1} \times H_1 \leq W_1 \rtimes \Gamma_1$, we get

$$|D| = k|W_1^{H_1} \times H_1| \leq kq^d = \frac{n}{d}q^d \leq q^n$$

as desired.

Thus we can assume for the rest of the proof that $G$ is a maximal primitive irreducible solvable subgroup of $\mathrm{GL}(V) = \mathrm{GL}_n(q)$. We note the following.

**Claim 2: $|D|$ is a power of $p$**

The basic argument that we use is similar to the one given by Jordan in [52, §76–§82]. Write $H = H_1 \times H_2$, such that $p \nmid |H_1|$ and $|H_2|$ a power of $p$. Suppose that $H_1 \neq \{1\}$. We will show that the abelian subgroup $V^{H_2} \times H_2$ of $V \rtimes G$ has larger order than $D$, which contradicts the maximality of $|D|$.

Since $p \nmid |H_1|$, by Maschke's theorem we can write

$$V = W_1^{r_1} \oplus \cdots \oplus W_t^{r_t} \oplus V^{H_1}, \tag{6.5}$$

where $W_i$ are nontrivial irreducible and pairwise nonisomorphic $\mathbb{F}_q[H_1]$-modules. Here $r_1, \ldots, r_t > 0$ are integers and $t \geq 1$. Denote $\nu_i = \dim W_i$. Since $H_1$ is abelian and acts irreducibly on $W_i$, we have $\mathrm{Hom}_{H_1}(W_i, W_i) \cong \mathbb{F}_{q^{\nu_i}}$ for all $1 \leq i \leq t$. Then the $\mathbb{F}_q$-algebra centralizer of $H_1$ in $\mathrm{Mat}_n(q)$ is isomorphic to

$$\mathrm{Mat}_{r_1}(\mathbb{F}_{q^{\nu_1}}) \oplus \cdots \oplus \mathrm{Mat}_{r_t}(\mathbb{F}_{q^{\nu_t}}) \oplus (\mathbb{F}_q)^d,$$

where $d = \dim V^{H_1}$. Since $H_1$ is abelian, its action on $W_i^{r_i}$ is by scalar matrices from $\mathrm{Mat}_{r_i}(\mathbb{F}_{q^{\nu_i}})$, and therefore

$$|H_1| \leq (q^{\nu_1} - 1) \cdots (q^{\nu_t} - 1). \tag{6.6}$$

Now $H_2$ commutes with $H_1$, so for all $1 \leq i \leq t$ the homogeneous summand $W_i^{r_i}$ is a nonzero $\mathbb{F}_{q^{\nu_i}}[H_2]$-module of dimension $r_i$. Because $H_2$ is a $p$-group, it has a nonzero fixed point space on $W_i^{r_i}$, and thus fixes at least $q^{\nu_i}$ vectors in $W_i^{r_i}$. Moreover $V^H = V^{H_1} \cap V^{H_2}$, so from the decomposition (6.5) we get a bound

$$|V^{H_2}| \geq q^{\nu_1} \cdots q^{\nu_t} |V^H|. \tag{6.7}$$

With (6.6) and (6.7) we get

$$\begin{aligned}
|D| &= |V^H| \cdot |H_1| \cdot |H_2| \\
&\leq (q^{\nu_1} - 1) \cdots (q^{\nu_t} - 1) |V^H| \cdot |H_2| \\
&< q^{\nu_1} \cdots q^{\nu_t} |V^H| \cdot |H_2| \\
&\leq |V^{H_2}| \cdot |H_2| = |V^{H_2} \times H_2|
\end{aligned}$$

which contradicts the maximality of $|D|$. Thus $|D|$ is a power of $p$, and in particular $H$ is a $p$-group.

Let $F_0 \trianglelefteq G$ be a maximal abelian normal subgroup of $G$. We proceed as in the setup of Remark 2.5.6. Then the $\mathbb{F}_q$-algebra generated by $F_0$ in $\mathrm{End}(V)$ is isomorphic to $\mathbb{K} = \mathbb{F}_{q^\nu}$, and $F_0 = \mathbb{K}^\times \cong C_{q^\nu-1}$. This makes $V$ into a $\mathbb{K}$-vector space of dimension $\mu = n/\nu$, and $C_{\mathrm{GL}(V)}(F_0)$ is the set of $\mathbb{K}$-linear maps in $\mathrm{GL}(V)$.

Let $H' = H \cap C_G(F_0)$.

**Claim 3: We have** $|V^{H'} \times H'| \geq |D|$
The following argument is similar to [52, §84]. If $H' = H$, there is nothing to prove. Suppose then that $H'$ is a proper subgroup of $H$.

It is clear that $V^H \subseteq V^{H'}$. Since $H' \leq C_G(F_0)$, the space $V^{H'}$ is $F_0$-invariant, so it is a vector space over $\mathbb{K}$. Let $x_1, \ldots, x_m$ be a $\mathbb{K}$-basis for $V^{H'}$, and extend this to $\mathbb{K}$-basis $x_1, \ldots, x_\mu$ of $V$.

Having fixed a $\mathbb{K}$-basis of $V$, we can define the map $\psi : V \to V$ and maps $g_A : V \to V$ for $A \in \mathrm{GL}_\mu(\mathbb{K})$ as in Remark 2.5.6.

We have $H = \langle H', g_A\psi^\delta \rangle$ for some $0 < \delta < \nu$ with $\delta \mid \nu$ and $A \in \mathrm{GL}_\mu(\mathbb{K})$. Now $V^H$ consists of vectors of the form $v = \sum_{i=1}^m \alpha_i x_i$ with $\alpha_1, \ldots, \alpha_m \in \mathbb{K}$ such that $g_A\psi^\delta v = v$. We have

$$g_A\psi^\delta v = g_A \sum_{i=1}^m \alpha_i^{q^\delta} x_i,$$

so $v \in V^H$ if and only if for all $1 \leq i \leq m$ we have

$$\sum_{j=1}^m A_{ji}\alpha_j^{q^\delta} = \begin{cases} \alpha_i, & \text{if } 1 \leq i \leq m. \\ 0, & \text{if } m+1 \leq i \leq \mu. \end{cases} \tag{6.8}$$

It follows from (6.8) that

$$\alpha_1 - A_{11}\alpha_1^{q^\delta} = \sum_{j=2}^m A_{ji}\alpha_j^{q^\delta}.$$

This is a polynomial equation of degree $q^\delta$, so for each $\alpha_2, \ldots, \alpha_m \in \mathbb{K}$ there are at most $q^\delta$ values of $\alpha_1$ that permit $v \in V^H$. Hence $|V^H| \leq q^{\nu(m-1)+\delta}$.

On the other hand $|V^{H'}| = q^{\nu m}$, so $|V^{H'}/V^H| \geq q^{\nu-\delta}$. Thus

$$\begin{aligned} |V^{H'} \times H'| &= |V^{H'}| \cdot |H'| \\ &\geq q^{\nu-\delta}|V^H| \cdot |H'| \\ &= q^{\nu-\delta}\frac{\nu}{\delta}|V^H| \cdot |H| \\ &\geq |D|, \end{aligned}$$

since $q^{\nu-\delta} \geq \frac{\nu}{\delta}$.

Therefore we may replace $D$ by the abelian subgroup $V^{H'} \times H'$. In other words, we can and will assume that $D = V^H \times H$, where $H \le C_G(F_0)$. Clearly we can also assume that $H \neq \{1\}$.

If $\mu < n$, then since $C_G(F_0)$ is an absolutely irreducible solvable subgroup of $\mathrm{GL}(W'_\lambda) = \mathrm{GL}_\mu(\mathbb{K})$ (Lemma 5.2.2 (i)), it follows by induction that

$$|V^H \times H| \le (q^\nu)^\mu = q^n$$

as desired. We can assume then that $\mu = n$, in which case $\nu = 1$ and $F_0 \cong \mathbb{F}_q^\times$ is the group of scalar matrices in $\mathrm{GL}_n(q)$.

Let $A = \mathrm{Fit}(C_G(F_0)) = \mathrm{Fit}(G)$, which is absolutely irreducible (Lemma 4.1.4, Theorem 4.1.16). Let $n = r_1^{\ell_1} \cdots r_k^{\ell_k}$ be the prime factorization of $n$, where $r_1 < \cdots < r_k$ are primes and $\ell_i > 0$ for all $1 \le i \le k$.

We have $A = F_0 R_1 \cdots R_k$, where $R_i \trianglelefteq G$ is extraspecial of order $r_i^{1+2\ell_i}$ and exponent $r_i \gcd(r_i, 2)$ (Proposition 4.1.11). Furthermore $Z(R_i) = F_0 \cap R_i$ is cyclic of order $r_i$, and $r_i \mid q - 1$ for all $1 \le i \le k$ (Proposition 4.1.18).

We have a homomorphism

$$\pi : G \to \mathrm{Sp}_{2\ell_1}(r_1) \times \cdots \times \mathrm{Sp}_{2\ell_k}(r_k)$$

defined by $\pi(g) = (g_1, \ldots, g_k)$, where $g_i$ is the action of $g$ on $R_i/Z(R_i)$. If $r_1 = 2$, we also have $g_1 \in O^\pm_{2\ell_1}(2)$.

Note that $\mathrm{Ker}\,\pi = A$ (Lemma 4.2.1). We have $\gcd(|H|, |A|) = 1$ since $H$ is a $p$-group, so $H \cap A = \{1\}$ and $H$ embeds into $\mathrm{Sp}_{2\ell_1}(r_1) \times \cdots \times \mathrm{Sp}_{2\ell_k}(r_k)$.

The primes $r_i$ are coprime to $p$ since $q \equiv 1 \mod r_i$ for all $1 \le i \le k$, so by Lemma 6.3.1 the image of $H$ in $\mathrm{Sp}_{2\ell_i}(r_i)$ has order at most $(r_i + 1)^{\ell_i}$. Therefore

$$|H| \le (r_1 + 1)^{\ell_1} \cdots (r_k + 1)^{\ell_k}.$$

Moreover by Theorem 6.2.4 we have $|V^H| \le q^{3n/4}$. Then

$$\begin{aligned} |D| = |V^H \times H| &\le (r_1 + 1)^{\ell_1} \cdots (r_k + 1)^{\ell_k} \cdot q^{3n/4} \\ &\le q^{\ell_1 + \cdots + \ell_k + 3n/4} \le q^n \end{aligned}$$

when $n > 8$.

Suppose then that $n \le 8$. In this case we have

$$|D| = |V^H \times H| \le (r_1 + 1)^{\ell_1} \cdots (r_k + 1)^{\ell_k} \cdot q^{3n/4} \le q^n \tag{6.9}$$

when $q > 8$. When $q \le 8$, the inequality (6.9) still holds, except for a few cases. We can rule out most of these values using the fact that $q \equiv 1 \mod r_i$ for the prime factors $r_i$ of $n$. This leaves us with the following cases:

- $q = 7, n = 2, 4$.
- $q = 5, n = 2, 4, 8$.
- $q = 4, n = 3$.
- $q = 3, n = 2, 4, 8$.

For these cases, when $q$ is odd we can apply Theorem 6.2.3 to conclude that $|V^H| \leq q^{n/2}$. Then

$$|D| = |V^H \times H| \leq (r_1 + 1)^{\ell_1} \cdots (r_k + 1)^{\ell_k} \cdot q^{n/2},$$

which implies $|D| \leq q^n$.

It remains to consider the case where $q = 4$ and $n = 3$. Then $H$ embeds into $\mathrm{Sp}_2(3)$ via the map $\pi$. Every abelian 2-subgroup of $\mathrm{Sp}_2(3)$ has order $\leq 4$, so $|H| \leq 4$. On the other hand $H$ is nontrivial, so $\dim V^H \leq 2$ and hence $|V^H| \leq 4^2$. Thus $|V^H \times H| \leq 4^3$, which completes the proof of the theorem. □

**Remark 6.3.4** As observed by Jordan in [52, §90], if $D \leq V \rtimes G$ is abelian and equality $|D| = q^n$ holds, then $D \cap V$ is nontrivial. Indeed, in this case $D$ is a $p$-group, which implies that $V^D \neq 0$. Since $D$ commutes with $V^D$, we must have $V^D \leq D$, as otherwise $\langle V^D, D\rangle$ would be an abelian subgroup of order $> q^n$, contradicting Theorem 6.3.3. Thus $V^D \leq D$, and in particular $D \cap V$ is nontrivial.

**Remark 6.3.5** It seems that at the end of proof given by Jordan, there is a small flaw which occurs in [52, §88–§89]. In [52, §88–§89], Jordan seems to claim that for $g$ acting on $V$ and $W$ as a unipotent linear map, we have $(V \otimes W)^g = V^g \otimes W^g$. However, this is not true. Suppose, for example, that $g$ acts on both $V$ and $W$ with a single unipotent Jordan block. Then $\dim(V \otimes W)^g = \min\{\dim V, \dim W\}$, while $\dim V^g = \dim W^g = 1$.

**Remark 6.3.6** More generally, if $G \leq \mathrm{GL}_n(q)$ is solvable and completely reducible, then every abelian subgroup of $\mathbb{F}_q^n \rtimes G$ has order at most $q^n$.

Indeed, in this case $G$ is isomorphic to a subgroup of $(\mathbb{F}_q^{n_1} \rtimes G_1) \times \cdots \times (\mathbb{F}_q^{n_k} \rtimes G_k)$, where $n_1 + \cdots + n_k = n$ and $G_i$ is a solvable and irreducible subgroup of $\mathrm{GL}_{n_i}(q)$. It follows then from Theorem 6.3.3 that every abelian subgroup of $\mathbb{F}_q^n \rtimes G$ has order at most $q^{n_1} \cdots q^{n_k} = q^n$.

**Remark 6.3.7** The assumption that $G$ is irreducible (or more generally, completely reducible) is necessary in Theorem 6.3.3. One reason is that we can find abelian $p$-subgroups of $\mathrm{GL}_n(q)$ of order $> q^n$; in general the maximal order of an abelian $p$-subgroup of $\mathrm{GL}_n(q)$ is $q^{\lfloor n^2/4 \rfloor}$ [27].

For example, let $k = \lceil n/2 \rceil$ and consider the subgroup $H \leq \mathrm{GL}_n(q)$ consisting of matrices of the form

$$\begin{pmatrix} I_k & * \\ 0 & I_{n-k} \end{pmatrix}.$$

Then $H$ is elementary abelian of order $|H| = q^{\lfloor n^2/4 \rfloor}$, and $|V^H| = q^k$. Thus $|V^H \times H| > |V|$ if $n \geq 3$, and $|H| > |V|$ if $n \geq 5$.

**Proposition 6.3.8** *Let $G \leq \Delta(V, \kappa)$ be an irreducible solvable group of the form $G = G^{\mathcal{B}}_{\mu,\nu}(X_1, \ldots, X_k)$, with corresponding subgroup $A = F_0 R_1 \cdots R_k$. Let $H \leq K \leq G$ be such that all of the following hold:*

*(i) $H \leq C_{G^\circ}(F_0)$ and $[H, H] \leq F_0$,*
*(ii) For every prime $r \mid \mu$, the action of $K$ on $O_r(A/F_0)$ is metrically completely reducible.*

*Then $|H| \leq |H \cap F_0| \cdot \mu^2$. If additionally*

*(iii) $H \trianglelefteq K$ and $|H| = |A|$,*

*then $H = A$.*

***Proof*** Replacing $K$ with $KF_0$, we can assume that $F_0 \leq K$. Suppose that (i) and (ii) hold. We will prove that $|H| \leq |H \cap F_0| \cdot \mu^2$. For this we may assume that $F_0 \leq H$; the general case follows from the fact that $HF_0$ has order $|H|[F_0 : H \cap F_0]$.

Recall that $\mu = r_1^{\ell_1} \cdots r_k^{\ell_k}$ is the prime factorization of $\mu$. By (ii), for all $1 \leq i \leq k$ we have an orthogonal decomposition

$$O_{r_i}(A/F_0) = Z_1^{(i)} \perp \cdots \perp Z_{t_i}^{(i)},$$

where $Z_j^{(i)}$ is an irreducible $\mathbb{F}_{r_i}[K]$-module for all $1 \leq j \leq t_i$ (Lemma 2.3.2). For all $1 \leq j \leq t_i$, let $H_j^{(i)}$ be the action of $H$ on $Z_j^{(i)}$.

The action of $K$ on $O_{r_i}(A/F_0)$ is completely reducible and $H_j^{(i)}$ is abelian, so by Theorem 6.3.3 (see Remark 6.3.6) we have

$$\left| \left(Z_j^{(i)}\right)^H \times H_j^{(i)} \right| \leq \left| Z_j^{(i)} \right| \tag{6.10}$$

for all $1 \leq j \leq t_i$.

Now $g \in H$ acts trivially on $A/F_0$ if and only if $g \in A$ (Lemma 4.2.1), so

$$|H/H \cap A| \leq \prod_{i,j} \left| H_j^{(i)} \right|.$$

Moreover $[H, H] \leq F_0$, so

$$H \cap A/F_0 \leq \prod_{i,j} \left(Z_j^{(i)}\right)^H.$$

Thus we get inequalities

$$\begin{aligned}[H : F_0] &= |H/H \cap A| \cdot |H \cap A/F_0| \\ &\leq \prod_{i,j} \left|H_j^{(i)}\right| \cdot \prod_{i,j} \left|\left(Z_j^{(i)}\right)^H\right| \\ &= \prod_{i,j} \left|\left(Z_j^{(i)}\right)^H \times H_j^{(i)}\right| \\ &\leq \prod_{i,j} \left|Z_j^{(i)}\right| && \text{by (6.10)} \\ &= \prod_{i=1}^{k} |O_{r_i}(A/F_0)| = \mu^2.\end{aligned}$$

This completes the proof of the first claim. Suppose then that (i)–(iii) hold. We have $|F_0 H| \leq |A| = |H|$ since $F_0 H$ satisfies (i) and (ii), so $H = F_0 H$ which gives $F_0 \leq H$. Because $|H| = |A|$, we have $[H : F_0] = \mu^2$, so the inequalities above imply

$$\left|\left(Z_j^{(i)}\right)^H \times H_j^{(i)}\right| = \left|Z_j^{(i)}\right| \tag{6.11}$$

for all $1 \leq i \leq k$ and $1 \leq j \leq t_i$.

As noted in Remark 6.3.4, it follows from (6.11) that $\left(Z_j^{(i)}\right)^H \neq 0$ for all $i$ and $j$. Now $K$ acts on $\left(Z_j^{(i)}\right)^H$ because $H \trianglelefteq K$, so $\left(Z_j^{(i)}\right)^H = Z_j^{(i)}$ since $Z_j^{(i)}$ is an irreducible $\mathbb{F}_{r_i}[K]$-module. Therefore $H$ acts trivially on $Z_j^{(i)}$ for all $i$ and $j$, which implies that $H$ acts trivially on $A/F_0$. Consequently $H \leq A$ (Lemma 4.2.1), which combined with $|H| = |A|$ gives $H = A$. □

**Corollary 6.3.9** *Let $G \leq \Delta(V, \kappa)$ be an irreducible solvable group of the form $G = G^{\mathcal{B}}_{\mu,\nu}(X_1, \ldots, X_k)$. If $H \leq C_{G^\circ}(F_0)$ and $[H, H] \leq F_0$, then*

$$|H| \leq |H \cap F_0| \cdot \mu^2.$$

*In particular,*

$$|H| \leq |A| = |F_0|\mu^2 = \begin{cases} (q^\nu - 1)\mu^2, & \textit{if } G \textit{ is of type } \mathcal{B}_0 \textit{ or } \mathcal{B}_1. \\ (q^\nu + 1)\mu^2, & \textit{if } G \textit{ is of type } \mathcal{B}_2. \\ 2(q - 1), & \textit{if } G \textit{ is of type } \mathcal{B}_3. \end{cases}$$

***Proof*** By Lemma 5.3.1 (i), this follows from Proposition 6.3.8 with $K = G$. □

**Lemma 6.3.10** *Let $G \leq \Delta(V, \kappa)$ be an irreducible solvable group of the form $G = G^{\mathcal{B}}_{\mu,\nu}(X_1, \ldots, X_k)$. Then $A = F_0 R_1 \cdots R_k$ contains an abelian subgroup of order $\mu|F_0|$.*

***Proof*** Let $\mu = r_1^{\ell_1} \cdots r_k^{\ell_k}$ be the prime factorization of $\mu$, so $A = F_0 R_1 \cdots R_k$ is a central product with $R_i = (r_i)^{1+2\ell_i}_{\varepsilon_i}$. Each $R_i$ contains a maximal abelian subgroup $H_i \leq R_i$ of order $C_{r_i}^{\ell_i+1}$ (Lemma 3.1.2). Then $F_0 H_1 \cdots H_k$ is an abelian subgroup of $A$ of order $|F_0| r_1^{\ell_1} \cdots r_k^{\ell_k} = |F_0|\mu$. □

The following result is proved by Jordan in [52, §92]. We will not need it in the proof of our main results, but we include it here for completeness.

**Proposition 6.3.11** *Let $G \leq O^{\varepsilon}_{2\ell}(2)$ be irreducible and solvable. If $H \leq G$ is an elementary abelian 2-group, then $|H| \leq 2^{\ell}$.*

***Proof*** *([52, §92])* We proceed by induction on $\ell$. The case $\ell = 1$ is clear, since $|O_2^+(2)| = 2$ and $|O_2^-(2)| = 6$.

Suppose then that $\ell > 1$. It is clear that we can assume that $G$ is maximal irreducible solvable. We first consider the case where $G$ is metrically imprimitive, so we have $2\ell = 2\ell'\ell''$ for $\ell'' > 1$ such that

$$G = X \wr Y = (X_1 \times \cdots \times X_{\ell''}) \rtimes Y,$$

where $X \leq O^{\varepsilon'}_{2\ell'}(2)$ is maximal metrically primitive irreducible solvable, and $Y \leq S_{\ell''}$ is maximal transitive solvable.

As in the proof of Theorem 6.3.3 (Claim 1), by induction we can reduce to the case where the image of $H$ in $Y$ is transitive. In this case, as seen in the proof of Theorem 6.3.3, we have $|H| = \ell''|H_1|$, where $H_1$ is the projection of $H \cap (X_1 \times \cdots \times X_{\ell''})$ into $X_1$. By induction we have $|H_1| \leq 2^{\ell'}$, so

$$|H| \leq \ell'' 2^{\ell'} \leq 2^{\ell'\ell''}$$

as desired.

Therefore we can assume that $G$ is metrically primitive. Since we are working in characteristic two, either $G$ is of type $\mathcal{B}_1$ or $\mathcal{B}_2$. Let $F_0$ and $A = \mathrm{Fit}(C_G(F_0))$ (Lemma 4.1.4) be the subgroups satisfying properties (F1)–(F3) and (A1)–(A3), respectively.

Let $\mu = r_1^{\ell_1} \cdots r_k^{\ell_k}$ is the prime factorization of $\mu$. We have $|F_0| = 2^{\nu} - 1$ in type $\mathcal{B}_1$ and $|F_0| = 2^{\nu} + 1$ in type $\mathcal{B}_2$. Furthermore the prime factors $r_i$ must divide $|F_0|$ (Proposition 4.1.18), so they are all odd. Then $G = G^{\mathcal{B}}_{\mu,\nu}(X_1, \ldots, X_k)$, where $n = 2\mu\nu$ and $X_i \leq \mathrm{GSp}_{2\ell_i}(r_i)$ is metrically completely reducible maximal solvable for all $1 \leq i \leq k$ (Lemma 4.3.30).

Let $\pi : G \to \prod_{i=1}^{k} \mathrm{GSp}_{2\ell_i}(r_i)$ be the usual map $\pi(g) = (g_1, \ldots, g_k)$, where $g_i$ is the action of $g$ on $R_i/Z(R_i)$. We have $\pi(H) \leq H_1 \times \cdots \times H_k$, where $H_i \leq$

$X_i$ is the action of $H$ on $R_i/Z(R_i)$. Since $|H_i|$ is coprime to $r_i$, it follows from Lemma 6.3.1 that $H_i$ is an elementary abelian 2-group generated by $\leq \ell_i$ elements, so $|H_i| \leq 2^{\ell_i}$.

Because $|A| = |F_0|\mu^2$ is coprime to $|H|$, the intersection $H \cap \operatorname{Ker}\pi$ embeds into $\operatorname{Ker}\pi/A$. The structure of $\operatorname{Ker}\pi/A$ is described in Lemmas 4.3.10 and 4.3.19.

From Lemma 4.3.19, we see that if $G$ is of type $\mathcal{B}_2$, every elementary abelian 2-subgroup of $\operatorname{Ker}\pi/A$ has order $\leq 2$, so

$$|H| = |H/H \cap \operatorname{Ker}\pi| \cdot |\pi(H)| \leq 2\mu \leq 2^{\nu\mu} = 2^{\ell}.$$

Suppose then that $G$ is of type $\mathcal{B}_1$, in which case $\nu \geq 3$, since $(q, \nu) = (2, 1)$ and $(2, 2)$ are not applicable for groups of type $\mathcal{B}_1$. By Lemma 4.3.10 every elementary abelian 2-subgroup of $\operatorname{Ker}\pi/A$ has order $\leq 4$, so we have

$$|H| \leq 4\mu \leq 2^{\nu\mu} = 2^{\ell},$$

as required. □

**Remark 6.3.12** Similarly to Theorem 6.3.3, in Proposition 6.3.11 the assumption that $G$ is irreducible (or more generally completely reducible) is necessary. For example, there is an elementary abelian subgroup of order 16 in $O_6^{\pm}(2)$.

# Chapter 7
# Maximality of the Groups Constructed

In this chapter, we will complete the classification of maximal irreducible solvable subgroups of $\Delta(V, \kappa)$ and $I(V, \kappa)$. When $\kappa \neq 0$, we will more generally classify metrically completely reducible maximal solvable subgroups of $\Delta(V, \kappa)$ and $I(V, \kappa)$. Furthermore, in the case where $q$ is even and $\kappa$ is a quadratic form, we will also classify metrically completely reducible maximal solvable subgroups of $\Omega(V, \kappa)$.

## 7.1 Maximality of $G^{\mathcal{B}}_{\mu,\nu}(X_1, \ldots, X_k)$

Throughout this section, we consider two metrically primitive irreducible solvable subgroups $G, \overline{G} \leq \Delta(V, \kappa)$ of the form $G = G^{\mathcal{B}}_{\mu,\nu}(X_1, \ldots, X_k)$ and $\overline{G} = G^{\overline{\mathcal{B}}}_{\overline{\mu},\overline{\nu}}(Y_1, \ldots, Y_\ell)$. Recall that we have shown that any maximal metrically primitive irreducible solvable subgroup is of this form (Lemma 4.3.30), and moreover we know precisely when $G$ is metrically primitive and irreducible (Lemma 5.2.2, Theorem 5.6.9).

In this section, we will see that with a few exceptions, assuming $G \leq \overline{G}$ implies $G = \overline{G}$. This gives us a classification of maximal metrically primitive irreducible solvable subgroups of $\Delta(V, \kappa)$ (Theorem 7.1.12).

With further analysis, we will also get a classification of the maximal metrically primitive irreducible solvable subgroups of $I(V, \kappa)$ (Theorem 7.1.13), and thus for every group $H$ with $I(V, \kappa) \leq H \leq \Delta(V, \kappa)$ (Lemma 2.1.3). In the case where $q$ is even and $\kappa$ is a quadratic form, we will similarly classify maximal metrically primitive irreducible solvable subgroups of $\Omega(V, \kappa)$ (Theorem 7.1.14).

**Remark 7.1.1** Note that for $\mu > 1$, the group $G^{\mathcal{B}}_{\mu,\nu}(X_1, \ldots, X_k)$ is described in terms of smaller metrically completely reducible maximal solvable subgroups $X_1$, …, $X_k$. Thus for a complete solution, we will have prove similar classification results for metrically imprimitive maximal irreducible solvable subgroups, and more

 
M. Korhonen, *Maximal Solvable Subgroups of Finite Classical Groups*,
Lecture Notes in Mathematics 2346, https://doi.org/10.1007/978-3-031-62915-0_7

generally for metrically completely reducible maximal solvable subgroups. These results will be obtained in Sects. 7.4 and 7.5.

One of the exceptional cases where $G^{\mathcal{B}}_{\mu,\nu}(X_1, \ldots, X_k)$ is not maximal solvable appears when $X_i^\circ$ is not maximal solvable in $\mathrm{Sp}_{2\ell_i}(r_i)$, together with some conditions on $q$ (Theorem 7.1.12 (ii)). Since $X_i^\circ$ is metrically completely reducible (Lemma 2.3.10 (i)), it follows that we will also have to classify metrically completely reducible maximal solvable subgroups of $\mathrm{Sp}_{2\ell_i}(r_i)$. As mentioned above, in the metrically primitive irreducible case, such a classification follows from results obtained in this section (Theorem 7.1.13). Later in Sect. 7.6 (Theorem 7.6.5), we will give a precise description of the cases where $X_i$ is maximal solvable, but $X_i^\circ$ is not maximal solvable in $\mathrm{Sp}_{2\ell_i}(r_i)$.

In the classification of maximal irreducible solvable subgroups of $I(V, \kappa)$, there are some exceptions to maximality which arise from metrically completely reducible maximal solvable subgroups $X \leq O^{\varepsilon}_{2\ell}(2)$ with the property that $X^{\Omega}$ is not maximal solvable in $\Omega^{\varepsilon}_{2\ell}(2)$ (Theorem 7.1.13 (iv)). It follows that we will also require a classification of metrically completely reducible maximal solvable subgroups of $\Omega^{\varepsilon}_{2\ell}(2)$, which in the metrically primitive irreducible case is obtained in this section (Theorem 7.1.14). In general this will be obtained in Sect. 7.6 (Theorems 7.6.3 and 7.6.4).

We will denote by $F_0$ and $\overline{F_0}$ subgroups satisfying (F1)–(F3) in $G$ and $\overline{G}$, respectively. Similarly we denote $A = \mathrm{Fit}(C_{G^\circ}(F_0))$ and $\overline{A} = \mathrm{Fit}(C_{\overline{G^\circ}}(\overline{F_0}))$, which are subgroups satisfying (A1)–(A3) in $G$ and $\overline{G}$, respectively (Corollary 5.3.6). We have $A = F_0 R_1 \cdots R_k$ and $\overline{A} = \overline{F_0}\ \overline{R_1} \cdots \overline{R_\ell}$ for extraspecial groups $R_i$ and $\overline{R_j}$ used in the construction of $G$ and $\overline{G}$.

We choose a generator $F_0 = \langle f \rangle$ and denote by $\alpha$ the number of eigenvalues of $f$ on $V' := \mathbb{K} \otimes_{\mathbb{F}_q} V$. For $\overline{F_0}$ and $\overline{G}$, we define $\overline{f}$ and $\overline{\alpha}$ similarly.

Our general approach is similar to the one used by Jordan in [52, Section VIII]. We begin with an elementary lemma, which appears as part of the proof of Jordan in [52, §122].

**Lemma 7.1.2** *Let $\mathbb{F}$ be a finite field. Let $S, T \in \mathrm{GL}_2(\mathbb{F})$ such that* $S = \begin{pmatrix} \gamma & 0 \\ 0 & \delta \end{pmatrix}$ *and* $T = \begin{pmatrix} x & y \\ z & w \end{pmatrix}$. *Suppose that $S$ normalizes $\langle T \rangle$, and that $T$ has order coprime to* $\mathrm{char}\,\mathbb{F}$. *Then one of the following holds:*

*(i)* $\gamma = \pm\delta$;
*(ii)* $y = z = 0$.

***Proof*** Since $S$ normalizes $T$, the conjugate $S^{-1}TS$ is centralized by $T$. Therefore $U := S^{-1}TS - T = \begin{pmatrix} 0 & (\gamma^{-1}\delta - 1)y \\ (\gamma\delta^{-1} - 1)z & 0 \end{pmatrix}$ is also centralized by $T$. It is readily checked that $TU = UT$ if and only if the following hold:

$$\begin{cases} (\gamma^{-1}\delta - 1)yw = (\gamma^{-1}\delta - 1)yx \\ (\gamma\delta^{-1} - 1)zx = (\gamma\delta^{-1} - 1)zw \\ (\gamma^{-1}\delta - 1)yz = (\gamma\delta^{-1} - 1)yz \end{cases}$$

Suppose that (i) does not hold, so $\gamma \neq \pm\delta$. Then the equalities above imply that $yw = yx$, $zx = zw$, and $yz = 0$. If $y \neq 0$, then we have $z = 0$ and $x = w$, so $T = \begin{pmatrix} x & y \\ 0 & x \end{pmatrix}$. But in this case $T$ has order divisible by char $\mathbb{F}$, contrary to what is assumed. Therefore we must have $y = 0$, and the same argument shows that $z = 0$. In other words (ii) holds, and the proof of the lemma is complete. □

**Lemma 7.1.3** *Let $H = G^{\Omega}$ if $q$ is even and $\kappa$ is a quadratic form, and $H = G^{\circ}$ otherwise. Suppose that $H \leq \overline{G^{\circ}}$. Then one of the following holds:*

*(i) $\overline{F_0}$ centralizes $F_0$;*
*(ii) $G$ is of type $\mathcal{B}_1$, with $q = 5$, $\nu = 1$, and $X_1 \leq \Omega^{+}_{2\ell_1}(2)$ if $\mu > 1$.*
*(iii) $G$ is of type $\mathcal{B}_2$, with $q = 3$, $\nu = 1$, and $X_1 \leq \Omega^{+}_{2\ell_1}(2)$ if $\mu > 1$.*

***Proof*** If $G$ is of type $\mathcal{B}_3$, the subgroup $F_0 = \langle -I_V \rangle$ centralizes $\overline{G^{\circ}}$, so (i) holds. Thus we will assume that $G$ is of type $\mathcal{B}_0$, $\mathcal{B}_1$, or $\mathcal{B}_2$.

A key observation in the proof is that $\overline{F_0}$ centralizes $[H, H]$, which follows since $\overline{G^{\circ}}/C_{\overline{G^{\circ}}}(\overline{F_0})$ is abelian (Lemma 4.3.33). In particular, it follows from Lemma 5.7.2 (for $H = G^{\circ}$) and Lemma 5.7.3 (for $H = G^{\Omega}$) that $\overline{F_0}$ centralizes $R_1 \cdots R_k$.

Moreover, we note the following observations which will be used throughout the proof. Suppose that $f^t \in [H, H]$, in which case $\overline{F_0}$ centralizes $f^t$. In particular $\overline{F_0}$ acts on the eigenspaces of $f^t$ on $V' = \mathbb{K} \otimes_{\mathbb{F}_q} V$. If $f$ and $f^t$ have the same eigenspaces on $V'$, then $\overline{F_0}$ centralizes $f$, and consequently $\overline{F_0}$ centralizes $F_0$.

Let $\lambda \in \mathbb{K}$ be an eigenvalue of $f$ on $V'$. For types $\mathcal{B}_0$ and $\mathcal{B}_2$, the eigenspaces of $f$ and $f^t$ differ if and only if there exists $0 < \alpha < [\mathbb{K} : \mathbb{F}_q]$ such that $\lambda^t = \lambda^{q^{\alpha}t}$. Similarly for type $\mathcal{B}_1$, the eigenspaces of $f$ and $f^t$ differ if and only if $\lambda^t = \lambda^{q^{\alpha}t}$ for some $0 < \alpha < [\mathbb{K} : \mathbb{F}_q]$, or $\lambda^t = \lambda^{-q^{\alpha}t}$ for some $0 \leq \alpha < [\mathbb{K} : \mathbb{F}_q]$.

We now consider the different types of $G$ in turn.

**Case 1: $G$ is of type $\mathcal{B}_0$**

In this case $H = G = G^{\circ}$, and $\mathbb{K} = \mathbb{F}_{q^{\nu}}$ with $|f| = q^{\nu} - 1$. If $\nu = 1$, then $F_0$ is the group of scalar matrices, so it is centralized by $\overline{F_0}$. We assume then that $\nu > 1$, and let $\lambda \in \mathbb{K}$ be an eigenvalue of $f$ on $V'$.

By Lemma 4.3.5, there exists an element of the form $g_A\psi^2$ in $G$. Thus $[G, G]$ contains

$$[f, g_A\psi^2] = [f, \psi^2] = f^{q^2-1}.$$

Suppose that the eigenspaces of $f$ and $f^{q^2-1}$ differ, so $\lambda^{q^2-1} = \lambda^{(q^2-1)q^\alpha}$ for some $0 < \alpha < \nu$. Then

$$(q^2 - 1)(q^\alpha - 1) \equiv 0 \mod q^\nu - 1. \tag{7.1}$$

Suppose first that $\nu > 2$. It follows from (7.1) that $q^\nu - 1$ cannot have a primitive prime divisor, so by Zsigmondy's theorem (Theorem 2.1.21) we have $q = 2$ and $\nu = 6$. However, it is readily seen that in this case (7.1) cannot hold for $0 < \alpha < \nu$, so we have contradiction.

Thus we can assume that $\nu = 2$, in which case $V' = W'_\lambda \oplus W'_{\lambda^q}$. By construction the $f$-eigenspaces $W'_\lambda$ and $W'_{\lambda^q}$ are both absolutely irreducible $\mathbb{K}[R_1 \cdots R_k]$-modules. In this case, consider first the possibility that $W'_\lambda \ncong W'_{\lambda^q}$ as $\mathbb{K}[R_1 \cdots R_k]$-modules. Then $\overline{F_0}$ acts on $W'_\lambda$ and $W'_{\lambda^q}$ since $\overline{F_0}$ centralizes $R_1 \cdots R_k$, so then $\overline{F_0}$ centralizes $F_0$.

Thus we can further assume that $W'_\lambda \cong W'_{\lambda^q}$ as $\mathbb{K}[R_1 \cdots R_k]$-modules. Because $W'_\lambda$ is absolutely irreducible and $\overline{F_0} = \langle \overline{f} \rangle$ centralizes $R_1 \cdots R_k$, after choosing suitable bases on $W'_\lambda \cong W'_{\lambda^q}$ we have

$$\overline{f} = \begin{pmatrix} xI & yI \\ zI & wI \end{pmatrix}$$

for some scalars $x, y, z, w \in \mathbb{K}$. Here the block matrix is written with respect to the decomposition $V' = W'_\lambda \oplus W'_{\lambda^q}$. Note that

$$f = \begin{pmatrix} \lambda I & 0 \\ 0 & \lambda^q I \end{pmatrix}.$$

Thus by applying Lemma 7.1.2 with $S = f$ and $T = \overline{f}$, it follows that either $\overline{f}$ centralizes $f$, or $\lambda = -\lambda^q$. If $\lambda = -\lambda^q$, then $\lambda^{2(q-1)} = 1$, implying $q^2 - 1 \mid 2(q-1)$, which is impossible. Therefore $\overline{f}$ centralizes $f$, which completes the proof of the lemma for $G$ of type $\mathcal{B}_0$.

**Case 2: $G$ is of type $\mathcal{B}_1$**
As in the previous case, we have $\mathbb{K} = \mathbb{F}_{q^\nu}$ and $|f| = q^\nu - 1$. Let $\lambda \in \mathbb{K}$ be an eigenvalue of $f$ on $V'$. We split the proof into three cases.

**Case 2.1: $\nu > 2$**
It follows from Lemma 4.3.13 (together with Lemma 4.3.12 (i) when $H = G^\Omega$) that there exists an element of the form $g_A \psi^2$ in $H$. As in the $\mathcal{B}_0$ case, we find that $[H, H]$ contains

$$[g_A \psi^2, f] = f^{q^2-1}.$$

Suppose that the eigenspaces of $f$ and $f^{q^2-1}$ differ on $V'$. Then either $\lambda^{q^2-1} = \lambda^{(q^2-1)q^\alpha}$ for some $0 < \alpha < \nu$, or $\lambda^{q^2-1} = \lambda^{-(q^2-1)q^\alpha}$ for some $0 \leq \alpha < \nu$. Thus one of the following equations holds:

$$(q^2-1)(q^\alpha - 1) \equiv 0 \mod q^\nu - 1 \text{ for some } 0 < \alpha < \nu \tag{7.2}$$

$$(q^2-1)(q^\alpha + 1) \equiv 0 \mod q^\nu - 1 \text{ for some } 0 \leq \alpha < \nu. \tag{7.3}$$

If $q^\nu - 1$ has no primitive prime divisor, by Zsigmondy's theorem (Theorem 2.1.21) we have $q = 2$ and $\nu = 6$, in which case it is readily seen that (7.2) and (7.3) have no solutions.

Therefore we can assume that $q^\nu - 1$ has a primitive prime divisor, in which case (7.2) cannot hold since $\nu > 2$. Furthermore (7.3) cannot hold for $0 \leq \alpha < \nu/2$, since $q^{2\alpha} - 1 = (q^\alpha - 1)(q^\alpha + 1)$ is not divisible by a primitive prime divisor of $q^\nu - 1$.

Hence (7.3) holds and $\nu/2 \leq \alpha < \nu$. If $\nu/2 < \alpha < \nu$, then $\gcd(\alpha, \nu) = 1$ and by Lemma 2.1.19 we have $\gcd(q^\alpha + 1, q^\nu - 1) \mid q + 1$. Then (7.3) implies $(q^2-1)(q+1) \equiv 0 \mod q^\nu - 1$, which is not possible when $\nu > 2$ and $q^\nu - 1$ has a primitive prime divisor.

It remains to consider the case where $\nu$ is even and $\alpha = \nu/2$. Then (7.3) implies that $q^{\nu/2} - 1$ divides $q^2 - 1$, so $\nu = 4$ and $\alpha = 2$. In this case the $f^{q^2-1}$-eigenspaces on $V'$ are the following:

$$W'_\lambda \oplus W'_{\lambda^{-q^2}}, \qquad W'_{\lambda^q} \oplus W'_{\lambda^{-q^3}},$$
$$W'_{\lambda^{q^2}} \oplus W'_{\lambda^{-1}}, \qquad W'_{\lambda^{q^3}} \oplus W'_{\lambda^{-q}}.$$

Since $\overline{f}$ centralizes $f^{q^2-1}$, it acts on each of these eigenspaces, all of which are sums of two absolutely irreducible $\mathbb{K}[R_1 \cdots R_k]$-modules. As at the end of the proof for groups of type $\mathcal{B}_0$, we find that either $\overline{F_0}$ centralizes $F_0$, or $\lambda = -\lambda^{-q^2}$. If $\lambda = -\lambda^{-q^2}$, then $\lambda^{2(q^2+1)} = 1$, implying $q^4 - 1 \mid 2(q^2+1)$, which is impossible. Thus $\overline{F_0}$ centralizes $F_0$, which completes the proof in the case where $\nu > 2$.

**Case 2.2:** $\nu = 2$

Suppose that $\nu = 2$. Note that it follows from Lemmas 5.2.2 and 5.2.4 that $H$ is irreducible. Therefore there exists an element of the form $g_A\varphi$ or $g_A\varphi\psi$ in $H$. We consider first the case where there exists an element of the form $g_A\varphi\psi$ in $H$. Then $[H, H]$ contains

$$[g_A\varphi\psi, f] = [\varphi\psi, f] = f^{q+1}.$$

We prove that

$$W'_\lambda \oplus W'_{\lambda^q} \qquad W'_{\lambda^{-1}} \oplus W'_{\lambda^{-q}}$$

are the eigenspaces of $f^{q+1}$ on $V'$. Since $\lambda^{q+1} = \lambda^{q(q+1)}$, the action of $f^{q+1}$ on both spaces is by scalar multiplication. To check that they are eigenspaces, it suffices to prove that $\lambda^{q+1} \neq \lambda^{-(q+1)}$. If $\lambda^{q+1} = \lambda^{-(q+1)}$, then $q^2 - 1 \mid 2(q+1)$, which implies $q = 2$ or $q = 3$. However $(q, \nu) = (2, 2)$ is excluded from the definition of groups of type $\mathcal{B}_1$, and $(q, \nu) = (3, 2)$ is not applicable for in that case $G$ is not metrically primitive (Lemma 5.6.5).

Since $\overline{F_0}$ centralizes $f^{q+1}$, it acts on the eigenspaces $W'_\lambda \oplus W'_{\lambda^q}$ and $W'_{\lambda^{-1}} \oplus W'_{\lambda^{-q}}$. As at the end of the proof for groups of type $\mathcal{B}_0$, we find that either $\overline{F_0}$ centralizes $F_0$, or $\lambda = -\lambda^{-q}$. If $\lambda = -\lambda^{-q}$, then $\lambda^{2(q-1)} = 1$, implying $q^2 - 1 \mid 2(q-1)$ which is impossible. Thus $\overline{F_0}$ centralizes $F_0$.

Next we consider the case where there exists an element of the form $g_A\varphi$ in $H$. Then $[H, H]$ contains

$$[g_A\varphi, f] = [\varphi, f] = f^2,$$

so $\overline{f}$ centralizes $f^2$. If $q$ is even, then $|f|$ is odd, so it follows that $f$ centralizes $\overline{f}$. If $q$ is odd, then $\overline{f}$ centralizes $(f^2)^{(q+1)/2} = f^{q+1}$, so as in the previous paragraph it follows that $\overline{F_0}$ centralizes $F_0$.

**Case 2.3:** $\nu = 1$

In this case $V' = W'_\lambda \oplus W'_{\lambda^{-1}}$. As at the end of the proof for groups of type $\mathcal{B}_0$, we find that either $\overline{F_0}$ centralizes $F_0$, or $\lambda = -\lambda^{-1}$. If $\lambda = -\lambda^{-1}$, then $\lambda^2 = -1$. In this case $q$ must be odd, so $H = G^\circ$. Now $\lambda$ is of order 4, so $|f| = q^\nu - 1 = 4$, and thus $q = 5$ and $\nu = 1$.

Because $|f| = 4$, we have $\mu = 2^{\ell_1}$ for some $\ell_1 \geq 0$. Suppose that $\mu > 1$ and $X_1 \not\leq \Omega^{\varepsilon_1}_{2\ell_1}(2)$. Then it follows from Lemma 5.5.2 that there exist $\varphi_0, g \in G^\circ$ such that $[\varphi_0, g] = f^{\pm 1}$, so $\overline{F_0}$ centralizes $F_0$. We may assume then that $\mu = 1$ or $X_1 \leq \Omega^{\varepsilon_1}_{2\ell_1}(2)$. If $\mu > 1$, we have $\varepsilon_1 = +$ by Lemma 5.1.4, so we are precisely in the exceptional case (ii) of the lemma.

**Case 3:** ***G* is of type $\mathcal{B}_2$**

In this case $\mathbb{K} = \mathbb{F}_{q^{2\nu}}$ and $|f| = q^\nu + 1$. Let $\lambda \in \mathbb{K}$ be an eigenvalue of $f$ on $V'$.

By Lemma 4.3.22 (together with Lemma 4.3.21 (i) when $H = G^\Omega$), there exists an element of the form $g_A\psi^2$ in $H$. As in the $\mathcal{B}_0$ case, we find that $[H, H]$ contains

$$[g_A\psi^2, f] = f^{q^2-1}.$$

Suppose that the eigenspaces of $f$ and $f^{q^2-1}$ differ, so $\lambda^{q^2-1} = \lambda^{(q^2-1)q^\alpha}$ for some $0 < \alpha < 2\nu$. Then

$$(q^2 - 1)(q^\alpha - 1) \equiv 0 \mod q^\nu + 1. \tag{7.4}$$

We consider first the case where $\nu > 1$. A primitive prime divisor of $q^{2\nu} - 1$ divides $q^{\nu} + 1$, so by (7.4) there is no primitive prime divisor for $q^{2\nu} - 1$. It follows then from Zsigmondy's theorem (Theorem 2.1.21) that $q = 2$ and $\nu = 3$. But this case is not applicable, since then $G$ is metrically imprimitive (Lemma 5.6.6).

Therefore the only possibility is that $\nu = 1$, in which case (7.4) does hold. Then $V' = W'_{\lambda} \oplus W'_{\lambda^q}$ and $\mathbb{K} = \mathbb{F}_{q^2}$. By construction $W'_{\lambda}$ and $W'_{\lambda^q}$ are both absolutely irreducible $\mathbb{K}[R_1 \cdots R_k]$-modules. Thus by arguing as at the end of the proof for groups of type $\mathcal{B}_0$, we find that either $\overline{F_0}$ centralizes $F_0$, or $\lambda = -\lambda^q$.

Thus we can assume that $\lambda = -\lambda^q$. In this case $\lambda^{2(q-1)} = 1$, so $q + 1 \mid 2(q-1)$, which implies $q = 3$. In particular $H = G^{\circ}$.

We have $|f| = 3 + 1 = 4$, so $\mu = 2^{\ell_1}$ for some $\ell_1 \geq 0$. Suppose that $\mu > 1$ and $X_1 \not\leq \Omega^{\varepsilon_1}_{2\ell_1}(2)$. Then it follows from Lemma 5.5.4 that there exist $g, \psi_0 \in G^{\circ}$ such that $[\psi_0, g] = f^{\pm 1}$, so $\overline{F_0}$ centralizes $F_0$. We may assume then that $\mu = 1$ or $X_1 \leq \Omega^{\varepsilon_1}_{2\ell_1}(2)$. If $\mu > 1$, then $\varepsilon_1 = +$ by Lemma 5.1.4, so we are in the exceptional case (iii) of the lemma. With this the proof of the lemma is complete. □

**Lemma 7.1.4** *Suppose that $G = G^{\mathcal{B}}_{\mu,\nu}(X_1, \ldots, X_k)$ and $\overline{G} = G^{\overline{\mathcal{B}}}_{\overline{\mu},\overline{\nu}}(Y_1, \ldots, Y_\ell)$ are such that $|\overline{F_0}|$ divides $|F_0|$. Then $G$ and $\overline{G}$ are as in one of the entries of Table 7.1.*

***Proof*** We first prove that the types of $G$ and $\overline{G}$ must be as in one of the entries of Table 7.1. In other words, we will rule out the following cases:

- $\overline{G}$ is of type $\mathcal{B}_1$ and $G$ is of type $\mathcal{B}_2$ or $\mathcal{B}_3$;
- $\overline{G}$ is of type $\mathcal{B}_2$ and $G$ is of type $\mathcal{B}_3$.

If $\overline{G}$ is of type $\mathcal{B}_1$ and $G$ is of type $\mathcal{B}_2$, then $|\overline{F_0}| = q^{\overline{\nu}} - 1$ and $|F_0| = q^{\nu} + 1$. In this case Lemma 2.1.20 (iv) implies that $(q, \overline{\nu})$ equals (2, 1), (3, 1), or (2, 2), but this is not applicable when $\overline{G}$ is of type $\mathcal{B}_1$. If $\overline{G}$ is of type $\mathcal{B}_1$ and $G$ is of type $\mathcal{B}_3$, we have $q^{\nu} - 1 \mid 2$, implying $(q, \overline{\nu}) = (2, 1)$ or $(3, 1)$ which is again not applicable for type $\mathcal{B}_1$. In the case where $\overline{G}$ is of type $\mathcal{B}_2$, we have $|\overline{F_0}| = q^{\overline{\nu}} + 1 > 2$, so $G$ cannot be of type $\mathcal{B}_3$.

**Table 7.1** For $G, \overline{G} \leq \Delta(V, \kappa)$, the possible configurations in the proof of Lemma 7.1.5 when $|\overline{F_0}|$ divides $|F_0|$

| Type of $G$ | Type of $\overline{G}$ | $\lvert F_0\rvert$ | $\alpha$ | $\lvert\overline{F_0}\rvert$ | $\overline{\alpha}$ |
|---|---|---|---|---|---|
| $\mathcal{B}_0$ | $\mathcal{B}_0$ | $q^{m\overline{\nu}} - 1$ | $m\overline{\nu}$ | $q^{\overline{\nu}} - 1$ | $\overline{\nu}$ |
| $\mathcal{B}_1$ | $\mathcal{B}_1$ | $q^{m\overline{\nu}} - 1$ | $2m\overline{\nu}$ | $q^{\overline{\nu}} - 1$ | $2\overline{\nu}$ |
| | $\mathcal{B}_2$ | $q^{2m\overline{\nu}} - 1$ | $4m\overline{\nu}$ | $q^{\overline{\nu}} + 1$ | $2\overline{\nu}$ |
| | $\mathcal{B}_3$ | $q^{\nu} - 1$ | $2\nu$ | 2 | 1 |
| $\mathcal{B}_2$ | $\mathcal{B}_2$ | $q^{(2m-1)\overline{\nu}} + 1$ | $(2m-1)2\overline{\nu}$ | $q^{\overline{\nu}} + 1$ | $2\overline{\nu}$ |
| | $\mathcal{B}_3$ | $q^{\nu} + 1$ | $\nu$ | 2 | 1 |
| $\mathcal{B}_3$ | $\mathcal{B}_3$ | 2 | 1 | 2 | 1 |

Thus the types of $G$ and $\overline{G}$ must be as in Table 7.1. It follows from Lemma 2.1.20 that in all cases the values of $|F_0|$ and $|\overline{F_0}|$ are as in Table 7.1, which completes the proof of (i). □

We continue with our analysis of the possible inclusions $G^\circ \leq \overline{G^\circ}$, and refine Lemma 7.1.3 to the following result.

**Lemma 7.1.5** *Let $H = G^\Omega$ if $q$ is even and $\kappa$ is a quadratic form, and $H = G^\circ$ otherwise. Suppose that $H \leq \overline{G^\circ}$. Then one of the following holds:*

*(i)* $\overline{F_0} = F_0$;
*(ii)* *$G$ is of type $\mathcal{B}_0$, with $q = 3$ and $\nu = 2$;*
*(iii)* *$G$ is of type $\mathcal{B}_1$, with $q = 5$ and $\nu = 2$;*
*(iv)* *$G$ is of type $\mathcal{B}_1$ or $\mathcal{B}_2$, $\overline{G}$ is of type $\mathcal{B}_3$, and $q^\nu \in \{3, 3^2, 3^4, 5, 5^2, 7\}$.*
*(v)* *$G$ is of type $\mathcal{B}_1$, with $q = 5$, $\nu = 1$, and $X_1 \leq \Omega^+_{2\ell_1}(2)$ if $\mu > 1$;*
*(vi)* *$G$ is of type $\mathcal{B}_2$, with $q = 3$, $\nu = 1$, and $X_1 \leq \Omega^+_{2\ell_1}(2)$ if $\mu > 1$.*

***Proof*** We can assume that (v) and (vi) do not hold. Then it follows from Lemma 7.1.3 that $\overline{F_0}$ centralizes $F_0$. On the other hand, as noted in the beginning of the proof of Lemma 7.1.3, we have $R_1 \cdots R_k \leq [H, H]$, so $\overline{F_0}$ centralizes $R_1 \cdots R_k$. Thus $\overline{F_0}$ centralizes $A = F_0 R_1 \cdots R_k$, which implies $\overline{F_0} \leq A$ by Lemma 4.3.32. Furthermore $Z(A) = F_0$, so we conclude that $\overline{F_0} \leq F_0$.

Since $\overline{F_0} \leq F_0$, the order $|\overline{F_0}|$ divides $|F_0|$. It follows from Lemma 7.1.4 that the possible configurations are as listed in Table 7.1. In particular, we find from Table 7.1 that in all cases $\overline{\alpha}$ divides $\alpha$. As a consequence, we deduce the following.

*Claim* $[A, A] \leq \overline{F_0}$.

Note that we have $n = \mu\alpha = \overline{\mu}\,\overline{\alpha}$, so $\overline{\mu} = \mu\alpha/\overline{\alpha}$. Thus for all $1 \leq i \leq k$, the prime $r_i$ divides $\overline{\mu}$. Furthermore, every prime divisor of $\overline{\mu}$ divides $|\overline{f}|$. Therefore $Z(R_i) = R_i \cap F_0$ is contained in $\overline{F_0}$, since $F_0$ has a unique subgroup of order $r_i$.

Now $A$ is a central product of $F_0$ and the extraspecial groups $R_i$, so we conclude that $[A, A] = Z(R_1) \cdots Z(R_k) \leq \overline{F_0}$.

Since $A$ centralizes $\overline{F_0}$ and $[A, A] \leq \overline{F_0}$, it follows from Corollary 6.3.9 that

$$|F_0|\mu^2 \leq |\overline{F_0}|\overline{\mu}^2.$$

Because $\overline{\mu} = \mu\alpha/\overline{\alpha}$, we conclude

$$|F_0| \leq |\overline{F_0}|(\alpha/\overline{\alpha})^2. \tag{7.5}$$

As we shall see next, the inequality (7.5) forces $F_0 = \overline{F_0}$, except for the cases (ii)–(iv) listed in the lemma. We consider the different types of $G$ in turn, using the notation from Table 7.1 throughout.

**Case 1: $G$ is of type $\mathcal{B}_0$**
In this case $H = G^\circ = G$, and inequality (7.5) becomes

$$(q^{m\overline{\nu}} - 1) \leq (q^{\overline{\nu}} - 1)m^2.$$

Since $q^{\overline{\nu}} > 2$ for a group of type $\mathcal{B}_0$, for $m > 1$ this inequality is only possible for $q = 3, \overline{\nu} = 1$, and $m = 2$, which is case (ii) of the lemma. If $m = 1$, then $F_0$ and $\overline{F_0}$ have the same order, so $\overline{F_0} = F_0$.

**Case 2: $G$ is of type $\mathcal{B}_1$**
If $\overline{G}$ is also of type $\mathcal{B}_1$, then inequality (7.5) implies

$$(q^{m\overline{\nu}} - 1) \leq (q^{\overline{\nu}} - 1)m^2.$$

As in the previous case, $m = 1$ implies $\overline{F_0} = F_0$. For $m > 1$ the inequality is only possible for $q^{\overline{\nu}} \in \{2, 3\}$, but for $\overline{G}$ of type $\mathcal{B}_1$ we must have $q^{\overline{\nu}} > 3$.

If $\overline{G}$ is of type $\mathcal{B}_2$, then inequality (7.5) becomes

$$(q^{2m\overline{\nu}} - 1) \leq (q^{\overline{\nu}} + 1)4m^2.$$

In this case $\overline{\mu} = \mu\alpha/\overline{\alpha} = 2m\mu$ is even, so $2 \mid q^{\overline{\nu}} + 1$, which implies that $q$ is odd. Hence one finds that the inequality is only possible for $q = 3, \overline{\nu} = 1, m = 1$; and $q = 5, \overline{\nu} = 1, m = 1$. For $(q, \overline{\nu}, m) = (3, 1, 1)$ we have $q = 3$ and $\nu = 2$, so this case is not applicable, since $G$ is not metrically primitive (Lemma 5.6.5). For $(q, \overline{\nu}, m) = (5, 1, 1)$ we have $q = 5$ and $\nu = 2$, so case (iii) of the lemma holds.

Next we consider the case where $\overline{G}$ is of type $\mathcal{B}_3$, so inequality (7.5) becomes

$$(q^\nu - 1) \leq 8\nu^2.$$

Since we have a group of type $\mathcal{B}_3$, in this case $q^\nu$ must be odd and $n$ is a power of 2. In particular $\nu$ is also a power of 2, so one finds that the inequality is only possible for $q^\nu \in \{3, 3^2, 3^4, 5, 5^2, 7\}$. All of these possibilities belong to case (iv) of the lemma.

**Case 3: $G$ is of type $\mathcal{B}_2$**
As seen in Table 7.1, in this case $\overline{G}$ is either of type $\mathcal{B}_2$ or $\mathcal{B}_3$.

If $\overline{G}$ is of type $\mathcal{B}_2$, then inequality (7.5) becomes

$$q^{(2m-1)\overline{\nu}} + 1 \leq (q^{\overline{\nu}} + 1)(2m - 1)^2.$$

If $m = 1$, then we have $\overline{F_0} = F_0$. In the case $m > 1$, the inequality is only possible for $q = 2, \overline{\nu} = 1, m \in \{2, 3, 4\}$ and $q = 3, \overline{\nu} = 1, m = 2$. However, taking into account the fact that $\overline{\mu} = (2m - 1)\mu$ and that every prime divisor of $\overline{\mu}$ must divide $|\overline{F_0}| = q^{\overline{\nu}} + 1$, we see that for $m > 1$ the only solution is $q = 2, \overline{\nu} = 1, m = 2$. In this case $q = 2$ and $\nu = 3$, so this case is not applicable since $G$ is not metrically primitive (Lemma 5.6.6).

Suppose then that $\overline{G}$ is of type $\mathcal{B}_3$, so inequality (7.5) becomes

$$(q^\nu + 1) \leq 8\nu^2.$$

We have a group of type $\mathcal{B}_3$, so $q^\nu$ must be odd and $n$ is a power of 2, and in particular $\nu$ is a power of 2. Thus we find that the inequality is only possible for $q^\nu \in \{3, 3^2, 3^4, 5, 5^2, 7\}$, so case (iv) of the lemma holds.

**Case 4: $G$ is of type $\mathcal{B}_3$**
Since in this case $|F_0| = 2$, it is immediate from $\overline{F_0} \leq F_0$ that $\overline{F_0} = F_0$.

We have proved that in all situations either $F_0 = \overline{F_0}$, or one of the exceptions (ii)–(vi) holds, so the proof of the lemma is complete. □

In the next result, we rule out most of the cases in Lemma 7.1.5 (iv).

**Lemma 7.1.6** *Let $G^\circ \leq \overline{G^\circ}$. Suppose that $G$ is of type $\mathcal{B}_1$ or $\mathcal{B}_2$, $\overline{G}$ is of type $\mathcal{B}_3$, and $q^\nu \in \{3, 3^2, 3^4, 5, 5^2, 7\}$. Then one of the following holds:*

*(i) $G$ is of type $\mathcal{B}_1$, $q \in \{5, 7, 9\}$, $\nu = 1$, and $e + \sum_{i=1}^k u_i \equiv 1 \mod 2$;*
*(ii) $G$ is of type $\mathcal{B}_2$, $q \in \{3, 7\}$, $\nu = 1$, and $e + \sum_{i=1}^k u_i \equiv 1 \mod 2$;*
*(iii) $G$ is of type $\mathcal{B}_1$, $q = 5$, and $\nu = 2$;*
*(iv) $G$ is of type $\mathcal{B}_1$, with $q = 5$, $\nu = 1$, and $X_1 \leq \Omega^+_{2\ell_1}(2)$ if $\mu > 1$.*
*(v) $G$ is of type $\mathcal{B}_2$, with $q = 3$, $\nu = 1$, and $X_1 \leq \Omega^+_{2\ell_1}(2)$ if $\mu > 1$.*

***Proof*** We can assume that (iv) and (v) do not hold. It follows then from Corollary 5.6.13 that $G^\circ$ is metrically primitive. We will prove the lemma using ideas similar to [52, §131–§141].

Since $\overline{G}$ is of type $\mathcal{B}_3$, we have $q$ odd and $n$ is a power of 2. Then $\overline{A}$ an extraspecial 2-group such that $Z(\overline{A}) = \overline{F_0} = \langle -I_V \rangle$, and $V$ is an absolutely irreducible $\mathbb{F}_q[\overline{A}]$-module.

We have $n = 2\nu\mu$, where $|F_0| = q^\nu - 1$ if $G$ is of type $\mathcal{B}_1$, and $|F_0| = q^\nu + 1$ if $G$ is of type $\mathcal{B}_2$. Then $\mu = 2^{\ell_1}$ for some $\ell_1 \geq 0$, since $n$ is a power of 2. If $\mu > 1$, we have $G = G^{\mathcal{B}}_{\mu,\nu}(X_1)$, where $X_1 \leq O^{\varepsilon_1}_{2\ell_1}(2)$ is metrically completely reducible maximal solvable, and $A = F_0 R_1$ with $R_1 = 2^{1+2\ell_1}_{\varepsilon_1}$. Note that $Z(R_1) = \overline{F_0} = \langle -I_V \rangle$. When $\mu = 1$, define $X_1 = 1$, $R_1 = 1$, and $\varepsilon_1 = +$.

We begin with the following observations.

**Claim 1: The action of $G^\circ$ on $\overline{A}/Z(\overline{A})$ is metrically completely reducible**
Suppose that $F/Z(\overline{R_1})$ is a nonzero, totally isotropic $G^\circ$-invariant subspace of $\overline{R_1}/Z(\overline{R_1})$. Then $F$ is a noncyclic abelian group normalized by $G^\circ$, which contradicts the fact that $G^\circ$ is metrically primitive (Lemma 2.4.3).

**Claim 2: $R_1 \leq \overline{A}$**
By Claim 1, the action of $G^\circ$ on $\overline{A}/\overline{F_0}$ is completely reducible (Lemma 2.3.2). Then $\overline{A}/\overline{F_0}$ is a completely reducible $\mathbb{F}_2[R_1]$-module by Clifford's theorem. Because $R_1$

is a 2-group, this implies that $R_1$ acts trivially on $\overline{A}/\overline{F_0}$, and thus $R_1 \leq \overline{A}$ by Lemma 4.3.25.

Now $R_1/\overline{F_0}$ is a non-degenerate subspace of $\overline{A}/\overline{F_0}$ by Claim 1 and Claim 2, so $\overline{A}$ is a central product $\overline{A} = R_1 R_1'$, where $R_1' := C_{\overline{A}}(R_1)$. (Alternatively, apply Lemma 4.1.1.) Since $n = 2\nu\mu$, here $R_1' = 2^{1+2\rho}_{\varepsilon_1'}$, where $2\nu = 2^\rho$.

By Claim 1, the action of $G^\circ$ on $R_1'/\overline{F_0}$ is metrically completely reducible. Let $Y_1' \leq O^{\varepsilon_1'}_{2\rho}(2)$ be a metrically completely reducible maximal solvable subgroup containing the action of $G^\circ$ on $R_1'/\overline{F_0}$.

The action of $\overline{A}$ on $V$ is absolutely irreducible, and we have a tensor decomposition

$$V = W \otimes U,$$

where $W$ is an absolutely irreducible $\mathbb{F}_q[R_1]$-module, and $U$ is an absolutely irreducible $\mathbb{F}_q[R_1']$-module (Theorem 3.2.1, Lemma 2.1.16).

Furthermore, the bilinear form $b$ corresponding to $\kappa$ can be written as

$$b = b_W \otimes b_U,$$

where $b_W$ is a non-degenerate $R_1$-invariant bilinear form on $W$ with $\operatorname{sgn}(b_W) = \varepsilon_1$, and $b_U$ is a non-degenerate $R_1'$-invariant bilinear form on $U$ with $\operatorname{sgn}(b_U) = \varepsilon_1'$ (see for example Sects. 3.2.3 and 3.2.6).

It follows from Lemma 2.1.16 that

$$\begin{aligned} N_{\mathrm{GL}(V)}(\overline{A}) &= N_{\mathrm{GL}(W)}(R_1) \otimes N_{\mathrm{GL}(U)}(R_1'), \\ C_{\mathrm{GL}(V)}(R_1) &= I_W \otimes \mathrm{GL}(U), \end{aligned} \tag{7.6}$$

so as seen in Lemma 4.3.29, we have $G^\circ \leq (G')^\circ$, where

$$G' := G^{\mathcal{B}_3}_{\mu,1}(X_1) \otimes G^{\mathcal{B}_3}_{2\nu,1}(Y_1').$$

Denote $\Lambda = C_{G^\circ}(R_1)$ and $\Lambda' = C_{(G')^\circ}(R_1)$. Note that by (7.6), we can identify $\Lambda'$ with $G^{\mathcal{B}_3}_{2\nu,1}(Y_1')^\circ$. By construction $G^{\mathcal{B}_3}_{2\nu,1}(Y_1')^\circ$ is contained in the normalizer of $R_1' = 2^{1+2\rho}_{\varepsilon_1'}$ in $\mathrm{GL}(U)$, so

$$\Lambda \leq \Lambda' \leq 2^{1+2\rho}_{\varepsilon_1'}.O^{\varepsilon_1'}_{2\rho}(2).$$

Next we will describe generators for $\Lambda$, which will be used to prove that one of the configurations listed in the lemma must hold.

When $G$ is of type $\mathcal{B}_1$, we define

$$\varphi_0 := \begin{cases} \varphi, & \text{if } \mu = 1. \\ \left(B_1^{(1)}\right)^{u_1} \varphi, & \text{if } \mu > 1. \end{cases}$$

$$\psi_0 := \psi.$$

Then $\varphi_0, \psi_0 \in G^\circ$ as seen in Remark 5.5.1, and moreover $\varphi_0, \psi_0$ centralize $R_1$. We have $\Lambda \leq (\operatorname{Ker} \pi)^\circ$ and $(\operatorname{Ker} \pi)^\circ = \langle A, \varphi_0, \psi_0 \rangle$ by Lemma 4.3.10. Since $C_A(R_1) = Z(A) = F_0$, it follows that $\Lambda = \langle f, \varphi_0, \psi_0 \rangle$.

Similarly when $G$ is of type $\mathcal{B}_2$, we see from the relations (4.4) described in Sect. 4.3.3 that the map $\psi \in G^\circ$ acts trivially on $R_1/Z(R_1)$. Similarly to Remark 5.5.3, we see from (4.4) that

$$\psi_0 := \begin{cases} \psi, & \text{if } \mu = 1 \text{ or } q \equiv 1 \mod 4. \\ \left(B_1^{(1)}\right)^{u_1} \psi, & \text{if } \mu > 1 \text{ and } q \equiv 3 \mod 4. \end{cases}$$

centralizes $R_1$. By Lemma 4.3.19 we have $(\operatorname{Ker} \pi)^\circ = \langle A, \psi_0 \rangle$, so $\Lambda = \langle f, \psi_0 \rangle$.

We will now proceed to consider the different possibilities for $G$ and $q^\nu$ case by case, ruling out all configurations except those stated in (i) and (ii). Note that $\nu \in \{1, 2, 4\}$.

**Case 1:** $\nu = 4$

**Case 1.1: $G$ is of type $\mathcal{B}_1$ and $q = 3$**
In this case $|F_0| = 3^4 - 1 = 80$. We have $\rho = 3$, so $\Lambda'$ is a subgroup of $2_\pm^{1+6}.O_6^\pm(2)$. Now $f \in \Lambda$ has order 80, so the image of $f$ in $O_6^\pm(2)$ has order divisible by $80/4 = 20$. But there is no element of order 20 in $O_6^\pm(2)$, so this is a contradiction.

**Case 1.2: $G$ is of type $\mathcal{B}_2$ and $q = 3$**
In this case $|F_0| = 3^4 + 1 = 2 \cdot 41$. As in the previous case $\Lambda'$ is a subgroup of $2_\pm^{1+6}.O_6^\pm(2)$. Now $f^2 \in \Lambda$ is an element of order 41, but there is no such element in $O_6^\pm(2)$, so we have a contradiction.

**Case 2:** $\nu = 2$

**Case 2.1: $G$ is of type $\mathcal{B}_1$ and $q = 9$**
In this case $|F_0| = 9^2 - 1 = 80$. We have $\rho = 2$, so $\Lambda'$ is a subgroup of $2_\pm^{1+4}.O_4^\pm(2)$. Now $f \in \Lambda \leq \Lambda'$ is an element of order 80, so the image of $f$ in $O_4^\pm(2)$ would have order divisible by $80/4 = 20$. This is a contradiction, since there is no element of order 20 in $O_4^\pm(2)$.

**Case 2.2: $G$ is of type $\mathcal{B}_2$ and $q = 9$**
In this case $|F_0| = 9^2 + 1 = 2 \cdot 41$, so $f^2 \in \Lambda$ is an element of order 41. We have $\rho = 2$, so as in Case 1.2 we have a contradiction, since $O_4^{\pm}(2)$ does not contain elements of order 41.

**Case 2.3: $G$ is of type $\mathcal{B}_1$ and $q = 3$**
This case is not applicable, since $G$ would be metrically imprimitive by Lemma 5.6.5.

**Case 2.4: $G$ is of type $\mathcal{B}_2$ and $q = 3$**
In this case $|F_0| = 3^2 + 1 = 10$. We have $\rho = 2$, so $\Lambda'$ is a subgroup of $2_{\pm}^{1+4}.X$, where $X$ is maximal irreducible solvable in $O_4^{\pm}(2)$. Now $g = f^2 \in \Lambda$ has order 5 and there is no such element in $O_4^{+}(2)$, so $\Lambda'$ is a subgroup of $2_{-}^{1+4}.X$, where $X \leq O_4^{-}(2)$ is metrically completely reducible maximal solvable. The only possibility is $X = G^{\mathcal{B}_2}_{1,2}$, which is metrically primitive irreducible of type $\mathcal{B}_2$, with $|X| = 20$.

From the basic properties of $X$ (for example Lemma 2.7.3, Table 2.2), we know that $X = \langle a, b\rangle = \langle a\rangle \rtimes \langle b\rangle$, where $a, b \in O_4^{-}(2)$ satisfy $|a| = 5$, $|b| = 4$, and $b^{-1}ab = a^2$. On the other hand, the image of $g$ in $O_4^{-}(2)$ has order 5 and $\psi_0^{-1} g \psi_0 = g^3 = g^{-2}$. From this it follows that the image of $\langle \psi_0, g\rangle$ in $O_4^{-}(2)$ must be equal to all of $X$.

In particular, the image of $\Lambda'$ in $O_4^{-}(2)$ is equal to all of $X$. However 2 is not a square modulo 3, so by Lemma 4.3.24 (ii) the image of $\Lambda'$ in $O_4^{-}(2)$ is contained in $X^{\Omega}$. By Lemma 5.1.3 we have $X \not\leq \Omega_4^{-}(2)$, so $X^{\Omega}$ has order 10, and we have a contradiction.

**Case 2.5: $G$ is of type $\mathcal{B}_1$ and $q = 5$**
In this case, statement (iii) of the lemma holds.

**Case 2.6: $G$ is of type $\mathcal{B}_2$ and $q = 5$**
In this case $|F_0| = 5^2 + 1 = 2 \cdot 13$, so $f^2 \in \Lambda$ is an element of order 13. In this case $\rho = 2$ and as in Case 1.2 we have a contradiction, since $O_4^{\pm}(2)$ does not contain an element of order 13.

**Case 3: $\nu = 1$**
Before considering the various cases, we make a general observation. We have $\overline{A} = R_1 R_1'$, where $\overline{R_1} = 2_{\varepsilon_1'}^{1+2}$. The action of $G^{\circ}$ on $R_1'/Z(R_1')$ is metrically completely reducible (Claim 1) and $O_2^{+}(2)$ is not metrically completely reducible, so we must have $\varepsilon_1' = -$. Recall that $\mathrm{sgn}(b) = (-1)^e$ by definition. From the decomposition $b = b_W \otimes b_U$ we get

$$\mathrm{sgn}(b) = \varepsilon_1 \varepsilon_1' = (-1)^{1+\sum_{i=1}^{k} u_i},$$

so

$$e + \sum_{i=1}^{k} u_i \equiv 1 \mod 2.$$

Thus for $G$ of type $\mathcal{B}_1$, for $q \in \{5, 7, 9\}$ case (i) of the lemma holds. Groups of type $\mathcal{B}_1$ are not defined for $(q, \nu) = (3, 1)$, so to prove the lemma in this case, it remains to rule out $q = 25$. Similarly when $G$ is of type $\mathcal{B}_2$, for $q \in \{3, 7\}$ case (ii) holds, so we need to rule out $q \in \{5, 9, 25\}$. We will next consider these remaining cases.

**Case 3.1: $G$ is of type $\mathcal{B}_1$ and $q = 25$**
In this case $|F_0| = 25-1 = 24$. We have $\rho = 1$, so $\Lambda'$ is a subgroup of $2_-^{1+2}.O_2^-(2)$. Now $f \in \Lambda$ is an element of order 24, so its image in $O_2^-(2)$ would have to be an element of order divisible by $24/4 = 6$. There is no such element, so we have a contradiction.

**Case 3.2: $G$ is of type $\mathcal{B}_2$ and $q = 5$**
In this case $|F_0| = 5 + 1 = 6$, and again $\Lambda'$ is a subgroup of $2_-^{1+2}.O_2^-(2)$.

Now $g = f^2 \in \Lambda$ has order 3, and $\psi_0^{-1} g \psi_0 = g^{-1}$. Therefore the image of $\langle g, \psi_0 \rangle \leq \Lambda$ in $O_2^-(2)$ must be equal to $O_2^-(2)$. However, since 2 is not a square modulo 5, it follows from Lemma 4.3.24 (ii) that the image of $\Lambda'$ in $O_2^-(2)$ is contained in $\Omega_2^-(2)$. Thus we have a contradiction.

**Case 3.3: $G$ is of type $\mathcal{B}_2$ and $q = 9$**
In this case $|F_0| = 9 + 1 = 10$. Now $f^2 \in \Lambda$ has order 5, but clearly there can be no such element in $2_-^{1+2}.O_2^-(2)$, so as in previous cases we have a contradiction.

**Case 3.4: $G$ is of type $\mathcal{B}_2$ and $q = 25$**
In this case $|F_0| = 25 + 1 = 26$. Then $f^2 \in \Lambda$ has order 13, so as in the previous case we have a contradiction since there is no such element in $2_-^{1+2}.O_2^-(2)$.

In all cases we have either found a contradiction, or shown that one of the configurations in the lemma holds; thus the proof of the lemma is complete. □

We can now refine Lemma 7.1.5 to the following.

**Lemma 7.1.7** *Suppose that $G^\circ \leq \overline{G^\circ}$. Then one of the following holds:*

*(i)* $\overline{F_0} = F_0$*;*
*(ii)* *$G$ is of type $\mathcal{B}_0$, with $q = 3$ and $\nu = 2$;*
*(iii)* *$G$ is of type $\mathcal{B}_1$, with $q = 5$ and $\nu = 2$;*
*(iv)* *$G$ is of type $\mathcal{B}_1$, $\overline{G}$ is of type $\mathcal{B}_3$, $q \in \{5, 7, 9\}$, $\nu = 1$, and $e + \sum_{i=1}^{k} u_i \equiv 1 \mod 2$;*
*(v)* *$G$ is of type $\mathcal{B}_2$, $\overline{G}$ is of type $\mathcal{B}_3$, $q \in \{3, 7\}$, $\nu = 1$, and $e + \sum_{i=1}^{k} u_i \equiv 1 \mod 2$;*
*(vi)* *$G$ is of type $\mathcal{B}_1$, with $q = 5$, $\nu = 1$, and $X_1 \leq \Omega_{2\ell_1}^+(2)$ if $\mu > 1$.*
*(vii)* *$G$ is of type $\mathcal{B}_2$, with $q = 3$, $\nu = 1$, and $X_1 \leq \Omega_{2\ell_1}^+(2)$ if $\mu > 1$.*

***Proof*** Follows from Lemmas 7.1.5 and 7.1.6. □

**Lemma 7.1.8** *Suppose that $G \leq \overline{G}$. Then one of the following holds:*

*(i)* $\overline{F_0} = F_0$;
*(ii)* *$G$ is of type $\mathcal{B}_0$, with $q = 3$ and $\nu = 2$;*
*(iii)* *$G$ is of type $\mathcal{B}_1$, with $q = 5$ and $\nu = 2$;*
*(iv)* *$G$ is of type $\mathcal{B}_1$, with $q = 5$, $\nu = 1$, and $e + \sum_{i=1}^{k} u_i \equiv 1 \mod 2$;*
*(v)* *$G$ is of type $\mathcal{B}_2$, with $q = 3$, $\nu = 1$, and $e + \sum_{i=1}^{k} u_i \equiv 1 \mod 2$;*

***Proof*** Suppose that (ii)–(v) do not hold. It follows from Lemma 7.1.7 that $\overline{F_0} = F_0$, except possibly in the following cases:

(a) $G$ is of type $\mathcal{B}_1$, with $q = 5$ and $\nu = 1$.
(b) $G$ is of type $\mathcal{B}_2$, with $q = 3$ and $\nu = 1$.
(c) $G$ is of type $\mathcal{B}_1$, $\overline{G}$ is of type $\mathcal{B}_3$, $q \in \{7, 9\}$, $\nu = 1$.
(d) $G$ is of type $\mathcal{B}_2$, $\overline{G}$ is of type $\mathcal{B}_3$, $q = 7$, $\nu = 1$.

Here in cases (c) and (d) we have $G$ of multiplier 1 and $\overline{G}$ of multiplier 2, as noted in the proofs of Lemmas 5.5.12, 5.5.13, and 5.5.14. Thus (c) and (d) cannot hold.

It remains to consider (a) and (b). In both cases $|F_0| = 4$, so $\mu = 2^{\ell_1}$ for some $\ell_1 \geq 0$. If $\mu > 1$, we have $G = G^{\mathcal{B}}_{\mu,1}(X_1)$ with $X_1 \leq O^{\varepsilon_1}_{2\ell_1}(2)$ metrically completely reducible maximal solvable, and $A = F_0 R_1$ with $R_1 = 2^{1+2\ell_1}_{\varepsilon_1}$. When $\mu = 1$, we define $R_1 = 1$ and $X_1 = 1$. We first observe the following.

**Claim 1: $\overline{F_0}$ centralizes $F_0$**
In case (a), we see as in Remark 5.5.1 that $\varphi \in G^\circ$. As seen in Sect. 4.3.2, we have $G = G^\circ \rtimes \langle \eta \rangle$, and it follows from (2.3) that $[\eta, \varphi] = f$. Since $\overline{G}/C_{\overline{G}}(\overline{F_0})$ is abelian (Lemma 4.3.33), it follows that $\overline{F_0}$ is centralized by $f$. Therefore $\overline{F_0}$ is centralized by $F_0$.

In the case (b), we have similarly $\psi \in G^\circ$ (Remark 5.5.3). As seen in Sect. 4.3.3, we have $G = \langle G^\circ, \eta \rangle$. Then $[\eta, \psi] = \eta^2 = f$, so we find that $\overline{F_0}$ is centralized by $F_0$.

**Claim 2: $\overline{F_0} \leq F_0$**
By Lemma 5.7.2 we have $R_1 \leq [G^\circ, G^\circ]$, so $\overline{F_0}$ centralizes $R_1$, since $\overline{G}/C_{\overline{G}}(\overline{F_0})$ is abelian (Lemma 4.3.33). Then $\overline{F_0}$ centralizes $A = F_0 R_1$, so as in the beginning of the proof of Lemma 7.1.5, we conclude that $\overline{F_0} \leq F_0$.

Next note that since $|F_0| = 4$ in both cases, if $F_0 \neq \overline{F_0}$, we must have $|\overline{F_0}| = 2$, in which case $\overline{G}$ is of type $\mathcal{B}_3$. Because $G$ is metrically primitive, arguing as in the beginning of the proof of Lemma 7.1.6 (Claim 1, Claim 2) we find that $R_1 \leq \overline{A}$. Similar arguments as in the proof of Lemma 7.1.6 (after Claim 2) show that we have a central product $\overline{A} = R_1 R_1'$, where $R_1' = 2^{1+2}_{\varepsilon_1'}$ is extraspecial.

Now arguing as in the proof of Lemma 7.1.6 (Case 3, $\nu = 1$) shows that $\varepsilon'_1 = -$, and

$$e + \sum_{i=1}^{k} u_i \equiv 1 \mod 2.$$

But then (iii) or (iv) holds, contrary to what we have assumed. Thus in both cases we conclude that $\overline{F_0} = F_0$, which completes the proof of the lemma. □

**Lemma 7.1.9** *Suppose that $G^\circ \leq \overline{G^\circ}$ and that $F_0 = \overline{F_0}$. Then $G$ and $\overline{G}$ are of the same type $\mathcal{B}_i$, $\alpha = \overline{\alpha}$, and $\mu = \overline{\mu}$. Furthermore, one of the following holds:*

*(i)* $A = \overline{A}$.
*(ii)* $G^\circ$ *is metrically imprimitive.*

***Proof*** The fact that $G$ and $\overline{G}$ are of the same type $\mathcal{B}_i$ follows easily from the fact that $|F_0| = |\overline{F_0}|$, see for example the proof of Lemma 5.8.1. Then $|F_0| = |\overline{F_0}|$ implies $\alpha = \overline{\alpha}$, and thus $\mu = \overline{\mu}$, since $n = \alpha\mu = \overline{\alpha}\,\overline{\mu}$. Note that $|A| = |\overline{A}|$, since $|A| = |F_0|\mu^2$ and similarly $|\overline{A}| = |\overline{F_0}|\overline{\mu}^2$.

Suppose that $G^\circ$ is metrically primitive. Then it follows from Lemma 2.4.3 that $G^\circ$ does not normalize any noncyclic abelian subgroups of $I(V, \kappa)$. In particular if $G^\circ$ normalizes an abelian subgroup $N$ of $\overline{A}$, we must have $N \leq F_0$.

Thus for every prime $r$ dividing $\mu$, the action of $G^\circ$ on $O_r(\overline{A}/F_0)$ is metrically completely reducible. Applying Proposition 6.3.8 with $H = A$, it follows that $A = \overline{A}$. □

**Lemma 7.1.10** *Suppose that $G \leq \overline{G}$ and that $F_0 = \overline{F_0}$. Then $G$ and $\overline{G}$ are of the same type $\mathcal{B}_i$, $\alpha = \overline{\alpha}$, $\mu = \overline{\mu}$, and $A = \overline{A}$.*

***Proof*** Since $G$ is metrically primitive, the result follows with the same argument as Lemma 7.1.9. □

**Lemma 7.1.11** *Assume that $n$ and $q$ are even, $\kappa$ is a quadratic form, and that $G^\Omega$ is irreducible. Suppose that $G^\Omega \leq \overline{G^\Omega}$. Then $G$ and $\overline{G}$ are of the same type $\mathcal{B}_i$, $\alpha = \overline{\alpha}$, $\mu = \overline{\mu}$, with $A = \overline{A}$ and $F_0 = \overline{F_0}$.*

***Proof*** It follows from Lemma 7.1.5 that $F_0 = \overline{F_0}$. In particular $|F_0| = |\overline{F_0}|$, so $G$ and $\overline{G}$ are of the same type $\mathcal{B}_i$ and $\alpha = \overline{\alpha}$. As in the beginning of the proof of Lemma 7.1.9, we conclude then that $\mu = \overline{\mu}$ and $|A| = |\overline{A}|$.

Since $G^\Omega$ is irreducible and metrically primitive (Corollary 5.6.14), arguing as in Lemma 7.1.9 proves that $A = \overline{A}$. □

**Theorem 7.1.12** *Let $G = G^{\mathcal{B}}_{\mu,\nu}(X_1, \ldots, X_k)$ be metrically primitive irreducible. Let $I$ be the set of indices $1 \leq i \leq k$ such that $X_i$ is of multiplier 2. Then one of the following holds.*

*(i)* $G$ *is maximal solvable in* $\Delta(V, \kappa)$.
*(ii) There exists* $1 \leq i \leq k$ *such that all of the following hold:*

*(a)* *$G$ is not of type $\mathcal{B}_3$;*
*(b)* *$r_i$ is odd;*
*(c)* *$X_i^\circ$ is not maximal solvable in $\mathrm{Sp}_{2\ell_i}(r_i)$;*
*(d)* *If $G$ is of type $\mathcal{B}_0$ or $\mathcal{B}_2$, then one of the following holds:*

- $\left(\frac{q}{r_i}\right) = +1.$
- $\left(\frac{q}{r_j}\right) = -1$ *for some* $j \in I$.

*(e)* *If $G$ is of type $\mathcal{B}_1$, then one of the following holds:*

- $\left(\frac{q}{r_i}\right) = \left(\frac{-1}{r_i}\right) = +1$, *and* $\left(\frac{q}{r_j}\right) = \left(\frac{-1}{r_j}\right) = +1$ *for all* $j \in I$.
- $\left(\frac{-1}{r_i}\right) = \left(\frac{-1}{r_j}\right) = +1$ *for all* $j \in I$, *and* $\left(\frac{q}{r_j}\right) = -1$ *for some* $j \in I$.
- $\left(\frac{-q}{r_i}\right) = \left(\frac{-q}{r_j}\right) = +1$ *for all* $j \in I$, *and* $\left(\frac{-1}{r_j}\right) = -1$ *for some* $j \in I$.

*(iii)* *$G$ is of type $\mathcal{B}_0$, with $q = 3$ and $\nu = 2$.*
*(iv)* *$G$ is of type $\mathcal{B}_1$, with $q = 5$ and $\nu = 2$.*
*(v)* *$G$ is of type $\mathcal{B}_1$, with $q = 5$, $\nu = 1$, and $e + \sum_{i=1}^{k} u_i \equiv 1 \mod 2$.*
*(vi)* *$G$ is of type $\mathcal{B}_2$, with $q = 3$, $\nu = 1$, and $e + \sum_{i=1}^{k} u_i \equiv 1 \mod 2$.*

*Furthermore, in cases (ii)–(vi) $G$ is not maximal solvable in $\Delta(V, \kappa)$.*

***Proof*** We will first check that in cases (ii)–(vi) $G$ is not maximal solvable in $\Delta(V, \kappa)$. In case (ii), it follows from Lemma 2.3.9 and Lemma 2.3.10 (i) that $X_i^\circ$ is metrically completely reducible. Thus there exists $Y_i \leq \mathrm{GSp}_{2\ell_i}(r_i)$ such that $Y_i$ is metrically completely reducible maximal solvable, and $X_i^\circ \lneqq Y_i^\circ$. For $1 \leq i \leq k$, denote the group of scalar matrices in $\mathrm{GSp}_{2\ell_i}(r_i)$ by $Z_i$. In (ii)(d) and (ii)(e), we have $\pi_i(G) \leq X_i^\circ Z_i \leq Y_i$ by Lemma 4.3.36. Therefore $G \leq \overline{G}$, where

$$\overline{G} := G^{\mathcal{B}}_{\mu,\nu}(X_1, \ldots, X_{i-1}, Y_i, X_{i+1}, \ldots, X_k).$$

Here $G^\circ \lneqq \overline{G^\circ}$ since $\pi_i(G)^\circ = X_i^\circ \lneqq Y_i^\circ = \pi_i(\overline{G})^\circ$, so $G$ is not maximal solvable.

In cases (iii)–(vi), the fact that $G$ is not maximal solvable follows from Lemmas 5.5.9, 5.5.11, 5.5.5, and 5.5.7.

Suppose then that (ii)–(vi) do not hold. We will prove that $G$ is maximal solvable. To this end, let $G \leq \overline{G} \leq \Delta(V, \kappa)$ with $\overline{G}$ maximal solvable. Then $\overline{G}$ is metrically primitive irreducible solvable since $G$ is, so $\overline{G}$ is of the form $\overline{G} = G^{\overline{\mathcal{B}}}_{\overline{\mu},\overline{\nu}}(Y_1, \ldots, Y_\ell)$ (Lemma 4.3.30).

Since (iii)–(vi) do not hold, it follows from Lemma 7.1.8 that $F_0 = \overline{F_0}$. Then by Lemma 7.1.10 we have $\mathcal{B} = \overline{\mathcal{B}}$, $\mu = \overline{\mu}$, $\nu = \overline{\nu}$, and $A = \overline{A}$. We have $k = \ell$ since $\mu = \overline{\mu}$, and $R_i = \overline{R_i}$ for all $1 \leq i \leq k$ by Theorem 5.7.9.

Let $\pi$ be the usual homomorphism

$$\pi : N_{\Delta(V,\kappa)}(F_0, R_1, \ldots, R_k) \to \prod_{i=1}^{k} \mathrm{GSp}_{2\ell_i}(r_i).$$

By definition, we have

$$G = \pi^{-1}(X_1 \times \cdots \times X_k),$$
$$\overline{G} = \pi^{-1}(Y_1 \times \cdots \times Y_k).$$

We will prove that $X_i = Y_i$ for all $1 \leq i \leq k$, which implies $G = \overline{G}$ and completes the proof of the theorem.

Let $1 \leq i \leq k$. Since $G \leq \overline{G}$, we have $\pi_i(G) \leq \pi_i(\overline{G})$. In particular

$$\pi_i(G)^\circ = X_i^\circ \leq Y_i^\circ = \pi_i(\overline{G})^\circ.$$

If $r_i = 2$, then $X_i^\circ = X_i$ and $Y_i^\circ = Y_i$, so $X_i = Y_i$ since $X_i$ is maximal solvable. Suppose then that $r_i > 2$. If $X_i^\circ$ is maximal solvable in $\mathrm{Sp}_{2\ell_i}(r_i)$, then $X_i^\circ = Y_i^\circ$ and so $X_i = Y_i$ since $X_i = N_{\mathrm{GSp}_{2\ell_i}(r_i)}(X_i^\circ)$ (Lemma 2.1.5).

Therefore we can assume that $r_i > 2$ and $X_i^\circ$ is not maximal solvable in $\mathrm{Sp}_{2\ell_i}(r_i)$. Note that $G$ is not of type $\mathcal{B}_3$ since $r_i$ is odd. Because we have assumed that (ii) does not hold, it follows from Lemma 4.3.36 that $\pi_i(G) \not\leq X_i^\circ Z_i$. On the other hand $X_i^\circ \leq \pi_i(G)$, so we have $\pi_i(G)Z_i = X_i$. Similarly $\pi_i(\overline{G})Z_i = Y_i$, so $X_i \leq Y_i$ and by maximality of $X_i$ we conclude that $X_i = Y_i$. □

**Theorem 7.1.13** *Let $G = G^{\mathcal{B}}_{\mu,\nu}(X_1, \ldots, X_k)$ be metrically primitive irreducible. Let $I$ be the set of indices $1 \leq i \leq k$ such that $X_i$ is of multiplier 2. Then one of the following holds.*

*(i) $G^\circ$ is maximal solvable in $I(V, \kappa)$.*
*(ii) $G^\circ$ is metrically imprimitive.*
*(iii) There exists $1 \leq i \leq k$ such that all of the following hold:*

*(a) $G$ is not of type $\mathcal{B}_3$;*
*(b) $r_i$ is odd;*
*(c) $X_i^\circ$ is not maximal solvable in $\mathrm{Sp}_{2\ell_i}(r_i)$;*
*(d) If $G$ is of type $\mathcal{B}_0$ or $\mathcal{B}_2$, then one of the following holds:*

- $\left(\dfrac{q}{r_i}\right) = +1.$
- $\left(\dfrac{q}{r_j}\right) = -1$ *for some $j \in I$.*

*(e) If $G$ is of type $\mathcal{B}_1$, then one of the following holds:*

- $\left(\frac{q}{r_i}\right)=\left(\frac{-1}{r_i}\right)=+1$, *and* $\left(\frac{q}{r_j}\right)=\left(\frac{-1}{r_j}\right)=+1$ *for all* $j\in I$.
- $\left(\frac{-1}{r_i}\right)=\left(\frac{-1}{r_j}\right)=+1$ *for all* $j\in I$, *and* $\left(\frac{q}{r_j}\right)=-1$ *for some* $j\in I$.
- $\left(\frac{-q}{r_i}\right)=\left(\frac{-q}{r_j}\right)=+1$ *for all* $j\in I$, *and* $\left(\frac{-1}{r_j}\right)=-1$ *for some* $j\in I$.

*(iv)* *$G$ is of type $\mathcal{B}_3$, $n>1$, and all of the following hold:*

*(a)* *2 is not a square in $\mathbb{F}_q$;*
*(b)* *$X_1^{\Omega}$ is not maximal solvable in $\Omega^{\varepsilon_1}_{2\ell_1}(2)$.*

*(v)* *$G$ is of type $\mathcal{B}_0$, with $q=3$ and $\nu=2$.*
*(vi)* *$G$ is of type $\mathcal{B}_1$, with $q=5$ and $\nu=2$.*
*(vii)* *$G$ is of type $\mathcal{B}_1$, with $q\in\{5,9\}$, $\nu=1$, and $e+\sum_{i=1}^k u_i\equiv 1 \mod 2$.*
*(viii)* *$G$ is of type $\mathcal{B}_1$, with $n$ a power of 2, $q=7$, $\nu=1$, and $e+\sum_{i=1}^k u_i\equiv 1 \mod 2$.*
*(ix)* *$G$ is of type $\mathcal{B}_2$, with $q\in\{3,7\}$, $\nu=1$, and $e+\sum_{i=1}^k u_i\equiv 1 \mod 2$.*

*Furthermore, in cases (iii)–(ix) $G^\circ$ is not maximal solvable in $I(V,\kappa)$.*

***Proof*** We first check that in cases (iii)–(ix) $G^\circ$ is not maximal solvable in $I(V,\kappa)$. In case (iii), the beginning of the proof of Theorem 7.1.12 shows that $G^\circ \lneqq \overline{G^\circ}$ for

$$\overline{G}:=G^{\mathcal{B}}_{\mu,\nu}(X_1,\ldots,X_{i-1},Y_i,X_{i+1},\ldots,X_k),$$

where $Y_i\le \mathrm{GSp}_{2\ell_i}(r_i)$ is metrically completely reducible maximal solvable with $X_i^\circ \lneqq Y_i^\circ$.

For (iv), suppose that $G$ is of type $\mathcal{B}_3$, $n>1$, and that 2 is not a square in $\mathbb{F}_q$. Then $n=2^{\ell_1}$ for some $\ell_1>0$, and $G=G^{\mathcal{B}_3}_{n,1}(X_1)$ with $X_1\le O^{\varepsilon_1}_{2\ell_1}(2)$. Furthermore, assume that $X_1^{\Omega}$ is not maximal solvable in $\Omega^{\varepsilon_1}_{2\ell_1}(2)$.

Because 2 is not a square and $G^\circ$ is metrically primitive, it follows from Lemma 5.3.7 that $X_1^{\Omega}$ is metrically completely reducible. Thus there exists a metrically completely reducible maximal solvable $Y_1\le O^{\varepsilon_1}_{2\ell_1}(2)$ such that $X_1^{\Omega}\lneqq Y_1^{\Omega}$. Define $\overline{G}=G^{\mathcal{B}_3}_{n,1}(Y_1)$ with $\overline{A}=A$. Then it follows from Lemma 4.3.27 (iii) that

$$\pi(G^\circ)=X_1^{\Omega}\lneqq Y_1^{\Omega}=\pi(\overline{G^\circ}),$$

so $G^\circ\lneqq\overline{G^\circ}$. Hence $G^\circ$ is not maximal solvable in $I(V,\kappa)$.

For (v)–(ix), we see from the following results that $G^\circ$ is not maximal solvable:

| | |
|---|---|
| Case (v): | Lemma 5.5.9. |
| Case (vi): | Lemma 5.5.11. |
| Case (vii): | Lemmas 5.5.5 and 5.5.14. |
| Case (viii): | Lemma 5.5.12. |
| Case (ix): | Lemmas 5.5.7 and 5.5.13. |

Suppose then that (ii)–(ix) do not hold. We will prove that $G^\circ$ is maximal solvable in $I(V, \kappa)$. To this end, let $G^\circ \leq \overline{G^\circ}$ with $\overline{G}$ maximal solvable. Then $\overline{G^\circ}$ is metrically primitive irreducible since $G$ is, so $\overline{G}$ is of the form $\overline{G} = G^{\overline{\mathcal{B}}}_{\overline{\mu},\overline{\nu}}(Y_1, \ldots, Y_\ell)$ (Lemma 4.3.30).

We will next see that $F_0 = \overline{F_0}$ by using Lemma 7.1.7 and our previous results. First note that in Lemma 7.1.7 (vi), case (vii) holds for $e = 1$, and $G^\circ$ is metrically primitive for $e = 0$ by Lemma 5.6.7 (iv). Similarly in Lemma 7.1.7 (vii), case (ix) holds for $e = 1$, and $G^\circ$ is metrically primitive for $e = 0$ by Lemma 5.6.8 (iv).

Hence Lemma 7.1.7 (vi) and (vii) do not hold, since we are assuming that $G^\circ$ is metrically primitive and (vi), (ix) do not hold. Moreover Lemma 7.1.7 (ii)–(v) do not hold, since we are assuming that (iv)–(ix) do not hold. Thus $F_0 = \overline{F_0}$ follows from Lemma 7.1.7.

Because $G^\circ$ is metrically primitive, by Lemma 7.1.9 we have $\mathcal{B} = \overline{\mathcal{B}}$, $\mu = \overline{\mu}$, $\nu = \overline{\nu}$, and $A = \overline{A}$. We have $k = \ell$ by $\mu = \overline{\mu}$, and $R_i = \overline{R_i}$ for all $1 \leq i \leq k$ by Theorem 5.7.9.

Then $G = \pi^{-1}(X_1 \times \cdots \times X_k)$ and $\overline{G} = \pi^{-1}(Y_1 \times \cdots \times Y_k)$, where $\pi$ is the usual homomorphism

$$\pi : N_{\Delta(V,\kappa)}(F_0, R_1, \ldots, R_k) \to \prod_{i=1}^{k} \mathrm{GSp}_{2\ell_i}(r_i).$$

We will prove that $X_i = Y_i$ for all $1 \leq i \leq k$, which implies that $G = \overline{G}$ and completes the proof of the theorem.

Suppose first that $G$ is not of type $\mathcal{B}_3$. Then $\pi(G) = \pi(G^\circ)$ and $\pi(\overline{G}) = \pi(\overline{G^\circ})$. Since (iii) does not hold, it follows by arguing as in the proof of Theorem 7.1.12 that $X_i = Y_i$ for all $1 \leq i \leq k$.

We can assume then that $G$ is of type $\mathcal{B}_3$ with $n > 1$, so $n = 2^{\ell_1}$ for some $\ell_1 > 0$ and $A = \overline{A} = 2^{1+2\ell_1}_{\varepsilon_1}$. Then $G = G^{\mathcal{B}_3}_{n,1}(X_1)$ and $\overline{G} = G^{\mathcal{B}_3}_{n,1}(Y_1)$ with $X_1, Y_1 \leq O^{\varepsilon_1}_{2\ell_1}(2)$ metrically completely reducible maximal solvable. If 2 is a square in $\mathbb{F}_q$, then by Lemma 4.3.27 we have $\pi(G) = X_1 \leq Y_1 = \pi(\overline{G})$. Therefore $X_1 = Y_1$, since $X_1$ is maximal solvable.

Suppose then that 2 is not a square in $\mathbb{F}_q$. Then $\pi(G) = X_1^\Omega \leq Y_1^\Omega = \pi(\overline{G})$ by Lemma 4.3.27. Because (iv) does not hold, we have $X_1^\Omega$ maximal solvable in $\Omega^{\varepsilon_1}_{2\ell_1}(2)$, so $X_1^\Omega = Y_1^\Omega$. Then $X_1 = Y_1$, since $X_1 = N_{O^{\varepsilon_1}_{2\ell_1}(2)}(X_1^\Omega)$ by Lemma 2.1.5.

□

**Theorem 7.1.14** *Assume that $n$ and $q$ are even and that $\kappa$ is a quadratic form. Let $G = G^{\mathcal{B}}_{\mu,\nu}(X_1, \ldots, X_k)$ be metrically primitive irreducible. Then one of the following holds.*

*(i) $G^{\Omega}$ is maximal solvable in $\Omega(V, \kappa)$.*
*(ii) $G^{\Omega}$ is not irreducible and $n > 2$.*
*(iii) There exists $1 \leq i \leq k$ such that $X_i^{\circ}$ is not maximal solvable in $\mathrm{Sp}_{2\ell_i}(r_i)$, and one of the following holds:*

*(a) $G$ is of type $\mathcal{B}_1$ and $\left(\frac{q}{r_i}\right) = \left(\frac{-1}{r_i}\right) = +1$.*
*(b) $G$ is of type $\mathcal{B}_2$.*

*Furthermore, in cases (ii) and (iii) $G^{\Omega}$ is not maximal solvable.*

***Proof*** If $n = 2$, then either $G = G^{\mathcal{B}_1}_{1,1} = \mathrm{GO}_2^+(q)$ or $G = G^{\mathcal{B}_2}_{1,1} = \mathrm{GO}_2^-(q)$, so $G^{\Omega} = \Omega_2^{\pm}(q) = \Omega(V, \kappa)$ is maximal solvable. Note that (ii) and (iii) do not apply when $n = 2$. We can assume then for the rest of the proof that $n > 2$.

First we will check that in cases (ii) and (iii) $G^{\Omega}$ is not maximal solvable in $\Omega(V, \kappa)$. In case (ii), this follows from Lemma 5.2.4 and Remark 5.2.5. For (iii), let $1 \leq i \leq k$ be such that $X_i^{\circ}$ is not maximal solvable in $\mathrm{Sp}_{2\ell_i}(r_i)$. It follows from Lemma 2.3.9 and Lemma 2.3.10 (i) that $X_i^{\circ}$ is metrically completely reducible. Thus there exists $Y_i \leq \mathrm{GSp}_{2\ell_i}(r_i)$ such that $Y_i$ is metrically completely reducible maximal solvable, and $X_i^{\circ} \lneqq Y_i^{\circ}$. For $1 \leq i \leq k$, denote the group of scalar matrices in $\mathrm{GSp}_{2\ell_i}(r_i)$ by $Z_i$.

In cases (iii)(a) and (iii)(b) we have $\pi_i(G^{\Omega}) \leq X_i^{\circ} Z_i$ by Lemma 4.3.36 and Lemma 5.1.8 (ii). Therefore $G^{\Omega} \leq \overline{G}^{\Omega}$, where

$$\overline{G} = G^{\mathcal{B}}_{\mu,\nu}(X_1, \ldots, X_{i-1}, Y_i, X_{i+1}, \ldots, X_k).$$

Then $G^{\Omega} \lneqq \overline{G^{\Omega}}$, because $\pi_i(G^{\Omega})^{\circ} = X_i^{\circ} \lneqq Y_i^{\circ} \leq \pi_i(\overline{G}^{\Omega})$ by Lemma 4.3.35 (iii).

Suppose that (ii) and (iii) do not hold. We will prove that $G^{\Omega}$ is maximal solvable in $\Omega(V, \kappa)$. Suppose that $G^{\Omega} \leq \overline{G}^{\Omega}$, where $\overline{G}$ is maximal solvable. Because $G^{\Omega}$ is irreducible, it follows from Theorem 5.6.11 that $G^{\Omega}$ is metrically primitive irreducible. Hence the same is true for $\overline{G}$, so $\overline{G}$ is of the form $\overline{G} = G^{\overline{\mathcal{B}}}_{\overline{\mu},\overline{\nu}}(Y_1, \ldots, Y_{\ell})$ (Lemma 4.3.30).

We first consider the case where $G$ is of type $\mathcal{B}_1$. By irreducibility of $G^{\Omega}$, it follows from Lemma 5.2.4 that $n/2$ is even, in which case $G^{\circ} = G^{\Omega}$ by Lemma 5.1.2. Furthermore, we have $\left(\frac{q}{r_j}\right) = +1$ for all $1 \leq j \leq k$ such that $X_j$ is of multiplier 2 by Lemma 5.1.6. Then because $G$ is irreducible, it follows from Lemma 4.3.14 that $\left(\frac{-1}{r_j}\right) = +1$ for all $1 \leq j \leq k$ such that $X_j$ is of multiplier 2. Since (iii)(a) does not hold, it follows from Theorem 7.1.13 that $G^{\circ}$ is maximal solvable in $I(V, \kappa)$. Thus $G^{\Omega} = G^{\circ}$ is maximal solvable in $\Omega(V, \kappa)$.

We assume then for the remainder of the proof that $G$ is of type $\mathcal{B}_2$. Since $G^\Omega$ is irreducible, by Lemma 7.1.11 we have $F_0 = \overline{F_0}$. Furthermore by Lemma 7.1.11 we have $\mathcal{B} = \overline{\mathcal{B}}$, $\mu = \overline{\mu}$, $\nu = \overline{\nu}$, and $A = \overline{A}$. We have $k = \ell$ since $\mu = \overline{\mu}$. Moreover $|F_0| = q^\nu + 1$ is odd, so $r_i$ is odd for all $1 \leq i \leq k$. Then as in the beginning of the proof of Theorem 5.7.9, we see that

$$R_i = \{x \in A : x^{r_i} = 1\} = \overline{R_i}$$

for all $1 \leq i \leq k$.

By definition $G = \pi^{-1}(X_1 \times \cdots \times X_k)$ and $\overline{G} = \pi^{-1}(Y_1 \times \cdots \times Y_k)$, where $\pi$ is the usual homomorphism

$$\pi : N_{\Delta(V,\kappa)}(F_0, R_1, \ldots, R_k) \to \prod_{i=1}^{k} \mathrm{GSp}_{2\ell_i}(r_i).$$

We will prove that $X_i = Y_i$ for all $1 \leq i \leq k$, which implies $G = \overline{G}$ and completes the proof of the theorem.

Let $1 \leq i \leq k$. Since $G^\Omega \leq \overline{G^\Omega}$, it follows from Lemma 4.3.22 (ii) and Lemma 4.3.21 (i) that $X_i^\circ \leq Y_i^\circ$. Because $G$ is of type $\mathcal{B}_2$ and (iii) does not hold, we have $X_i^\circ$ maximal solvable in $\mathrm{Sp}_{2\ell_i}(r_i)$. Thus $X_i^\circ = Y_i^\circ$, which implies $X_i = Y_i$ since $X_i = N_{\mathrm{GSp}_{2\ell_i}(r_i)}(X_i^\circ)$ (Lemma 2.1.5). □

In the case $\mu = 1$, with Theorems 7.1.12–7.1.14 we get the following result.

**Theorem 7.1.15** *Let $\nu \geq 1$ be an integer. Then all of the following hold:*

*(i) $G_{1,\nu}^{\mathcal{B}}$ is metrically primitive maximal irreducible solvable in $\Delta(V, \kappa)$, except in the following cases:*

- *(a) Type $\mathcal{B}_0$, with $(q, \nu) = (3, 2)$.*
- *(b) Type $\mathcal{B}_1$, with $(q, \nu) = (3, 2)$ or $(5, 2)$.*
- *(c) Type $\mathcal{B}_1$, $e = 1$, with $(q, \nu) = (5, 1)$.*
- *(d) Type $\mathcal{B}_2$, with $(q, \nu) = (2, 3)$.*
- *(e) Type $\mathcal{B}_2$, $e = 1$, with $(q, \nu) = (3, 1)$.*

*(ii) $\left(G_{1,\nu}^{\mathcal{B}}\right)^\circ$ is metrically primitive maximal irreducible solvable in $I(V, \kappa)$, except in the following cases:*

- *(a) Type $\mathcal{B}_0$, with $(q, \nu) = (3, 2)$.*
- *(b) Type $\mathcal{B}_1$, with $(q, \nu) = (3, 2)$ or $(5, 2)$.*
- *(c) Type $\mathcal{B}_1$, $e = 1$, with $(q, \nu) = (5, 1)$, $(7, 1)$, or $(9, 1)$.*
- *(d) Type $\mathcal{B}_2$, with $(q, \nu) = (2, 3)$.*
- *(e) Type $\mathcal{B}_2$, $e = 1$, with $(q, \nu) = (3, 1)$ or $(7, 1)$.*

*(iii) Assume n and q even, with $\kappa$ a quadratic form. Then $\left(G^{\mathcal{B}}_{1,\nu}\right)^{\Omega}$ is metrically primitive maximal irreducible solvable in $\Omega(V, \kappa)$, except in the following cases:*

*(a) Type $\mathcal{B}_1$, with $\nu$ odd.*
*(b) Type $\mathcal{B}_2$, with $(q, \nu) = (2, 3)$.*

***Proof*** Claim (i) follows from Lemma 5.2.2, Theorems 5.6.9, and 7.1.12. Similarly (ii) follows from Lemma 5.2.2, Theorems 5.6.10, and 7.1.13. For (iii), the result follows from Lemma 5.2.4, Theorems 5.6.11, and 7.1.14. □

**Remark 7.1.16** In the classification given by Jordan [52], we note that in the proof there are some untreated cases. These issues appear both in the metrically primitive case treated in this section, and the cases of metrically imprimitive groups and metrically completely reducible groups discussed in Sects. 7.4 and 7.5.

More specifically, in his proof that the groups he constructed are maximal irreducible solvable subgroups, the basic strategy Jordan outlines in [52, §47] is the following. Let $L$ and $\overline{L}$ be two groups given by the construction [51], either in $\mathrm{GL}_n(p)$ ($p$ prime), $\mathrm{GSp}_n(p)$ ($n$ even, $p$ odd prime), or $O_n^{\pm}(2)$ ($n$ even).

In [52, §47] Jordan states that the goal is to prove that except for the *cas d'exclusion* listed in [52, §21, §35–47], the following hold:

- If $L \leq \overline{L}$, then $L = \overline{L}$.
- In the case of $\mathrm{GSp}_n(p)$ ($p$ odd prime), if $L^{\circ} \leq \overline{L^{\circ}}$, then $L^{\circ} = \overline{L^{\circ}}$.

There are a few issues that come up here and in the proof given by Jordan. For example:

- In the case of $O_n^{\pm}(2)$, Jordan does not consider the maximality of $L^{\Omega}$ in $\Omega_n^{\pm}(2)$, which is necessary in view of Theorem 7.1.13 (iv). Considering the maximality of $L^{\Omega}$ only needs a few additional arguments. Later in Theorem 7.6.3, we will see that $L \leq O_n^{\pm}(2)$ is maximal irreducible solvable and $L^{\Omega}$ is irreducible, then $L^{\Omega}$ is maximal solvable in $\Omega_n^{\pm}(2)$.
- As pointed out by Jordan in [52, §45–47], there are cases where $L$ is maximal solvable and $L^{\circ}$ is not maximal solvable, and these need to be taken into account—see Theorem 7.1.12 (ii) and Theorem 7.1.13 (iii). However, Jordan seems to consider the maximality of $\overline{L}$ only in the metrically primitive case.

  For example, in the case where $L = (H \wr K) \cap \Delta(V, \kappa)$ is maximal irreducible solvable as in Theorem 2.2.14, it is possible that $L^{\circ}$ is not maximal solvable. One case where this happens is when $H$ is maximal solvable in $\Delta(W, \kappa')$, but $H^{\circ}$ is not maximal solvable in $I(W, \kappa')$.

  For the completely reducible case, Jordan only considers groups of the form $L^{\circ} = H_1^{\circ} \times \cdots \times H_t^{\circ}$ with $H_i \leq \Delta(W_i, \kappa)$ as in Lemma 2.3.7. (This is a *complexe* in the terminology of Jordan from [52, §21].) However, there are examples of metrically completely reducible $L = (H_1 \times \cdots \times H_t) \cap \Delta(V, \kappa)$ (with $t \geq 2$) such that $L$ is maximal solvable, but $H_i^{\circ}$ are not maximal solvable in $I(W_i, \kappa)$ (Remark 7.5.5). See also Remark 7.5.6.

- In the case where $L$ is metrically imprimitive, Jordan does not consider the case where $L$ is of semiprimary type (Example 2.2.7).

## 7.2 Completely Reducible Subgroups of $G^{\mathcal{B}}_{\mu,\nu}(X_1, \ldots, X_k)$

Throughout this section, we denote by $G$ a metrically primitive irreducible solvable group of the form $G = G^{\mathcal{B}}_{\mu,\nu}(X_1, \ldots, X_k)$, where $G \leq \Delta(V, \kappa)$. In Sect. 7.1, we obtained a classification that describes when $G$ is maximal solvable in $\Delta(V, \kappa)$, and when $G^\circ$ is maximal solvable in $I(V, \kappa)$. We will now move towards a classification of metrically imprimitive maximal irreducible solvable subgroups, and metrically completely reducible maximal solvable subgroups. As noted in Remark 7.1.1, this is also necessary to fully classify metrically primitive maximal irreducible solvable subgroups.

We consider the question of when $G^\circ$ can contain completely reducible groups of the form $H_1^\circ \times \cdots \times H_t^\circ$, such that $V = W_1 \perp \cdots \perp W_t$ with $H_i \leq \Delta(W_i, \kappa)$ maximal irreducible solvable and $H_i \neq 1$ for all $1 \leq i \leq t$. In the case where $n$ and $q$ are even and $\kappa$ is a quadratic form, we will obtain similar results for $G^\Omega$.

Note that if $H_i$ is metrically imprimitive, then it follows from Theorem 2.2.14 that $H_i^\circ$ contains $H_1^{(i)} \times \cdots \times H_d^{(i)}$, such that $W_i = Z_1 \perp \cdots \perp Z_d$ with $H_j^{(i)} \leq \Delta(Z_j, \kappa)$ metrically primitive maximal irreducible solvable for all $1 \leq j \leq d$. Thus we will mostly consider the case where the groups $H_i$ are metrically primitive. Under this assumption, there are very few cases where $H_1^\circ \times \cdots \times H_t^\circ$ can be contained in $G$, so the general case will be easy to deal with afterwards.

As a corollary, we will be able to describe when metrically imprimitive groups of the form in Theorem 2.2.14 (i)–(ii) can be contained in a metrically primitive solvable subgroup of $\Delta(V, \kappa)$ (Theorems 7.2.19 and 7.2.21). This is the first step towards the classification of metrically imprimitive maximal irreducible solvable subgroups, which will be obtained in Sect. 7.4.

For reference, we record the usual setup that we consider in this section in the following example.

**Example 7.2.1** The general setup and notation that we will use throughout this section is as follows. Let $V = W_1 \perp \cdots \perp W_t$ be an orthogonal decomposition, where $t \geq 2$ and $\dim W_i \geq 1$ for all $1 \leq i \leq t$. (Note that for $\kappa = 0$, orthogonal decomposition just means a direct sum.)

For all $1 \leq i \leq t$, let $H_i$ be a maximal metrically primitive irreducible solvable subgroup of $\Delta(W_i, \kappa)$. Assume that $H_i \neq 1$, in which case it follows from Lemma 2.1.6 that $H_i^\circ \neq 1$. (Note that the only case where $H_i = 1$ can hold is when $\kappa = 0$ and $H_i = \mathrm{GL}_1(2)$.)

Let $L' = H_1 \times \cdots \times H_t$ be the completely reducible subgroup of $\mathrm{GL}(V)$ corresponding to $H_1, \ldots, H_t$ and the decomposition $V = W_1 \perp \cdots \perp W_t$. Denote

$L := L' \cap \Delta(V, \kappa)$. Then if $\kappa \neq 0$, we have

$$L = \{(h_1, \ldots, h_t) \in L' : \tau_i(h_i) = \tau_j(h_j) \text{ for all } 1 \leq i, j \leq t\},$$

where $\tau_i$ is the usual homomorphism $\tau_i : \Delta(W_i, \kappa) \to \mathbb{F}_q^{\times}$. We have

$$L^{\circ} = H_1^{\circ} \times \cdots \times H_t^{\circ}.$$

We will denote $n_i = \dim W_i$ for all $1 \leq i \leq t$.

For all $1 \leq i \leq t$, let $F_0^{(i)}$ be the normal subgroup of $H_i$ satisfying properties (F1)–(F3). Then $F_0^{(i)}$ is cyclic (Lemma 2.4.3), so choose a generator $f_i$. Let $\mathbb{K}_i/\mathbb{F}_q$ be the splitting field of the characteristic polynomial of $f_i$. Following the usual notation established in Sect. 2.4, we denote by $\mu_i$ the multiplicity of an eigenvalue of $f_i$ on $\mathbb{K}_i \otimes_{\mathbb{F}_q} V$ (Lemma 2.4.9). We denote by $\alpha_i$ the number of eigenvalues of $f_i$ on $\mathbb{K}_i \otimes_{\mathbb{F}_q} V$, so $n_i = \alpha_i \mu_i$. We define $\nu_i := \alpha_i$ if $H_i$ is not of type $\mathcal{B}_1$ or $\mathcal{B}_2$, and $\nu_i = \alpha_i/2$ otherwise. Furthermore, set $\omega_i := |f_i|$.

We use similar notation for $G = G^{\mathcal{B}}_{\mu,\nu}(X_1, \ldots, X_k)$, so $F_0 = \langle f \rangle$ is a subgroup of $G$ satisfying properties (F1)–(F3), with splitting field $\mathbb{K}$. Furthermore, we denote by $\alpha$ the number of eigenvalues of $f$ on $\mathbb{K} \otimes_{\mathbb{F}_q} V$, and $\omega := |f|$.

**Remark 7.2.2** Suppose that $L^{\circ} = H_1^{\circ} \times \cdots \times H_t^{\circ}$ is as in Example 7.2.1. Each $H_i^{\circ}$ contains an abelian subgroup of order $\omega_i \mu_i$ (Lemma 6.3.10). Therefore if $L^{\circ} \leq G^{\circ}$, it follows from Corollary 6.3.9 that

$$\prod_{i=1}^{t} \omega_i \mu_i \leq \omega \mu^2.$$

Since $n = \alpha\mu$ and $n_i = \alpha_i \mu_i$ for all $i$, we get

$$\alpha \cdot \prod_{i=1}^{t} \frac{\omega_i}{\alpha_i} n_i \leq \frac{\omega}{\alpha} \cdot n^2, \tag{7.7}$$

called *l'inégalité fondamentale* by Jordan [52, §100]. Following the approach of Jordan, throughout this section we will use (7.7) to limit the types of configurations that can occur.

In the case where $n$ and $q$ are even and $\kappa$ is a quadratic form, it follows from Lemma 6.3.10 that $H_i^{\Omega}$ contains an abelian subgroup of order $\omega_i \mu_i$ for all $1 \leq i \leq t$. Thus if $L^{\Omega} \leq G^{\Omega}$, the inequality (7.7) holds.

**Lemma 7.2.3** *Let $L^{\circ} = H_1^{\circ} \times \cdots \times H_t^{\circ}$ as in Example 7.2.1. Suppose that $L^{\circ} \leq G^{\circ}$. Then for all $1 \leq i \leq t$, one of the following holds:*

*(i) $F_0$ acts on $W_i$ and $\sum_{j \neq i} W_j$;*
*(ii) $n_i = 1$.*

***Proof*** Let $1 \leq i \leq t$ and suppose that $n_i > 1$. We argue similarly to [52, §94] to prove that (i) holds. First we note the following.

**Claim: If** $1 \neq z \in [H_i^\circ, H_i^\circ]$ **is such that** $z \in F_0^{(i)}$**, then** $F_0$ **acts on** $W_i$ **and** $\sum_{j \neq i} W_j$

Such a $z$ centralizes $F_0$, since $G/C_{G^\circ}(F_0)$ is abelian (Lemma 4.3.33). Since $F_0^{(i)}$ is cyclic, the subgroup generated by $z$ is normal in $H_i$, and thus the action of $\langle z \rangle$ on $V$ is completely reducible by Clifford's theorem.

Furthermore, by Lemma 2.4.3 we have $W_i^z = 0$. Then because $F_0$ centralizes $z$, it acts on $V^z = \sum_{j \neq i} W_j$, and also on $W_i$, which is the sum of the nontrivial irreducible $\mathbb{F}_q[\langle z \rangle]$-submodules in $V$.

With the claim proved, we now proceed prove the lemma by finding such a $z$. If $\mu_i > 1$, there exists an extraspecial $r$-subgroup $R \trianglelefteq H_i^\circ$ such that $Z(R) = R \cap F_0^{(i)}$ (Propositions 4.1.17, 4.1.11). We have $Z(R) = [R, R]$, so there exists $x, y \in R$ with $[x, y] \in F_0^{(i)}$ of order $r$.

Suppose then that $\mu_i = 1$, so $H_i$ is of type $\mathcal{B}_0$, $\mathcal{B}_1$, or $\mathcal{B}_2$, since $\dim W_i > 1$. If $[\mathbb{K}_i : \mathbb{F}_q] > 1$, then $|f_i| > q - 1$ and for each of the types there exists $\psi_i \in H_i^\circ$ such that $\psi_i^{-1} f_i \psi_i = f_i^q$. Therefore $[\psi_i, f_i] = f_i^{1-q} \neq 1$.

Thus we can assume that $\mu_i = [\mathbb{K}_i : \mathbb{F}_q] = 1$, in which case $H_i$ must be of type $\mathcal{B}_1$. Here there exists $\varphi_i \in H_i^\circ$ such that $\varphi_i^{-1} f_i \varphi_i = f_i^{-1}$, so $[\varphi_i, f_i] = f_i^2 \neq 1$. This completes the proof of the lemma. □

**Lemma 7.2.4** *Assume that* $n$ *and* $q$ *are even, and that* $\kappa$ *is a quadratic form. Let* $L = (H_1 \times \cdots \times H_t) \cap \Delta(V, \kappa)$ *as in Example 7.2.1. Suppose that* $L^\Omega \leq G^\Omega$*. Then* $F_0$ *acts on* $W_i$ *for all* $1 \leq i \leq t$.

***Proof*** Note that since $q$ is even, we have $n_i > 1$ for all $1 \leq i \leq t$. As in the proof of Lemma 7.2.3, it will suffice to find for all $1 \leq i \leq t$ an element $1 \neq z \in [L^\Omega, L^\Omega]$ such that $z \in F_0^{(i)}$. If $H_i^\circ \leq \Omega(W_i, \kappa)$, it follows from the proof of Lemma 7.2.3 that we can find such a $z$ in $[H_i^\circ, H_i^\circ] = [H_i^\Omega, H_i^\Omega]$, and therefore $F_0$ acts on $W_i$.

Suppose then that $H_i^\circ \not\leq \Omega(W_i, \kappa)$. If $H_j^\circ \leq \Omega(W_j, \kappa)$ for all $j \neq i$, then as earlier it follows from the proof of Lemma 7.2.3 that $F_0$ acts on $W_j$ for all $j \neq i$. Then $F_0$ acts on $\sum_{j \neq i} W_j$, and therefore on $(\sum_{j \neq i} W_j)^\perp = W_i$.

Thus we can assume that there exists $j \neq i$ such that $H_j^\circ \not\leq \Omega(W_j, \kappa)$. Choose $h_j \in H_j^\circ \setminus H_j^\Omega$. As seen in the proof of Lemma 7.2.3, there exists $h_i, h_i' \in H_j^\circ$ such that $1 \neq z = [h_i, h_i'] \in F_0^{(i)}$. Then

$$z = [h_i, h_j] = [h_i h_j, h_i' h_j] = [h_i h_j, h_i'] = [h_i, h_i' h_j].$$

Now $h_i \in L^\Omega$ or $h_i h_j \in L^\Omega$, and moreover $h_i' \in L^\Omega$ or $h_i' h_j \in L^\Omega$. Thus we conclude that $z \in [L^\Omega, L^\Omega]$. □

**Lemma 7.2.5** *Let* $L^\circ = H_1^\circ \times \cdots \times H_t^\circ$ *be as in Example 7.2.1. Suppose that* $L^\circ \leq G^\circ$*, and that* $F_0$ *acts on* $W_i$ *for all* $1 \leq i \leq t$*. Then* $f = x_1 \cdots x_t$*, where*

*$x_i \in Z(H_i^\circ) \leq F_0^{(i)}$, and $|x_i| = |f|$ for all $1 \leq i \leq t$. In particular $\omega \mid \omega_i$ for all $1 \leq i \leq t$.*

***Proof*** *([52, §96])* We can write $f = x_1 \cdots x_t$, where $x_i \in I(V, \kappa)$ acts on $W_i$, and acts trivially on $W_j$ for $j \neq i$. Clearly $|x_i| \leq |f|$ for all $1 \leq i \leq t$. If equality does not hold, then there exists $d > 0$ such that $x_i^d = 1$ and $x^d \neq 1$. But then $f^d$ acts trivially on $W_i$, contradicting the fact that $V^g = 0$ for all $g \in F_0 \setminus \{1\}$ (Lemma 2.4.3).

Therefore $|x_i| = |f|$ for all $1 \leq i \leq t$. Next we will prove that $x_i$ centralizes $H_i^\circ$ for all $1 \leq i \leq t$. To this end, let $g \in H_i^\circ$. Since $F_0$ is normalized by $L$, there exists $d > 0$ such that $gfg^{-1} = f^d$. Thus

$$x_1 \cdots x_{i-1}(gx_ig^{-1})x_{i+1} \cdots x_t = x_1^d \cdots x_t^d.$$

It follows that $gx_ig^{-1} = x_i^d$, and $x_j^d = x_j$ for all $j \neq i$. Because $t > 1$ and $|x_j| = |x_i|$ for $j \neq i$, we conclude that $x_i^d = x_i$. Therefore $gx_ig^{-1} = x_i$.

We have proved that $x_i$ centralizes $H_i^\circ$. The centralizer of $H_i^\circ$ in $I(V, \kappa)$ is contained in $F_0^{(i)}$ (Lemma 4.3.32), so $x_i \in Z(H_i^\circ) \leq F_0^{(i)}$. Since $|x_i| = |f|$, it follows that $\omega \mid \omega_i$ for all $1 \leq i \leq t$. □

**Lemma 7.2.6** *Assume that $n$ and $q$ are even, and that $\kappa$ is a quadratic form. Let $L = (H_1 \times \cdots \times H_t) \cap \Delta(V, \kappa)$ as in Example 7.2.1. Suppose that $L^\Omega \leq G^\Omega$. Then $f = x_1 \cdots x_t$, where $x_i \in Z(H_i^\Omega) \leq F_0^{(i)}$, and $|x_i| = |f|$ for all $1 \leq i \leq t$. In particular $\omega \mid \omega_i$ for all $1 \leq i \leq t$.*

***Proof*** It follows from Lemma 7.2.4 that $F_0$ acts on $W_i$ for all $1 \leq i \leq t$, so the result follows with same proof as Lemma 7.2.5. □

In the conclusion of Lemmas 7.2.5 and 7.2.6, we have $\omega \mid \omega_i$ for all $1 \leq i \leq t$. This places various restrictions on the values of $\omega$, $\omega_i$ and the types of $G$ and $H_i$, which we record in the next lemma, similarly to [52, §97].

**Lemma 7.2.7** *Suppose that $\overline{G} = G^{\overline{\mathcal{B}}}_{\overline{\mu},\overline{\nu}}(Y_1, \ldots, Y_\ell) \leq \Delta(W, \kappa)$ is metrically primitive and irreducible, where $W$ is a non-degenerate subspace of $V$. Denote the corresponding subgroup $F_0$ and integers $\omega$, $\alpha$ in $\overline{G}$ by $\overline{F_0}$, $\overline{\omega}$, $\overline{\alpha}$. Then the following statements hold.*

*(i) If $\omega$ divides $\overline{\omega}$, then $G$ and $\overline{G}$ are as in one of the entries of Table 7.2.*
*(ii) If $\omega$ divides $\overline{\omega}$, then $\overline{\omega}/\overline{\alpha} \geq \omega/\alpha$.*

***Proof*** For claim (i), it follows from the proof of Lemma 7.1.4 and Table 7.1 that the possible configurations are as listed in Table 7.2. For (ii), we will see that the ratio

$$\overline{\omega}/\overline{\alpha} : \omega/\alpha \geq 1$$

in all cases.

**Table 7.2** For metrically primitive groups of the form $G = G^{\mathcal{B}}_{\mu,\nu}(X_1, \ldots, X_k) \leq \Delta(V, \kappa)$ and $\overline{G} = G^{\overline{\mathcal{B}}}_{\overline{\mu},\overline{\nu}}(Y_1, \ldots, Y_\ell) \leq \Delta(W, \kappa)$ where $W$ is a non-degenerate subspace of $V$, the possible configurations when $|F_0|$ divides $|\overline{F_0}|$; see Lemma 7.2.7

| Type of $G$ | $\lvert F_0\rvert$ | $\alpha$ | Type of $\overline{G}$ | $\lvert\overline{F_0}\rvert$ | $\overline{\alpha}$ |
|---|---|---|---|---|---|
| $\mathcal{B}_0$ | $q^\nu - 1$ | $\nu$ | $\mathcal{B}_0$ | $q^{m\nu} - 1$ | $m\nu$ |
| $\mathcal{B}_1$ | $q^\nu - 1$ | $2\nu$ | $\mathcal{B}_1$ | $q^{m\nu} - 1$ | $2m\nu$ |
| $\mathcal{B}_2$ | $q^\nu + 1$ | $2\nu$ | $\mathcal{B}_1$ | $q^{2m\nu} - 1$ | $4m\nu$ |
| | | | $\mathcal{B}_2$ | $q^{(2m-1)\nu} + 1$ | $(2m-1)2\nu$ |
| $\mathcal{B}_3$ | 2 | 1 | $\mathcal{B}_1$ | $q^{\overline{\nu}} - 1$ | $2\overline{\nu}$ |
| | | | $\mathcal{B}_2$ | $q^{\overline{\nu}} + 1$ | $2\overline{\nu}$ |
| | | | $\mathcal{B}_3$ | 2 | 1 |

If $G$ is of type $\mathcal{B}_0$ or type $\mathcal{B}_1$ as in Table 7.2, then

$$\overline{\omega}/\overline{\alpha} : \omega/\alpha = \frac{q^{m\nu} - 1}{m(q^\nu - 1)} \geq 1$$

for all $m, \nu \geq 1$. This follows since the ratio increases with $m$, and for $m = 1$ it is equal to 1.

If $G$ is of type $\mathcal{B}_2$ and $\overline{G}$ is of type $\mathcal{B}_1$ as in Table 7.2, then similarly

$$\overline{\omega}/\overline{\alpha} : \omega/\alpha = \frac{q^{2m\nu} - 1}{2m(q^\nu + 1)} \geq 1,$$

except for $(q, \nu, m) = (2, 1, 1)$, but this case is not applicable for $\overline{G}$ of type $\mathcal{B}_1$.

If $G$ is of type $\mathcal{B}_2$ and $\overline{G}$ is of type $\mathcal{B}_2$ as in Table 7.2, then

$$\overline{\omega}/\overline{\alpha} : \omega/\alpha = \frac{q^{(2m-1)\nu} + 1}{(2m - 1)(q^\nu + 1)} \geq 1$$

for all $m, \nu \geq 1$.

If $G$ is of type $\mathcal{B}_3$, then $q$ must be odd. Thus if $\overline{G}$ is of type $\mathcal{B}_1$ as in Table 7.2, then

$$\overline{\omega}/\overline{\alpha} : \omega/\alpha = \frac{q^{\overline{\nu}} - 1}{4\overline{\nu}} \geq 1$$

except when $(q, \overline{\nu}) = (3, 1)$, but this is not applicable for $\overline{G}$ of type $\mathcal{B}_1$. If $\overline{G}$ is of type $\mathcal{B}_2$ as in Table 7.2, then

$$\overline{\omega}/\overline{\alpha} : \omega/\alpha = \frac{q^{\overline{\nu}} + 1}{4\overline{\nu}} \geq 1$$

since $q > 2$.

Finally in the case where $G$ and $\overline{G}$ are both of type $\mathcal{B}_3$, we have $\overline{\omega}/\overline{\alpha} = \omega/\alpha = 2$ and the ratio equals 1. This completes the proof of (ii) and the lemma. □

**Lemma 7.2.8** *Suppose that $G = G^{\mathcal{B}}_{\mu,\nu}(X_1, \ldots, X_k)$ is irreducible and solvable. If $|G^\circ|$ is a power of 2, then $\mu = 1$ and $n$ is a power of 2. Furthermore, one of the following holds:*

*(i)* $\nu = 1$.
*(ii)* $q = 3$, $\nu = 2$, *and* $|F_0| = 3^2 - 1 = 2^3$.

***Proof*** We begin similarly to [52, §99]. Suppose that $\mu > 1$. We will prove that $|G^\circ|$ is divisible by an odd prime.

If $\mu$ has an odd prime divisor $r$, then $A = \mathrm{Fit}(C_{G^\circ}(F_0))$ contains an extraspecial $r$-group of order $r^{1+2\ell}$ for some $\ell > 0$. Therefore we can assume that $\mu = 2^{\ell_1}$ for some $\ell_1 > 0$.

We have $G = G^{\mathcal{B}}_{\mu,\nu}(X_1)$, where $X_1 \leq O^{\varepsilon_1}_{2\ell_1}(2)$ is metrically completely reducible maximal solvable, and $A = F_0 R_1$ with $R_1 = 2^{1+2\ell_1}_{\varepsilon_1}$. By Lemma 5.7.1 the action of $G^\circ$ on $R_1/Z(R_1)$ has no nonzero fixed points, so $|G^\circ|$ cannot be a power of two.

We have proved that if $|G^\circ|$ is a power of 2, then $\mu = 1$. In this case $|G^\circ| = n|F_0|$ (Table 2.2), so $n$ must also be a power of 2. Now it remains to show that either (i) or (ii) holds. If $G$ is of type $\mathcal{B}_3$, then $\nu = 1$ by definition. For the other types, we have $|F_0| = q^\nu \pm 1$, which must be a power of two. Thus it follows from Lemma 2.1.22 that either (i) or (ii) holds. □

**Lemma 7.2.9** *Let $L^\circ = H_1^\circ \times \cdots \times H_t^\circ$ be as in Example 7.2.1, and suppose that $L^\circ \leq G^\circ$. Then $t \leq 4$. If $t = 4$, then $n = 4$, $n_1 = n_2 = n_3 = n_4 = 1$, and one of the following holds:*

*(i)* $\kappa = 0$, $q = 3$, *and* $H_1 = H_2 = H_3 = H_4 = \mathrm{GL}_1(3)$;
*(ii)* $\kappa \neq 0$ *is a quadratic form,* $q$ *is odd, and* $H_1 = H_2 = H_3 = H_4 = \mathrm{GO}_1(q)$;

***Proof*** Since elements of $H_i^\circ$ act trivially on $\sum_{j \neq i} W_j$, it follows from Theorem 6.2.3 that $n - n_i \leq 3n/4$, so $n_i \geq n/4$ for all $1 \leq i \leq t$. Because $n = n_1 + \cdots + n_t$, we conclude that $t \leq 4$.

Suppose then that $t = 4$. Since $n_i \geq n/4$ for all $i$, it follows that $n_1 = n_2 = n_3 = n_4 = n/4$. Every element of $H_i^\circ$ has a fixed point space of dimension $\geq 3n/4$ on $V$, so by Theorem 6.2.3 the order $|H_i^\circ|$ is a power of two. Thus by Lemma 7.2.8 we have $\mu_i = 1$ for all $1 \leq i \leq 4$.

Moreover, by Lemma 7.2.8 for all $1 \leq i \leq 4$ either $\nu_i = 1$, or $(q, \nu_i) = (3, 2)$. If $(q, \nu_i) = (3, 2)$, we must have $|F_0^{(i)}| = 3^2 - 1 = 8$, so $H_i$ is of type $\mathcal{B}_0$ or $\mathcal{B}_1$. Then either $H_i$ is not maximal solvable (Lemma 5.5.9) or $H_i$ is metrically imprimitive (Lemma 5.6.5), so this case does not apply. Hence $\nu_i = 1$ for all $1 \leq i \leq 4$.

First we consider the case where $n_i > 1$ for all $i$. Since $\mu_i = \nu_i = 1$, in this case $H_i$ cannot be of type $\mathcal{B}_0$ or $\mathcal{B}_3$. Therefore $H_i$ is of type $\mathcal{B}_1$ or $\mathcal{B}_2$, and $n_i = 2\nu_i\mu_i = 2$ for all $i$. Hence $n = n_1 + n_2 + n_3 + n_4 = 8$. Note that $\omega_i > 2$ because $H_i$ is of type $\mathcal{B}_1$ or $\mathcal{B}_2$. Thus $\omega_i \geq 4$ for all $i$, since $\omega_i$ must be a power of two.

Now $G^\circ$ contains the abelian group $F_0^{(1)} \times F_0^{(2)} \times F_0^{(3)} \times F_0^{(4)}$, so it follows from Corollary 6.3.9 that

$$\omega_1\omega_2\omega_3\omega_4 \leq \omega\mu^2. \tag{7.8}$$

By Lemmas 7.2.3 and 7.2.5, we have $\omega \mid \omega_i$ for all $i$, and in particular $\omega \leq \omega_i$. Therefore if $\mu < 8$, inequality (7.8) gives

$$\omega_1\omega_2\omega_3 \leq \mu^2 \leq 7^2,$$

which is a contradiction since $\omega_1\omega_2\omega_3 \geq 4^3$. Thus we must have $\mu = n = 8$, in which case $G$ is of type $\mathcal{B}_3$. In this case $\omega = 2$, so (7.8) becomes

$$\omega_1\omega_2\omega_3\omega_4 \leq 2^7,$$

which is a contradiction since $\omega_1\omega_2\omega_3\omega_4 \geq 4^4$.

It remains to consider the case where $n_i = 1$ for all $1 \leq i \leq 4$, so $n = 4$. If $\kappa \neq 0$, it follows from $\dim W_i = 1$ that $\kappa$ is a quadratic form and (ii) holds. Suppose then that $\kappa = 0$, in which case $H_i = \mathrm{GL}_1(q)$ for all $1 \leq i \leq 4$. Note that $q > 2$, since $H_i \neq 1$ for all $i$. Because $H_1 \times H_2 \times H_3 \times H_4$ is abelian, it follows from Corollary 6.3.9 that

$$(q-1)^4 \leq (q^\nu - 1)\mu^2.$$

Note that $\mu \in \{1, 2, 4\}$ since $n = 4$. For $\mu = 4$ and $\mu = 2$, we get inequalities $(q-1)^4 \leq (q-1) \cdot 4^2$ and $(q-1)^4 \leq (q^2-1) \cdot 2^2$ respectively, which can only hold for $q \leq 3$. Since $q > 2$, for $\mu = 4$ and $\mu = 2$ we must have $q = 3$, so (i) holds.

Finally in the case where $\mu = 1$, we have $|G| = 4(q^4 - 1)$ (Table 2.2). Thus $(q-1)^4$ divides $4(q^4-1)$, and it is straightforward to check that this is only possible for $q = 2, 3$. Since $q > 2$, it follows that $q = 3$ and (i) holds. This completes the proof of the lemma. □

**Lemma 7.2.10** *Assume that $n$ and $q$ are even, and that $\kappa$ is a quadratic form. Let $L = (H_1 \times \cdots \times H_t) \cap \Delta(V, \kappa)$ as in Example 7.2.1. Suppose that $L^\Omega \leq G^\Omega$. Then $t \leq 3$.*

***Proof*** Since elements of $H_i^\Omega$ act trivially on $\sum_{j \neq i} W_j$, it follows as in the beginning of the proof of Lemma 7.2.9 that $t \leq 4$ and $n_i \geq n/4$ for all $1 \leq i \leq t$.

Suppose that $t = 4$. Since $n_i \geq n/4$ for all $i$, it follows that $n_1 = n_2 = n_3 = n_4 = n/4$. Every element of $H_i^\Omega$ has a fixed point space of dimension $\geq 3n/4$ on $V$, so by Theorem 6.2.3 the order $|H_i^\Omega|$ is a power of two. Then $H_i^\circ$ is a 2-group, so $V^{H_i^\circ} \neq 0$ since $q$ is even. This is impossible since $H_i^\circ$ is irreducible, so we must have $t \leq 3$. □

**Lemma 7.2.11** *Let $L = H_1^\circ \times H_2^\circ \times H_3^\circ$ be as in Example 7.2.1 with $t = 3$, and suppose that $L^\circ \leq G^\circ$. Then $n_i = 1$ for some $1 \leq i \leq 3$.*

***Proof*** Suppose that $n_i > 1$ for all $i$. We will assume, without loss of generality, that $1 < n_1 \leq n_2 \leq n_3$.

It follows from Lemma 7.2.3 that $F_0$ acts on $W_i$ for all $1 \leq i \leq 3$. Then by Lemma 7.2.5 we have $f = x_1x_2x_3$, where $x_i \in Z(H_i^\circ) \leq F_0^{(i)}$, and $|x_i| = |f|$ for all $1 \leq i \leq 3$. In particular $\omega \mid \omega_i$ for all $1 \leq i \leq 3$.

As noted in Remark 7.2.2, the following inequality holds:

$$\alpha \cdot \frac{\omega_1}{\alpha_1} \cdot \frac{\omega_2}{\alpha_2} \cdot \frac{\omega_3}{\alpha_3} \cdot n_1n_2n_3 \leq \frac{\omega}{\alpha} \cdot n^2. \tag{7.9}$$

Since $n = n_1+n_2+n_3$, we have $n_1 \leq n/3$. By Theorem 6.2.3, the fixed point space of every non-identity element of $G^\circ$ has dimension $\leq 3n/4$. Since elements of $H_1^\circ$ act trivially on $W_2 \perp W_3$, we must have $n_1 \geq n/4$.

Now $n_1 \leq n_2, n_3 \leq n - 2n_1$ and $n_2n_3 = n_2(n - n_1 - n_2)$, so it follows that

$$n_1(n - 2n_1) \leq n_2n_3 \leq \left(\frac{n - n_1}{2}\right)^2.$$

Because $n/4 \leq n_1 \leq n/3$, we have

$$\frac{n^2}{9} \leq n_1(n - 2n_1) \leq \frac{n^2}{8}.$$

Therefore $n^2 \leq 9n_2n_3$. Furthermore $\omega_2/\alpha_2, \omega_3/\alpha_3 \geq \omega/\alpha$ since $\omega \mid \omega_2, \omega_3$ (Lemma 7.2.7 (ii)). Thus (7.9) implies that

$$\omega \cdot \omega_1 \cdot \mu_1 \leq 9. \tag{7.10}$$

Since $\omega \mid \omega_1$ and $\omega, \omega_1 \geq 2$, the only possible solutions to the inequality (7.10) are the following:

$$\begin{cases} \omega = \omega_1 = 3 & \mu_1 = 1 \\ \omega = 2, \omega_1 = 4 & \mu_1 = 1 \\ \omega = \omega_1 = 2 & \mu_1 = 1 \\ \omega = \omega_1 = 2 & \mu_1 = 2 \end{cases} \tag{7.11}$$

We first consider the case where $\omega = \omega_1 = 3$ and $\mu_1 = 1$. Since $|f| = |x_1| = |f_1| = 3$ and $x_1$ centralizes $H_1^\circ$, it follows that $F_0^{(1)}$ is central in $H_1^\circ$. Because $\mu_1 = 1$, we have $C_{H_1^\circ}(F_0) = F_0$ (Lemma 2.9.3), so $H_1^\circ = F_0$. But then we must have $n_1 = 1$, which is a contradiction.

Therefore $\omega = 2$, in which case $\mu$ is a power of two. Furthermore either $G$ is of type $\mathcal{B}_3$ with $q$ odd, or $G$ is of type $\mathcal{B}_0$ with $q = 3$, $\nu = 1$. In both cases $n$ is a power of two, so by Theorem 6.2.3 every element of prime order $r > 2$ in $G$ has a fixed point space of dimension $\leq n/2$. Since every element of $H_1^\circ$ has a fixed point space of dimension $\geq n - n_1 > n/2$, it follows that $|H_1^\circ|$ is a power of two, and thus $\mu_1 = 1$ by Lemma 7.2.8.

If $\omega_1 = 2$, then either $H_1$ is of type $\mathcal{B}_3$ with $q$ odd, or $H_1$ is of type $\mathcal{B}_0$ with $q = 3$, $\nu_1 = 1$. Since $\mu_1 = 1$, in both cases $n_1 = 1$, contrary to what we have assumed.

Thus the only possibility that remains from (7.11) is $\omega = 2$, $\omega_1 = 4$, and $\mu_1 = 1$. In this case $G$ cannot be of type $\mathcal{B}_0$, since $\omega = q^\nu - 1$ implies $q = 3$ and $\omega_1 = q^{\nu_1} - 1$ implies $q = 5$. Therefore $\kappa \neq 0$, so $G$ is of type $\mathcal{B}_3$.

Since $\omega_1 = 4$, either $H_1$ is of type $\mathcal{B}_1$ with $q = 5$, $\nu_1 = 1$, or $H_1$ is of type $\mathcal{B}_2$ with $q = 3$, $\nu_1 = 1$. In both cases $\nu_1 = 1$, so $n_1 = 2\nu_1\mu_1 = 2$. Because $n/4 \leq n_1 \leq n/3$ and $n$ is a power of two, it follows that $n = 8$. Moreover $n_2, n_3$ are even with $n_2 + n_3 = 6$, so $n_2 = 2$ and $n_3 = 4$. The same argument as for $H_1$ shows that $\mu_2 = 1$, $\nu_2 = 1$, and $\omega_2 = 4$.

Because $H_1$ is of type $\mathcal{B}_1$ or $\mathcal{B}_2$ with $\mu_1 = 1$, we have $|H_1^\circ| = \omega_1 n_1 = 8$ (Table 2.2), and similarly $|H_2| = 8$. Thus $|H_1^\circ \times H_2^\circ \times H_3^\circ| = 2^6 |H_3^\circ|$.

We have $G = G_{8,1}^{\mathcal{B}_3}(X)$, where $X \leq O_6^{\varepsilon_1}(2)$ is metrically completely reducible maximal solvable. The possibilities for $X$ and $|X|$ are as follows, see Sect. 5.4 and Table 5.5.

| $X$ | $\lvert X\rvert$ | $\varepsilon_1$ |
|---|---|---|
| $O_4^+(2) \times O_2^-(2)$ | $2^4 \cdot 3^3$ | $-$ |
| $G_{1,2}^{\mathcal{B}_2} \times O_2^-(2)$ | $2^3 \cdot 3 \cdot 5$ | $+$ |
| $O_2^-(2) \wr S_3$ | $2^4 \cdot 3^4$ | $-$ |
| $G_{1,3}^{\mathcal{B}_1}$ | $2 \cdot 3 \cdot 7$ | $+$ |
| $G_{1,3}^{\mathcal{B}_2}$ | $2 \cdot 3^3$ | $-$ |
| $G_{3,1}^{\mathcal{B}_2}(\mathrm{GSp}_2(3))$ | $2^4 \cdot 3^4$ | $-$ |

Note that $X \not\leq \Omega_6^{\varepsilon_1}(2)$ in all cases. We have $q \in \{3, 5\}$, so 2 is not a square in $\mathbb{F}_q$. Thus by Lemma 4.3.27 (iii)

$$|G^\circ| = 2^7 \cdot |X^\Omega| = 2^6 \cdot |X|.$$

It follows that $|H_3^\circ|$ must divide $|X|$. We will now see that in most cases this is not possible.

If $H_3$ is of type $\mathcal{B}_1$ or $\mathcal{B}_2$, we have $4 = n_3 = 2\nu_2\mu_2$, so $\mu_2 \in \{1, 2\}$. If $\mu_2 = 1$, then $|H_3^\circ|$ is equal to $(5^2 \pm 1) \cdot 4$ or $(3^2 \pm 1) \cdot 4$ (Table 2.2). If $\mu_2 = 2$, then $|H_3^\circ|$ equals $(5 \pm 1) \cdot 2^2 \cdot 2 \cdot |O_2^-(2)|$ or $(3 + 1) \cdot 2^2 \cdot 2 \cdot |O_2^-(2)|$. In the case where $H_3$ is of type $\mathcal{B}_3$, since 2 is not a square in $\mathbb{F}_q$, we have $|H_3^\circ| = 2^5 \cdot 72/2$ or $|H_3^\circ| = 2^5 \cdot 20/2$. Therefore the possibilities for $|H_3^\circ|$ are as follows:

| | | | | |
|---|---|---|---|---|
| Type $\mathcal{B}_1$ or $\mathcal{B}_2$, $q = 5$: | $2^5 \cdot 3$ | $2^3 \cdot 13$ | $2^6 \cdot 3$ | $2^5 \cdot 3^2$ |
| Type $\mathcal{B}_1$ or $\mathcal{B}_2$, $q = 3$: | $2^5$ | $2^3 \cdot 5$ | $2^5 \cdot 3$ | $2^6 \cdot 3$ |
| $\mathcal{B}_3$: | $2^7 \cdot 3^2$ | $2^6 \cdot 5$ | | |

For $|H_3^\circ|$ to divide $|X|$, the only possibility is that $|H_3^\circ| = 2^3{\cdot}5$, and $|X| = 2^3{\cdot}3{\cdot}5$, with $X = G^{\mathcal{B}_2}_{1,2} \times O_2^-(2) < O_6^+(2)$. In this case $G^\circ$ is a subgroup of $2_+^{1+6}.O_6^+(2)$, and the image of $G^\circ$ in $O_6^+(2)$ is equal to $X^\Omega$. Here $2_+^{1+6} = A = \mathrm{Fit}(G^\circ)$.

Since $H_3^\circ$ is centralized by $H_1^\circ \times H_2^\circ$, it follows that there is an element of order 5 in $G^\circ$ with centralizer of order divisible by $2^6$. We will see that this is not possible. If $x \in G^\circ$ has order 5, then its image $x'$ in $O_6^+(2)$ is contained in $X$. The centralizer of $x'$ in $X$ is equal to $\langle x' \rangle \times O_2^-(2)$. The action of $x'$ on the 4-dimensional summand has no nonzero fixed points (Lemma 2.4.3), so the fixed point space of $x'$ has dimension 2. It follows that $A \cap C_{G^\circ}(x)$ has order $\leq 2^3$, so $|C_{G^\circ}(x)|$ divides $2^3 \cdot |\langle x' \rangle \times O_2^-(2)| = 2^4 \cdot 3 \cdot 5$. In particular $|C_{G^\circ}(x)|$ is not divisible by $2^6$, so we have a contradiction. □

**Lemma 7.2.12** *Assume that $n$ and $q$ are even, and that $\kappa$ is a quadratic form. Let $L = (H_1 \times H_2 \times H_3) \cap \Delta(V, \kappa)$ as in Example 7.2.1, with $t = 3$. Then $L^\Omega \not\leq G^\Omega$.*

***Proof*** Suppose for the sake of contradiction that $L^\Omega \leq G^\Omega$. It follows from Lemma 7.2.4 that $F_0$ acts on $W_i$ for all $1 \leq i \leq 3$. Furthermore, by Lemma 7.2.6 we have $\omega \mid \omega_i$ for all $1 \leq i \leq 3$, and $f = x_1 x_2 x_3$ with $x_i \in Z(H_i^\Omega) \leq F_0^{(i)}$.

Since $q$ is even, we have $n_i$ even for all $1 \leq i \leq t$. As in the proof of Lemma 7.2.11, we can assume that $1 < n_1 \leq n_2 \leq n_3$, in which case it follows from Theorem 6.2.3 that $n/4 \leq n_1 \leq n/3$. Arguing as in the proof of Lemma 7.2.11 shows that the inequality (7.9) holds and implies

$$\omega \cdot \omega_1 \cdot \mu_1 \leq 9. \tag{7.12}$$

We have $\omega \mid \omega_1$ and $\omega, \omega_1 > 1$ are odd since $q$ is even, so the only possibility is that $\omega = \omega_1 = 3$ and $\mu_1 = 1$.

In this case one of the following holds:

- $H$ and $H_1$ are of type $\mathcal{B}_1$ with $q = 4$ and $\nu = \nu_1 = 1$.
- $H$ and $H_1$ are of type $\mathcal{B}_2$ with $q = 2$ and $\nu = \nu_1 = 1$.

In both cases $\nu_1 = 1$, so $n_1 = 2\nu_1\mu_1 = 2$. Since $n/4 \leq n_1 \leq n/3$, we have $6 \leq n \leq 8$. Since $n$ is even, we have $n = 6$ or $n = 8$. Here $n = 2\nu\mu$ is not possible, since then $\mu = 4$, contradicting the fact that $\omega = 3$ is odd.

Therefore $n = 6$, in which case $n_1 = n_2 = n_3 = 2$. Furthermore, either $H_1 = H_2 = H_3 = \mathrm{GO}_2^+(4)$ or $H_1 = H_2 = H_3 = \mathrm{GO}_2^-(2)$. In both cases we have $H_i^\circ \not\leq \Omega(W_i, \kappa)$ for all $1 \leq i \leq 3$.

We will next prove that $x_1 \in Z(H_1^\circ)$. We know that $x_1 \in Z(H_1^\Omega)$, so it will suffice to check that $x_1$ centralizes any $g_1 \in H_1^\circ \setminus H_1^\Omega$. Let $g_2 \in H_2^\circ \setminus H_2^\Omega$ and denote $g = g_1 g_2$. Then $g \in L^\Omega$ and $g$ normalizes $F_0$, so there exists $d > 0$ such that $gfg^{-1} = f^d = x_1^d x_2^d x_3^d$. Then

$$gfg^{-1} = (g_1 x_1 g_1^{-1})(g_2 x_2 g_2^{-1}) x_3,$$

so $x_3^d = x_3$. Because $|x_1| = |x_2| = |x_3| = 3$, it follows that $g_1 x_1 g_1^{-1} = x_1^d = x_1$. Therefore we conclude that $x_1 \in Z(H_1^\circ)$.

We have $|x_1| = \omega_1 = 3$ and $x_1 \in Z(H_1^\circ)$, so $F_0^{(1)} = \langle x_1 \rangle$ is central in $H_1^\circ$. Because $\mu_1 = 1$, we have $C_{H_1^\circ}(F_0) = F_0$ (Lemma 2.9.3), so $H_1^\circ = F_0^{(1)}$. But $H_1^\circ$ is not cyclic when $H_1 = \mathrm{GO}_2^-(2)$ or $H_1 = \mathrm{GO}_2^+(4)$, so we have a contradiction. □

**Lemma 7.2.13** *Let $L = H_1^\circ \times H_2^\circ \times H_3^\circ$ be as in Example 7.2.1 with $t = 3$, and suppose that $L^\circ \leq G^\circ$. Then one of the following holds:*

*(i)* $n = 3$, $\kappa = 0$, $q = 4$, *and* $L = \mathrm{GL}_1(4) \times \mathrm{GL}_1(4) \times \mathrm{GL}_1(4)$.
*(ii)* $n = 4$, $\kappa \neq 0$ *is a quadratic form,* $q = 3$, *and* $L = O_1(3) \times O_1(3) \times O_2^-(3)$.
*(iii)* $n = 4$, $\kappa \neq 0$ *is a quadratic form,* $q = 5$, *and* $L = O_1(5) \times O_1(5) \times O_2^+(5)$.

***Proof*** It follows from Lemma 7.2.13 that $n_i = 1$ for some $i$, say $n_1 = 1$ and assume that $n_1 \leq n_2 \leq n_3$. As in the beginning of the proof of Lemma 7.2.13, it follows from Theorem 6.2.3 that $n_1 \geq n/4$, so $n = 3$ or $n = 4$. We consider these two possibilities.

**Case 1:** $n = 3$
If $\kappa \neq 0$, there are no metrically primitive groups in odd dimension (Proposition 4.1.20). Therefore we must have $\kappa = 0$ in this case, with $n_1 = n_2 = n_3 = 1$ and $H_1 = H_2 = H_3 = \mathrm{GL}_1(q)$, where $q > 2$.

We have $\mu \in \{1, 3\}$. Suppose first that $\mu = 1$. It is straightforward to check that $|G| = 3 \cdot (q^3 - 1)$ is divisible by $|H_1 \times H_2 \times H_3| = (q-1)^3$ only if $q = 2, 4$, so $q = 4$ and (i) holds.

In the case where $\mu = 3$, it follows from Corollary 6.3.9 that

$$(q-1)^3 \leq \omega \cdot \mu^2 = 9(q-1).$$

This inequality only holds for $q = 3$ and $q = 4$. However $\mu = 3$ implies that $\omega = q - 1$ is divisible by 3, so $q = 4$ and (i) holds. This completes the proof of the lemma in this case.

**Case 2:** $n = 4$
In this case, we must have $n_1 = n_2 = 1$ and $n_3 = 2$. It follows from Lemma 7.2.3 that $F_0$ acts on $W_3$ and $W_1 \perp W_2$. We can write $f = x_1 x_2$, where $x_1$ is the action of $f$ on $W_1 \perp W_2$, and $x_2$ is the action of $f$ on $W_3$. Arguing as in the proof of Lemma 7.2.5, we find that $|x_1| = |x_2| = |f|$ and $x_2 \in Z(H_3^\circ) \leq F_0^{(3)}$. In particular $|f| = \omega$ divides $|f_3| = \omega_3$.

We now consider the two different possibilities for $\kappa$.

**Case 2.1:** $\kappa = 0$
We have $L = \mathrm{GL}_1(q) \times \mathrm{GL}_1(q) \times X$, where $X$ is maximal primitive irreducible solvable in $\mathrm{GL}_2(q)$ and $q > 2$. In this case either $\omega_3 = q^2 - 1$ and $|X| = 2(q^2 - 1)$, or $\omega_3 = q - 1$ and $|X| = 24(q - 1)$ (Table 5.1). Because $\omega = q^\nu - 1$ must divide $\omega_3$, we have $\nu \in \{1, 2\}$. If $\nu = 2$, then $\omega = \omega_3 = q^2 - 1$. By Corollary 6.3.9 we have

$$(q-1)^2 \cdot (q^2 - 1) \leq (q^2 - 1) \cdot 2^2,$$

and this inequality can only hold for $q \leq 3$. Therefore $q = 3$, but then $H_3$ is not maximal solvable. This follows from Lemma 5.5.9, but is also obvious since $\mathrm{GL}_2(3)$ is solvable and $H_3 = \Gamma\mathrm{L}_1(3^2) \lneqq \mathrm{GL}_2(3)$.

Thus $\nu = 1$, so $\mu = 4$. Note that because prime factors of $\mu$ divide $\omega$ (Proposition 4.1.18), it follows that $q$ is odd. If $\nu_3 = 2$, by Corollary 6.3.9

$$(q-1)^2 \cdot (q^2 - 1) \leq (q - 1) \cdot 4^2,$$

which implies $q \leq 3$. Hence $q = 3$, which as before is not applicable, since $H_3$ is not maximal solvable in $\mathrm{GL}_2(3)$.

Thus we must have $\nu_3 = 1$, in which case Corollary 6.3.9 implies that

$$(q-1)^3 \leq (q-1) \cdot 4^2,$$

which implies $q \leq 5$. Because $q$ is odd, the only possibilities are $q = 3$ and $q = 5$.

Suppose that $q = 3$. In this case $H_3 = \mathrm{GL}_2(3)$ contains an element $x$ of order 3, which is unipotent and thus has a fixed point space of dimension 1 on $W_3$. But then $x$ has a fixed point space of dimension 3 on $V$, which contradicts Theorem 6.2.3.

It remains to consider $q = 5$. In this case $|L| = 4 \cdot 4 \cdot |H_3| = 2^9 \cdot 3$. Since $|O_4^-(2)|$ is not divisible by 9, it follows that $G = (\mathbb{F}_5^\times \circ 2_+^{1+4}).O_4^+(2)$, in which case $|G| = 4 \cdot 4^2 \cdot 72 = 2^9 \cdot 3^2$. Hence if $L \leq G$, a Sylow 2-subgroup of $L$ is a Sylow 2-subgroup of $G$. Now $G$ contains an absolutely irreducible subgroup of the form $2_+^{1+4}$, so in particular a Sylow 2-subgroup of $G$ is irreducible. Clearly a Sylow 2-subgroup of $L$ is not irreducible, so we have a contradiction.

**Case 2.2:** $\kappa \neq 0$
Since $n_1 = 1$, in this case $\kappa$ must be a quadratic form, with $q$ odd and $H_1 = H_2 = \mathrm{GO}_1(q)$. Orthogonal groups in dimension two are solvable, so $H_3 = \mathrm{GO}_2^{\varepsilon'}(q)$. Then $L^\circ = O_1(q) \times O_1(q) \times O_2^{\varepsilon'}(q)$, with $\varepsilon = (-1)^{(q-1)/2}\varepsilon'$. In this case $H_3$ is of type $\mathcal{B}_1$ if $\varepsilon' = +$, and of type $\mathcal{B}_2$ if $\varepsilon' = -$. In both cases $\omega_3 = q - \varepsilon'$.

We consider first the case where $G$ is of type $\mathcal{B}_1$ of $\mathcal{B}_2$. Then $\omega = q^\nu \pm 1$, with $\nu \in \{1, 2\}$. Since $q > 2$ and $\omega$ divides $\omega_3 = q - \varepsilon'$, it follows that $\nu = 1$. If $\omega \neq \omega_3$, then $q = 3$, with $\omega = q - 1$ and $\omega_3 = q + 1$. In this case, statement (ii) of the lemma holds. Therefore we must have $\omega = \omega_3$. Now $f = x_1x_2$ with $|f| = |x_2|$, so we conclude that $\langle x_2 \rangle = F_0^{(3)}$. Since $x_2 \in Z(H_3^\circ)$, it follows that $F_0^{(3)}$ is central in

$H_3^\circ$. However, this is not the case. We have $Z(O_2^{\varepsilon'}(q)) = \{\pm I_2\}$ if $(q, \varepsilon') \neq (3, +)$, and $(q, \varepsilon') = (3, +)$ is not applicable since we require $q^{\nu_3} > 3$ for groups of type $\mathcal{B}_1$.

Therefore $G$ must be of type $\mathcal{B}_3$, in which case $\omega = 2$ and $\mu = 4$. Note that $\varepsilon = +$, since $n > 2$ (Sect. 4.3.4).

It follows from Corollary 6.3.9 that

$$2 \cdot 2 \cdot (q - \varepsilon') \leq \omega\mu^2 = 2^5,$$

so $q \leq 9$ if $\varepsilon' = +$, and $q \leq 7$ if $\varepsilon' = -$.

Since $\varepsilon' = (-1)^{(q-1)/2}\varepsilon = (-1)^{(q-1)/2}$, for $q = 3$ and $q = 5$ statements (ii) and (iii) of the lemma hold, respectively. Furthermore $q$ is odd, so it remains to rule out $q = 7$ with $\varepsilon' = -$, and $q = 9$ with $\varepsilon' = +$. Note that in both cases $\omega_3 = 8$. Thus $f_3$ is an element of order 8 which has a fixed point space of dimension 2 on $V$, and centralizer of order $\geq |C_L(f_3)| = 2 \cdot 2 \cdot 8 = 2^5$. However, we will see that there is no element of order 8 in $G^\circ$ with these properties, so this is a contradiction.

Now 2 is a square in $\mathbb{F}_q$ for $q \in \{7, 9\}$, so $G^\circ = 2_+^{1+4}.O_4^+(2)$ (Lemma 4.3.27). Let $\overline{f_3}$ be the image of $f_3$ in $O_4^+(2)$. Then $\overline{f_3}$ is a unipotent matrix, so $|\overline{f_3}| \leq 4$. Furthermore $2_+^{1+4}$ has exponent 4, so $|\overline{f_3}| > 1$ and thus $|\overline{f_3}| \in \{2, 4\}$. If $|\overline{f_3}| = 2$, then $f_3^2$ is an element of order 4 in $2_+^{1+4}$ with a fixed point space of dimension $\geq 2$, contradicting Lemma 6.1.1 (ii).

Therefore $|\overline{f_3}| = 4$. A straightforward calculation (which we omit) shows that $O_4^+(2) = O_2^-(2) \wr S_2 \cong S_3 \wr S_2$ contains a unique conjugacy class of elements of order 4, and that $C_{O_4^+(2)}(x) = \langle x \rangle$ for every $x \in O_4^+(2)$ with $|x| = 4$. This can also be deduced from more general results such as [60, Theorem 4.2, Lemma 6.13, Theorem 7.3].

In any case, it follows that $C_{G^\circ}(f_3)$ is contained in $C_A(f_3).\langle\overline{f_3}\rangle$, where $A = 2_+^{1+4}$. On the other hand since $|\overline{f_3}| = 4$, the action of $\overline{f_3}$ on $\mathbb{F}_2^4$ consists of a single unipotent Jordan block, and thus the fixed point space of $\overline{f_3}$ has dimension 1. This implies that $|C_A(f_3)| \leq 4$, so we conclude that $|C_{G^\circ}(f_3)| \leq 4 \cdot 4 = 2^4$, contradicting the fact that $|C_{G^\circ}(f_3)| \geq |C_L(f_3)| = 2^5$. This completes the proof of the lemma. □

**Lemma 7.2.14** *Let $L^\circ = H_1^\circ \times H_2^\circ$ be as in Example 7.2.1 with $t = 2$, and suppose that $L^\circ \leq G^\circ$. Then all of the following statements hold:*

*(i) $n$ is a power of 2 and $q$ is odd;*
*(ii) $n_1 = n_2 = n/2$;*
*(iii) If $G$ is of type $\mathcal{B}_0$, then one of the following holds:*

*(a) $q = 5$, $\nu = \nu_1 = \nu_2 = 1$.*
*(b) $q = 3$, $\nu = 1$, and $\nu_i \in \{1, 2\}$.*

*(iv) If $G$ is of type $\mathcal{B}_1$, then $H_1$ and $H_2$ are of type $\mathcal{B}_1$, with $q^\nu = q^{\nu_1} = q^{\nu_2} = 5$;*
*(v) If $G$ is of type $\mathcal{B}_2$, then $H_2$ and $H_2$ are of type $\mathcal{B}_2$, with $q^\nu = q^{\nu_1} = q^{\nu_2} = 3$;*

(vi) *If $G$ is of type $\mathcal{B}_3$, then one of the following holds, up to interchanging $H_1$ and $H_2$:*

(a) *$H_1$ is of type $\mathcal{B}_1$, $H_2$ is of type $\mathcal{B}_1$, with $q^{\nu_1} = q^{\nu_2} = 5$;*
(b) *$H_1$ is of type $\mathcal{B}_1$, $H_2$ is of type $\mathcal{B}_2$, with $q^{\nu_1} = q^{\nu_2} = 5$;*
(c) *$H_1$ is of type $\mathcal{B}_2$, $H_2$ is of type $\mathcal{B}_2$, with $q^{\nu_1}, q^{\nu_2} \in \{3, 3^2\}$;*
(d) *$H_1$ is of type $\mathcal{B}_1$, $H_2$ is of type $\mathcal{B}_3$, with $q^{\nu_1} \in \{5, 7, 9\}$;*
(e) *$H_1$ is of type $\mathcal{B}_2$, $H_2$ is of type $\mathcal{B}_3$, with $q^{\nu_1} \in \{3, 5, 7, 9\}$;*
(f) *$H_1$ and $H_2$ are of type $\mathcal{B}_3$.*

***Proof*** We first consider $n_1 = n_2 = 1$, in which case $n = 2$.

If $\kappa \neq 0$, then $\kappa$ must be a quadratic form with $q$ odd, and $H_1^\circ = H_2^\circ = O_1(q)$. Thus (i), (ii) and (vi)(f) hold in this case, as required.

Suppose then that $\kappa = 0$, in which case $L = L^\circ = \mathrm{GL}_1(q) \times \mathrm{GL}_1(q)$. We have $n = \mu\nu = 2$, so $\mu \in \{1, 2\}$. If $\mu = 1$, then $|L| = (q-1)^2$ divides $|G| = 2(q^2 - 1)$ (Table 2.2), which implies $q \in \{2, 3, 5\}$. Since $q > 2$, we have $q \in \{3, 5\}$, so (i)–(iii) hold. If $\mu = 2$, then 2 divides $\omega = q - 1$, so $q$ is odd. Furthermore, by Corollary 6.3.9 we have

$$|L| = (q-1)^2 \leq \omega\mu^2 = 4(q-1),$$

so $q \leq 5$. Hence $q \in \{3, 5\}$ and statements (i)–(iii) of the lemma hold.

Thus for the rest of the proof we will assume that $n_1 > 1$ or $n_2 > 1$. In this case, it follows from Lemma 7.2.3 that $F_0$ acts on $W_1$ and $W_2$. By Lemma 7.2.5 we have $f = x_1x_2$, where $x_i \in Z(H_i^\circ) \leq F_0^{(i)}$ and $|x_i| = |f|$ for all $1 \leq i \leq 2$. In particular $\omega \mid \omega_i$, so $\omega_i/\alpha_i \geq \omega/\alpha$ for all $1 \leq i \leq 2$ (Lemma 7.2.7 (ii)).

As noted in Remark 7.2.2, the following inequality holds:

$$\alpha \cdot \frac{\omega_1}{\alpha_1} \cdot \frac{\omega_2}{\alpha_2} \cdot n_1n_2 \leq \frac{\omega}{\alpha} \cdot n^2. \tag{7.13}$$

We have $n = n_1 + n_2$. By Theorem 6.2.3, the fixed point space of every non-identity element of $G^\circ$ has dimension $\leq 3n/4$. Since elements of $H_i^\circ$ act trivially on the other summand of dimension $n - n_i$, we must have $n_i \geq n/4$. Therefore

$$\frac{n}{4} \leq n_i \leq \frac{3}{4}n$$

for all $1 \leq i \leq 2$. Because $n_1n_2 = n_1(n - n_1)$, we conclude that

$$n_1n_2 \geq \frac{3}{16}n^2.$$

Combining this with (7.13) and the fact that $\omega_1/\alpha_1, \omega_2/\alpha_2 \geq \omega/\alpha$ (Lemma 7.2.7 (ii)) gives

$$\alpha \cdot \frac{\omega_i}{\alpha_i} \leq \frac{16}{3} \tag{7.14}$$

for all $1 \leq i \leq 2$.

We will now consider the different types of $G$ case-by-case, and use (7.14) to rule out all configurations except for those listed in the lemma. Note that by Lemma 7.2.7 (i), the types of $G$ and $H_i$ and the values of $\omega$ and $\omega_i$ must be as in Table 7.2.

**Case 1: $G$ is of type $\mathcal{B}_0$**
Note that in this case $q^\nu > 2$. Since $\omega \mid \omega_i$, we have $\omega = q^\nu - 1$ and $\omega_i = q^{m_i \nu} - 1$ for some integers $m_i, \nu \geq 1$ (Table 7.2). Then (7.14) becomes

$$\frac{q^{m_i \nu} - 1}{m_i} \leq \frac{16}{3}. \tag{7.15}$$

It is readily seen that for $q^\nu \geq 7$ the inequality (7.15) cannot hold. Thus $2 < q^\nu < 7$, in which case the only solutions to (7.15) are the following:

- $q = 2$ and $\nu = \nu_i = 2$;
- $q = 3$, $\nu = 1$ and $\nu_i \in \{1, 2\}$;
- $q = 4$ and $\nu = \nu_i = 1$;
- $q = 5$ and $\nu = \nu_i = 1$;

Consider first the case where $q = 2$ and $\nu = \nu_i = 2$. Here $\omega = \omega_1 = \omega_2 = 3$, so $\mu$ and $\mu_i$ must be equal to a power of 3, say $\mu = 3^\rho$ and $\mu_i = 3^{\rho_i}$. Then $n = \nu\mu = 2 \cdot 3^\rho$ and $n_i = \nu_i \mu_i = 2 \cdot 3^{\rho_i}$. But now $n = n_1 + n_2$ implies $3^\rho = 3^{\rho_1} + 3^{\rho_2}$, which is impossible. Similarly when $q = 4$ and $\nu = \nu_i = 1$, we find that $n = 3^\rho$ and $n_i = 3^{\rho_i}$, which is not possible when $n = n_1 + n_2$.

When $q = 3$, the case $\nu_i = 2$ is ruled out, since by Lemma 5.5.9 the group $H_i$ would not be maximal solvable. Thus we are left with the following cases:

- $q = 3$ and $\nu = \nu_1 = \nu_2 = 1$.
- $q = 5$ and $\nu = \nu_1 = \nu_2 = 1$.

In both cases $\omega = \omega_i = q^\nu - 1$ is a power of 2. Thus $n = \mu$ and $n_i = \mu_i$ must be powers of 2, which combined with $n = n_1 + n_2$ implies that $n_1 = n_2 = n/2$. We have shown that claims (i)–(iii) of the lemma hold, which completes the proof for groups of type $\mathcal{B}_0$.

**Case 2: $G$ is of type $\mathcal{B}_1$**
Since $\omega \mid \omega_i$, in this case both $H_1$ and $H_2$ must be of type $\mathcal{B}_1$ (Table 7.2), with $\omega = q^\nu - 1$ and $\omega_i = q^{m_i \nu} - 1$ for some integers $m_i, \nu \geq 1$.

Then inequality (7.14) becomes

$$\frac{q^{m_i \nu} - 1}{m_i} \leq \frac{16}{3}. \tag{7.16}$$

as in Case 1. Because $q^\nu > 4$ or $(q, \nu) = (4, 1)$ for groups of type $\mathcal{B}_1$, the only possible solutions to (7.16) are the following:

- $q = 4$ and $\nu = \nu_i = 1$.
- $q = 5$ and $\nu = \nu_i = 1$.

Thus if $q = 4$, we have $\nu = \nu_1 = \nu_2 = 1$ and $\omega = \omega_1 = \omega_2 = 3$. As in Case 1, we see that $n = 2 \cdot 3^\rho$ and $n_i = 2 \cdot 3^{\rho_i}$. By $n = n_1 + n_2$ we have $3^\rho = 3^{\rho_1} + 3^{\rho_2}$, which is not possible.

Therefore we must have $q = 5$ and $\nu = \nu_1 = \nu_2 = 1$, in which case $\omega = \omega_1 = \omega_2 = 4$. Then $\mu$ and $\mu_i$ must be powers of 2, so $n = 2\nu\mu = 2\mu$ and $n_i = 2\nu_i\mu_i = 2\mu_i$ are powers of 2. Since $n = n_1 + n_2$, it follows that $n_1 = n_2 = n/2$. We have proved that claims (i), (ii), and (iv) of the lemma hold, which completes the proof in this case.

**Case 3: $G$ is of type $\mathcal{B}_2$**

When $G$ is of type $\mathcal{B}_2$, it follows from $\omega \mid \omega_i$ that $H_i$ is of type $\mathcal{B}_1$ or $\mathcal{B}_2$ (Table 7.2).

We will first narrow down the possible values of $q, \nu, \nu_i$ when $H_i$ is of type $\mathcal{B}_1$. In this case $\omega = q^\nu + 1$ and $\omega_i = q^{2m_i\nu} - 1$ for some integers $m_i, \nu \geq 1$ (Table 7.2). Then (7.14) becomes

$$\frac{q^{2m_i \nu} - 1}{2m_i} \leq \frac{16}{3}. \tag{7.17}$$

The inequality (7.17) has no solutions for $q^\nu \geq 7$, and for $2 \leq q^\nu < 7$ the only solutions arc thc following:

- $q = 2$, $\nu = 1$, and $\nu_i = 2$.
- $q = 2$, $\nu = 1$, and $\nu_i = 4$.
- $q = 3$, $\nu = 1$, and $\nu_i = 2$.

Here $q = 2$ and $\nu_i = 2$ is not possible, since $H_i$ is of type $\mathcal{B}_1$. Furthermore $q = 3$ and $\nu_i = 2$ is excluded, since in this case $H_i$ is metrically imprimitive (Lemma 5.6.5). Therefore if $H_i$ is of type $\mathcal{B}_1$, we must have $q = 2$, $\nu = 1$, and $\nu_i = 4$.

Next we will narrow down the possibilities for $q, \nu, \nu_i$ when $H_i$ is of type $\mathcal{B}_2$. In this case $\omega = q^\nu + 1$ and $\omega_i = q^{(2m_i-1)\nu} + 1$ for some integers $m_i, \nu \geq 1$ (Table 7.2). Thus (7.14) becomes

$$\frac{q^{(2m_i-1)\nu} + 1}{2m_i - 1} \leq \frac{16}{3}. \tag{7.18}$$

Among the solutions to the inequality (7.18), the solution $q = 2$, $\nu = 1$, $\nu_i = 3$ must be excluded, since then $H_i$ would be metrically imprimitive (Lemma 5.6.6). The remaining solutions to the inequality (7.18) are the following:

- $q = 2$ and $\nu = \nu_i = 1$.
- $q = 2$ and $\nu = \nu_i = 2$.
- $q = 3$ and $\nu = \nu_i = 1$.
- $q = 4$ and $\nu = \nu_i = 1$.

We will now consider the two possible cases: when $H_1$ or $H_2$ is of $\mathcal{B}_1$, and when they are both of type $\mathcal{B}_2$.

**Case 3.1: $H_1$ or $H_2$ is of type $\mathcal{B}_1$**
Suppose, without loss of generality, that $H_1$ is of type $\mathcal{B}_1$. As we have already seen, in this case we must have $q = 2$, $\nu = 1$, and $\nu_1 = 4$. Then $\omega = 3$, so $\mu = 3^\rho$, and $n = 2\nu\mu = 2 \cdot 3^\rho$.

We have $\omega_1 = 2^4 - 1 = 15$, so the only possible prime factors of $\mu_1$ are 3 and 5, say $\mu_1 = 3^{\rho_1} \cdot 5^{\sigma_1}$. Then $n_1 = 2\nu_1\mu_1 = 2^3 \cdot 3^{\rho_1} \cdot 5^{\sigma_1}$. In particular $n_1$ is a multiple of 8. Hence $H_2$ cannot be of type $\mathcal{B}_1$, as otherwise $n_2$ would also be a multiple of 8, but $n = n_1 + n_2 = 2 \cdot 3^\rho$ is not.

Therefore $H_2$ must be of type $\mathcal{B}_2$. Because $q = 2$ and $\nu = 1$, from the possibilities for $H_2$ we have narrowed down earlier, we must have $\nu_2 = 1$. Then $\omega_2 = 3$, so $\mu_2 = 3^{\rho_2}$ and $n_2 = 2\nu_2\mu_2 = 2 \cdot 3^{\rho_2}$. In particular $n_1 \neq n_2$, so either $n_1 < n/2$ or $n_2 < n/2$.

If $n_1 < n/2$, then since $H_1^\circ$ acts trivially on $W_2$, every element of $H_1^\circ$ has its fixed point space of dimension $> n/2$. But then $f_1^3$ is an element of order 5 with a fixed point space of dimension $> n/2$, contradicting Theorem 6.2.3.

Therefore $n_2 < n/2 < n_1$. Consider first the case where $\mu_2 > 1$. Let $R \trianglelefteq H_2^\circ$ be the extraspecial 3-subgroup $R = 3_+^{1+2\rho_2}$ used in the construction of $H_2^\circ$. Then $x \in R \setminus Z(R)$ is an element of order 3 with a fixed point space of dimension $n_2/3$ on $W_1$ (Lemma 6.1.1 (iii)), and thus a fixed point space of dimension $n_2/3 + n_1$ on $V$. But $n_2/3 + n_1 = n/3 + 2n_1/3 > 2n/3$, so we have a contradiction by Theorem 6.2.3.

Thus $\mu_2 = 1$. In this case $n_2 = 2\nu_2\nu_2 = 2$. Since $n/4 \leq n_2 < n/2$ and $n = 2 \cdot 3^\rho$, we must have $n = 6$. But then $n_1 = n - n_2 = 6$ is not a multiple of 8, so we have a contradiction. Thus it is not possible that $H_1$ or $H_2$ is of type $\mathcal{B}_1$.

**Case 3.2: $H_1$ and $H_2$ are both of type $\mathcal{B}_2$**
Suppose first that $q = 2$ and $\nu = 1$. Then $\nu_1 = \nu_2 = 1$, so $\omega = \omega_1 = \omega_2 = 3$. Similarly to Case 1 and Case 2, we see that $n = 2 \cdot 3^\rho$ and $n_i = 2 \cdot 3^{\rho_i}$; this implies $3^\rho = 3^{\rho_1} + 3^{\rho_2}$ which is not possible.

If $q = 2$ and $\nu = 2$, then $\nu_1 = \nu_2 = 2$, so $\omega = \omega_1 = \omega_2 = 5$. Then $n = 4 \cdot 5^\rho$ and $n_i = 4 \cdot 5^{\rho_i}$, implying $5^\rho = 5^{\rho_1} + 5^{\rho_2}$ which is not possible. Similarly when $q = 4$ and $\nu = 1$, we have $\nu = \nu_1 = \nu_2 = 1$, so $\omega = \omega_1 = \omega_2 = 5$ which again leads to the contradiction $5^\rho = 5^{\rho_1} + 5^{\rho_2}$.

Therefore the only possibility is that $q = 3$ and $\nu = 1$, in which case $\nu_1 = \nu_2 = 1$. Here $\omega = \omega_1 = \omega_2 = 4$, so $n$, $n_1$, and $n_2$ are powers of two. Then $n = n_1 + n_2$

implies that $n_1 = n_2 = n/2$. We conclude that statements (i), (ii), and (v) of the lemma hold, as desired.

**Case 4:** ***G* is of type $\mathcal{B}_3$**
In this case $n$ is a power of 2, and $q$ must be odd, so (i) holds. We split the proof into two cases.

**Case 4.1:** $n_1 < n/2$ **or** $n_2 < n/2$
Suppose without loss of generality that $n_1 < n/2$. Then $n_2 > n/2$, so every element of $H_1^\circ$ has a fixed point space of dimension $> n/2$ on $V$. Because $n$ is not divisible by 3, it follows from Theorem 6.2.3 that $|H_1^\circ|$ is a power of two. Therefore $\mu_1 = 1$ and $n_1$ is a power of two by Lemma 7.2.8. Furthermore, by Lemma 7.2.8 either $\nu_1 = 1$, or $(q, \nu_1) = (3, 2)$ and $H_1$ is of type $\mathcal{B}_1$. However in the latter case $H_1$ would be metrically imprimitive (Lemma 5.6.5), so we must have $\nu_1 = 1$.

Since $n$ and $n_1$ are both powers of two and $n/4 \leq n_1 < n/2$, it follows that $n_1 = n/4$. Then $n_2 = 3n/4$, and in particular $n_2$ is not a power of 2. Therefore $H_2$ cannot be of type $\mathcal{B}_3$. Thus $H_2$ is of type $\mathcal{B}_1$ or $\mathcal{B}_2$, with $\omega_2 = q^{\nu_2} \pm 1$, $\alpha_2 = 2\nu_2$, and $n_2 = 2\nu_2\mu_2$.

If $H_1$ is of type $\mathcal{B}_3$, it follows from $\mu_1 = 1$ that $n_1 = 1$. But then $n_2 = 3$, contradicting the fact that $n_2$ is even. Therefore $H_1$ is of type $\mathcal{B}_1$ or $\mathcal{B}_2$, and it follows from $\nu_1 = \mu_1 = 1$ that $\omega_1 = q \pm 1$, $\alpha_1 = 2$, and $n_1 = 2$. On the other hand $n_1 = n/4$, so $n = 8$ and $n_2 = 6$. Because $n_2 = 2\nu_2\mu_2$, we have $\nu_2 \in \{1, 3\}$.

Plugging the values into the inequality (7.13) gives

$$\frac{\omega_1}{2} \cdot \frac{\omega_2}{2\nu_2} = \frac{q \pm 1}{2} \cdot \frac{q^{\nu_2} \pm 1}{2\nu_2} \leq \frac{32}{3}. \tag{7.19}$$

For $q > 7$ there are no solutions to the inequality (7.19). For $q \leq 7$, the possible solutions are the following:

- $q = 3$, $\omega_1 = 3 \pm 1$, and $\omega_2 = 3 \pm 1$.
- $q = 3$, $\omega_1 = 3 \pm 1$, and $\omega_2 = 3^3 \pm 1$.
- $q = 5$, $\omega_1 = 5 \pm 1$, and $\omega_2 = 5 \pm 1$.
- $q = 7$, $\omega_1 = 7 - 1$, and $\omega_2 = 7 - 1$.

Most of these solutions can be ruled out. We cannot have $\omega_1 = 3 - 1$, since $(q, \nu_1) = (3, 1)$ is excluded from the definition of groups of type $\mathcal{B}_1$. Furthermore, we know that $\omega_1$ must be a power of two. For the solutions with $\nu_2 = 1$ we have $\mu_2 = 3$, so $\omega_2$ must be a multiple of 3 when $\nu_2 = 1$. Taking these restrictions into consideration, we are left with the following solutions to the inequality (7.19).

- $q = 3$, $\omega_1 = 3 + 1$, and $\omega_2 = 3^3 + 1$.
- $q = 3$, $\omega_1 = 3 + 1$, and $\omega_2 = 3^3 - 1$.
- $q = 5$, $\omega_1 = 5 - 1$, and $\omega_2 = 5 + 1$.

Before considering these cases, note that since $G$ is of type $\mathcal{B}_3$, we have $G^\circ \leq 2_{\pm}^{1+6}.O_6^{\pm}(2)$.

Consider the case where $q = 3$, $\omega_1 = 3 + 1$, and $\omega_2 = 3^3 + 1 = 2^2 \cdot 7$. Since $7 \nmid |O_6^-(2)|$, we have $G^\circ \leq 2_+^{1+6}.O_6^+(2)$. The image of $f_2$ in $O_6^+(2)$ must have order divisible by 7. In $O_6^+(2)$ an element of order 7 generates a subgroup which is its own centralizer, so in fact the image of $f_2$ in $O_6^+(2)$ has order 7. Then $f_2^7$ is an element of order 4 in $2_+^{1+6}$ with a fixed point space of dimension 2, but this is a contradiction by Lemma 6.1.1 (ii).

Suppose that $q = 3$, $\omega_1 = 3 + 1$, and $\omega_2 = 3^3 - 1 = 2 \cdot 13$. In this case $f_2^2$ is an element of order 13, but there is no such element in $G^\circ$ since $13 \nmid |O_6^{\pm}(2)|$, so we have a contradiction.

It remains to consider the case where $q = 5$, $\omega_1 = 5 - 1$, and $\omega_2 = 5 + 1$. Here $\nu_2 = 1$ and $\mu_2 = 3$, so the order of $H_2^\circ$ (Lemma 4.3.23) is given by

$$|H_2^\circ| = 6 \cdot 3^2 \cdot |\mathrm{GSp}_2(3)| = 2^5 \cdot 3^4.$$

In particular $H_2^\circ$ contains a Sylow 3-subgroup $Q$ of order $3^4$, which is centralized by $H_1^\circ$, with $|H_1^\circ| = 8$. We will see that there is no such subgroup in $G^\circ$.

Since $3^4 \nmid |O_6^+(2)|$, the group $G^\circ$ is contained in $2_-^{1+6}.O_6^-(2)$. Let $\overline{Q}$ be the image of $Q$ in $O_6^-(2)$. We have $|O_6^-(2)| = 2^7 \cdot 3^4 \cdot 5$, so $\overline{Q}$ is a Sylow 3-subgroup of $O_6^-(2)$. A 3-Sylow of $O_6^-(2)$ can be constructed as a wreath product $\overline{Q} = \Omega_2^-(2) \wr C_3$, which is irreducible (Lemma 2.2.6). Furthermore, it is straightforward to see that the centralizer of $\Omega_2^-(2) \wr C_3$ in $O_6^-(2)$ has order 3, so $C_{O_6^-(2)}(\overline{Q}) = Z(\overline{Q}) \cong C_3$.

Therefore $C_{G^\circ}(Q) = C_A(Q)Z(Q)$, where $A = 2_-^{1+6}$. Because $\overline{Q}$ is irreducible, the fixed point space of $\overline{Q}$ is zero, and thus $|C_A(Q)| = 2$. We conclude that $|C_{G^\circ}(Q)| = 2 \cdot 3 = 6$, contradicting the fact that $Q$ is centralized by $H_1^\circ$ with $|H_1^\circ| = 8$.

**Case 4.2:** $n_1 = n_2 = n/2$

Suppose that $n_1 = n_2 = n/2$. We will show that the configurations listed in (vi) are the only ones possible. Since $n_1 = n_2 = n/2$ and $\omega = 2$, the inequality (7.13) gives

$$\frac{\omega_1}{\alpha_1} \cdot \frac{\omega_2}{\alpha_2} \leq 8. \tag{7.20}$$

Consider first the case where neither $H_1$ nor $H_2$ is of type $\mathcal{B}_3$. In this case $n_i = 2\nu_i\mu_i$, $\alpha_i = 2\nu_i$ and (7.20) becomes

$$\frac{q^{\nu_1} \pm 1}{2\nu_1} \cdot \frac{q^{\nu_2} \pm 1}{2\nu_2} \leq 8. \tag{7.21}$$

It is straightforward to see that the inequality (7.21) cannot hold for $q \geq 7$. For $q < 7$, we have the following solutions to inequality (7.21), up to exchanging $H_1$ and $H_2$:

- $\omega_1 = 5 - 1, \omega_2 = 5 - 1.$
- $\omega_1 = 5 - 1, \omega_2 = 5 + 1.$

- $\omega_1 = 3 + 1, \omega_2 = 3 + 1.$
- $\omega_1 = 3 + 1, \omega_2 = 3^2 + 1.$
- $\omega_1 = 3^2 + 1, \omega_2 = 3^2 + 1.$

Note that here we have taken into account that (a) $q$ must be odd; (b) $n_i = 2\nu_i\mu_i$ is a power of two, so $\nu_i$ is a power of two; (c) $\omega_i = 3 - 1$ is not possible since $(q, \nu_i) \neq (3, 1)$ for groups of type $\mathcal{B}_1$; (d) $\omega_i = 3^2 - 1$ is not possible since groups of type $\mathcal{B}_1$ with $(q, \nu_i) = (3, 2)$ are metrically imprimitive (Lemma 5.6.5). All of the solutions above correspond to case (vi)(a), (vi)(b), or (vi)(c) of the lemma.

Next we consider the case where exactly one of $H_1$ or $H_2$ is of type $\mathcal{B}_3$. To this end, suppose without loss of generality that $H_1$ is of type $\mathcal{B}_1$ or $\mathcal{B}_2$, and $H_2$ is of type $\mathcal{B}_3$. Then $\omega_2/\alpha_2 = 2$, so inequality (7.20) becomes

$$\frac{q^{\nu_1} \pm 1}{2\nu_1} \leq 4. \tag{7.22}$$

If $q > 9$, there are no solutions to the inequality (7.22). For $q \leq 9$ we have the following solutions:

- $q = 9, \nu_1 = 1, \omega_1 = 9 \pm 1.$
- $q = 7, \nu_1 = 1, \omega_1 = 7 \pm 1.$
- $q = 7, \nu_1 = 1, \omega_1 = 5 \pm 1.$
- $q = 3, \nu_1 = 1, \omega_1 = 3 + 1.$
- $q = 3, \nu_2 = 2, \omega_1 = 3^2 + 1.$

Here we have again taken into account that $q$ is odd, and that $(q, \nu_1) = (3, 1)$ or $(q, \nu_1) = (3, 2)$ is not applicable for $H_1$ of type $\mathcal{B}_1$ (Lemma 5.6.5). All of the solutions above to correspond to case (vi)(d) or (vi)(e) of the lemma.

Finally if $H_1$ and $H_2$ are both of type $\mathcal{B}_3$, then (vi)(f) of the lemma holds. This completes the proof of the lemma. □

**Lemma 7.2.15** *Assume that $n$ and $q$ are even, and that $\kappa$ is a quadratic form. Let $L = (H_1 \times H_2) \cap \Delta(V, \kappa)$ as in Example 7.2.1, with $t = 2$. Then $L^\Omega \not\leq G^\Omega$.*

***Proof*** Suppose for the sake of contradiction that $L^\Omega \leq G^\Omega$. It follows from Lemmas 7.2.4 and 7.2.6 that $F_0$ acts on $W_1$ and $W_2$, and that $\omega \mid \omega_1, \omega_2$.

As in the proof of Lemma 7.2.14, the inequality (7.13) holds, from which it follows that

$$\alpha \cdot \frac{\omega_i}{\alpha_i} \leq \frac{16}{3} \tag{7.23}$$

for all $1 \leq i \leq 2$. We will now go through the arguments used in the proof of Lemma 7.2.14 to find a contradiction. Since $q$ is even, the group $G$ is of type $\mathcal{B}_1$ or type $\mathcal{B}_2$.

**Case 1: $G$ is of type $\mathcal{B}_1$**
Since $\omega \mid \omega_i$, in this case both $H_1$ and $H_2$ must be of type $\mathcal{B}_1$ (Table 7.2). Since $q$ is even, as in the proof of Lemma 7.2.14 (Case 2) it follows from (7.23) that $q = 4$ and $\nu = \nu_1 = \nu_2 = 1$. Then $n = 2 \cdot 3^\rho$ and $n_i = 2 \cdot 3^{\rho_i}$, which by $n = n_1 + n_2$ leads to the contradiction $3^\rho = 3^{\rho_1} + 3^{\rho_2}$.

**Case 2: $G$ is of type $\mathcal{B}_2$**
Because $q$ is even, as in the proof of Lemma 7.2.14 (Case 3) it follows from (7.23) that the possibilities for $H_i$ are as follows:

- $H_i$ is of type $\mathcal{B}_1$, $q = 2$, $\nu = 1$, and $\nu_i = 4$.
- $H_i$ is of type $\mathcal{B}_2$, $q = 2$, and $\nu = \nu_i = 1$.
- $H_i$ is of type $\mathcal{B}_2$, $q = 2$, and $\nu = \nu_i = 2$.
- $H_i$ is of type $\mathcal{B}_2$, $q = 4$, and $\nu = \nu_i = 1$.

Here if $H_1$ or $H_2$ is of type $\mathcal{B}_1$, the exact same argument as in Lemma 7.2.14 (Case 3.1) leads to a contradiction. The other possibility is that $H_1$ and $H_2$ are both of type $\mathcal{B}_2$. Here arguing as in Lemma 7.2.14 (Case 3.2) leads to a contradiction.

In both cases we have a contradiction, which completes the proof of the lemma. □

**Lemma 7.2.16** *Let $L^\circ = H_1^\circ \times H_2^\circ$ be as in Example 7.2.1 with $t = 2$, and suppose that $L^\circ \leq G^\circ$. Then one of the following holds:*

*(i) $n = 2$, $n_1 = n_2 = 1$, and one of the following holds:*

*(a) $\kappa = 0$, $q \in \{3, 5\}$, and $H_1 = H_2 = \mathrm{GL}_1(q)$.*
*(b) $\kappa \neq 0$ is a quadratic form, $q$ is odd, and $H_1 = H_2 = \mathrm{GO}_1(q)$.*

*(ii) $n = 4$, $n_1 = n_2 = 2$, and one of the following holds:*

*(a) $\kappa \neq 0$ is a quadratic form, $q = 5$, and $H_1 = H_2 = \mathrm{GO}_2^+(5)$.*
*(b) $\kappa \neq 0$ is a quadratic form, $q = 3$, and $H_1 = H_2 = \mathrm{GO}_2^-(3)$.*

***Proof*** It follows from Lemma 7.2.14 that $q$ is odd, $n$ is a power of two, and $n_1 = n_2 = n/2$. Furthermore, by Lemma 7.2.14 we have $\nu = 1$, so $n = \alpha\mu$ with $\alpha \in \{1, 2\}$ and $\mu = 2^s$ for some $s \geq 0$.

We consider first the case where $n = 2$. If $\kappa = 0$, then we must have $H_1 = H_2 = \mathrm{GL}_1(q)$. Then by Lemma 7.2.14 (iii) we have $q \in \{3, 5\}$. In the case where $\kappa \neq 0$, it is clear that $\kappa$ must be a quadratic form and $H_1 = H_2 = \mathrm{GO}_1(q)$. Thus for $n = 2$ claim (i) holds.

We assume then for the remainder of the proof that $n > 2$. Note that since $\alpha \in \{1, 2\}$, we have $s \geq 1$, so $A = F_0 H$ with $H = 2_{\varepsilon_1}^{1+2s}$ and $G = G_{\mu,\nu}^{\mathcal{B}}(X_1)$ with $X_1 \leq O_{2s}^{\varepsilon_1}(2)$ metrically completely reducible maximal solvable.

Denote $\theta = -I_V$ and $\theta_i = -I_{W_i}$, so $\theta = \theta_1\theta_2$ and $\theta \in G^\circ$, $\theta_i \in H_i^\circ$. Define $\Lambda := C_H(\theta_1)$, which is the stabilizer of $W_1$ in $H$. We first estimate the order of $\Lambda$, similarly to [52, §112–114].

**Claim 1: If $\theta_1$ acts nontrivially on $A/F_0$, then $|\Lambda| \geq 2^s$**
Suppose that $\theta_1$ acts nontrivially on $A/F_0$. Since $\theta_1$ has order 2, the action of $\theta_1$ on $A/F_0 \cong \mathbb{F}_2^{2s}$ is as a unipotent linear map with all Jordan blocks of size 1 or 2. Therefore, the fixed point space of $\theta_1$ on $A/F_0$ has dimension $\geq s$.

The fixed point space of $\theta_1$ on $A/F_0$ is equal to $QF_0/F_0$, where $Q = \{x \in H : [x, \theta_1] \in F_0\}$. Thus $|QF_0/F_0| \geq 2^s$, which gives

$$|Q| \geq 2^s |Q \cap F_0| = 2^{s+1} \tag{7.24}$$

since $H \cap F_0 = \langle \theta \rangle$.

We have a homomorphism $\rho : Q \to F_0$ defined by $x \mapsto [x, \theta_1]$. Because $H \trianglelefteq G$, the image of $\rho$ is contained in $H \cap F_0 = \langle \theta \rangle$ for all $x \in H$. Hence $\operatorname{Ker} \rho = \Lambda$ has index $\leq 2$ in $Q$, so it follows from (7.24) that $|\Lambda| \geq 2^s$.

**Claim 2: If $\theta_1$ acts trivially on $A/F_0$, then $|\Lambda| = 2^{2s}$**
Suppose that $\theta_1$ acts trivially on $A/F_0$. Then $[x, \theta_1] \in F_0$ for all $x \in H$, so we have a homomorphism $\rho : H \to F_0$ defined by $x \mapsto [x, \theta_1]$. Since $x^2 = \pm 1$ for all $x \in H$, the image of $\rho$ is contained in $\langle \theta \rangle$.

If $\rho$ is trivial, then $H$ is centralized by $\theta_1$. In that case $\theta_1$ centralizes $A = F_0 H$, which would imply $\theta_1 \in F_0$ by Lemma 4.3.32. But this is a contradiction, since $\theta_1$ acts trivially on $W_2$ and no non-identity element of $F_0$ has nonzero fixed points (Lemma 2.4.3). Therefore $\rho$ must be nontrivial. It follows that $\operatorname{Ker} \rho = \Lambda$ has index 2 in $H$, so $|\Lambda| = 2^{2s}$.

For $1 \leq i \leq 2$, let $\Lambda_i$ be the image of $\Lambda$ in $\mathrm{GL}(W_i)$. Then each $x \in \Lambda$ can be written in the form $x = x_1 x_2$ with $x_i \in \Lambda_i$. We will next describe the order of $|\Lambda_i|$.

**Claim 3: $|\Lambda| = |\Lambda_i|$ if $\theta_1 \notin H$, and $|\Lambda| = 2|\Lambda_i|$ if $\theta_1 \in H$**
By symmetry it will suffice to prove the claim for $i = 1$. We have a surjective homomorphism $\rho : \Lambda \to \Lambda_1$ defined by $x \mapsto x_1$, to prove the claim we will determine the kernel of $\rho$.

If $x_1 = 1$, then $x$ acts trivially on $W_1$. In particular $x$ has nonzero fixed points on $V$, so $x^2 = -1$ is not possible and we must have $x^2 = 1$. If $x \notin Z(H) = \langle \theta \rangle$, then $x$ is conjugate to $-x = \theta x$ in $H$ and the eigenspaces of $x$ on $V$ are of equal dimension. Thus either $x = 1$, or $x$ acts as $-I_{W_2}$ on $W_2$. Hence

$$\operatorname{Ker} \rho = \begin{cases} 1, & \text{if } \theta_2 \notin H. \\ \langle \theta_2 \rangle, & \text{if } \theta_2 \in H. \end{cases} \tag{7.25}$$

Because $\theta = \theta_1 \theta_2 \in H$, we have $\theta_1 \in H$ if and only if $\theta_2 \in H$. Thus we conclude from (7.25) that $|\Lambda| = |\Lambda_1|$ if $\theta_1 \notin H$, and $|\Lambda| = 2|\Lambda_1|$ if $\theta_1 \in H$.

**Claim 4: Suppose that $\theta_1 \notin H$. Then $|\Lambda_i| \leq 4$ if $4 \mid \omega_i$, and $|\Lambda_i| \leq 2$ if $4 \nmid \omega_i$**
Since $\theta_1 \in H$ if and only if $\theta_2 \in H$, by symmetry it will suffice to prove the claim for $i = 1$. Let $g \in H_1^\circ$. Then for all $x \in \Lambda$, we have $[g, x] \in H$ since $H \trianglelefteq G$.

On the other hand $[g, x] = [g, x_1]$ fixes all the points in $W_2$, so as in the proof of Claim 3 it follows that $[g, x_1] = 1$ or $[g, x_1] = \theta_1$. Since $\theta_1 \notin H$, we conclude that $[g, x_1] = 1$ for all $g \in H_1^\circ$, and thus $x_1 \in F_0^{(1)}$ by Lemma 4.3.32.

We have proved that $\Lambda_1 \leq F_0^{(1)} = \langle f_1 \rangle$. Since every element of $\Lambda_1$ has order $\leq 4$, it follows that $|\Lambda_1| \leq 4$ in all cases, and $|\Lambda_1| \leq 2$ if $4 \nmid \omega_1$.

**Claim 5: Suppose that $\theta_1 \in H$. Then $|\Lambda_i| \leq 8$ if $4 \mid \omega_i$, and $|\Lambda_i| \leq 4$ if $4 \nmid \omega_i$**
As in Claim 4, we can assume $i = 1$ without loss of generality. Denote $A^{(1)} := \mathrm{Fit}(C_{H_1^\circ}(F_0^{(1)}))$, which is the subgroup satisfying properties (A1)–(A3) in $H_1^\circ$ (Lemma 4.1.4). Since $n_1$ is a power of two, we have $A^{(1)} = F_0^{(1)} R_1^{(1)}$ with $R_1^{(1)} \trianglelefteq H_1$ an extraspecial 2-group.

As in the proof of Claim 4, for all $g \in H_1^\circ$ and $x \in \Lambda$ we have $[g, x] \in \langle \theta_1 \rangle$. It follows that for all $x \in \Lambda$, the subgroup $N = C_{H_1^\circ}(x_1)$ is a normal subgroup of index $\leq 2$ in $H_1^\circ$. By Lemma 5.7.2, the subgroup $R_1^{(1)}$ is contained in $N$, so we conclude that $\Lambda_1$ centralizes $R_1^{(1)}$.

Let $\Lambda_1'$ be the centralizer of $F_0^{(1)}$ in $\Lambda_1$. Because $[f_1, x] \in \langle \theta_1 \rangle$ for all $x \in \Lambda$, we have $[\Lambda_1 : \Lambda_1'] \leq 2$, and thus

$$|\Lambda_1| \leq 2|\Lambda_1'|.$$

Now $\Lambda_1'$ centralizes $F_0^{(1)}$ and $R_1^{(1)}$, so it follows from Lemma 4.3.32 that $\Lambda_1' \leq Z(A^{(1)}) = F_0^{(1)}$. Therefore $|\Lambda_1'| \leq 4$ in all cases, and $|\Lambda_1'| \leq 2$ if $4 \nmid \omega_1$. From this the claim follows.

Putting all the claims proved so far together, we can bound the value of $s$.

**Claim 6: We have $s \leq 2$ in all cases. Furthermore, if $4 \nmid \omega_i$ for some $i$, then $s = 1$**
Suppose first that $\theta_1$ acts nontrivially on $A/F_0$, in which case we must have $\theta_1 \notin H$. It follows from Claim 1, Claim 3, and Claim 4 that

$$2^s \leq |\Lambda| = |\Lambda_i| \leq 4,$$

so $s \leq 2$. If $4 \nmid \omega_i$ then $|\Lambda_i| \leq 2$ by Claim 4, so we must have $s = 1$.

Next we consider the case where $\theta_1$ acts trivially on $A/F_0$. It follows from Claim 2, Claim 3, and Claim 5 that

$$2^{2s} = |\Lambda| \leq 2|\Lambda_i| \leq 16,$$

so $s \leq 2$. If $4 \nmid \omega_i$ then $2|\Lambda_i| \leq 8$ by Claim 4, so we must have $s = 1$.

Therefore $\mu \in \{2, 4\}$, and $\mu = 4$ is only possible when $4 \mid \omega_1, \omega_2$. We will now consider the different possibilities for $G$ case-by-case.

**Case 1: $G$ is of type $\mathcal{B}_0$**
By Lemma 7.2.14 we have $\alpha = \nu = 1$. Therefore if $\mu = 2$, then $n = \nu\mu = 2$, a case which has already been ruled out from consideration. Suppose then that $\mu = 4$, in which case we must have $4 \mid \omega_i$, so it follows from Lemma 7.2.14 that $q = 5$ and $\nu = \nu_1 = \nu_2 = 1$. Note that $n = 4$ in this case, and $n_i = 2$ with $\mu_i = 2$. The order of $G = G^{\mathcal{B}_0}_{4,1}(X_1)$ is given by

$$|G| = 4 \cdot 4^2 \cdot 1 \cdot 72 = 2^9 \cdot 3^2$$

if $X_1 = O_4^+(2)$, and

$$|G| = 4 \cdot 4^2 \cdot 1 \cdot 20 = 2^8 \cdot 5$$

if $X_1 = G^{\mathcal{B}_2}_{1,1} < O_4^-(2)$ (Lemma 4.3.7).

Since $n_i = 2$ and $\mu_i = 2$, we have $|H_i| = 4 \cdot 2^2 \cdot |O_2^-(2)| = 2^5 \cdot 3$. Thus $|H_1 \times H_2| > |G|$, contradicting $H_1 \times H_2 \leq G$.

**Case 2: $G$ is of type $\mathcal{B}_1$ or $\mathcal{B}_2$**
In this case, it follows from Lemma 7.2.14 (iv) and (v) that one of the following holds:

- $G$, $H_1$, and $H_2$ are of type $\mathcal{B}_1$, with $q = 5$ and $\nu_1 = \nu_2 = 1$.
- $G$, $H_1$, and $H_2$ are of type $\mathcal{B}_2$, with $q = 3$ and $\nu_1 = \nu_2 = 1$.

In both cases $|F_0| = |F_0^{(1)}| = |F_0^{(2)}| = 4$. We first consider $\mu = 2$. In this case

$$|G^\circ| = 4 \cdot 2^2 \cdot 2 \cdot |O_2^-(2)| = 2^6 \cdot 3.$$

Moreover $n_i = n/2$, so $\mu_i = 1$ and thus $|H_i^\circ| = 2^3$.

Therefore if $H_1^\circ \times H_2^\circ$ is contained in $G^\circ$, it is equal to a Sylow 2-subgroup of $G$. However, we claim that a Sylow 2-subgroup of $G^\circ$ is irreducible, so $H_1^\circ \times H_2^\circ \not\leq G^\circ$. If $G$ is of type $\mathcal{B}_1$, then $V' := \mathbb{K} \otimes_{\mathbb{F}_q} V$ decomposes as $V' = W'_\lambda \oplus W'_{\lambda^{-1}}$, where $W'_{\lambda^{\pm 1}}$ are absolutely irreducible and nonisomorphic $\mathbb{K}[A]$-modules (Lemma 5.2.2). The linear map $\varphi \in \operatorname{Ker}\pi$ swaps the summands $W'_\lambda$ and $W'_{\lambda^{-1}}$, so $\operatorname{Ker}\pi = \langle A, \varphi \rangle$ acts irreducibly on $V'$ by Lemma 2.1.15. Thus $\operatorname{Ker}\pi$ is an irreducible 2-subgroup of $G^\circ$, so a Sylow 2-subgroup of $G^\circ$ is irreducible. Arguing similarly when $G$ is of type $\mathcal{B}_2$, one finds that $\operatorname{Ker}\pi = \langle A, \psi \rangle$ is irreducible, so a Sylow 2-subgroup of $G^\circ$ is irreducible.

It remains to consider the case where $\mu = 4$. The order of $G$ is given by

$$|G| = 4 \cdot 4^2 \cdot 2 \cdot 72 = 2^{10} \cdot 3^2$$

if $X_1 = O_4^+(2)$, and

$$|G| = 4 \cdot 4^2 \cdot 2 \cdot 20 = 2^9 \cdot 5$$

if $X_1 < O_4^-(2)$ with $|X_1| = 20$ (see Lemmas 4.3.15 and 4.3.23). We have $n_i = n/2 = 4$, so $\mu_i = 2$. Thus $|H_i^\circ| = 4 \cdot 2^2 \cdot 2 \cdot |O_2^-(2)| = 2^6 \cdot 3$. Therefore $|H_1 \times H_2| > |G|$, so $H_1 \times H_2 \leq G$ is impossible.

**Case 3: $G$ is of type $\mathcal{B}_3$**
If $\mu = 2$, then $n = 2$, which we have already ruled out. Thus we must have $\mu = 4$. In this case $4 \mid \omega_1, \omega_2$, so it follows from Lemma 7.2.14 (vi) that one of the following cases holds:

- $H_1$ and $H_2$ are of type $\mathcal{B}_1$, with $q = 5$ and $\nu_1 = \nu_2 = 1$.
- $H_1$ and $H_2$ are of type $\mathcal{B}_2$, with $q = 3$ and $\nu_1 = \nu_2 = 1$.

If $e = 1$, then it follows from Lemmas 5.5.5 and 5.5.7 that $H_i$ is not maximal solvable, contrary to what we have assumed. Thus we must have $e = 0$.

Furthermore, in both cases we have $|F_0^{(i)}| = 4$, $\mu_i = 1$, and $n_i = 2$, so $|H_i^\circ| = 8$. Since $e = 0$, it follows that $H_i^\circ = O_2^+(5)$ if $q = 5$, and $H_i^\circ = O_2^-(3)$ if $q = 3$. Thus either $q = 5$ and $H_i = \mathrm{GO}_2^+(5)$, or $q = 3$ and $H_i = \mathrm{GO}_2^-(3)$. These are precisely the exceptions listed in case (ii) of the lemma.

With the result proved in all these cases, the proof of the lemma is complete. □

Summarizing the results we have obtained so far, we have the following theorem.

**Theorem 7.2.17** *Let $L^\circ = H_1^\circ \times \cdots \times H_t^\circ$ as in Example 7.2.1. Suppose that $L^\circ \leq G^\circ$. Then $t \in \{2, 3, 4\}$ and one of the following holds:*

*(i) $t = 4$, $n = 4$, and one of the following holds:*

*(a) $\kappa = 0$, $q = 3$, and $L^\circ = \mathrm{GL}_1(3) \times \mathrm{GL}_1(3) \times \mathrm{GL}_1(3) \times \mathrm{GL}_1(3)$;*
*(b) $\kappa \neq 0$ quadratic form, $q$ is odd, and $L^\circ = O_1(q) \times O_1(q) \times O_1(q) \times O_1(q)$;*

*(ii) $t = 3$, $n \in \{3, 4\}$, and one of the following holds:*

*(a) $\kappa = 0$, $q = 4$, and $L^\circ = \mathrm{GL}_1(4) \times \mathrm{GL}_1(4) \times \mathrm{GL}_1(4)$;*
*(b) $\kappa \neq 0$ quadratic form, $q = 3$, and $L^\circ = O_1(3) \times O_1(3) \times O_2^-(3)$;*
*(c) $\kappa \neq 0$ quadratic form, $q = 5$, and $L^\circ = O_1(5) \times O_1(5) \times O_2^+(5)$;*

*(iii) $t = 2$, $n \in \{2, 4\}$, and one of the following holds:*

*(a) $\kappa = 0$, $q = 3$, and $L^\circ = \mathrm{GL}_1(3) \times \mathrm{GL}_1(3)$;*
*(b) $\kappa = 0$, $q = 5$, and $L^\circ = \mathrm{GL}_1(5) \times \mathrm{GL}_1(5)$;*
*(c) $\kappa \neq 0$ quadratic form, $q$ is odd, and $L^\circ = O_1(q) \times O_1(q)$;*
*(d) $\kappa \neq 0$ quadratic form, $q = 3$, and $L^\circ = O_2^-(3) \times O_2^-(3)$;*
*(e) $\kappa \neq 0$ quadratic form, $q = 5$, and $L^\circ = O_2^+(5) \times O_2^+(5)$;*

***Proof*** The theorem follows from Lemmas 7.2.9, 7.2.13, and 7.2.16. □

**Theorem 7.2.18** *Assume that $n$ and $q$ are even, and that $\kappa$ is a quadratic form. Let $L = (H_1 \times \cdots \times H_t) \cap \Delta(V, \kappa)$ as in Example 7.2.1. Then $L^\Omega \not\leq G^\Omega$.*

***Proof*** The theorem follows from Lemmas 7.2.10, 7.2.12, and 7.2.15. □

**Theorem 7.2.19** *Let $X = (H \wr K) \cap \Delta(V, \kappa)$ be as in Theorem 2.2.14 (i). If $X^\circ \leq G^\circ$, then one of the following holds:*

*(i) $\kappa = 0$, $d = 1$, $H = \mathrm{GL}_1(q)$, and one of the following holds:*

- *(a) $q \in \{3, 5\}$ and $X = \mathrm{GL}_1(q) \wr S_2$.*
- *(b) $q = 4$ and $X = \mathrm{GL}_1(4) \wr S_3$.*
- *(c) $q = 3$ and $X = \mathrm{GL}_1(3) \wr S_4$.*

*(ii) $\kappa \neq 0$ is a quadratic form, $q$ is odd, and one of the following holds:*

- *(a) $d = 1$ and $X^\circ = O_1(q) \wr S_2$.*
- *(b) $d = 1$ and $X^\circ = O_1(q) \wr S_4$.*
- *(c) $d = 2$, $q = 3$, and $X^\circ = O_2^-(3) \wr S_2$.*
- *(d) $d = 2$, $q = 5$, and $X^\circ = O_2^+(5) \wr S_2$.*

***Proof*** We have $X = L^\circ \rtimes K$, where $L^\circ = H_1^\circ \times \cdots \times H_t^\circ$ is as in Example 7.2.1. Thus it follows from Theorem 7.2.17 that $L^\circ$ must be as in one of the cases listed in Theorem 7.2.17 (i)–(iii). In all of these cases $t \in \{2, 3, 4\}$, so $K = S_t$ since $K$ is maximal solvable. Furthermore $H_i \cong H_j$ as linear groups for all $i$, $j$, so cases (ii)(b) and (ii)(c) of Theorem 7.2.17 (i)–(iii) are not applicable. Since $X = H \wr S_t$, the remaining cases of Theorem 7.2.17 give precisely the configurations listed in (i) and (ii). □

**Theorem 7.2.20** *Assume that $n$ and $q$ are even, and that $\kappa$ is a quadratic form. Let $X = (H \wr K) \cap \Delta(V, \kappa)$ be as in Theorem 2.2.14 (i). Then $X^\Omega \not\leq G^\Omega$.*

***Proof*** Similarly to the proof of Theorem 7.2.19, the result follows from Theorem 7.2.18. □

**Theorem 7.2.21** *Let $X = \mathrm{semiwr}(H^\circ) \leq \Delta(V, \kappa)$ a semiprimary imprimitive subgroup constructed as in Example 2.2.7, where $H \leq \Delta(W, \kappa')$ is maximal irreducible solvable of multiplier 2. If $X \leq G$, then all of the following hold:*

*(i) $\kappa$ is a quadratic form, $n = 2$, and $q > 3$;*
*(ii) $H = \mathrm{GO}_1(q)$;*
*(iii) $G = \mathrm{GO}_2^\varepsilon(q)$, where $\varepsilon = (-1)^{(q+1)/2}$.*

*Furthermore, when (i)–(iii) hold, we have $X \lneq G$.*

***Proof*** When (i)–(iii) hold, it follows from Remark 2.2.13 that $X \lneq G$. Conversely, suppose that $X \leq G$. We will prove that (i)–(iii) hold.

Let $Z$ be the subgroup of scalar matrices in $\Delta(W, \kappa')$. Since $H$ is maximal irreducible solvable of multiplier 2, we have $H = (Y^\circ \wr K)Z$, where $Y^\circ$ is metrically primitive maximal irreducible solvable, and $K \leq S_k$ maximal transitive solvable with $k \geq 1$ (Theorem 2.2.14). We have $X^\circ = H_1^\circ \times H_2^\circ$ with respect to an orthogonal decomposition $V = Z_1 \perp Z_2$, where $H_i^\circ \leq \Delta(Z_i, \kappa)$ is similar to $H^\circ = Y^\circ \wr K$ for all $1 \leq i \leq 2$. Then for $i = 1, 2$ we have a decomposition $Z_i = W_1^{(i)} \perp \cdots \perp W_k^{(i)}$, where $\{W_1^{(i)}, \ldots, W_k^{(i)}\}$ is the orthogonal system of imprimitivity defining $H_i^\circ$ if $k > 1$.

Now $H_i^\circ$ contains a subgroup $\left(Y_1^{(i)}\right)^\circ \times \cdots \times \left(Y_k^{(i)}\right)$ with respect to the decomposition $Z_i = W_1^{(i)} \perp \cdots \perp W_k^{(i)}$, where $\left(Y_j^{(i)}\right)^\circ$ is isometric to $Y^\circ$ and metrically primitive maximal irreducible solvable in $I(W_j^{(i)}, \kappa)$ for all $1 \leq j \leq k$.

Then $\left(Y_1^{(1)}\right)^\circ \times \cdots \times \left(Y_k^{(1)}\right) \times \left(Y_1^{(2)}\right)^\circ \times \cdots \times \left(Y_k^{(2)}\right)^\circ$ is contained in $G^\circ$, so by Theorem 7.2.17 we have $k = 1$ or $k = 2$.

If $k = 2$, it follows from Theorem 7.2.17 that $\kappa$ is a quadratic form, $n = 4$, and $H = (O_1(q) \wr S_2)Z$. Because $H$ is of multiplier 2, it is properly contained in $\Delta(V, \kappa) = \mathrm{GO}_2^\varepsilon(q)$, contradicting the assumption that $H$ is maximal solvable.

Therefore $k = 1$, so $X^\circ = H_1^\circ \times H_2^\circ$ with $H_i^\circ$ metrically primitive maximal irreducible solvable. We have $X^\circ \leq G^\circ$ and $H$ is of multiplier 2, so it follows from Theorem 7.2.17 that $\kappa$ is a quadratic form, $n = 2$, and $H = \mathrm{GO}_1(q)$. We must have $q > 3$, because $G$ is metrically primitive and for $q = 3$ the group $\Delta(V, \kappa) = \mathrm{GO}_2^+(3)$ is not (Remark 2.2.13). This proves that (i) and (ii) hold.

For (iii), note that $\varepsilon = (-1)^{(q+1)/2}$ as seen in the construction of Example 2.2.7. Furthermore by $n = 2$ it is clear that $G = G_{1,1}^{\mathcal{B}_1} = \mathrm{GO}_2^+(q)$ if $\varepsilon = +$, and $G = G_{1,1}^{\mathcal{B}_2} = \mathrm{GO}_2^-(q)$ if $\varepsilon = -$. Therefore (iii) holds, which completes the proof of the theorem. □

**Theorem 7.2.22** *Suppose that $X = (H \wr K) \cap \Delta(V, \kappa)$ is as in Theorem 7.2.19 (i) or (ii). Then there exists a metrically primitive group $G \leq \Delta(V, \kappa)$ such that $X \lneqq G$. Furthermore $X^\circ \lneqq G^\circ$, except in the following cases:*

*(i) $\kappa \neq 0$ is a quadratic form, $q = 3$, and $X^\circ = O_1(3) \wr S_2 = O_2^-(3)$.*
*(ii) $\kappa \neq 0$ is a quadratic form, $q = 5$, and $X^\circ = O_1(5) \wr S_2 = O_2^+(5)$.*

***Proof*** We first suppose that $\kappa = 0$ and consider the examples in Theorem 7.2.19 (i). We will consider the cases (a)–(c) in turn. In each case, it turns out that the group $X$ is contained in the normalizer $N_{\mathrm{GL}_n(q)}(R)$ of an extraspecial group $R$.

**Case (i)(a): $n = 2$ and $q \in \{3, 5\}$**
For $q = 3$ the result is clear, since $X \lneqq G$ for $G = \mathrm{GL}_2(3)$, and moreover $\mathrm{GL}_2(3) = 2_-^{1+2}.O_2^-(2)$ is solvable and primitive.

Suppose that $q = 5$. From the construction of Sect. 3.2.4, we have an absolutely irreducible subgroup $R = 2_-^{1+2}$ in $\mathrm{GL}_2(5)$ with generators

$$A = \begin{pmatrix} 2 & \\ & -2 \end{pmatrix}, \quad B = \begin{pmatrix} & -1 \\ 1 & \end{pmatrix}.$$

By Theorem 3.2.5, the normalizer $G = N_{\mathrm{GL}_2(5)}(R) = \langle A, B, C, E \rangle$, where

$$C = \begin{pmatrix} 2 & -1 \\ 1 & -2 \end{pmatrix}, \quad E = \begin{pmatrix} -2 & \\ & -1 \end{pmatrix}.$$

Note that $G = 2_-^{1+2}.O_2^-(2)$ is primitive irreducible solvable (Theorem 5.6.9).

We have $\langle A, E\rangle = \mathrm{GL}_1(5) \times \mathrm{GL}_1(5)$, the group of diagonal matrices in $\mathrm{GL}_2(5)$. Furthermore

$$E^2 B = \begin{pmatrix} & 1 \\ 1 & \end{pmatrix},$$

so we conclude that $\mathrm{GL}_1(5) \wr S_2 = \langle A, E, B\rangle \lneq G$.

**Case (i)(b):** $n = 3$ **and** $q = 4$
Let $\theta \in \mathbb{F}_4^\times$ be a primitive element, so $|\theta| = 3$. From the construction of Sect. 3.2.1, we have an absolutely irreducible subgroup $R = 3_+^{1+2}$ in $\mathrm{GL}_3(4)$ with generators

$$A = \begin{pmatrix} 1 & & \\ & \theta & \\ & & \theta^2 \end{pmatrix}, \quad B = \begin{pmatrix} & & 1 \\ 1 & & \\ & 1 & \end{pmatrix}.$$

By Theorem 3.2.2, the normalizer $G = N_{\mathrm{GL}_4(3)}(R) = \langle A, B, C, E\rangle$, where

$$C = \begin{pmatrix} 1 & 1 & 1 \\ 1 & \theta & \theta^2 \\ 1 & \theta^2 & \theta \end{pmatrix}, \quad E = \begin{pmatrix} 1 & & \\ & 1 & \\ & & \theta \end{pmatrix}.$$

Note that by Theorem 3.2.2 we have $G = 3_+^{1+2}.\mathrm{Sp}_2(3)$. Furthermore $G$ is primitive irreducible solvable by Theorem 5.6.9. We next find generators for the subgroup $\mathrm{GL}_1(4) \wr S_3$ in $G$.

We have

$$C^2 = \begin{pmatrix} 1 & & \\ & & 1 \\ & 1 & \end{pmatrix},$$

so $B$ and $C^2$ generate all permutations of the basis vectors. By conjugating $E$ with these permutations, we get

$$BEB^{-1} = \begin{pmatrix} \theta & & \\ & 1 & \\ & & 1 \end{pmatrix}, \quad B^2EB^{-2} = \begin{pmatrix} 1 & & \\ & \theta & \\ & & 1 \end{pmatrix}$$

so $\langle E, BEB^{-1}, B^2EB^{-2}\rangle = \mathrm{GL}_1(4) \times \mathrm{GL}_1(4) \times \mathrm{GL}_1(4)$. Therefore $\mathrm{GL}_1(4) \wr S_3 = \langle E, B, C^2\rangle \lneq G$, as required.

**Case (i)(c):** $n = 4$ **and** $q = 3$
By the construction of Sect. 3.2.1, we have an absolutely irreducible subgroup $R = 2_+^{1+4}$ of $\mathrm{GL}_4(3)$ with generators $R = \langle A_1, B_1, A_2, B_2 \rangle$, where

$$A_1 = \begin{pmatrix} 1 & & & \\ & 1 & & \\ & & -1 & \\ & & & -1 \end{pmatrix}, \qquad B_1 = \begin{pmatrix} & & 1 & \\ & & & 1 \\ 1 & & & \\ & 1 & & \end{pmatrix},$$

$$A_2 = \begin{pmatrix} 1 & & & \\ & -1 & & \\ & & 1 & \\ & & & -1 \end{pmatrix}, \qquad B_2 = \begin{pmatrix} & 1 & & \\ 1 & & & \\ & & & 1 \\ & & 1 & \end{pmatrix}.$$

Here the matrices are written with respect to the basis $\{v_{\xi_1,\xi_2}\} = \{v_{0,0}, v_{1,0}, v_{0,1}, v_{1,1}\}$ used in Sect. 3.2.1.

By Theorem 3.2.2, the normalizer $G = N_{\mathrm{GL}_4(3)}(R)$ is generated by $R$ together with

$$C_1 = \begin{pmatrix} 1 & & 1 & \\ & 1 & & 1 \\ 1 & & -1 & \\ & 1 & & -1 \end{pmatrix}, \qquad C_2 = \begin{pmatrix} 1 & 1 & & \\ 1 & -1 & & \\ & & 1 & 1 \\ & & 1 & -1 \end{pmatrix},$$

$$D_{12} = \begin{pmatrix} 1 & & & \\ & 1 & & \\ & & 1 & \\ & & & -1 \end{pmatrix}.$$

Furthermore $G = 2_+^{1+4}.O_4^+(2)$, so $G$ is solvable since $O_4^+(2)$ is, and $G$ is primitive irreducible solvable by Theorem 3.2.2.

We will now check that $G$ contains $\mathrm{GL}_1(3) \wr S_4$. The generators $B_1$ and $B_2$ correspond to permutations $(13)(24)$ and $(14)(23)$ of the basis vectors, while

$$C_1 D_{12} C_1^{-1} = \begin{pmatrix} 1 & & & \\ & & & 1 \\ & & 1 & \\ & 1 & & \end{pmatrix}, \quad C_2 D_{12} C_2^{-1} = \begin{pmatrix} 1 & & & \\ & 1 & & \\ & & & 1 \\ & & 1 & \end{pmatrix}$$

correspond to permutations $(234)$ and $(34)$. These generate all of $S_4$, so $K = \langle B_1, B_2, C_1 D_{12} C_1^{-1}, C_2 D_{12} C_2^{-1} \rangle \cong S_4$. Because $K$ acts transitively on the basis vectors, conjugating $D_{12}$ with elements of $K$ provides generators for $\mathrm{GL}_1(3) \times \mathrm{GL}_1(3) \times \mathrm{GL}_1(3) \times \mathrm{GL}_1(3)$. Thus

$$\mathrm{GL}_1(3) \wr S_4 = \langle D_{12}, B_1, B_2, C_1D_{12}C_1^{-1}, C_2D_{12}C_2^{-1}\rangle$$

which completes the proof that $\mathrm{GL}_1(3) \wr S_4 \lneqq G$.

Next we will consider the examples in Theorem 7.2.19 (ii), in which case $\kappa \neq 0$ is a quadratic form and $q$ is odd. We consider each case in turn.

**Case (ii)(a):** $d = 1, n = 2$
In this case, the claimed result follows from the discussion in Example 2.2.11 and Remark 2.2.13, which shows that $X \lneqq \mathrm{GO}_2^{\varepsilon}(q)$. Furthermore $X^{\circ} \lneqq O_2^{\varepsilon}(q)$, except for $q \in \{3, 5\}$.

**Case (ii)(b):** $d = 1, n = 4$
As seen in Sect. 3.2.3, we have an absolutely irreducible subgroup $R = 2_+^{1+4}$ in $\mathrm{GO}_4^+(q)$ with $R = \langle A_1, B_1, A_2, B_2\rangle$, where the matrices $A_1, B_1, A_2, B_2$ are as in case (i)(c). Note that the basis vectors form an orthonormal basis.

Let $\zeta \in \mathbb{F}_q^{\times}$ be a primitive element. By Theorem 3.2.4, the normalizer $G = N_{\mathrm{GO}_4^+(q)}(R)$ has generators

$$G = \langle R, C_1, C_2, D_{12}, \zeta I_4\rangle,$$

where the matrices $C_1$, $C_2$, $D_{12}$ are as in case (i)(c). Furthermore $G = (\mathbb{F}_q^{\times} \circ 2_+^{1+4}).O_4^+(2)$ by Theorem 3.2.4, and $G$ is metrically primitive irreducible solvable (Theorem 5.6.9).

It follows as in case (i)(c) that

$$O_1(q) \wr S_4 = \langle D_{12}, B_1, B_2, C_1D_{12}C_1^{-1}, C_2D_{12}C_2^{-1}\rangle.$$

Furthermore $O_1(q) \wr S_4 \lneqq G^{\circ}$. We have $X = \langle O_1(q) \wr S_4, \zeta I_4\rangle$ so $X \lneqq G$, as required.

**Case (ii)(c):** $d = 2, n = 4$
Recall that $O_2^-(3) = O_1(3) \wr S_2$ (Remark 2.2.13). Thus

$$X^{\circ} = O_2^-(3) \wr S_2 = O_1(3) \wr S_2 \wr S_2 \lneqq O_1(3) \wr S_4 \lneqq G^{\circ}$$

with $G$ as in case (ii)(b). With generators as in case (ii)(b), we have

$$X^{\circ} = O_1(3) \wr S_2 \wr S_2 = \langle D_{12}, B_1, B_2, C_2D_{12}C_2^{-1}\rangle.$$

We have $\mathrm{GO}_2^-(3) = \langle O_1(3) \wr S_2, \eta\rangle$, where

$$O_1(3) \wr S_2 = \left\langle \begin{pmatrix} -1 & 0 \\ 0 & 1 \end{pmatrix}, \begin{pmatrix} 0 & 1 \\ 1 & 0 \end{pmatrix} \right\rangle$$

$$\eta = \begin{pmatrix} 1 & 1 \\ 1 & -1 \end{pmatrix}.$$

Therefore $X = \langle X^\circ, C_2\rangle \lneqq G$.

**Case (ii)(d):** $d = 2, n = 4$
Note that $O_2^+(5) = O_1(5) \wr S_2$ (Remark 2.2.13). Thus we find that

$$X^\circ = O_1(5) \wr S_2 \wr S_2 \lneqq G^\circ$$

with $G$ as in the previous case.

We have $\mathrm{GO}_2^+(5) = \langle O_1(5) \wr S_2, \eta\rangle$, where

$$O_1(5) \wr S_2 = \left\langle \begin{pmatrix} -1 & 0 \\ 0 & -1 \end{pmatrix}, \begin{pmatrix} 0 & 1 \\ 1 & 0 \end{pmatrix} \right\rangle$$

$$\eta = \begin{pmatrix} 2 & 0 \\ 0 & 3 \end{pmatrix}.$$

Now

$$C_2^2 A_2 = \begin{pmatrix} 2 & & & \\ & 3 & & \\ & & 2 & \\ & & & 3 \end{pmatrix},$$

so $X = \langle X^\circ, C_2^2 A_2\rangle \lneqq G$.

With all cases handled, the proof of the theorem is complete. □

We end this section with the following result, where we consider metrically primitive irreducible $G = G^{\mathcal{B}}_{\mu,\nu}(X_1, \ldots, X_k)$ such that $G^\circ$ is metrically imprimitive. It turns out that in this case $G^\circ$ is not maximal solvable, except when $G = \mathrm{GO}_2^-(3)$ or $G = \mathrm{GO}_2^+(5)$. (This gives a slight refinement of Theorem 7.1.13, see Theorem 7.1.13 (ii).)

**Proposition 7.2.23** *Let $G \le \Delta(V, \kappa)$ be metrically primitive irreducible solvable of the form $G = G^{\mathcal{B}}_{\mu,\nu}(X_1, \ldots, X_k)$. Suppose that $G^\circ$ is metrically imprimitive. Then one of the following holds:*

*(i) $G^\circ$ is not maximal solvable in $I(V, \kappa)$.*
*(ii) $\kappa \neq 0$ is a quadratic form, $q = 3$, $n = 2$, and $G = \mathrm{GO}_2^-(3)$.*
*(iii) $\kappa \neq 0$ is a quadratic form, $q = 5$, $n = 2$, and $G = \mathrm{GO}_2^+(5)$.*

***Proof*** Suppose that $G^\circ$ is maximal solvable in $I(V, \kappa)$. Since $G^\circ$ is metrically imprimitive, it follows as in the proof of Theorem 2.2.14 that $G^\circ = H^\circ \wr K$ for some $H \leq \Delta(W, \kappa')$ and $K \leq S_t$, where $n = t \dim W$.

Since $G$ is metrically primitive, it follows from by Theorem 7.2.22 that $H^\circ \wr K$ is as Theorem 7.2.22 (ii). Since $G^\circ$ is maximal solvable in $I(V, \kappa)$, by Theorem 7.2.22 $\kappa \neq 0$ is a quadratic form, and either $G^\circ = O_2^-(3)$ or $G^\circ = O_2^+(5)$. Consequently either (ii) or (iii) holds, as required. □

## 7.3 Systems of Imprimitivity for Completely Reducible Subgroups

Suppose that $L = (H_1 \times \cdots \times H_t) \cap \Delta(V, \kappa)$ with respect to a decomposition $V = W_1 \perp \cdots \perp W_t$, where $H_i \leq \Delta(W_i, \kappa)$ is metrically primitive maximal irreducible solvable for all $1 \leq i \leq t$. In this section, we will consider the possible orthogonal systems of imprimitivity for $L$. In Sects. 7.4 and 7.5, we will apply these results in the classification of metrically imprimitive maximal irreducible solvable subgroups, and metrically completely reducible maximal solvable subgroups.

In this section we will prove that in most cases, if $V = Z_1 \perp \cdots \perp Z_s$ such that $L$ acts on $\{Z_1, \ldots, Z_s\}$, then $Z_i$ is $L^\circ$-invariant for all $1 \leq i \leq s$ (Lemma 7.3.6). We will also obtain similar results for $L^\circ$ (Lemma 7.3.7). In the case where $q$ is even and $\kappa$ is a quadratic form, we will prove a similar result for $L^\Omega$ (Lemma 7.3.9).

We begin with some results on $G^{\mathcal{B}}_{\mu,\nu}(X_1, \ldots, X_k)$-orbits of nonzero vectors in $V$.

**Lemma 7.3.1** *Let $G$ be a finite irreducible subgroup of* $\mathrm{GL}(V)$, *where* $\dim V = n$. *Then $|Gv| \geq n + 1$ for all $v \in V \setminus \{0\}$, unless $n = 1$ and $G = 1$.*

***Proof*** If $n = 1$, then $|Gv| = |G|$ and the claim follows. Suppose then that $n > 1$ and let $v \in V \setminus \{0\}$. The subspace spanned by $Gv$ is $G$-invariant, so $V$ is spanned by $Gv$. Hence $Gv$ contains a basis of $V$, and in particular $|Gv| \geq n$. The vector $\sum_{g \in G} gv$ is fixed by the action of $G$, so by irreducibility we must have $\sum_{g \in G} gv = 0$. Thus $Gv$ is not linearly independent, so $|Gv| > n$ as required. □

The next two lemmas are given by Jordan in [52, §50, §61].

**Lemma 7.3.2** *Let $G = G^{\mathcal{B}}_{\mu,\nu}(X_1, \ldots, X_k) \leq \Delta(V, \kappa)$ be irreducible. Then the $A$-orbit of any nonzero vector in $V' = \mathbb{K} \otimes_{\mathbb{F}_q} V$ contains $\geq |F_0|\mu$ vectors.*

***Proof*** Let $v \in V' \setminus \{0\}$. Since $V'$ decomposes into a direct sum of $f$-eigenspaces, and each eigenspace is $A$-invariant, it will suffice to consider the case where $v$ is an $f$-eigenvector. As a $\mathbb{K}[A]$-module, each $f$-eigenspace is isomorphic to $W'_\lambda \downarrow A$ or $(W'_\lambda)^* \downarrow A$ twisted by an automorphism of $A$ (as noted in the proof of Lemma 5.2.2). Therefore it will suffice to consider the case where $v \in W'_\lambda$.

The action of $A$ on $W_\lambda$ is irreducible (Lemma 5.2.2). Thus $Av$ spans $W'_\lambda$, and in particular it contains a basis $v_1, \ldots, v_\mu$ of $W'_\lambda$. Since $f$ acts on $W'_\lambda$ by multiplication with $\lambda$, the vectors $f^d v_i$ are distinct for all $0 \le d < |F_0|$ and $1 \le i \le \mu$. Hence $|Av| \ge |F_0|\mu$. □

**Lemma 7.3.3** *Let $G = G^{\mathcal{B}}_{\mu,\nu}(X_1, \ldots, X_k) \le \Delta(V, \kappa)$ be irreducible. Then the following statements hold:*

*(i) The A-orbit of any nonzero vector in $V' = \mathbb{K} \otimes_{\mathbb{F}_q} V$ contains $\ge n+1$ vectors, except when $G = \mathrm{GL}_1(2)$ is of type $\mathcal{B}_0$ with $q = 2$, $n = 1$.*
*(ii) The A-orbit of any nonzero vector in $V' = \mathbb{K} \otimes_{\mathbb{F}_q} V$ contains $> n+1$ vectors, except in the following cases:*

*(a) G is of type $\mathcal{B}_0$, $q = 2$, $\nu = 1$, $\mu = 1$, $n = 1$.*
*(b) G is of type $\mathcal{B}_0$, $q = 3$, $\nu = 1$, $\mu = 1$, $n = 1$.*
*(c) G is of type $\mathcal{B}_0$, $q = 2$, $\nu = 2$, $\mu = 1$, $n = 2$.*
*(d) G is of type $\mathcal{B}_1$, $q = 2$, $\nu = 3$, $\mu = 1$, $n = 6$.*
*(e) G is of type $\mathcal{B}_1$, $q = 4$, $\nu = 1$, $\mu = 1$, $n = 2$.*
*(f) G is of type $\mathcal{B}_2$, $q = 2$, $\nu = 1$, $\mu = 1$, $n = 2$.*
*(g) G is of type $\mathcal{B}_2$, $q = 2$, $\nu = 2$, $\mu = 1$, $n = 4$.*
*(h) G is of type $\mathcal{B}_3$, q odd, $n = 1$.*

***Proof*** It follows from Lemma 7.3.2 that the $A$-orbit of every nonzero vector in $V' = \mathbb{K} \otimes_{\mathbb{F}_q} V$ contains $\ge |F_0|\mu$ vectors. For the different types, the values of $|F_0|\mu$ and $n+1$ are as follows:

| Type | $\lvert F_0\rvert\mu$ | $n+1$ |
|---|---|---|
| $\mathcal{B}_0$ | $(q^\nu - 1)\mu$ | $\mu\nu + 1$ |
| $\mathcal{B}_1$ | $(q^\nu - 1)\mu$ | $2\mu\nu + 1$ |
| $\mathcal{B}_2$ | $(q^\nu + 1)\mu$ | $2\mu\nu + 1$ |
| $\mathcal{B}_3$ | $2n$ | $n+1$ |

Recall that for type $\mathcal{B}_0$ we require $q^\nu > 2$, type $\mathcal{B}_1$ we require $q^\nu > 4$ or $(q, \nu) = (4, 1)$, and type $\mathcal{B}_3$ requires $q$ odd. Among the applicable values, it is readily checked that $|F_0|\mu > n+1$, except for the values listed in (ii). Thus claim (ii) follows. Furthermore, among the cases in (ii) we always have $|F_0|\mu \ge n+1$, except when $G$ is of type $\mathcal{B}_0$ with $q = 2$, $\nu = 1$, $\mu = 1$, so we conclude that claim (i) holds. □

The following lemma is similar to [57, Lemma 3.2] in the case $\kappa = 0$, and follows with the same proof.

**Lemma 7.3.4** *Let $H_1 \times \cdots \times H_k \le M \le \Delta(V, \kappa)$ such that all of the following hold:*

*(i) $V = W_1 \perp \cdots \perp W_k$, where $W_i$ is a nontrivial irreducible $\mathbb{F}_q[H_i]$-module for all $1 \le i \le k$;*
*(ii) M acts on $W_i$ and the action is metrically primitive for all $1 \le i \le k$;*
*(iii) the direct factors $H_j$ act trivially on $W_i$ for all $j \ne i$.*

*If $V = Q_1 \perp \cdots \perp Q_\ell$ and $M$ acts on $\{Q_1, \ldots, Q_\ell\}$, then we have $\ell \leq k$.*

***Proof*** Replacing $\oplus$ with $\perp$, proceeding by induction on $k$, and arguing exactly as in [57, Lemma 3.2] works. □

**Lemma 7.3.5** *Suppose $q$ is odd, and let $g \in \mathrm{GL}(V)$ be such that $g^2 = 1$. If $V = Z_1 \oplus Z_2$ and $g$ acts nontrivially on $\{Z_1, Z_2\}$, then* $\dim V^g = n/2$.

***Proof*** Since $q$ is odd and $g^2 = 1$, the action of $g$ is diagonalizable and we have $V = V_1 \oplus V_{-1}$, where $V_c$ denotes the $g$-eigenspace for eigenvalue $c$. If $g$ acts nontrivially on $\{Z_1, Z_2\}$, it is straightforward to see that

$$V_1 = \{v + g(v) : v \in Z_1\}$$
$$V_{-1} = \{v - g(v) : v \in Z_1\},$$

so $\dim V_1 = \dim V_{-1} = \dim Z_1 = n/2$. □

**Lemma 7.3.6** *Assume that $q$ is odd and $\kappa \neq 0$. Let $L = (H_1 \times \cdots \times H_t) \cap \Delta(V, \kappa)$ with respect to a decomposition $V = W_1 \perp \cdots \perp W_t$, such that $H_i \leq \Delta(W_i, \kappa)$ is metrically primitive maximal irreducible solvable for all $1 \leq i \leq t$.*

*Let $L^\circ \leq G \leq \Delta(V, \kappa)$ be such that the action of $N_G(W_i)$ on $W_i$ is metrically primitive irreducible for all $1 \leq i \leq t$. Suppose $V = Z_1 \perp \cdots \perp Z_s$ such that $G$ acts transitively on $\{Z_1, \ldots, Z_s\}$. Then one of the following holds:*

*(i) $L^\circ$ acts on $Z_m$ for all $1 \leq m \leq s$.*
*(ii) $\kappa$ is a quadratic form and* $\dim Z_m = 1$ *for all $1 \leq m \leq s$.*

***Proof*** Let $1 \leq i \leq t$. It will suffice to prove that either $H_i^\circ$ acts on $Z_m$ for all $1 \leq m \leq s$, or (ii) holds.

We consider first the case where $\dim W_i = 1$. In this case $\kappa$ must be a quadratic form and $H_i^\circ = O_1(q)$. Then $H_i^\circ = \langle g \rangle$, where $g^2 = 1$ and $\dim V^g = n - 1$.

If $g$ does not act on $Z_m$ for all $m$, there is an orbit $\{Z_\alpha, Z_\beta\}$ of size 2 in the action of $g$ on $\{Z_1, \ldots, Z_s\}$. It follows from Lemma 7.3.5 that $\dim(Z_\alpha \perp Z_\beta)^g = d$, where $d = \dim Z_\alpha = \dim Z_\beta$. On the other hand, by $\dim V^g = n - 1$ we must have $\dim(Z_\alpha \perp Z_\beta)^g = 2d - 1$, so $d = 1$. Then $\dim Z_m = 1$ for all $1 \leq m \leq s$ since $G$ acts transitively on $\{Z_1, \ldots, Z_s\}$, so (ii) holds.

Suppose then that $\dim W_i > 1$. There exists a summand $Z_{m_0}$ which has nonzero projection to $W_i$. Let $v \in Z_{m_0}$ such that $v = w + w'$, where $w \in W_i \setminus \{0\}$, and $w' \in \sum_{j \neq i} W_j$. Because $q$ is odd and $\dim W_i > 1$, by Lemma 7.3.3 the $H_i^\circ$-orbit of $v$ contains $> \dim W_i + 1$ vectors. Since the orbit is contained in $W_i \oplus \langle w' \rangle$, it cannot be linearly independent. Thus there exist $v_1 \neq v_2$ in the $H_i^\circ$-orbit of $v$ that are contained in the same summand, say $v_1, v_2 \in Z_{m'}$ for some $m'$. Then $v_1 - v_2 \in W_i \cap Z_{m'}$, so $W_i \cap Z_{m'} \neq 0$.

Thus $\sum_{m=1}^{s} W_i \cap Z_m$ is a nonzero $H_i^\circ$-invariant subspace of $W_i$, so by irreducibility

$$W_i = \sum_{m=1}^{s} W_i \cap Z_m.$$

The action of $N_G(W_i)$ on $W_i$ is irreducible and metrically primitive, so we must have $W_i = W_i \cap Z_{m'}$.

In other words, we have $W_i \subseteq Z_{m'}$. Because $W_i$ is $H_i^\circ$-invariant, it follows that $H_i^\circ$ acts on $Z_{m'}$. Furthermore $W_i$ is the unique nontrivial irreducible $\mathbb{F}_q[H_i^\circ]$-submodule in $V$, so the action of $H_i^\circ$ on $\sum_{m \neq m'} Z_m$ is trivial. This completes the proof of the lemma. □

**Lemma 7.3.7** *Let $L = (H_1 \times \cdots \times H_t) \cap \Delta(V, \kappa)$ with respect to a decomposition $V = W_1 \perp \cdots \perp W_t$, and assume that $H_i \leq \Delta(W_i, \kappa)$ is metrically primitive maximal irreducible solvable for all $1 \leq i \leq t$. Let $L^\circ \leq G \leq \Delta(V, \kappa)$ be such that $G$ acts metrically primitively on $W_i$ for all $1 \leq i \leq t$. Suppose $V = Z_1 \perp \cdots \perp Z_s$ such that $G$ acts transitive on $\{Z_1, \ldots, Z_s\}$. Then one of the following holds:*

*(i)* $s = 1$.
*(ii)* $s = 2$, $t = 2$, $n = 2$, *and one of the following holds:*

- *(a)* $\kappa = 0$, $q = 3$, *and* $L^\circ = \mathrm{GL}_1(3) \times \mathrm{GL}_1(3)$;
- *(b)* $\kappa$ *is a quadratic form,* $q$ *is odd, and* $L^\circ = O_1(q) \times O_1(q)$.

***Proof*** Throughout we will denote $d_i := \dim W_i$ for all $1 \leq i \leq t$.

Suppose first that $W_i \cap Z_j \neq 0$ for some $j$. Since $G$ acts on $\{Z_1, \ldots, Z_s\}$, the subspace $\sum_{j=1}^{s} W_i \cap Z_j$ is a nonzero $G$-submodule of $W_i$, so by irreducibility

$$W_i = (W_i \cap Z_1) \perp \cdots \perp (W_i \cap Z_s).$$

On the other hand $G$ acts transitively on $\{Z_1, \ldots, Z_s\}$ and the action of $G$ on $W_i$ is metrically primitive, so it follows that $s = 1$ and (i) holds.

Therefore we can assume that $W_i \cap Z_j = 0$ for all $i$ and $j$. We will prove that (ii) holds in this case. Let $1 \leq i \leq t$. There exists some summand $Z_{j_0}$ which has nonzero projection to $W_i$. Since $Z_{j_0} \cap W_i = 0$, we can find $v \in Z_{j_0}$ such that $v = w_i + w_{i_0} + v'$, where $w_i \in W_i$ and $w_{i_0} \in W_{i_0}$ are nonzero with $i \neq i_0$, and $v' \in \sum_{i' \neq i, i_0} W_{i'}$. We note the following.

**Claim:** $s \geq d_i + 1$ **and** $|H_i^\circ v| = d_i + 1$
If the $H_i^\circ$-orbit of $v$ contains vectors $v_1 \neq v_2$ in the same summand $Z_j$, then $v_1 - v_2 \in W_i \cap Z_j$, which contradicts $W_i \cap Z_j = 0$.

Therefore $s \geq |H_i^\circ v|$. Furthermore, the orbit $H_i^\circ v$ is linearly independent, since the vectors are contained in distinct $Z_j$'s. Thus $|H_i^\circ v| \leq d_i + 1$, since $H_i^\circ v$ is contained in the subspace $W_i \oplus \langle w_{i_0} + v' \rangle$ of dimension $d_i + 1$. On the other hand, by Lemma 7.3.1 we have $|H_i^\circ v| \geq d_i + 1$, so $|H_i^\circ v| = d_i + 1$.

We will next prove that $s = 2$. The orbit of $v$ under $H_i^\circ \times H_{i_0}^\circ$ is contained in $W_i \oplus W_{i_0} \oplus \langle v' \rangle$, which has dimension $d_i + d_{i_0} + 1$. On the other hand, by Lemma 7.3.1 the $(H_i^\circ \times H_{i_0}^\circ)$-orbit contains at least

$$(d_i + 1)(d_{i_0} + 1) > d_i + d_{i_0} + 1$$

vectors. Thus the $(H_i^\circ \times H_{i_0}^\circ)$-orbit of $v$ cannot be linearly independent, so it contains vectors $v_1 \neq v_2$ which are contained in the same summand $Z_{j_0}$. Then $v_1 - v_2 \in (W_i \oplus W_{i_0}) \cap Z_{j_0}$, and in particular $(W_i \oplus W_{i_0}) \cap Z_{j_0} \neq 0$.

Therefore $\sum_{j=1}^{s} (W_i \oplus W_{i_0}) \cap Z_j$ is a nonzero $G$-submodule of $W_i \oplus W_{i_0}$. It cannot be equal to $W_i$ or $W_{i_0}$ since $W_i \cap Z_j = W_{i_0} \cap Z_j = 0$, so by Lemma 2.1.15 we have

$$W_i \oplus W_{i_0} = Z_1' \perp \cdots \perp Z_s',$$

where $Z_j' = (W_i \oplus W_{i_0}) \cap Z_j$ for all $1 \leq j \leq s$. The action of $G$ on $\{Z_1, \dots, Z_s\}$ is transitive, so by Lemma 7.3.4 we have $s \leq 2$. We cannot have $s = 1$ since $W_i \cap Z_j = 0$, so $s = 2$ and $V = Z_1 \perp Z_2$.

By the claim proved earlier, we have $s \geq d_i + 1$ and $|H_i^\circ v| = d_i + 1$. Since $s = 2$, we conclude that $d_i = 1$ and $H_i^\circ = \{\pm 1\}$ for all $1 \leq i \leq t$. In particular $q$ must be odd.

To complete the proof of the lemma, let $h_i$ be a generator of $H_i^\circ$. If $h_i$ leaves $Z_1$ and $Z_2$ invariant, then either $Z_1$ or $Z_2$ must contain a $h_i$-eigenvector with eigenvalue $-1$, but this is not possible since $W_i \cap Z_1 = W_i \cap Z_2 = 0$. Therefore $h_i$ acts nontrivially on $\{Z_1, Z_2\}$, so it follows from Lemma 7.3.5 that $\dim V^{h_i} = n/2$. On the other hand $\dim V^{h_i} = n - 1$, so $n = 2$. We have proved that $s = t = n = 2$, and it follows from $H_i^\circ = \{\pm 1\}$ that (ii)(a) or (ii)(b) holds. □

**Lemma 7.3.8** *Assume that $n$ and $q$ are even, and that $\kappa$ is a quadratic form. Let $L = (H_1 \times \cdots \times H_t) \cap \Delta(V, \kappa)$ with respect to a decomposition $V = W_1 \perp \cdots \perp W_t$, and assume that $H_i \leq \Delta(W_i, \kappa)$ is metrically primitive maximal irreducible solvable for all $1 \leq i \leq t$. Suppose $V = Z_1 \perp \cdots \perp Z_s$ such that $L^\Omega$ acts on $\{Z_1, \dots, Z_s\}$. Then $s \leq t$.*

***Proof*** We will argue similarly to the proof in [57, Lemma 3.2], with a few necessary adjustments. If $t = 1$, then $L^\Omega = H_1^\Omega$ is metrically primitive by Theorem 5.6.11, so $s = 1$. We assume then that $t > 1$ and proceed by induction on $t$.

Suppose first that the action of $L^\Omega$ on $\{Z_1, \dots, Z_s\}$ is not transitive, with orbits $\{Z_1^{(1)}, \dots, Z_{s_1}^{(1)}\}, \dots, \{Z_1^{(m)}, \dots, Z_{s_m}^{(m)}\}$. Then for all $1 \leq i \leq m$ the sum $Z_1^{(i)} \perp \cdots \perp Z_{s_i}^{(i)}$ is a non-degenerate $L^\Omega$-submodule, so by Lemma 2.3.12 we have

$$Z_1^{(i)} \perp \cdots \perp Z_{s_i}^{(i)} = W_1^{(i)} \perp \cdots \perp W_{t_i}^{(i)}$$

for some subset $\{W_1^{(i)}, \ldots, W_{t_i}^{(i)}\}$ of $\{W_1, \ldots, W_t\}$. By applying induction on

$$(H_1^{(i)} \times \cdots \times H_{t_i}^{(i)})^{\Omega},$$

it follows that $s_i \leq t_i$ for all $1 \leq i \leq m$. Since $s = s_1 + \cdots + s_m$ and $t = t_1 + \cdots + t_m$, we conclude that $s \leq t$.

Therefore we can assume that $L^{\Omega}$ acts transitively on $\{Z_1, \ldots, Z_s\}$. Let $t_0 > 0$ be minimal such that

$$Z_{i_0} \cap (W_{i_1} \perp \cdots \perp W_{i_{t_0}}) \neq 0$$

for some $1 \leq i_0 \leq s$ and $1 \leq i_1 < \cdots < i_{t_0} \leq t$.

Denote $Z'_j := Z_j \cap (W_{i_1} \perp \cdots \perp W_{i_{t_0}})$ for $1 \leq j \leq s$. Then $L^{\Omega}$ acts transitively on $\{Z'_1, \ldots, Z'_s\}$, and

$$Z'_1 \perp \cdots \perp Z'_s$$

is an $L^{\Omega}$-submodule of $W_{i_1} \perp \cdots \perp W_{i_{t_0}}$.

We consider the case where $Z'_1$ is degenerate, so $Z'_1 \cap (Z'_1)^{\perp} \neq 0$. Then $Z = \sum_{j=1}^{s} Z'_j \cap (Z'_j)^{\perp}$ is a nonzero totally isotropic $L^{\Omega}$-submodule of $V$. It follows from Lemma 2.3.12 that there exists $1 \leq i \leq t$ such that $Z \subseteq W_i$. Furthermore, by Theorem 5.6.11 the action of $H_i^{\Omega}$ on $Z$ is metrically primitive, so it follows that $s = 1$.

Suppose then that $Z'_1$ is non-degenerate, in which case $Z'_1 \perp \cdots \perp Z'_s$ is a non-degenerate $L^{\Omega}$-submodule of $W_{i_1} \perp \cdots \perp W_{i_{t_0}}$. Hence $Z'_1 \perp \cdots \perp Z'_s$ is a sum of some $W_i$'s (Lemma 2.3.12), so by minimality of $t_0$ we must have

$$Z'_1 \perp \cdots \perp Z'_s = W_{i_1} \perp \cdots \perp W_{i_{t_0}}.$$

If $t_0 < t$, then by induction $s \leq t_0 < t$. Therefore we can assume that $t_0 = t$, which implies that

$$Z_i \cap (W_{i_1} \perp \cdots \perp W_{i_{t-1}}) = 0 \tag{7.26}$$

for all $1 \leq i \leq \ell$ and $1 \leq i_1 < \cdots < i_{t-1} \leq t$. In particular the projection of $Z_i$ into $W_j$ is injective for all $i$ and $j$, so

$$\dim Z_i \leq \dim W_j \tag{7.27}$$

for all $i$ and $j$.

Next let $s_0 > 0$ be minimal such that $W_i \cap (Z_{j_1} \perp \cdots \perp Z_{j_{s_0}}) \neq 0$ for some $1 \leq i \leq t$ and $1 \leq j_1 < \cdots < j_{s_0} \leq s$. We will first prove that $H_j^{\Omega}$ acts nontrivially on $Z_{j_1} \perp \cdots \perp Z_{j_{s_0}}$ for all $j \neq i$. To this end, let $h \in H_j^{\Omega}$. We have $h(Z_{j_1} \perp \cdots \perp$

$Z_{j_{s_0}}) = Z_{j'_1} \perp \cdots \perp Z_{j'_{s_0}}$ for some $j'_1 < \cdots < j'_{s_0}$. On the other hand $H_j$ acts trivially on $W_i$, so for any $v \in W_i \cap (Z_{j_1} \perp \cdots \perp Z_{j_{s_0}})$ we have

$$hv = v \in (Z_{j_1} \perp \cdots \perp Z_{j_{s_0}}) \cap (Z_{j'_1} \perp \cdots \perp Z_{j'_{s_0}}).$$

By the minimality of $s_0$ we must have $Z_{j_1} \perp \cdots \perp Z_{j_{s_0}} = Z_{j'_1} \perp \cdots \perp Z_{j'_{s_0}}$, so $H_j^{\Omega}$ acts on $Z_{j_1} \perp \cdots \perp Z_{j_{s_0}}$.

Next we check that the action of $H_j^{\Omega}$ on $Z_{j_1} \perp \cdots \perp Z_{j_{s_0}}$ is nontrivial. To this end, let $v \in Z_{j_1} \setminus \{0\}$. Then by (7.26) we have $v = w_1 + \cdots + w_t$ with $w_i \in W_i$ and $w_i \neq 0$ for all $1 \leq i \leq t$. Because the action of $H_j^{\Omega}$ is nontrivial on $W_j$, there exists $h \in H_j$ such that $hw_j \neq w_j$, so $hv \neq v$.

We have proved that $H_j^{\Omega}$ acts nontrivially on $Z_{j_1} \perp \cdots \perp Z_{j_{s_0}}$ for all $j \neq i$. The action of $H_j^{\Omega}$ on $W_j$ has no trivial submodules (Lemma 2.2.17) and $H_j^{\Omega}$ acts trivially on $W_{j'}$ for $j' \neq j$, so we must have $W_j \cap (Z_{j_1} \perp \cdots \perp Z_{j_{s_0}}) \neq 0$. Therefore by the arguments in the previous two paragraphs, we can conclude that $H_i^{\Omega}$ also acts nontrivially on $Z_{j_1} \perp \cdots \perp Z_{j_{s_0}}$.

Hence $Z_{j_1} \perp \cdots \perp Z_{j_{s_0}}$ is an $(H_1^{\Omega} \times \cdots \times H_t^{\Omega})$-invariant subspace, on which $H_i^{\Omega}$ acts nontrivially for all $1 \leq i \leq t$. It follows then from Lemmas 2.2.17 and 2.1.15 that

$$Z_{j_1} \perp \cdots \perp Z_{j_{s_0}} = W'_1 \perp \cdots \perp W'_t,$$

where $W'_i$ is a nonzero $\mathbb{F}_q[H_i^{\Omega}]$-submodule of $W_i$ for all $1 \leq i \leq t$.

On the other hand $Z_{j_1} \perp \cdots \perp Z_{j_{s_0}}$ is non-degenerate, so $W'_i = W_i$ for all $1 \leq i \leq t$ (Lemma 2.2.17). Therefore $Z_{j_1} \perp \cdots \perp Z_{j_{s_0}} = W_1 \perp \cdots \perp W_t$, which implies that $s_0 = s$. Then $W_i \cap (Z_{j_1} \perp \cdots \perp Z_{j_{s-1}}) = 0$ for all $1 \leq j_1 < \cdots < j_{s-1} \leq s$, so $\dim W_i \leq \dim Z_j$. Combining this with (7.27) gives $\dim W_i = \dim Z_j$ for all $i$ and $j$, so $s = t$ and the proof of the lemma is complete. □

**Lemma 7.3.9** *Assume that $n$ and $q$ are even, and that $\kappa$ is a quadratic form. Let $L = (H_1 \times \cdots \times H_t) \cap \Delta(V, \kappa)$ with respect to a decomposition $V = W_1 \perp \cdots \perp W_t$, and assume that $H_i \leq \Delta(W_i, \kappa)$ is metrically primitive maximal irreducible solvable for all $1 \leq i \leq t$. Suppose $V = Z_1 \perp \cdots \perp Z_s$ such that $L^{\Omega}$ acts transitively on $\{Z_1, \ldots, Z_s\}$. Then $s = 1$.*

***Proof*** Suppose first that $W_i \cap Z_j \neq 0$ for some $i$ and $j$. Then

$$Z = (W_i \cap Z_1) \perp \cdots \perp (W_i \cap Z_s)$$

is a nonzero $L^{\Omega}$-submodule of $W_i$. Note that $W_i \cap Z_j \neq 0$ for all $1 \leq j \leq s$, since $L^{\Omega}$ acts transitively on $\{Z_1, \ldots, Z_s\}$.

If $H_i^{\Omega}$ is irreducible, we have $Z = W_i$, and therefore $s = 1$ since $H_i^{\Omega}$ is metrically primitive by Theorem 5.6.11.

If $H_i^\Omega$ is not irreducible, then we have a totally singular decomposition $W_i = W_i' \oplus W_i''$, where $W_i'$ and $W_i''$ are nonisomorphic irreducible $\mathbb{F}_q[H_i^\Omega]$-modules (Theorem 5.6.11). Thus either $Z = W_i$, $Z = W_i'$, or $Z = W_i''$ (Lemma 2.1.15). By Theorem 5.6.11 (see also Remark 5.6.12) $H_i^\Omega$ acts primitively on $W_i'$ and $W_i''$, and metrically primitively on $W_i$. Therefore we must have $s = 1$ in this case as well.

Therefore we can assume that $W_i \cap Z_j = 0$ for all $i$ and $j$. Let $1 \leq i \leq t$. There exists a summand $Z_{j_0}$ that has nonzero projection to $W_i$. Because $W_i \cap Z_{j_0} = 0$, we can find $v \in Z_{j_0}$ such that $v = w_i + w_{i_0} + w'$, where $w_i \in W_i$ and $w_{i_0} \in W_{i_0}$ are nonzero with $i \neq i_0$, and $w' \in \sum_{i' \neq i, i_0} W_{i'}$.

By Lemma 7.3.3 we have $|H_i^\Omega w_i| \geq d_i + 1$. Thus by arguing as in the proof of Lemma 7.3.7 (Claim), it follows that $s \geq d_i + 1$ and $|H_i^\Omega w_i| = d_i + 1$. In this case $H_i$ is as in one the exceptional cases of Lemma 7.3.3, and in particular $H_i^\Omega$ is irreducible (Lemma 5.2.4).

The $(H_i^\Omega \times H_{i_0}^\Omega)$-orbit of $v$ has size

$$(d_i + 1)(d_{i_0} + 1) > d_i + d_{i_0} + 1,$$

by Lemma 7.3.3. Thus as in the proof of Lemma 7.3.7, we have $(W_i \perp W_{i_0}) \cap Z_j \neq 0$ for some $1 \leq j \leq s$. Therefore $\sum_{j=1}^{s} (W_i \perp W_{i_0}) \cap Z_j$ is a nonzero $L^\Omega$-submodule of $W_i \perp W_{i_0}$. Because $H_i^\Omega$ and $H_{i_0}^\Omega$ are irreducible and $W_i \cap Z_j = W_{i_0} \cap Z_j = 0$, by Lemma 2.1.15 we have

$$W_i \perp W_{i_0} = ((W_i \perp W_{i_0}) \cap Z_1) \perp \cdots \perp ((W_i \perp W_{i_0}) \cap Z_s).$$

It follows from Lemma 7.3.8 that $s \leq 2$. On the other hand $s \geq d_i + 1 > 2$ since $d_i$ is even, so we have a contradiction. This completes the proof of the lemma. □

## 7.4 Maximality of Metrically Imprimitive Subgroups

Suppose that $G = (H \wr K) \cap \Delta(V, \kappa)$ is as in Theorem 2.2.14 (i), so $H \leq \Delta(W, \kappa')$ is metrically primitive maximal irreducible solvable, and $K \leq S_k$ is maximal transitive solvable with $k > 1$, where $\dim V = k \dim W$. In this section we will:

- Classify the cases where $G$ is maximal solvable in $\Delta(V, \kappa)$ (Theorem 7.4.7).
- Classify the cases where $G^\circ$ is maximal solvable in $I(V, \kappa)$ (Theorem 7.4.6).
- For $q$ even and $\kappa$ is a quadratic form, prove that $G^\Omega$ is maximal solvable in $\Omega(V, \kappa)$ (Theorem 7.4.8).

We will also consider semiprimary groups semiwr($H^\circ$) as in Theorem 2.2.14 (ii), where $H \leq \Delta(W, \kappa')$ is maximal irreducible solvable of multiplier 2. In this case we will prove that semiwr($H^\circ$) is maximal solvable in $\Delta(V, \kappa)$, except for semiwr($O_1(q)$) with $\kappa$ a quadratic form and $q > 3$ (Theorem 7.4.9).

In Sect. 7.2 we have seen that in most cases, these groups cannot be contained in a metrically primitive solvable group. Thus it remains to determine when they can be contained in another metrically imprimitive group with structure as in Theorem 2.2.14 (i)–(ii). For a large part, this comes down to the question of possible systems of imprimitivity for the groups $(H \wr K) \cap \Delta(V, \kappa)$, $H^\circ \wr K$, and $(H \wr K)^\Omega$ as above. Some results in this direction were obtained in [57] in the case where $\kappa = 0$.

**Theorem 7.4.1** *Suppose that* $\dim V = k \dim W$, *where* $k > 1$. *Let* $H$ *be a nontrivial irreducible metrically primitive subgroup of* $I(W, \kappa')$ *and let* $K \leq S_k$ *transitive, so that* $G = H \wr K$ *as in Example 2.2.5 is an irreducible subgroup of* $I(V, \kappa)$ *(Lemma 2.2.6). Then* $G$ *has a unique nonrefinable orthogonal system of imprimitivity, except when* $\dim W = 1$, $n = k$ *is even,* $|H| = 2$, *and* $K$ *is a subgroup of* $S_2 \wr S_{n/2}$.

***Proof*** The result follows exactly as in [57, Proof of Theorem 1.1, pp. 633–634], replacing "primitive" with "metrically primitive", replacing $\oplus$ with $\perp$, and using Lemma 7.3.4 instead of [57, Lemma 3.2]. □

**Lemma 7.4.2** *Let* $G = (H \wr K) \cap \Delta(V, \kappa)$ *be as in Theorem 2.2.14 (i), with orthogonal system of imprimitivity* $V = W_1 \perp \cdots \perp W_k$. *Then one of the following holds.*

*(i)* $\{W_1, \ldots, W_k\}$ *is the unique nonrefinable system of imprimitivity for* $G^\circ$ *on* $V$.
*(ii)* $H^\circ$ *is metrically imprimitive.*
*(iii)* $\dim W_1 = 1$, $q$ *is odd,* $n = k$ *is even,* $K = S_2 \wr K'$ *for some* $K' \leq S_{k/2}$, *and one of the following holds:*

  *(a)* $\kappa = 0$, $q = 3$, *and* $H = \mathrm{GL}_1(3)$.
  *(b)* $\kappa \neq 0$ *is a quadratic form and* $H = \mathrm{GO}_1(q)$.

***Proof*** For the proof we can assume that (ii) does not hold, so $H^\circ$ is metrically primitive. Applying Theorem 7.4.1 on $G^\circ = H^\circ \wr K$ we see that (i) holds, except possibly when $\dim W_1 = 1$, $|H^\circ| = 2$, $n = k$ is even, and $K \leq S_2 \wr S_{n/2}$.

We check that (iii) holds in the exceptional case of Theorem 7.4.1. To this end, it follows from $\dim W_1 = 1$ and $|H^\circ| = 2$ that either $\kappa = 0$ and $H = \mathrm{GL}_1(3)$, or $\kappa \neq 0$ is a quadratic form and $H^\circ = O_1(q)$. In other words $q$ is odd, and either (iii)(a) or (iii)(b) holds. Next note that $K$ is a subgroup of $S_2 \wr K'$, where $K'$ is the image of $K$ in $S_{n/2}$. Thus $K = S_2 \wr K'$ since $K$ is maximal solvable, which completes the proof of the claims in (iii). □

**Lemma 7.4.3** *Let* $G = (H \wr K) \cap \Delta(V, \kappa)$ *be as in Theorem 2.2.14 (i), with orthogonal system of imprimitivity* $V = W_1 \perp \cdots \perp W_k$. *Then one of the following holds.*

*(i)* $\{W_1, \ldots, W_k\}$ *is the unique nonrefinable system of imprimitivity for* $G$ *on* $V$.
*(ii)* $\dim W_1 = 1$, $q$ *is odd,* $k$ *is even,* $K = S_2 \wr K'$ *for some* $K' \leq S_{k/2}$, *and one of the following holds:*

  *(a)* $\kappa = 0$, $q = 3$, *and* $H = \mathrm{GL}_1(3)$.
  *(b)* $\kappa \neq 0$ *is a quadratic form and* $H = \mathrm{GO}_1(q)$.

***Proof*** Let $\{Z_1, \ldots, Z_\ell\}$ be an orthogonal system of imprimitivity for $G$ on $V$, so

$$V = Z_1 \perp \cdots \perp Z_\ell$$

with $\ell \geq 2$ and $G$ acts on $\{Z_1, \ldots, Z_\ell\}$.

Suppose that $H$ is as in one of the exceptions of Lemma 7.3.3 (ii). It follows from Corollary 5.6.13 that in these cases $H^\circ$ is metrically primitive, so the result follows from Lemma 7.4.2.

Thus we can assume that $H$ is not among the exceptional cases listed in Lemma 7.3.3. Let $v \in Z_1 \setminus \{0\}$. Then there exists $1 \leq i \leq k$ such that $v = w_i + v'$, where $w_i \in W_i \setminus \{0\}$ and $v' \in \sum_{j \neq i} W_j$.

By Lemma 7.3.3 the $H_i^\circ$-orbit of $v$ contains $> d + 1$ vectors, where $d = \dim W_i$. On the other hand, the $H_i^\circ$ is contained in the subspace $W_i \oplus \langle v' \rangle$ of dimension $d + 1$. Thus vectors in the $H_i^\circ$-orbit of $v$ cannot be linearly independent, so we can find $v_1 \neq v_2$ in the orbit which are contained in the same summand $Z_j$. Then $v_1 - v_2 \in W_i \cap Z_j$, and in particular $W_i \cap Z_j \neq 0$.

Hence we may assume without loss of generality that $W_1 \cap Z_1 \neq 0$. In this case we can argue similarly to [57, Proof of Theorem 1.1, p. 633]. First note that

$$(W_1 \cap Z_1) \perp \cdots \perp (W_1 \cap Z_\ell)$$

is a nonzero $N_G(W_1)$-submodule of $W_1$. Because $G$ is irreducible, the action of $N_G(W_1)$ on $W_1$ must be irreducible (Lemma 2.2.4 (ii)), so $W_1 = (W_1 \cap Z_1) \perp \cdots \perp (W_1 \cap Z_\ell)$. Furthermore the action of $N_G(W_1)$ on $W_1$ is equal to $H_1$, which is metrically primitive, so $W_1 = W_1 \cap Z_1$.

Therefore $W_1 \subseteq Z_1$, and the $N_G(Z_1)$-submodule generated by $W_1$ is equal to $W_{i_1} \perp \cdots \perp W_{i_r}$ for some $1 = i_1 < \cdots < i_r$. Because $G$ is irreducible, the action of $N_G(Z_1)$ on $Z_1$ is irreducible (Lemma 2.2.4 (ii)), so $Z_1 = W_{i_1} \perp \cdots \perp W_{i_r}$. Consequently $\{W_1, \ldots, W_k\}$ is a refinement of $\{Z_1, \ldots, Z_\ell\}$, and the result follows. □

**Lemma 7.4.4** *Suppose that $n$ is even and $q$ is odd. Let $G = H \wr S_2 \wr K' \leq \mathrm{GL}(V)$, where $H = \{\pm 1\} \leq \mathrm{GL}_1(q)$, and $K' \leq S_{n/2}$ is transitive. Let $V = W_1 \oplus \cdots \oplus W_n$ be the system of imprimitivity defining $G$, where $W_i = \langle e_i \rangle$ for all $1 \leq i \leq n$ and $\{\{e_1, e_2\}, \ldots, \{e_{n-1}, e_n\}\}$ is a system of imprimitivity for $S_2 \wr K'$ on $\{e_1, \ldots, e_n\}$. Then any nonrefinable system of imprimitivity for $G$ is equal to one of the following:*

*(i)* $\{W_1, \ldots, W_n\}$.
*(ii)* $\{\langle e_1 + \lambda e_2 \rangle, \langle e_1 - \lambda e_2 \rangle, \ldots, \langle e_{n-1} + \lambda e_n \rangle, \langle e_{n-1} - \lambda e_n \rangle\}$, *where* $\lambda^2 = \pm 1$.

*Furthermore, the action of $G$ on the system of imprimitivity in (ii) is isomorphic to $S_2 \wr K'$.*

***Proof*** It follows from Lemma 1.3.3 that $\{\{W_1, W_2\}, \ldots, \{W_{n-1}, W_n\}\}$ is the unique nonrefinable system of imprimitivity for $S_2 \wr K'$ on $\{W_1, \ldots, W_n\}$. Thus it follows from [57, Remark 3.3] that every nonrefinable system of imprimitivity for $G$ must be as in (i) or (ii).

For the remaining claim, let $\{W_1', \ldots, W_n'\}$ be as in (ii), so $W_1' = \langle e_1+\lambda e_2\rangle$, $W_2' = \langle e_1 - \lambda e_2\rangle, \ldots, W_{n-1}' = \langle e_{n-1} + \lambda e_n\rangle$, $W_n' = \langle e_{n-1} - \lambda e_n\rangle$. Note that $G$ contains maps $h_i$ defined by $h_i(e_i) = -e_i$, and $h_i(e_j) = e_j$ for all $j \neq i$. For the action of $G$ on $\{W_1', \ldots, W_n'\}$, these maps generate the base group $S_2 \times \cdots \times S_2$ of the wreath product $S_2 \wr K'$. Furthermore, the action of $G$ on $\{\{W_1, W_2\}, \ldots, \{W_{n-1}, W_n\}\}$ is exactly the same as its action on $\{\{W_1', W_2'\}, \ldots, \{W_{n-1}', W_n'\}\}$, since $W_{2i-1} \oplus W_{2i} = W_{2i-1}' \oplus W_{2i}'$ for all $1 \leq i \leq n/2$. Thus the action of $G$ on $\{W_1', \ldots, W_n'\}$ is precisely $S_2 \wr K'$, as claimed. □

**Lemma 7.4.5** *Assume that $n$ and $q$ are even, and that $\kappa$ is a quadratic form. Let $G = (H \wr K) \cap \Delta(V, \kappa)$ be as in Theorem 2.2.14 (i), with orthogonal system of imprimitivity $V = W_1 \perp \cdots \perp W_k$. Then $\{W_1, \ldots, W_k\}$ is the unique nonrefinable system of imprimitivity for $G^\Omega$ on $V$.*

***Proof*** Let $V = Z_1 \perp \cdots \perp Z_\ell$ be an orthogonal system of imprimitivity for $G^\Omega$ on $V$. By construction $G$ contains $L = (H_1 \times \cdots \times H_k) \cap \Delta(V, \kappa)$ with respect to the decomposition $V = W_1 \perp \cdots \perp W_k$, where $H_i \leq \Delta(W_i, \kappa)$ is isometric to $H$ for all $1 \leq i \leq k$.

It follows from Lemma 7.3.9 that the $L^\Omega$-orbits on $\{Z_1, \ldots, Z_\ell\}$ must all be of size 1. Therefore $L^\Omega$ acts on $Z_i$ for all $i$, so by Lemma 2.3.12 we have

$$Z_i = W_1^{(i)} \perp \cdots \perp W_{k/\ell}^{(i)}$$

for some subset $\{W_1^{(i)}, \ldots, W_{k/\ell}^{(i)}\}$ of $\{W_1, \ldots, W_k\}$. In other words $\{W_1, \ldots, W_k\}$ is a refinement of $\{Z_1, \ldots, Z_\ell\}$. □

**Theorem 7.4.6** *Let $G = (H \wr K) \cap \Delta(V, \kappa)$ be as in Theorem 2.2.14 (i). Then one of the following holds.*

*(i) $G^\circ$ is maximal solvable in $I(V, \kappa)$.*
*(ii) $H^\circ$ is not maximal solvable in $I(W, \kappa)$.*
*(iii) $\kappa = 0$, $d = 1$, $H = \mathrm{GL}_1(q)$, and one of the following holds:*

*(a) $q \in \{3, 5\}$ and $G = \mathrm{GL}_1(q) \wr S_2 \wr K'$.*
*(b) $q = 4$ and $G = \mathrm{GL}_1(4) \wr S_3 \wr K'$.*
*(c) $q = 3$ and $G = \mathrm{GL}_1(3) \wr S_4 \wr K'$.*

*(iv) $\kappa \neq 0$ is a quadratic form, $q$ is odd, and one of the following holds:*

*(a) $d = 1$ and $G^\circ = O_1(q) \wr S_2 \wr K'$, with $q > 5$.*
*(b) $d = 1$ and $G^\circ = O_1(q) \wr S_4 \wr K'$.*
*(c) $d = 2$, $q = 3$, and $G^\circ = O_2^-(3) \wr S_2 \wr K'$.*
*(d) $d = 2$, $q = 5$, and $G^\circ = O_2^+(5) \wr S_2 \wr K'$.*

*Furthermore, in cases (ii)–(iv) $G^\circ$ is not maximal solvable in $I(V, \kappa)$.*

***Proof*** We first check that in cases (ii)–(iv) $G^\circ$ is not maximal solvable in $I(V, \kappa)$. In case (ii) this is clear, for if $H^\circ \lneq H' \leq I(V, \kappa)$ for some solvable $H'$, then $G^\circ = H^\circ \wr K \lneq H' \wr K$. In cases (iii) and (iv) we have $G^\circ = H^\circ \wr S_t \wr K'$ for some $t \in \{2, 3, 4\}$. For the configurations in (iii) and (iv), it follows from Theorem 7.2.22 that $H^\circ \wr S_t \lneq (G')^\circ$ for some irreducible metrically primitive solvable group $G'$. Thus $G^\circ \lneq (G')^\circ \wr K'$, so $G^\circ$ is not maximal solvable.

Suppose then that $G^\circ$ is not maximal solvable in $I(V, \kappa)$. We will verify that one of (ii)–(iv) must hold. First we consider the case where $H^\circ$ is metrically imprimitive. In this case it follows from Proposition 7.2.23 that $H^\circ$ is not maximal solvable in $I(V, \kappa)$, except when $\kappa \neq 0$ is a quadratic form and one of the following holds:

- $q = 3$, $d = 2$, and $H^\circ = O_2^-(3)$.
- $q = 5$, $d = 2$, and $H^\circ = O_2^+(5)$.

In both cases $n$ is even. Since $O_2^-(3) = O_1(3) \wr S_2$ and $O_2^+(5) = O_1(5) \wr S_2$, we have $G^\circ = O_1(q) \wr S_2 \wr K$ with $q \in \{3, 5\}$. If $S_2 \wr K$ is not maximal solvable in $S_n$, by Theorem 1.1.2 we have $K = S_2 \wr K'$ for some $K' \leq S_{n/2}$, so (iv)(c) or (iv)(d) holds. If $S_2 \wr K$ is maximal solvable in $S_n$, we reduce to the case where $H^\circ$ is metrically primitive, by replacing $H$ with $\mathrm{GO}_1(q)$ and $K$ with $S_2 \wr K'$.

Therefore we can assume for the rest of the proof that $H^\circ$ is metrically primitive. Since $G^\circ$ is irreducible but not maximal solvable, there exists a maximal irreducible solvable group $\overline{G} \leq \Delta(V, \kappa)$ such that $G^\circ \lneq \overline{G^\circ}$. If $\overline{G}$ is metrically primitive, it follows from Theorems 7.2.19 and 7.2.22 that $G^\circ$ is as in (ii) or (iii), with $K' = 1$.

Therefore we can assume that $\overline{G}$ is not metrically primitive. Then $\overline{G}$ is not semiprimary since $G^\circ$ is irreducible, so by Theorem 2.2.14, we have $\overline{G} = (\overline{H} \wr \overline{K}) \cap \Delta(V, \kappa)$ as in Theorem 2.2.14 (i). Let $V = W_1 \perp \cdots \perp W_t$ be the system of imprimitivity defining $G$, and $V = Z_1 \perp \cdots \perp Z_s$ be the system of imprimitivity defining $\overline{G}$.

We first assume that $G$ is not as in Lemma 7.4.2 (iii). Since $H^\circ$ is metrically primitive, it follows from Lemma 7.4.2 that $\{W_1, \ldots, W_t\}$ is the unique nonrefinable orthogonal system of imprimitivity for $G^\circ$. By uniqueness $\{W_1, \ldots, W_t\}$ must be a refinement of $\{Z_1, \ldots, Z_s\}$, so for all $1 \leq i \leq s$ we have $Z_i = W_1^{(i)} \perp \cdots \perp W_{d'}^{(i)}$ for some subset $\{W_1^{(i)}, \ldots, W_{d'}^{(i)}\}$ of $\{W_1, \ldots, W_t\}$. It follows that $G^\circ = H^\circ \wr K_1 \wr \overline{K}$, where $H^\circ \wr K_1 \lneq \overline{H^\circ}$. Applying Theorems 7.2.19 and 7.2.22 on $H^\circ \wr K_1$, it follows that that $G^\circ$ is as in (iii) or (iv), with $K' = \overline{K}$.

Suppose then that $G$ is as in Lemma 7.4.2 (iii). In this case $d = 1$, $n$ is even, and one of the following holds:

- $\kappa = 0$ and $G^\circ = \mathrm{GL}_1(3) \wr S_2 \wr K'$.
- $\kappa \neq 0$ is a quadratic form, $q$ is odd, and $G^\circ = O_1(q) \wr S_2 \wr K'$.

The first of these configurations corresponds to (iii)(a), and the second to (iv)(a) if $q > 5$. We will now consider the case where $G^\circ = O_1(q) \wr S_2 \wr K'$ for $q \in \{3, 5\}$.

In this case we have a system of imprimitivity $\{W_1, \ldots, W_n\}$ for $G^\circ$ such that

$$V = W_1 \perp \cdots \perp W_n,$$
$$W_i = \langle e_i \rangle \text{ for all } 1 \leq i \leq n,$$

where $S_2 \wr K'$ acts on $\{e_1, \ldots, e_n\}$ and $\{\{e_1, e_2\}, \ldots, \{e_{n-1}, e_n\}\}$ is a system of imprimitivity for $S_2 \wr K'$.

If $\{W_1, \ldots, W_n\}$ is a refinement of $\{Z_1, \ldots, Z_s\}$, it follows as earlier that $G^\circ = O_1(q) \wr S_2 \wr \overline{K}$, where $H^\circ \wr S_2 \leq \overline{H^\circ}$. But $O_1(3) \wr S_2 = O_2^-(3)$ and $O_1(5) \wr S_2 = O_2^+(5)$, so $O_1(q) \wr S_2 = \overline{H^\circ}$. Then $G^\circ = \overline{H^\circ} \wr \overline{K}$, which contradicts $G^\circ \lneqq \overline{G^\circ}$.

Therefore we can assume for the rest of the proof that $\{W_1, \ldots, W_n\}$ is not a refinement of $\{Z_1, \ldots, Z_s\}$. Denote $Q_1 = W_1 \oplus W_2, \ldots, Q_{n/2} = W_{n-1} \oplus W_n$. Then

$$V = Q_1 \perp \cdots \perp Q_{n/2}$$

and $\{Q_1, \ldots, Q_{n/2}\}$ is a system of imprimitivity for $G^\circ$ on $V$. We will prove the following.

**Claim:** $\{Q_1, \ldots, Q_{n/2}\}$ **is a refinement of** $\{Z_1, \ldots, Z_s\}$
Since $G^\circ$ acts on $\{Z_1, \ldots, Z_s\}$, we can refine $\{Z_1, \ldots, Z_s\}$ into a nonrefinable system of imprimitivity for $G^\circ$. Because $\{W_1, \ldots, W_n\}$ is not a refinement of $\{Z_1, \ldots, Z_s\}$, it follows from Lemma 7.4.4 that there exists $\lambda \in \mathbb{F}_q$ such that $\lambda^2 = \pm 1$ and the system $\{Z_1, \ldots, Z_s\}$ refines to $V = W_1' \oplus \cdots \oplus W_n'$, where $W_1' = \langle e_1 + \lambda e_2 \rangle$, $W_2' = \langle e_1 - \lambda e_2 \rangle, \ldots, W_{n-1}' = \langle e_{n-1} + \lambda e_n \rangle$, $W_n' = \langle e_{n-1} - \lambda e_n \rangle$.

For all $1 \leq i \leq s$, we have

$$Z_i = (W_1')^{(i)} \oplus \cdots \oplus (W_\ell')^{(i)}$$

for some subset $\{(W_1')^{(i)}, \ldots, (W_\ell')^{(i)}\}$ of $\{W_1', \ldots, W_n'\}$.

If $\ell = 1$, then $\{Z_1, \ldots, Z_s\} = \{W_1', \ldots, W_n'\}$, and so $\overline{G^\circ} = O_1(q) \wr \overline{K}$. By Lemma 7.4.4 the action of $G^\circ$ on $\{W_1', \ldots, W_n'\}$ is equal to $S_2 \wr K'$, with respect to the system of imprimitivity $\{\{W_1', W_2'\}, \ldots, \{W_{n-1}', W_n'\}\}$. Thus $S_2 \wr K' \leq \overline{K}$, and so $S_2 \wr K' = \overline{K}$ since $K = S_2 \wr K'$ is maximal solvable. But then $G^\circ = \overline{G^\circ}$, contradicting $G^\circ \lneqq \overline{G^\circ}$.

Therefore $\ell > 1$. Now

$$\{\{(W_1')^{(1)}, \ldots, (W_\ell')^{(1)}\}, \ldots, \{(W_1')^{(n/\ell)}, \ldots, (W_\ell')^{(n/\ell)}\}\}$$

forms a nontrivial system of imprimitivity for the action of $G^\circ$ on $\{W_1', \ldots, W_n'\}$, so by Lemma 1.3.3 it can be refined to $\{\{W_1', W_2'\}, \ldots, \{W_{n-1}', W_n'\}\}$. Then each $W_1' \oplus W_2' = Q_1, \ldots, W_{n-1}' \oplus W_n' = Q_{n/2}$ is contained in some $Z_i$, and so $\{Q_1, \ldots, Q_{n/2}\}$ is a refinement of $\{Z_1, \ldots, Z_s\}$.

We have proved that $\{Q_1, \ldots, Q_{n/2}\}$ is a refinement of $\{Z_1, \ldots, Z_s\}$. But $Q_i = W_{2i-1} \oplus W_{2i}$, so in fact $\{W_1, \ldots, W_n\}$ is a refinement of $\{Z_1, \ldots, Z_s\}$, contrary to what we have assumed. This completes the proof of the theorem. □

**Theorem 7.4.7** *Let $G = (H \wr K) \cap \Delta(V, \kappa)$ be as in Theorem 2.2.14 (i). Then one of the following holds.*

*(i) $G$ is maximal solvable in $\Delta(V, \kappa)$.*
*(ii) $\kappa = 0$, $d = 1$, $H = \mathrm{GL}_1(q)$, and one of the following holds:*

*(a) $q \in \{3, 5\}$ and $G = \mathrm{GL}_1(q) \wr S_2 \wr K'$.*
*(b) $q = 4$ and $G = \mathrm{GL}_1(4) \wr S_3 \wr K'$.*
*(c) $q = 3$ and $G = \mathrm{GL}_1(3) \wr S_4 \wr K'$.*

*(iii) $\kappa \neq 0$ is a quadratic form, $q$ is odd, and one of the following holds:*

*(a) $d = 1$ and $G^\circ = O_1(q) \wr S_2 \wr K'$.*
*(b) $d = 1$ and $G^\circ = O_1(q) \wr S_4 \wr K'$.*
*(c) $d = 2$, $q = 3$, and $G^\circ = O_2^-(3) \wr S_2 \wr K'$.*
*(d) $d = 2$, $q = 5$, and $G^\circ = O_2^+(5) \wr S_2 \wr K'$.*

*Furthermore, in cases (ii) and (iii) $G$ is not maximal solvable in $\Delta(V, \kappa)$.*

***Proof*** We will argue similarly to Theorem 7.4.6. First we note that in cases (ii) and (iii), it follows from Theorem 7.2.22 that $G$ is not maximal solvable.

Suppose then that $G$ is not maximal solvable in $\Delta(V, \kappa)$, so there exists a maximal irreducible solvable group $\overline{G} \leq \Delta(V, \kappa)$ such that $G \lneqq \overline{G}$. We will check that either (ii) or (iii) holds.

If $\overline{G}$ is metrically primitive, it follows from Theorem 7.2.19 that $G$ is as in (ii) or (iii), with $K' = 1$.

Suppose then that $\overline{G}$ is metrically imprimitive. Then $\overline{G}$ is not semiprimary since $G^\circ = H^\circ \wr K$ is irreducible (Lemma 2.2.6), so $\overline{G} = (\overline{H} \wr \overline{K}) \cap \Delta(V, \kappa)$ as in Theorem 2.2.14 (i). Let $V = W_1 \perp \cdots \perp W_t$ be the system of imprimitivity defining $G$, and let $V = Z_1 \perp \cdots \perp Z_s$ be the system of imprimitivity defining $\overline{G}$.

If $G$ is as in Lemma 7.4.3 (ii), then $d = 1$ and either (ii)(a) or (iii)(a) holds. Thus we can assume that $G$ is not as in Lemma 7.4.3 (ii), so $G$ has a unique nonrefinable orthogonal system of imprimitivity. Then $\{W_1, \ldots, W_t\}$ must be a refinement of $\{Z_1, \ldots, Z_s\}$. As in the proof of Theorem 7.4.6, it follows that $K = K_1 \wr \overline{K}$, where $\overline{K}$ is the action of $G$ on $\{Z_1, \ldots, Z_s\}$, and $H^\circ \wr K_1 \leq \overline{H^\circ}$. Applying Theorem 7.2.19 on $H^\circ \wr K_1$, we find that that $G^\circ$ is as in (ii) or (iii), with $K' = \overline{K}$. □

**Theorem 7.4.8** *Assume that $n$ and $q$ are even, and that $\kappa$ is a quadratic form. Let $G = (H \wr K) \cap \Delta(V, \kappa)$ be as in Theorem 2.2.14 (i). Then $G^\Omega$ is maximal solvable in $\Omega(V, \kappa)$.*

***Proof*** Let $V = W_1 \perp \cdots \perp W_k$ be the orthogonal system of imprimitivity defining $G$. Suppose for the sake of contradiction that $G^\Omega$ is not maximal solvable, and let $\overline{G} \leq \Delta(V, \kappa)$ be maximal solvable such that $G^\Omega \lneqq \overline{G^\Omega}$.

First note that $\overline{G^\Omega}$ is irreducible, because $G^\Omega$ is (Lemma 2.2.21). It follows from Theorem 7.2.18 that $\overline{G}$ cannot be metrically primitive. Therefore $\overline{G}$ is metrically imprimitive, so by Theorem 2.2.14 we have $\overline{G} = (\overline{H} \wr \overline{K}) \cap \Delta(V, \kappa)$ for some metrically primitive maximal irreducible solvable $\overline{H} \leq \Delta(\overline{W}, \overline{\kappa})$ and maximal transitive solvable $\overline{K} \leq S_\ell$, where $\ell > 1$.

Let $V = Z_1 \perp \cdots \perp Z_\ell$ be the orthogonal system of imprimitivity defining $\overline{G}$. Then $G^\Omega$ acts on $\{Z_1, \ldots, Z_\ell\}$, so by Lemma 7.4.5 the system $\{W_1, \ldots, W_k\}$ is a refinement of $\{Z_1, \ldots, Z_\ell\}$. Thus for all $1 \leq i \leq \ell$ we have

$$Z_i = W_1^{(i)} \perp \cdots \perp W_{t_i}^{(i)}$$

for some subset $\{W_1^{(i)}, \ldots, W_{t_i}^{(i)}\}$ of $\{W_1, \ldots, W_k\}$.

In this case $K \leq K' \wr K''$, where $K''$ is the action of $G^\Omega$ on $\{Z_1, \ldots, Z_\ell\}$, and $K'$ is the action of $N_{G^\Omega}(Z_1)$ on $\{W_1^{(1)}, \ldots, W_{t_1}^{(1)}\}$. Because $K$ is maximal solvable, we have $K = K' \wr K''$ with $K'$ and $K''$ maximal transitive solvable. Furthermore $K'' \leq \overline{K}$, so by maximality $K'' = \overline{K}$ and $K = K' \wr \overline{K}$.

Therefore $G^\circ = H^\circ \wr K' \wr \overline{K}$. Now the action of $N_{G^\Omega}(Z_1)$ on $Z_1$ is isometric to $H^\circ \wr K'$, while the action of $N_{\overline{G^\Omega}}(Z_1)$ on $Z_1$ is isometric to $\overline{H^\circ}$ (Lemma 2.3.12 (i)). Thus we have $H^\circ \wr K' \leq \overline{H^\circ}$. Since $\overline{H}$ is metrically primitive, by Theorem 7.2.18 we cannot have $K' \neq 1$, so $K' = 1$ and $H^\circ \leq \overline{H^\circ}$. Because $H^\circ$ is maximal solvable, it follows that $H^\circ = \overline{H^\circ}$, but then $G^\circ = \overline{G^\circ}$, contradicting $G^\Omega \lneq \overline{G^\Omega}$. □

**Theorem 7.4.9** *Let* $G = \operatorname{semiwr}(H^\circ) \leq \Delta(V, \kappa)$ *as in Example 2.2.7, where* $H \leq \Delta(W, \kappa')$. *Suppose that* $H$ *is maximal irreducible solvable in* $\Delta(W, \kappa')$, *and that* $H$ *is of multiplier 2. Then one of the following holds:*

*(i)* $G$ *is maximal irreducible solvable in* $\Delta(V, \kappa)$.
*(ii)* $n = 2$, $H = \mathrm{GO}_1(q)$, *and* $q > 3$.

*Furthermore, if (ii) holds, then* $G$ *is not maximal solvable.*

***Proof*** Note that in case (ii) we have $G \lneq \mathrm{GO}_2^\varepsilon(q)$ (Remark 2.2.13), so $G$ is not maximal solvable. By Theorem 2.2.19, the group $G$ is irreducible and solvable. Thus for the proof of the theorem, it will suffice to prove that either $G$ is maximal solvable, or that (ii) holds.

Since $H$ is of multiplier 2, it cannot be metrically imprimitive of semiprimary type (Theorem 2.2.19). Thus by Theorem 2.2.14, we can write $H = (X \wr Y) \cap \Delta(W, \kappa')$, where $X \leq \Delta(U, \kappa'')$ is metrically primitive maximal irreducible solvable and $Y \leq S_m$ is maximal transitive solvable, where $m \geq 1$ and $m \dim U = \dim W$. Note that here $X$ is of multiplier 2 since $H$ is (Theorem 2.2.18), so $X^\circ$ is metrically primitive maximal irreducible solvable in $I(U, \kappa'')$.

Let $V = W_1 \perp W_2$ be the decomposition used in the construction of $G$, so $G^\circ = H_1^\circ \times H_2^\circ$ with $H_i \leq \Delta(W_i, \kappa)$ similar to $H$ for all $1 \leq i \leq 2$. Then $H_i^\circ = X_i^\circ \wr Y$, where $X_i^\circ$ is similar to $X$ for all $1 \leq i \leq 2$.

Suppose that $G \leq \overline{G} \leq \Delta(V, \kappa)$ such that $\overline{G}$ is maximal solvable. We will verify that in all cases either $G = \overline{G}$, or (ii) holds, which will prove the theorem.

If $\overline{G}$ is metrically primitive, it follows from Theorem 7.2.21 that (ii) holds. Suppose then that $\overline{G}$ is metrically imprimitive. We consider the two possibilities for $\overline{G}$ given by Theorem 2.2.14.

**Case 1: $\overline{G}$ is an isometric imprimitive subgroup**

In this case $\overline{G} = (\overline{H} \wr \overline{K}) \cap \Delta(V, \kappa)$, as constructed in Example 2.2.5. Let $V = Z_1 \perp \cdots \perp Z_\ell$ be the orthogonal system of imprimitivity defining $\overline{G}$, where $\ell > 1$. Because $G$ is of multiplier 1, the same is true for $\overline{G}$. Therefore $\overline{H}$ is of multiplier 1 (Theorem 2.2.18), and in particular $\dim Z_i$ is even for all $1 \leq i \leq \ell$. We now separate into two cases.

**Case 1.1: $G^\circ$ acts intransitively on $\{Z_1, \ldots, Z_\ell\}$**

The only proper nonzero $\mathbb{F}_q[G^\circ]$-submodules of $V$ are $W_1$ and $W_2$ (Lemma 2.1.15), so in this case $G^\circ$ has two orbits on $\{Z_1, \ldots, Z_\ell\}$. Hence

$$W_1 = Z_1^{(1)} \perp \cdots \perp Z_s^{(1)},$$
$$W_2 = Z_1^{(2)} \perp \cdots \perp Z_s^{(2)},$$

where $\{Z_1^{(1)}, \ldots, Z_s^{(1)}\}$ and $\{Z_1^{(2)}, \ldots, Z_s^{(2)}\}$ are the $G^\circ$-orbits on $\{Z_1, \ldots, Z_\ell\}$.

Then $H_1^\circ \leq (\overline{H} \wr Y') \cap \Delta(W_1, \kappa)$, where $Y'$ is the action of $H_1^\circ$ on $\{Z_1^{(1)}, \ldots, Z_s^{(1)}\}$. Because $H_1$ is of multiplier 2, we have $H_1 = H_1^\circ Z$, where $Z$ is the group of scalar matrices in $\Delta(W_1, \kappa)$. Hence $H_1 \leq (\overline{H} \wr Y') \cap \Delta(W_1, \kappa)$, and so $H_1 = (\overline{H} \wr Y') \cap \Delta(W_1, \kappa)$ because $H_1$ is maximal solvable. But this is contradiction, because $H_1$ is of multiplier 2 and $(\overline{H} \wr Y') \cap \Delta(W_1, \kappa)$ is of multiplier 1 (Theorem 2.2.18).

**Case 1.2: $G^\circ$ acts transitively on $\{Z_1, \ldots, Z_\ell\}$**

We have decompositions $W_1 = U_1 \perp \cdots \perp U_m$ and $W_2 = U_1' \perp \cdots \perp U_m'$, which are the orthogonal systems of imprimitivity defining $H_1^\circ = X_1^\circ \wr Y$ and $H_2^\circ = X_2^\circ \wr Y$ if $m > 1$. Then $G^\circ$ has a normal subgroup

$$R^\circ := \left(X_1^{(1)}\right)^\circ \times \cdots \times \left(X_m^{(1)}\right)^\circ \times \left(X_1^{(2)}\right)^\circ \times \cdots \times \left(X_m^{(2)}\right)^\circ,$$

where $X_j^{(1)} \leq \Delta(U_j, \kappa)$ and $X_j^{(2)} \leq \Delta(U_j', \kappa)$ are similar to $X$ for all $1 \leq j \leq m$.

Since $X^\circ$ is metrically primitive maximal irreducible solvable, by Theorem 7.3.7 the $R^\circ$-orbits on $\{Z_1, \ldots, Z_\ell\}$ are all of size 1 or of size 2. If there is an $R^\circ$-orbit of size 2, it follows from Theorem 7.3.7 that $\dim Z_i = 1$, contradicting the fact that $\dim Z_i$ is even. Therefore $R^\circ$ acts on $Z_i$ for all $1 \leq i \leq \ell$.

Then by Lemma 2.1.15, each $Z_i$ must be an orthogonal direct sum with summands of the form $U_j$ or $U_j'$. Because $G^\circ$ acts transitively on $\{Z_1, \ldots, Z_\ell\}$, for all $1 \leq i \leq \ell$ we have

$$Z_i = U_1^{(i)} \perp \cdots \perp U_{m/\ell}^{(i)} \perp (U_1')^{(i)} \perp \cdots \perp (U_{m/\ell}')^{(i)}$$

for some subsets $\{U_1^{(i)}, \ldots, U_{m/\ell}^{(i)}\}$ of $\{U_1, \ldots, U_m\}$ and $\{(U_1')^{(i)}, \ldots, (U_{m/\ell}')^{(i)}\}$ of $\{U_1', \ldots, U_m'\}$.

Note that $H_1^\circ$ acts trivially on $(U_j')^{(i)}$ for all $i$ and $j$. Thus because $H_1^\circ$ acts on $\{Z_1, \ldots, Z_\ell\}$, it follows that $Z_i$ is $H_1^\circ$-invariant for all $1 \leq i \leq \ell$. Similarly each $Z_i$ is $H_2^\circ$-invariant, so $G^\circ = H_1^\circ \times H_2^\circ$ acts on $Z_i$ for all $1 \leq i \leq \ell$. But this is contradiction, because $G^\circ$ acts transitively on $\{Z_1, \ldots, Z_\ell\}$ and $\ell > 1$.

**Case 2: $\overline{G}$ is of semiprimary type**

In this case $\overline{G} = \text{semiwr}(\overline{H^\circ})$, where $\overline{H} \leq \Delta(Z, \overline{\kappa})$ is of multiplier 2, as constructed in Example 2.2.7. Let $V = Z_1 \perp Z_2$ be the orthogonal system of imprimitivity defining $\overline{G}$.

We have $G^\circ \leq \overline{G^\circ} = \overline{H_1^\circ} \times \overline{H_2^\circ}$. Thus $Z_1$ and $Z_2$ are $\mathbb{F}_q[G^\circ]$-submodules, so by Lemma 2.1.15 we have $\{Z_1, Z_2\} = \{W_1, W_2\}$. We can assume without loss of generality that $Z_1 = W_1$ and $Z_2 = W_2$, so $H_i^\circ \leq \overline{H_i^\circ}$. This implies $H_i^\circ = \overline{H_i^\circ}$ for $i = 1, 2$ since $H^\circ$ is maximal solvable, so $G^\circ = \overline{G^\circ}$. Thus $\overline{G^\circ} \leq G \leq \overline{G}$, from which we conclude $G = \overline{G}$ since $G$ is of multiplier 1. □

We end this section by describing when two metrically imprimitive maximal irreducible solvable subgroups are conjugate. For a group $G = (H \wr K) \cap \Delta(V, \kappa)$ of isometric type we have $G^\circ$ irreducible (Lemma 2.2.18), while for $G = \text{semiwr}(H^\circ)$ we have $G^\circ$ reducible. Thus is suffices to consider conjugacy among groups of the same type (isometric or semiprimary).

**Proposition 7.4.10** *Let $G = (H \wr K) \cap \Delta(V, \kappa)$ and $\overline{G} = (\overline{H} \wr \overline{K}) \cap \Delta(V, \kappa)$ be maximal irreducible solvable as in Theorem 2.2.14 (i). Then $G$ and $\overline{G}$ are conjugate in $\Delta(V, \kappa)$ if and only if all of the following conditions hold:*

*(i) $H$ is similar to $\overline{H}$;*
*(ii) $K \cong \overline{K}$ as permutation groups.*

***Proof*** *(Cf. Proposition 1.3.8)* Let $V = W_1 \perp \cdots \perp W_k$ be the orthogonal system of imprimitivity defining $G$, and $V = Z_1 \perp \cdots \perp Z_\ell$ the orthogonal system of imprimitivity defining $\overline{G}$.

Suppose first that $G$ and $\overline{G}$ are conjugate in $\Delta(V, \kappa)$, and let $g \in \Delta(V, \kappa)$ such that $gGg^{-1} = \overline{G}$. Now $\{W_1, \ldots, W_k\}$ is a nonrefinable system of imprimitivity for $G$, so $\{gW_1, \ldots, gW_k\}$ is a nonrefinable system of imprimitivity for $\overline{G}$.

Since $\overline{G}$ is maximal solvable, it follows from Theorem 7.4.7 that Lemma 7.4.3 (ii) does not hold. Therefore $\overline{G}$ has a unique nonrefinable orthogonal system of imprimitivity, so by Lemma 7.4.3 we have $k = \ell$ and $\{gW_1, \ldots, gW_k\} = \{Z_1, \ldots, Z_k\}$.

Relabeling the summands if necessary, we can assume that $gW_i = Z_i$ for all $1 \leq i \leq k$. The action of $N_G(W_1)$ on $W_1$ is similar to $H$, while the action of $N_{\overline{G}}(Z_1)$ is similar to $\overline{H}$. On the other hand $gN_G(W_1)g^{-1} = N_{\overline{G}}(Z_1)$, so $H$ and $\overline{H}$ are similar.

The action of $G$ on $\{W_1, \ldots, W_k\}$ is isomorphic as a permutation group to $K$. Since $gW_i = Z_i$ for all $1 \leq i \leq k$, it follows that the action of $gGg^{-1} = \overline{G}$ on

$\{Z_1, \ldots, Z_k\}$ is also isomorphic to $K$ as a permutation group. Therefore $K \cong \overline{K}$ as permutation groups, so (i) and (ii) hold.

Conversely, suppose that (i) and (ii) hold. In this case $k = \ell$, since $K \cong \overline{K}$ as permutation groups. The group $G$ contains $M = (H_1 \times \cdots \times H_k) \cap \Delta(V, \kappa)$, where $H_i \leq \Delta(W_i, \kappa)$ is isometric to $H$ for all $1 \leq i \leq k$. Similarly $\overline{G}$ contains $\overline{M} = (\overline{H_1} \times \cdots \times \overline{H_k}) \cap \Delta(V, \kappa)$, where $\overline{H_i} \leq \Delta(Z_i, \kappa)$ is isometric to $\overline{H}$ for all $1 \leq i \leq k$.

Because $H$ and $\overline{H}$ are similar, for all $1 \leq i \leq k$ there is a similarity $g_i : W_i \to Z_i$ such that $g_i H_i g_i^{-1} = \overline{H_i}$. Since $H_i$ is isometric to $H$, by multiplying $g_i$ with suitable scalars we can assume that $\tau(g_i) = \tau(g_j)$ for all $1 \leq i, j \leq k$. Then $g = g_1 \cdots g_k \in \Delta(V, \kappa)$, and $gMg^{-1} = \overline{M}$. Since $K \cong \overline{K}$ as permutation groups, it follows that $gGg^{-1} \leq \overline{M}\,\overline{K} = \overline{G}$. Because $gGg^{-1}$ is maximal solvable, we have $gGg^{-1} = \overline{G}$, which completes the proof of the proposition. □

**Proposition 7.4.11** *Let $G = \mathrm{semiwr}(H^\circ)$ and $\overline{G} = \mathrm{semiwr}(\overline{H^\circ})$ be semiprimary subgroups of $\Delta(V, \kappa)$, where $H \leq \Delta(W, \kappa')$ and $\overline{H} \leq \Delta(Z, \kappa'')$ are maximal irreducible solvable of multiplier 2. Assume that $G$ and $\overline{G}$ are maximal solvable in $\Delta(V, \kappa)$. Then $G$ and $\overline{G}$ are conjugate in $\Delta(V, \kappa)$ if and only if $H$ and $\overline{H}$ are similar.*

***Proof*** Suppose that $G$ and $\overline{G}$ are conjugate in $\Delta(V, \kappa)$. Since $G$ is of multiplier 1, it follows that $G$ and $\overline{G}$ are conjugate in $I(V, \kappa)$, so let $g \in I(V, \kappa)$ be such that $gGg^{-1} = \overline{G}$.

In particular $gG^\circ g^{-1} = \overline{G^\circ}$. We have $G^\circ = H_1^\circ \times H_2^\circ$ with $H_i^\circ$ similar to $H^\circ$, and $\overline{G^\circ} = \overline{H_1^\circ} \times \overline{H_2^\circ}$ with $\overline{H_i^\circ}$ similar to $\overline{H^\circ}$. Thus it follows from Lemma 2.3.14 that $H_1^\circ$ is isometric to $\overline{H_1^\circ}$ or $\overline{H_2^\circ}$. Consequently $H^\circ$ and $\overline{H^\circ}$ are similar. Because $H$ and $\overline{H}$ are of multiplier 2, we conclude that $H$ and $\overline{H}$ are similar.

Conversely, suppose that $H$ and $\overline{H}$ are similar. We have $G^\circ = H_1^\circ \times H_2^\circ$ and $\overline{G^\circ} = \overline{H_1^\circ} \times \overline{H_2^\circ}$ as in the previous paragraph.

If $\dim W = \dim Z$ is odd, then $H_i^\circ$ is isometric to $H^\circ$ and $\overline{H_i^\circ}$ is isometric to $\overline{H^\circ}$, so $G^\circ$ is conjugate to $\overline{G^\circ}$ in $I(V, \kappa)$ (Lemma 2.3.14). Since $N_{\Delta(V,\kappa)}(G^\circ) = G$ and $N_{\Delta(V,\kappa)}(\overline{G^\circ}) = \overline{G}$ (Lemma 2.1.5), it follows that $G$ is conjugate to $\overline{G}$ in $I(V, \kappa)$.

If $\dim W = \dim Z$ is even, then one of the $H_i^\circ$ is isometric to $H^\circ$, and the other one is similar but not isometric to $H^\circ$. Suppose without loss of generality that $H_1^\circ$ is isometric to $H^\circ$, and $H_2^\circ$ is similar but not isometric to $H^\circ$. Since $H$ is similar to $\overline{H}$, we can assume that $H_1^\circ$ is isometric to $\overline{H_1^\circ}$ and $H_2^\circ$ is isometric to $\overline{H_2^\circ}$. Then $G^\circ$ and $\overline{G^\circ}$ are conjugate in $I(V, \kappa)$ (Lemma 2.3.14), so as in the previous paragraph it follows that $G$ and $\overline{G}$ are conjugate in $I(V, \kappa)$. □

**Remark 7.4.12** In the case where $\kappa = 0$ and $\Delta(V, \kappa) = \mathrm{GL}(V)$, a proof of Proposition 7.4.10 is given in [74, Theorem 2.5.8]. The proof there is attributed to L. G. Kovács and is based on a result of Gross [30, Theorem 5.2].

## 7.5 Maximality of Metrically Completely Reducible Subgroups

Throughout this section, we will assume that $\kappa \neq 0$. In this section, we will classify metrically completely reducible maximal solvable subgroups of $\Delta(V, \kappa)$ and $I(V, \kappa)$. Furthermore, in the case where $q$ is even and $\kappa$ is a quadratic form, we will classify metrically completely reducible maximal solvable subgroups of $\Omega(V, \kappa)$.

More specifically, we know by Lemma 2.3.9 that every metrically completely reducible maximal solvable subgroup of $\Delta(V, \kappa)$ is of the form $L = (H_1 \times \cdots \times H_t) \cap \Delta(V, \kappa)$ with respect to a decomposition $V = W_1 \perp \cdots \perp W_t$, where $H_i \leq \Delta(W_i, \kappa)$ is maximal irreducible solvable for all $1 \leq i \leq t$. In Theorems 7.5.3 and 7.5.4, we prove that $L$ is maximal solvable in $\Delta(V, \kappa)$, with a few known exceptions. Furthermore, we will see that in the exceptional cases $L$ is not maximal solvable.

Similarly we know that in $I(V, \kappa)$, every metrically completely reducible maximal solvable subgroup is of the form $L^\circ = H_1^\circ \times \cdots \times H_t^\circ$ with respect to a decomposition $V = W_1 \perp \cdots \perp W_t$, where $H_i^\circ \leq I(W_i, \kappa)$ is maximal irreducible solvable for all $1 \leq i \leq t$. In Theorem 7.5.1, we will prove that $L^\circ$ is maximal, with a few exceptions. In the case where $q$ is even and $\kappa$ is a quadratic form, we will prove a similar result for $L^\Omega$ (Theorem 7.5.7).

**Theorem 7.5.1** *Assume that $\kappa \neq 0$. Let $L^\circ = H_1^\circ \times \cdots \times H_t^\circ$ with respect to a decomposition $V = W_1 \perp \cdots \perp W_t$, where $t \geq 1$. Suppose that $H_i^\circ$ is maximal irreducible solvable in $I(W_i, \kappa)$ for all $1 \leq i \leq t$. Then one of the following holds.*

*(i) $L^\circ$ is maximal solvable in $I(V, \kappa)$.*
*(ii) There exists $i \neq j$ such that one of the following holds:*

  *(a) $H_i^\circ$ is isometric to $H_j^\circ$.*
  *(b) $H_i^\circ$ is isometric to $H_j^\circ \wr S_2$.*
  *(c) $H_i^\circ$ is isometric to $H_j^\circ \wr S_3$.*

*Furthermore, if (ii) holds, then $L^\circ$ is not maximal solvable.*

***Proof*** If (ii) holds, then $L^\circ$ has a factor isometric to $H_j^\circ \times (H_j^\circ \wr S_k)$ for some $1 \leq k \leq 3$. This is properly contained in the solvable group $H_j^\circ \wr S_{k+1} \leq I(W_i \perp W_j, \kappa)$, so $L^\circ$ is not maximal solvable.

We will assume for the remainder of the proof that (ii) does not hold, and will prove that $L^\circ$ is maximal solvable in this case. If $t = 1$, then $L^\circ = H_1^\circ$ is maximal solvable by assumption, so we assume that $t > 1$ and proceed by induction on $t$. Suppose for the sake of contradiction that $L^\circ$ is not maximal solvable, in which case $L^\circ \lneq G^\circ$ for some maximal solvable subgroup $G^\circ \leq I(V, \kappa)$. We consider the different possibilities for $G^\circ$ in turn.

**Case 1: $G^\circ$ is irreducible and metrically primitive**
Since $H_i^\circ$ are maximal solvable, we have $H_i^\circ = X_i \wr K_i$ for $X_i \leq I(W_i', \kappa_i')$ maximal metrically primitive irreducible solvable, and $K_i \leq S_{k_i}$ maximal transitive solvable, where $k_i \geq 1$. We can write $W_i = W_1^{(i)} \perp \cdots \perp W_{k_i}^{(i)}$, where $\left\{W_1^{(i)}, \ldots, W_{k_i}^{(i)}\right\}$ is the orthogonal system of imprimitivity defining $H_i^\circ$ if $k_i > 1$. Then $H_i^\circ$ contains $\left(H_1^{(i)}\right)^\circ \times \cdots \times \left(H_{k_i}^{(i)}\right)^\circ$, where $\left(H_j^{(i)}\right)^\circ \leq I(W_j^{(i)}, \kappa)$ is isometric to $X_i$ for all $1 \leq j \leq k_i$.

It follows from Theorem 7.2.17 that $k_1 + \cdots + k_t \leq 4$, so $K_i = S_{k_i}$ with $k_i \leq 4$ for all $1 \leq i \leq t$. Furthermore, by Theorem 7.2.17, $\kappa$ is a quadratic form and $X_i = O_1(q)$ for all $i$. Therefore $L^\circ = (O_1(q) \wr S_{k_1}) \times \cdots \times (O_1(q) \wr S_{k_t})$. We have assumed that (ii)(a) does not hold, so $k_i \neq k_j$ for all $i \neq j$. On the other hand $k_1 + \cdots + k_t \leq 4$ and $t > 1$, so $t = 2$ and $\{k_1, k_2\}$ equals $\{1, 2\}$ or $\{1, 3\}$. But this is a contradiction, since we have assumed that (ii)(b) and (ii)(c) do not hold.

**Case 2: $G^\circ$ is irreducible and metrically imprimitive**
In this case $G^\circ = H^\circ \wr K$, where $H^\circ \leq I(Z, \kappa')$ is metrically primitive maximal irreducible solvable, and $K \leq S_\ell$ maximal transitive solvable with $\ell > 1$. Let $V = Z_1 \perp \cdots \perp Z_\ell$ be the orthogonal system of imprimitivity defining $G^\circ$. We consider the two different possibilities.

**Case 2.1: $L^\circ$ acts intransitively on $\{Z_1, \ldots, Z_\ell\}$**
Let $\{Z_1^{(1)}, \ldots, Z_{\ell_1}^{(1)}\}, \ldots, \{Z_1^{(u)}, \ldots, Z_{\ell_u}^{(u)}\}$ be the orbits of $L^\circ$ on $\{Z_1, \ldots, Z_\ell\}$, where $u > 1$. Then $Q_i := Z_1^{(i)} \perp \cdots \perp Z_{\ell_i}^{(i)}$ is an $L^\circ$-invariant subspace of $V$, so by Lemma 2.3.8 we have

$$Q_i = W_1^{(i)} \perp \cdots \perp W_{t_i}^{(i)}$$

for some subset $\{W_1^{(i)}, \ldots, W_{t_i}^{(i)}\}$ of $\{W_1, \ldots, W_t\}$.

The action of $L^\circ$ on $Q_i$ is contained in $H^\circ \wr K_i$, where $K_i$ is the action of $L^\circ$ on $\left\{Z_1^{(i)}, \ldots, Z_{\ell_i}^{(i)}\right\}$. On the other hand, the action of $L^\circ$ on $Q_i$ is equal to $\left(H_1^{(i)}\right)^\circ \times \cdots \times \left(H_{t_i}^{(i)}\right)^\circ$ which is maximal solvable in $I(Q_i, \kappa)$ by induction, so

$$\left(H_1^{(i)}\right)^\circ \times \cdots \times \left(H_{t_i}^{(i)}\right)^\circ = H^\circ \wr K_i.$$

The wreath product $H^\circ \wr K_i$ is irreducible by Lemma 2.2.6, so $t_i = 1$. It follows that $u = t$, so after relabeling we can assume that $Q_i = W_i$ and $H_i^\circ = H^\circ \wr K_i$ for all $1 \leq i \leq t$. Hence

$$L^\circ = H^\circ \wr K_1 \times \cdots \times H^\circ \wr K_t.$$

Then the action of $L^\circ$ on $\{Z_1, \ldots, Z_\ell\}$ is equal to $K_1 \times \cdots \times K_t \leq K$. Because statement (ii) does not hold, it follows from Theorem 1.1.2 that $K_1 \times \cdots \times K_t$ is maximal solvable, so $K_1 \times \cdots \times K_t = K$. But this is a contradiction, since $t > 1$ and $K$ is transitive.

**Case 2.2: $L^\circ$ acts transitively on $\{Z_1, \ldots, Z_\ell\}$**
As in the beginning of Case 1, write $H_i^\circ = X_i \wr K_i$ for $X_i$ metrically primitive maximal irreducible solvable, and $K_i \leq S_{k_i}$ maximal transitive solvable, where $k_i \geq 1$. Write

$$W_i = W_1^{(i)} \perp \cdots \perp W_{k_i}^{(i)}, \tag{7.28}$$

where $\left\{W_1^{(i)}, \ldots, W_{k_i}^{(i)}\right\}$ is the orthogonal system of imprimitivity defining $H_i^\circ$ if $k_i > 1$. Then $H_i^\circ$ contains $X_1^{(i)} \times \cdots \times X_{k_i}^{(i)}$, where $X_j^{(i)} \leq I(W_j^{(i)}, \kappa)$ is isometric to $X_i$ for all $1 \leq j \leq k_i$.

Denote

$$R^\circ = X_1^{(1)} \times \cdots \times X_{k_1}^{(1)} \times \cdots \times X_1^{(t)} \times \cdots \times X_{k_t}^{(t)}.$$

Because $L^\circ$ acts transitively on $\{Z_1, \ldots, Z_\ell\}$ and $R^\circ \trianglelefteq L^\circ$, we know that $L^\circ$ acts transitively on the $R^\circ$-orbits on $\{Z_1, \ldots, Z_\ell\}$. Thus by Lemma 7.3.7, the $R^\circ$-orbits on $\{Z_1, \ldots, Z_\ell\}$ are either all of size 1 or all of size 2.

We consider first the case where the $R^\circ$-orbits are of size 2. Then $\ell$ is even, and after relabeling we can assume that $\{Z_1, Z_2\}, \ldots, \{Z_{\ell-1}, Z_\ell\}$ are the $R^\circ$-orbits on $\{Z_1, \ldots, Z_\ell\}$. Because $Z_{2i-1} \perp Z_{2i}$ is $R^\circ$-invariant for all $1 \leq i \leq \ell/2$, it follows from Lemma 2.3.8 that $Z_{2i-1} \perp Z_{2i}$ is a sum of some $W_j^{(j')}$'s. Moreover the groups $X_j^{(j')}$ are metrically primitive maximal irreducible solvable in $I(W_j^{(j')}, \kappa)$, so by Lemma 7.3.7 we have

$$Z_{2i-1} \perp Z_{2i} = W_{t_i}^{(t_i')} \perp W_{s_i}^{(s_i')}$$

with $\dim Z_{2i-1} = \dim Z_{2i} = 1$ for all $i$ and $\dim W_j^{(j')} = 1$ for all $j$, $j'$, so in particular $\ell = n$. Furthermore $\kappa$ is a quadratic form, and $X_j^{(j')} = O_1(q)$ for all $j$ and $j'$.

Because $L^\circ$ acts transitively on the pairs $\{Z_{2i-1}, Z_{2i}\}$, after relabeling we can assume that the summands $W_{t_i}^{(t_i')}$ are on the same $L^\circ$-orbit. Note that $L^\circ$ does not act transitively on all $W_j^{(j')}$ because $t > 1$. Therefore we have

$$V = \left(W_{t_1}^{(t_1')} \perp \cdots \perp W_{t_{n/2}}^{(t_{n/2}')}\right) \perp \left(W_{s_1}^{(s_1')} \perp \cdots \perp W_{s_{n/2}}^{(s_{n/2}')}\right),$$

where $\left\{W_{t_1}^{(t_1')}, \ldots, W_{t_{n/2}}^{(t_{n/2}')}\right\}$ and $\left\{W_{s_1}^{(s_1')}, \ldots, W_{s_{n/2}}^{(s_{n/2}')}\right\}$ are the two $L^\circ$-orbits on the summands $W_j^{(j')}$.

Since $X_i = O_1(q)$ for all $i$, it follows that $L^\circ = (O_1(q) \wr \Gamma_1) \times (O_1(q) \wr \Gamma_2)$, where $\Gamma_1$ is the action of $L^\circ$ on $\left\{W_{t_1}^{(t_1')}, \ldots, W_{t_{n/2}}^{(t_{n/2}')}\right\}$, and $\Gamma_2$ is the action of $L^\circ$ on $\left\{W_{s_1}^{(s_1')}, \ldots, W_{s_{n/2}}^{(s_{n/2}')}\right\}$. We have $\Gamma_1 \cong \Gamma_2$ as permutation groups, since both actions are equivalent to the action of $L^\circ$ on the pairs $\{\{Z_1, Z_2\}, \ldots, \{Z_{n-1}, Z_n\}\}$. But this is a contradiction, since we have assumed that (ii)(a) does not hold.

We consider then the case where all the $R^\circ$-orbits are of size 1. In other words, in this case $R^\circ$ acts on $Z_i$ for all $1 \leq i \leq \ell$. By Lemma 2.3.8, each $Z_i$ is a sum of some $W_j^{(j')}$'s. Therefore (possibly after relabeling the summands $W_j^{(j')}$ in (7.28)) we can write

$$Z_1 = \left(W_1^{(1)} \perp \cdots \perp W_{m_1}^{(1)}\right) \perp \cdots \perp \left(W_1^{(t)} \perp \cdots \perp W_{m_t}^{(t)}\right),$$

where $\{W_1^{(i)}, \ldots, W_{m_i}^{(i)}\}$ are the summands of $W_i$ from (7.28) that appear in $Z_1$.

Because $L^\circ$ acts transitively on $\{Z_1, \ldots, Z_\ell\}$, we have $m_i > 0$ for all $1 \leq i \leq t$. Since $H_i^\circ$ acts trivially on $W_1^{(j)}$ for $j \neq i$, it follows that $Z_1$ is invariant under $H_i^\circ$ for all $1 \leq i \leq t$. But then $Z_1$ is $L^\circ$-invariant, contradicting the fact that $L^\circ$ acts transitively on $\{Z_1, \ldots, Z_\ell\}$ and $\ell > 1$.

**Case 3: $G^\circ$ is not irreducible**

Because $L^\circ$ is metrically completely reducible (Lemma 2.3.10), the same is true for $G^\circ$. Thus by Lemma 2.3.7 we have $G^\circ = \overline{H_1^\circ} \times \cdots \times \overline{H_s^\circ}$, with respect to a decomposition $V = Z_1 \perp \cdots \perp Z_s$ such that $s \geq 2$. Furthermore, here $\overline{H_i^\circ}$ is maximal irreducible solvable in $I(Z_i, \kappa)$ for all $1 \leq i \leq s$.

Since $L^\circ$ acts on $Z_i$, it follows from Lemma 2.1.15 that

$$Z_i = W_1^{(i)} \perp \cdots \perp W_{t_i}^{(i)}$$

for some subset $\{W_1^{(i)}, \ldots, W_{t_i}^{(i)}\}$ of $\{W_1, \ldots, W_t\}$. Then $\left(H_1^{(i)}\right)^\circ \times \cdots \times \left(H_{t_i}^{(i)}\right)^\circ \leq \overline{H_i^\circ}$. By induction $\left(H_1^{(i)}\right)^\circ \times \cdots \times \left(H_{t_i}^{(i)}\right)^\circ$ is maximal solvable, so

$$\left(H_1^{(i)}\right)^\circ \times \cdots \times \left(H_{t_i}^{(i)}\right)^\circ = \overline{H_i^\circ}.$$

Since $\overline{H_i^\circ}$ is irreducible, we have $t_i = 1$ and $\left(H_1^{(i)}\right)^\circ = \overline{H_i^\circ}$ for all $1 \leq i \leq t$. But then $L^\circ = G^\circ$, contrary to the assumption $L^\circ \lneqq G^\circ$.

In all cases we have found a contradiction, so we conclude that $L^\circ$ is maximal solvable. □

**Remark 7.5.2** Let $\kappa, \kappa' \neq 0$ be non-degenerate forms of the same type. It is possible that two maximal irreducible solvable subgroups $X \leq I(W, \kappa)$ and $Y \leq I(W', \kappa')$ are similar, but not isometric (see Lemma 2.1.10). In this case it follows from Theorem 7.5.1 that $X \times Y$ is a maximal solvable subgroup of $I(W \perp W', \kappa + \kappa')$.

For a concrete example, there is a maximal irreducible solvable subgroup $G = G_{2,1}^{\mathcal{B}_3}(O_2^-(2))$ of type $\mathcal{B}_3$ in $\mathrm{GSp}_2(7)$. Here $G$ is the normalizer of an extraspecial group $2_-^{1+2}$ in $\mathrm{GL}_2(7) = \mathrm{GSp}_2(7)$. Because 2 is a square in $\mathbb{F}_7$, it follows that $G$ has multiplier 2 (Lemma 5.1.1), and thus

$$\{xGx^{-1} \cap \mathrm{Sp}_2(7) : x \in \mathrm{GSp}_2(7)\}$$

splits into two $\mathrm{Sp}_2(7)$-classes (Lemma 2.1.10).

Representatives for these two conjugacy classes of subgroups are given by $X = G \cap \mathrm{Sp}_2(7)$ and $Y = xGx^{-1} \cap \mathrm{Sp}_2(7)$, where $\tau(x) = \det(x)$ is nonsquare (Lemma 2.1.10). Then by Theorem 7.5.1, the direct product $X \times Y$ is a maximal solvable subgroup of $\mathrm{Sp}_4(7)$.

Note that in this case the subgroup $(G \times xGx^{-1}) \cap \mathrm{GSp}_4(7)$ is not maximal solvable in $\mathrm{GSp}_4(7)$, since it is contained in the semiprimary group corresponding to $X$ (Example 2.2.7).

By Lemmas 2.3.7, 2.3.14, and Theorem 7.5.1, the classification of metrically completely reducible maximal solvable subgroups of $I(V, \kappa)$ is reduced to the irreducible case. We will next prove similar results for $\Delta(V, \kappa)$. We begin with the case where the subgroup is of multiplier 2, which for the most part will follow from Theorem 7.5.1.

**Theorem 7.5.3** *Assume that $\kappa \neq 0$. Let $L = (H_1 \times \cdots \times H_t) \cap \Delta(V, \kappa)$ with respect to a decomposition $V = W_1 \perp \cdots \perp W_t$, where $t \geq 1$. Suppose that $H_i^\circ$ is maximal irreducible solvable in $I(W_i, \kappa)$ for all $1 \leq i \leq t$, and that $H_i$ is maximal irreducible solvable in $\Delta(W_i, \kappa)$ for all $1 \leq i \leq t$. Assume that $L$ is of multiplier 2. Then one of the following holds:*

- *(i) $L$ is maximal solvable in $\Delta(V, \kappa)$.*
- *(ii) There exists $i \neq j$ such that one of the following holds:*
  - *(a) $H_i^\circ$ is isometric to $H_j^\circ$.*
  - *(b) $H_i^\circ$ is isometric to $H_j^\circ \wr S_2$.*
  - *(c) $H_i^\circ$ is isometric to $H_j^\circ \wr S_3$.*
- *(iii) There exists a permutation $u_1, v_1, \ldots, u_r, v_r, k_1, \ldots, k_s$ of $\{1, \ldots, t\}$ such that all of the following hold:*
  - *(a) $H_{u_i}^\circ$ is similar to $H_{v_i}^\circ$ for all $1 \leq i \leq r$.*
  - *(b) $H_{k_i}$ is of multiplier 1 for all $1 \leq i \leq s$.*

*Furthermore, if (ii) or (iii) holds, then $L$ is not maximal solvable in $\Delta(V, \kappa)$.*

***Proof*** Since $L$ is of multiplier 2, we have $q$ odd and $L = L^\circ Z$, where $Z \leq \mathrm{GL}(V)$ is the group of scalar matrices. If (ii) holds, it follows from Theorem 7.5.1 that $L^\circ$ is not maximal solvable in $I(V,\kappa)$, and thus $L$ is not maximal solvable in $\Delta(V,\kappa)$ (Lemma 2.1.2).

We will next prove that if (iii) holds, then $L$ is not maximal solvable. To this end, we first note that if $H^\circ_{u_i}$ is isometric to $H^\circ_{v_i}$ for some $1 \leq i \leq r$, it follows again from Theorem 7.5.1 that $L$ is not maximal solvable in $\Delta(V,\kappa)$. Thus we can assume that for all $1 \leq i \leq r$, we have $H^\circ_{u_i}$ similar but not isometric to $H^\circ_{v_i}$. Let $\varphi_i : W_{u_i} \to W_{v_i}$ be a similarity such that $\varphi_i H^\circ_{u_i} \varphi_i^{-1} = H_{v_i}$.

Let $c_i \in \mathbb{F}_q$ be the scalar such that $b(\varphi_i(v), \varphi_i(w)) = c_i b(v,w)$ for all $v, w \in W_{u_i}$. Because $H^\circ_{u_i}$ and $H^\circ_{v_i}$ are not isometric, the multiplier $c_i$ cannot be a square, as otherwise a scalar multiple of $\varphi_i$ would be an isometry. Thus $c_i$ is nonsquare for all $1 \leq i \leq r$, so by replacing $\varphi_i$'s with suitable scalar multiples we can assume that there is a primitive element $c \in \mathbb{F}_q$ such that $c_i = c$ for all $1 \leq i \leq r$. Because $H_{k_i}$ is of multiplier 1 for all $1 \leq i \leq s$, there exist $h_i \in H_{k_i}$ such that $\tau(h_i) = c$ for all $1 \leq i \leq s$.

Now we define for each $1 \leq i \leq r$ a similarity $\psi_i : W_{u_i} \perp W_{v_i} \to W_{u_i} \perp W_{v_i}$ by

$$w + w' \mapsto \varphi_i(w) + \varphi_i^{-1}(w')$$

for all $w \in W_{u_i}$ and $w' \in W_{v_i}$. It is readily checked that $\psi_i$ normalizes $H^\circ_{u_i} \times H^\circ_{v_i}$, and acts nontrivially on $\{W_{u_i}, W_{v_i}\}$ for all $1 \leq i \leq s$. (In fact $H^\circ_{u_i} \times H^\circ_{v_i}$ and $\psi_i$ generate a semiprimary group corresponding to $H^\circ_{u_i}$, as constructed in Example 2.2.7.) Furthermore, then $g = \psi_1 \ldots \psi_r h_1 \cdots h_s \in \Delta(V,\kappa)$ normalizes $L = L^\circ Z$ and acts nontrivially on $\{W_{u_i}, W_{v_i}\}$, so $L$ is not maximal solvable.

Next we will assume that (ii) and (iii) do not hold, and prove that $L$ is maximal solvable. Because (ii) does not hold, it follows from Theorem 7.5.1 that $L^\circ$ is maximal solvable in $I(V,\kappa)$. Suppose for the sake of contradiction that $L$ is not maximal solvable, so $L \lneqq G$ for some maximal solvable $G \leq \Delta(V,\kappa)$. We have $L^\circ \leq G^\circ$, so $G^\circ = L^\circ$ by maximality of $L^\circ$. Therefore $G \leq N_{\Delta(V,\kappa)}(L^\circ)$.

Since $L^\circ$ is maximal solvable in $I(V,\kappa)$, we have $N_{I(V,\kappa)}(L^\circ) = L^\circ$, so $L = L^\circ Z = N_{I(V,\kappa)}(L^\circ)Z$ has index 2 in $N_{\Delta(V,\kappa)}(L^\circ)$. Consequently $G = N_{\Delta(V,\kappa)}(L^\circ)$ and $[G : L] = 2$. Because $G$ normalizes $L$, it acts on the irreducible $L$-submodules of $V$, which are precisely the summands $W_i$. Let $g \in G \setminus L$, so $\tau(g) \in \mathbb{F}_q^\times$ is nonsquare since $G^\circ Z = L^\circ Z = L$. Then the action of $g$ on $\{W_1, \ldots, W_t\}$ is as a permutation of order 2, since $[G : L] = 2$. Thus there exists a permutation $u_1, v_1, \ldots, u_r, v_r, k_1, \ldots, k_s$ of $\{1, \ldots, t\}$ such that $g$ swaps $W_{u_i}$ and $W_{v_i}$ for all $1 \leq i \leq r$, and $g(W_{k_i}) = W_{k_i}$ for all $1 \leq i \leq s$.

Then $gH^\circ_{u_i}g^{-1} = H^\circ_{v_i}$, so $H^\circ_{u_i}$ is similar to $H^\circ_{v_i}$ for all $1 \leq i \leq r$. Furthermore $gH^\circ_{k_i}g^{-1} = H^\circ_{k_i}$, so the restriction

$$g_i := g_{|W_{k_i}} : W_{k_i} \to W_{k_i}$$

normalizes $H_{k_i}^\circ$ for all $1 \le i \le s$. Because $H_{k_i}$ is maximal solvable, it follows from Lemma 2.1.5 that $g_i \in H_{k_i}$. Now $\tau(g) = \tau(g_i)$ is nonsquare, so we conclude that $H_{k_i}$ is of multiplier 1 for all $1 \le i \le k$. Thus (iii) holds, contrary to what we have assumed. This contradiction proves that $L$ must be maximal solvable, and completes the proof of the theorem. □

**Theorem 7.5.4** *Assume that $\kappa \neq 0$. Let $L = (H_1 \times \cdots \times H_t) \cap \Delta(V, \kappa)$ with respect to a decomposition $V = W_1 \perp \cdots \perp W_t$, where $t \ge 1$. Suppose that $H_i$ is maximal irreducible solvable in $\Delta(W_i, \kappa)$ for all $1 \le i \le t$. Assume that $L$ is of multiplier 1. Then one of the following holds.*

*(i) $L$ is maximal solvable in $\Delta(V, \kappa)$.*
*(ii) There exists $i \neq j$ such that one of the following holds:*

*(a) $H_i^\circ$ is similar to $H_j^\circ$.*
*(b) $H_i^\circ$ is similar to $H_j^\circ \wr S_2$.*
*(c) $H_i^\circ$ is similar to $H_j^\circ \wr S_3$.*

*(iii) There exists $i \neq j$ such that $H_i = \mathrm{semiwr}(X_i^\circ)$, $H_j = \mathrm{semiwr}(X_j^\circ)$, and one of the following holds:*

*(a) $X_i^\circ$ is similar to $X_j^\circ$.*
*(b) $X_i^\circ$ is similar to $X_j^\circ \wr S_2$.*
*(c) $X_i^\circ$ is similar to $X_j^\circ \wr S_3$.*

*Furthermore, if (ii) or (iii) holds, then $L$ is not maximal solvable in $\Delta(V, \kappa)$.*

***Proof*** If $q$ is even, then $\Delta(V, \kappa) = I(V, \kappa)Z$, where $Z \le \mathrm{GL}(V)$ is the group of scalar matrices. Then $L$ is maximal solvable if and only if $L^\circ$ is maximal solvable (Lemma 2.1.3), so the result follows from Theorem 7.5.1. Thus we will assume for the rest of the proof that $q$ is odd.

For the claims, we will first prove that if (ii) or (iii) holds, then $L$ is not maximal solvable. Suppose first that (ii) holds, in which case there exists $i \neq j$ such that $H_i^\circ$ is similar to $H_j^\circ \wr S_k$ for some $1 \le k \le 3$.

Since $L$ is of multiplier 1, for all $1 \le j \le t$ the group $H_j$ is of multiplier 1 in $\Delta(W_j, \kappa)$. Thus it follows from Lemma 2.1.12 that $H_i$ is similar to $(H_j \wr S_k) \cap \Delta(W_j \perp \cdots \perp W_j, \kappa)$. Furthermore, if $\varphi \in \Delta(W_i, \kappa)$ with $\tau(\varphi) = c$, then for all $1 \le j \le t$ there exists $h_j \in H_j$ such that $\tau(h_j) = c$. Then conjugating $L$ with $g = h_1 \cdots h_{i-1} \varphi h_{i+1} \cdots h_t$ gives

$$gLg^{-1} = H_1 \times \cdots \times H_{i-1} \times \varphi H_i \varphi^{-1} \times H_{i+1} \times \cdots \times H_t.$$

Thus by replacing $L$ with a conjugate, we can assume that $H_i = (H_j \wr S_k) \cap \Delta(W_j \perp \cdots \perp W_j, \kappa)$. Then as in the beginning of the proof of Theorem 7.5.1, it follows that $L$ is not maximal solvable.

Suppose next that (iii) holds, with $H_i = \mathrm{semiwr}(X_i^\circ)$ and $H_j = \mathrm{semiwr}(X_j^\circ)$. Because $\mathrm{semiwr}(X)$ is similar to $\mathrm{semiwr}(Y)$ if $X$ and $Y$ are similar, we can assume that $X_i^\circ$ is isometric to $X_j^\circ \wr S_k$ for some $1 \leq k \leq 3$. We have

$$(H_i \times H_j) \cap \Delta(W_i \perp W_j, \kappa) = \text{semiwr}(X_i^\circ \times X_j^\circ).$$

It follows from Theorem 7.5.1 that $X_i^\circ \times X_j^\circ$ is not maximal solvable, so $L$ is not maximal solvable in $\Delta(V, \kappa)$.

We will assume then for the rest of the proof that (ii) and (iii) do not hold. We will see similarly to the proof of Theorem 7.5.1 that $L$ is maximal solvable. For $t = 1$, we have $L = H_1$ maximal solvable by assumption, so suppose $t > 1$ and proceed by induction on $t$. Suppose for the sake of contradiction that $L$ is not maximal solvable, in which case $L \lneq G$ for some maximal solvable subgroup $G \leq \Delta(V, \kappa)$. We consider the different possibilities for $G$ in turn.

**Case 1: $G$ is irreducible and metrically primitive**
We have $L^\circ = H_1^\circ \times \cdots \times H_t^\circ$ contained in $G^\circ$. Since $\kappa \neq 0$ and the $H_i$ are of multiplier 1, it follows from Theorem 7.2.17 that $\kappa$ is a quadratic form and one of the following holds:

- $q = 3, t = 2, H_1 = H_2 = \text{GO}_2^-(3)$.
- $q = 5, t = 2, H_1 = H_2 = \text{GO}_2^+(5)$.

In both cases $H_1$ and $H_2$ are similar, contradicting our assumption that (ii) does not hold.

**Case 2: $G$ is irreducible and metrically imprimitive**
We consider the two possibilities for $G$ given by Theorem 2.2.14.

**Case 2.1: $G$ is of semiprimary type**
We have $G = \text{semiwr}(\overline{H}^\circ)$, where $\overline{H} \leq \Delta(W, \kappa')$ is maximal irreducible solvable of multiplier 2. Let $V = Z_1 \perp Z_2$ be the orthogonal system of imprimitivity defining $G$. Then $G^\circ = \overline{H_1^\circ} \times \overline{H_2^\circ}$, with $\overline{H_i^\circ} \leq I(Z_i, \kappa)$ similar to $\overline{H}^\circ$.

Suppose first that $Z_1$ and $Z_2$ are $L$-invariant. Then it follows from Lemma 2.3.10 that

$$Z_1 = W_{i_1} \perp \cdots \perp W_{i_r}$$
$$Z_2 = W_{i'_1} \perp \cdots \perp W_{i'_r}$$

for some indices $i_m, i'_m$. In this case the action of $N_G(Z_1)$ on $Z_1$ contains

$$\left(H_{i_1} \times \cdots \times H_{i_r}\right) \cap \Delta(Z_1, \kappa),$$

which is of multiplier 1. But this is a contradiction, because the action of $N_G(Z_1)$ on $Z_1$ is similar to $\overline{H}$, and thus of multiplier 2.

Thus we can assume that $L$ acts transitively on $\{Z_1, Z_2\}$. Because $Z_1$ and $Z_2$ are $G^\circ$-invariant, it follows that $Z_1$ and $Z_2$ are $L^\circ$-invariant. Furthermore for all

$1 \leq i \leq t$, either $H_i^\circ$ is irreducible, or $H_i$ is semiprimary and $W_i$ is a direct sum of two nonisomorphic irreducible $\mathbb{F}_q[H_i^\circ]$-modules (Lemma 2.2.15, Theorem 2.2.18).

Therefore if some $H_i^\circ$ is irreducible, it follows from Lemma 2.1.15 that $W_i \subseteq Z_1$ or $W_i \subseteq Z_2$. But since $W_i$ is $L$-invariant and $L$ acts on $\{Z_1, Z_2\}$, this would imply that $Z_1$ and $Z_2$ are $L$-invariant, contradicting our assumption.

Hence $H_i^\circ$ is reducible for all $1 \leq i \leq t$, in which case $H_i$ is semiprimary for all $1 \leq i \leq t$. Then $H_i = \text{semiwr}(X_i^\circ)$ with $X_i$ maximal irreducible solvable of multiplier 2, and $W_i$ is an orthogonal direct sum of two nonisomorphic irreducible $\mathbb{F}_q[H_i^\circ]$-modules. Moreover we must have $W_i \not\subseteq Z_1$ and $W_i \not\subseteq Z_2$ for all $1 \leq i \leq t$, as noted in the previous paragraph. Therefore by applying Lemma 2.1.15 on $L^\circ$, we conclude that

$$Z_1 = W_1' \perp \cdots \perp W_t',$$
$$Z_2 = W_1'' \perp \cdots \perp W_t'',$$

where $W_i = W_i' \perp W_i''$ is the orthogonal system of imprimitivity defining $H_i = \text{semiwr}(X_i^\circ)$. Then

$$L = (\text{semiwr}(X_1^\circ) \times \cdots \times \text{semiwr}(X_t^\circ)) \cap \Delta(V, \kappa) = \text{semiwr}(X_1^\circ \times \cdots \times X_t^\circ).$$

The action of $L^\circ$ on $Z_1$ is isometric with $X_1^\circ \times \cdots \times X_t^\circ$, so we can identify

$$X_1^\circ \times \cdots \times X_t^\circ \leq \overline{H_1^\circ} \leq I(Z_1, \kappa).$$

Because (iii) does not hold, it follows from Theorem 7.5.1 that $X_1^\circ \times \cdots \times X_t^\circ$ is maximal solvable in $I(Z_1, \kappa)$. Therefore $X_1^\circ \times \cdots \times X_t^\circ = \overline{H_1^\circ}$, which is a contradiction since $\overline{H_1^\circ}$ is irreducible and $t > 1$.

**Case 2.2: *G* is of isometric type**

By Theorem 2.2.14 we have $G = (\overline{H} \wr \overline{K}) \cap \Delta(V, \kappa)$ as in Example 2.2.5, where $\overline{H} \leq \Delta(W, \kappa')$ is metrically primitive maximal irreducible solvable, and $\overline{K} \leq S_\ell$ is maximal transitive solvable with $\ell > 1$. Let $V = Z_1 \perp \cdots \perp Z_\ell$ be the orthogonal system of imprimitivity defining $G$.

Suppose first that $L$ does not act transitively on $\{Z_1, \ldots, Z_\ell\}$. Let

$$\{Z_1^{(1)}, \ldots, Z_{\ell_1}^{(1)}\}, \ldots, \{Z_1^{(m)}, \ldots, Z_{\ell_m}^{(m)}\}$$

be the orbits of $L$ on $\{Z_1, \ldots, Z_\ell\}$, where $m > 1$. Denote

$$Q_i := Z_1^{(i)} \perp \cdots \perp Z_{\ell_i}^{(i)}$$

for all $1 \leq i \leq m$.

It follows from Lemma 2.3.8 that for all $1 \leq i \leq m$, we have

$$Q_i = W_1^{(i)} \perp \cdots \perp W_{t_i}^{(i)}$$

for some subset $\{W_1^{(i)}, \ldots, W_{t_i}^{(i)}\}$ of $\{W_1, \ldots, W_t\}$.

Then for all $1 \leq i \leq m$, we have

$$\left(H_1^{(i)} \times \cdots \times H_{t_i}^{(i)}\right) \cap \Delta(Q_i, \kappa) \leq (\overline{H} \wr K_i) \cap \Delta(Q_i, \kappa)$$

where $K_i$ is the action of $L$ on $\{Z_1^{(i)}, \ldots, Z_{\ell_i}^{(i)}\}$.

By induction $\left(H_1^{(i)} \times \cdots \times H_{t_i}^{(i)}\right) \cap \Delta(Q_i, \kappa)$ is maximal solvable, so

$$\left(H_1^{(i)} \times \cdots \times H_{t_i}^{(i)}\right) \cap \Delta(Q_i, \kappa) = (\overline{H} \wr K_i) \cap \Delta(Q_i, \kappa).$$

Since $(\overline{H} \wr K_i) \cap \Delta(Q_i, \kappa)$ is irreducible for all $i$ (Lemma 2.2.18), it follows that $t_i = 1$ and $H_1^{(i)} = (\overline{H} \wr K_i) \cap \Delta(W_1^{(i)}, \kappa)$ for all $i$.

Hence

$$L = \left((\overline{H} \wr K_1) \times \cdots \times (\overline{H} \wr K_t)\right) \cap \Delta(V, \kappa),$$

which implies that $K_1 \times \cdots \times K_t$ is contained in $\overline{K}$. Because statement (ii) of the lemma does not hold, it follows from Theorem 1.1.2 that $K_1 \times \cdots \times K_t$ is maximal solvable in $S_k$, so $K_1 \times \cdots \times K_t = \overline{K}$. But this is a contradiction, since $\overline{K}$ is transitive and $t > 1$.

Therefore we can assume that $L$ acts transitively on $\{Z_1, \ldots, Z_\ell\}$. We now establish some notation, as follows.

For $1 \leq i \leq t$, if $H_i$ is not of semiprimary type, then $H_i = (X_i \wr Y_i) \cap \Delta(W_i, \kappa)$ where $X_i$ is metrically primitive maximal irreducible solvable, and $Y_i \leq S_{k_i}$ is maximal transitive solvable, with $k_i \geq 1$. Then we have an orthogonal decomposition

$$W_i = W_1^{(i)} \perp \cdots \perp W_{k_i}^{(i)},$$

where $\{W_1^{(i)}, \ldots, W_{k_i}^{(i)}\}$ is the orthogonal system of imprimitivity defining $H_i$ if $k_i > 1$.

If $H_i$ is of semiprimary type, then $H_i = \mathrm{semiwr}(X_i^\circ \wr Y_i)$, where $X_i$ is metrically primitive maximal irreducible solvable of multiplier 2, and $Y_i \leq S_{k_i'}$ is maximal transitive solvable, where $k_i' \geq 1$. Denote $k_i = 2k_i'$. We have $W_i = W_i' \perp W_i''$, where $W_i'$ and $W_i''$ are nonisomorphic irreducible $H_i^\circ$-modules. Here the action of

$H_i^\circ$ on $W_i'$ and $W_i''$ is isometric to $X_i^\circ \wr Y_i$. Thus we have orthogonal decompositions

$$W_i' = W_1^{(i)} \perp \cdots \perp W_{k_i'}^{(i)},$$
$$W_i'' = W_{k_i'+1}^{(i)} \perp \cdots \perp W_{k_i}^{(i)}$$

where $\{W_1^{(i)}, \ldots, W_{k_i'}^{(i)}\}$ and $\{W_{k_i'+1}^{(i)}, \ldots, W_{k_i}^{(i)}\}$ are orthogonal systems of imprimitivity on $W_i'$ and $W_i''$, respectively.

Now for all $1 \leq i \leq t$, the subgroup $H_i^\circ$ contains $\left(X_1^{(i)}\right)^\circ \times \cdots \times \left(X_{k_i}^{(i)}\right)^\circ$ with respect to the decomposition $W_i = W_1^{(i)} \perp \cdots \perp W_{k_i}^{(i)}$, where $\left(X_j^{(i)}\right)^\circ$ is similar to $X_i^\circ$ for all $1 \leq j \leq k_i$. Then $L^\circ$ contains

$$R^\circ := \left(X_1^{(1)}\right)^\circ \times \cdots \times \left(X_{k_1}^{(1)}\right)^\circ \times \cdots \times \left(X_1^{(t)}\right)^\circ \times \cdots \times \left(X_{k_t}^{(t)}\right)^\circ.$$

Note that for all $i$ and $j$, the action of $N_L(W_j^{(i)})$ on $W_j^{(i)}$ is equal to $X_j^{(i)}$, which is irreducible and metrically primitive. Furthermore we have assumed that $\kappa \neq 0$ and $q$ is odd, so we can apply Lemma 7.3.6. We have $\dim Z_m > 1$ for all $1 \leq m \leq s$ since $\overline{H}$ is of multiplier 1. Thus it follows from Lemma 7.3.6 that $R^\circ$ acts on $Z_m$ for all $1 \leq m \leq \ell$.

In particular $Z_1$ is $R^\circ$-invariant, so by Lemma 2.1.15 we have

$$Z_1 = W_{j_1}^{(i_1)} \perp \cdots \perp W_{j_u}^{(i_u)}$$

for some indices $i_1, j_1, \ldots, i_u, j_u$. We note the following.

**Claim: $W_i \not\subseteq Z_1$ and $Z_1 \not\subseteq W_i$ for all $1 \leq i \leq t$**
Consider the possibility that $W_i \subseteq Z_1$. Then because $W_i$ is $L$-invariant, it follows that $Z_1$ is $L$-invariant. But this is a contradiction, for we have assumed that $L$ acts transitively on $\{Z_1, \ldots, Z_\ell\}$ and $\ell > 1$. Similarly consider the possibility that $Z_1 \subseteq W_i$ for some $i$. Because $W_i$ is $L$-invariant and $L$ acts transitively on $\{Z_1, \ldots, Z_\ell\}$, we have $V = W_i$, contrary to the fact that $t > 1$. Therefore $Z_1 \not\subseteq W_i$ for all $1 \leq i \leq t$.

By the claim $Z_m \not\subseteq W_i$ for all $i$, so for every $i_\alpha$ we have $i_\alpha \neq i_\beta$ for some index $i_\beta$. Because $\left(H_{i_\alpha}\right)^\circ$ acts trivially on $W_{j_\beta}^{(i_\beta)}$, it follows that $Z_1$ is $\left(H_{i_\alpha}\right)^\circ$-invariant. If $H_{i_\alpha}$ is not semiprimary, then $\left(H_{i_\alpha}\right)^\circ$ is irreducible. In this case the $\left(H_{i_\alpha}\right)^\circ$-submodule generated by $W_{j_\alpha}^{(i_\alpha)}$ is equal to $W_{i_\alpha}$, contradicting the fact that $W_{i_\alpha} \not\subseteq Z_1$.

Thus $H_{i_\alpha}$ must be semiprimary for all $i_\alpha$, and $Z_1$ contains $W_{i_\alpha}'$, where

$$W_{i_\alpha} = W_{i_\alpha}' \perp W_{i_\alpha}''$$

with $W'_{i_\alpha}$ and $W''_{i_\alpha}$ nonisomorphic irreducible $\mathbb{F}_q[(H_{i_\alpha})^\circ]$-modules. There exists $g \in L$ with $g(W'_{i_\alpha}) = W''_{i_\alpha}$ since $L$ is of multiplier 1. Then $W_{i_\alpha} = W'_{i_\alpha} \perp W''_{i_\alpha}$ is contained in $Z_1 \perp Z_m$, where $g(Z_1) = Z_m$. Since $W_{i_\alpha}$ is $L$-invariant and $L$ acts transitively on $\{Z_1, \ldots, Z_\ell\}$, it follows that $\{Z_1, \ldots, Z_\ell\} = \{Z_1, Z_m\}$ and $\ell = 2$.

We conclude then that $H_i$ is semiprimary for all $i$, so $H_i = \text{semiwr}(D_i^\circ)$ with $D_i = (X_i \wr Y_i) \cap \Delta(W_i, \kappa)$ maximal irreducible solvable of multiplier 2 for all $1 \leq i \leq t$. Furthermore $\{Z_1, \ldots, Z_\ell\} = \{Z_1, Z_2\}$ such that

$$Z_1 = W'_1 \perp \cdots \perp W'_t,$$
$$Z_2 = W''_1 \perp \cdots \perp W''_t,$$

where $W_i = W'_i \perp W''_i$ with $W'_i$ and $W''_i$ nonisomorphic irreducible $\mathbb{F}_q[H_i^\circ]$-modules for all $1 \leq i \leq t$. Hence

$$L = \left(\text{semiwr}(D_1^\circ) \times \cdots \times \text{semiwr}(D_t^\circ)\right) \cap \Delta(V, \kappa) = \text{semiwr}(D_1^\circ \times \cdots \times D_t^\circ),$$

with $V = Z_1 \perp Z_2$ the orthogonal system of imprimitivity defining $L$ as a semiprimary group.

The action of $N_G(Z_1)$ on $Z_1$ is equal to a group $\overline{H_1}$ which is isometric to $H$. Since the action of $L^\circ$ on $Z_1$ is isometric with $D_1^\circ \times \cdots \times D_t^\circ$, we can identify

$$D_1^\circ \times \cdots \times D_t^\circ \leq \overline{H_1^\circ} \leq I(Z_1, \kappa).$$

Because we are assuming that (iii) does not hold, it follows from Theorem 7.5.1 that $D_1^\circ \times \cdots \times D_t^\circ$ is maximal solvable in $I(Z_1, \kappa)$. Therefore $D_1^\circ \times \cdots \times D_t^\circ = \overline{H_1^\circ}$, but this is a contradiction since $\overline{H_1^\circ}$ is irreducible and $t > 1$.

**Case 3: $G$ is not irreducible**
In this case $G = (\overline{H_1} \times \cdots \times \overline{H_s}) \cap \Delta(V, \kappa)$, with respect to a decomposition $V = Z_1 \perp \cdots \perp Z_s$ such that $s \geq 2$. Furthermore, here $\overline{H_i}$ is maximal irreducible solvable in $\Delta(Z_i, \kappa)$ for all $1 \leq i \leq s$.

Since $L$ contains $H_1^\circ \times \cdots \times H_t^\circ$ and since $L$ acts irreducibly on $W_i$ for all $1 \leq i \leq t$, it follows from Lemma 2.1.15 that all submodules of $L$ on $V$ are of the form $W_{i_1} \perp \cdots \perp W_{i_r}$ for some $1 \leq i_1 < \cdots < i_r \leq t$. Thus for all $1 \leq i \leq s$, we have

$$Z_i = W_1^{(i)} \perp \cdots \perp W_{t_i}^{(i)}$$

for some subset $\{W_1^{(i)}, \ldots, W_{t_i}^{(i)}\}$ of $\{W_1, \ldots, W_t\}$.

Now by arguing as in the proof of Theorem 7.5.1 (Case 3), it follows that $L = G$, contradicting the assumption $L \lneqq G$.

In all cases we have found a contradiction, so we conclude that $L$ is maximal solvable in $\Delta(V, \kappa)$. □

**Remark 7.5.5** In Theorem 7.5.4, it is not necessarily the case that $H_i^\circ$ are maximal solvable in $I(W_i, \kappa)$.

For example, the groups $X = G_{1,1}^{\mathcal{B}_1}$ and $Y = G_{1,1}^{\mathcal{B}_2}$ are both metrically primitive maximal irreducible solvable in $\mathrm{GSp}_2(7)$. Furthermore, both $X$ and $Y$ are of multiplier 1. By Lemmas 5.5.12 and 5.5.13 the subgroups $X^\circ$ and $Y^\circ$ are not maximal solvable in $\mathrm{Sp}_2(7)$. However, it is still the case by Theorem 7.5.4 that $(X \times Y) \cap \mathrm{GSp}_4(7)$ is maximal solvable in $\mathrm{GSp}_4(7)$.

**Remark 7.5.6** The exceptions of Theorem 7.5.4 (iii) are missed by Jordan in [52], who only considers the maximality of $L^\circ$. This then leads to additional examples where $G_{\mu,\nu}^{\mathcal{B}_0}(X_1, \ldots, X_k) < \mathrm{GL}_n(p)$ is not maximal irreducible solvable, and which are not covered by the *cas d'exclusion* listed in [52, §41–47]. (See also Remark 7.1.16.)

For example, let $X = G_{2,1}^{\mathcal{B}_3}(O_2^-(2)) < \mathrm{GSp}_2(7)$. Then $X$ is primitive maximal irreducible solvable in $\mathrm{GSp}_2(7)$ (Theorem 7.1.12), and $X$ is of multiplier 2 (Lemma 5.1.5). Then it follows from Theorem 7.4.9 that $\mathrm{semiwr}(X^\circ)$ is maximal irreducible solvable in $\mathrm{GSp}_4(7)$, and $\mathrm{semiwr}(X^\circ \wr S_2)$ is maximal irreducible solvable in $\mathrm{GSp}_8(7)$. However, it follows from Theorem 7.5.4 that $\mathrm{semiwr}(X^\circ) \times \mathrm{semiwr}(X^\circ \wr S_2)$ is not maximal solvable in $\mathrm{GSp}_{12}(7)$.

Note that for $\mathrm{semiwr}(X^\circ) \times \mathrm{semiwr}(X^\circ \wr S_2)$, the case of Theorem 7.5.4 (ii) does not apply.

**Theorem 7.5.7** *Assume that $n$ and $q$ are even, and that $\kappa$ is a quadratic form. Let $L = (H_1 \times \cdots \times H_t) \cap \Delta(V, \kappa)$ with respect to a decomposition $V = W_1 \perp \cdots \perp W_t$, where $t \geq 1$ and $H_i \leq \Delta(W_i, \kappa)$ is maximal irreducible solvable for all $1 \leq i \leq t$. Then one of the following holds.*

*(i) $L^\Omega$ is maximal solvable in $\Omega(V, \kappa)$.*
*(ii) There exists $1 \leq i \leq t$ such that $H_i^\Omega$ is not maximal solvable in $\Omega(W_i, \kappa)$, and $H_j^\circ \leq \Omega(W_j, \kappa)$ for all $j \neq i$.*
*(iii) $t > 1$, and there exists $1 \leq i \leq t$ such that $H_i^\circ = O_2^+(q)$ and $H_j^\circ \leq \Omega(W_j, \kappa)$ for all $j \neq i$.*
*(iv) There exists $i \neq j$ such that one of the following holds:*

*(a) $H_i^\circ$ is isometric to $H_j^\circ$.*
*(b) $H_i^\circ$ is isometric to $H_j^\circ \wr S_2$.*
*(c) $H_i^\circ$ is isometric to $H_j^\circ \wr S_3$.*

*Furthermore, if (ii), (iii), or (iv) holds, then $L^\Omega$ is not maximal solvable in $\Omega(V, \kappa)$.*

***Proof*** We first check that in cases (ii)–(iv), the group $L^\Omega$ is not maximal solvable in $\Omega(V, \kappa)$.

Suppose first that (ii) holds, and assume without loss of generality that $H_1^\Omega$ is not maximal solvable, and $H_j^\circ \leq \Omega(W_j, \kappa)$ for all $j \neq 1$. Let $H_1^\Omega \lneq X \leq \Omega(W_1, \kappa)$ with $X$ solvable. Then $L^\Omega = H_1^\Omega \times H_2^\circ \times \cdots \times H_t^\circ$ is properly contained in the solvable subgroup $X \times H_2^\circ \times \cdots \times H_t^\circ$ of $\Omega(V, \kappa)$.

For (iii), we can assume without loss of generality that $H_1^\circ = O_2^+(q)$, and $H_j^\circ \leq \Omega(W_j, \kappa)$ for all $j \neq 1$. In this case $W_1$ has a basis $v$, $w$ such that $Q(v) = Q(w) = 0$ and $b(v, w) = 1$. Let $\zeta \in \mathbb{F}_q^\times$ be a primitive element. Then $H_1^\Omega = \langle g \rangle$, where $gv = \zeta v$ and $gw = \zeta^{-1} w$.

Because $H_j^\circ \leq \Omega(W_j, \kappa)$, it follows from Lemma 2.2.20 (iii) and Lemma 5.1.3 that the quadratic form on $W_j$ is of plus type for all $j \neq 1$. Thus we can find a basis $v_1, \ldots, v_d, w_1, \ldots, w_d$ for $W_2 \perp \cdots \perp W_t$ such that $Q(v_i) = Q(w_j) = b(v_i, v_j) = b(w_i, w_j) = 0$ and $b(v_i, w_j) = \delta_{i,j}$ for all $1 \leq i, j \leq d$. With respect to the basis $v, v_1, \ldots, v_d, w_1, \ldots, w_d, w$, every $g \in L^\Omega = H_1^\Omega \times H_2^\circ \times \cdots \times H_t^\circ$ is of the form

$$\begin{pmatrix} \alpha & & \\ & X & \\ & & \alpha^{-1} \end{pmatrix},$$

where $\alpha \in \mathbb{F}_q^\times$, and $X$ is the action of $g$ on $W^\perp / W = W_2 \perp \cdots \perp W_t$.

Therefore $L^\Omega$ normalizes the subgroup $U \leq \Omega(V, \kappa)$ consisting of all matrices of the form

$$\begin{pmatrix} 1 & * & * \\ 0 & I_{2d} & * \\ 0 & 0 & 1 \end{pmatrix}.$$

Using the fact that $t > 1$, it follows that $U \neq 1$, so $L^\Omega$ is properly contained in the solvable group $L^\Omega U \leq \Omega(V, \kappa)$. Thus $L^\Omega$ is not maximal solvable.

For (iv), suppose that $H_i^\circ$ is isometric to $H_j^\circ \wr S_k$ for some $1 \leq k \leq 3$. We will show that $L^\Omega$ is not maximal solvable in $\Omega(V, \kappa)$. By replacing $L$ with an $I(V, \kappa)$-conjugate, we can assume that $H_i^\circ = H_j^\circ \wr S_k$. We have a proper inclusion

$$\left((H_j^\circ \wr S_k) \times H_j^\circ\right)^\Omega \lneq (H_j^\circ \wr S_{k+1})^\Omega \leq \Omega(W_i \perp W_j, \kappa)$$

since the permutations in $S_{k+1}$ are contained in $\Omega(W_i \perp W_j, \kappa)$ (Lemma 2.2.20 (i)). From this it follows that $L^\Omega$ is not maximal solvable in $\Omega(V, \kappa)$.

We will assume then for the rest of the proof that (ii)–(iv) do not hold, and prove that $L^\Omega$ is maximal solvable in $\Omega(V, \kappa)$. When $t = 1$, we have $L^\Omega = H_1^\Omega$ maximal solvable since (ii) does not hold. Hence we can assume $t > 1$ and proceed by induction. We first note the following.

**Claim: $L^\Omega$ is metrically completely reducible**

If $L^\Omega$ is not metrically completely reducible, by Lemma 2.3.12 there exists $i$ such that $H_i^\Omega$ is not irreducible, and $H_j^\circ \leq \Omega(W_j, \kappa)$ for all $j \neq i$. Because $H_i^\circ$ is maximal irreducible solvable, it follows from Lemma 2.2.21 that $H_i^\circ$ is metrically primitive. Then $H_i^\circ$ is of type $\mathcal{B}_1$ by Lemma 5.2.4, and we have a totally singular decomposition $W_i = W_i' \oplus W_i''$, where $W_i'$ and $W_i''$ are nonisomorphic irreducible

$\mathbb{F}_q[H_i^\Omega]$-modules. If $\dim W_i > 2$, then $H_i^\Omega$ is not maximal solvable as noted in Remark 5.2.5, so (ii) holds, contrary to what we have assumed. If $\dim W_i = 2$, then $H_i^\circ = O_2^+(q)$ and (iii) holds, again contrary to what we have assumed. Therefore we conclude that $L^\Omega$ is metrically completely reducible.

For rest of the proof we suppose for the sake of contradiction that $L^\Omega$ is not maximal solvable. Then $L^\Omega \lneqq G^\Omega$ for some maximal solvable $G \leq \Delta(V, \kappa)$, such that $G^\Omega$ is maximal solvable in $\Omega(V, \kappa)$. It follows from Theorem 7.2.18 that $G$ is not metrically primitive irreducible. We consider the remaining possibilities.

**Case 1: $G$ is irreducible and metrically imprimitive**
We have $G^\circ = \overline{H^\circ} \wr \overline{K}$, where $\overline{H^\circ}$ is metrically primitive maximal irreducible solvable, and $\overline{K} \leq S_\ell$ is maximal transitive solvable with $\ell > 1$. Let $V = Z_1 \perp \cdots \perp Z_\ell$ be the orthogonal system of imprimitivity defining $G$. We split into two cases.

**Case 1.1: $L^\Omega$ acts intransitively on $\{Z_1, \ldots, Z_\ell\}$**
Let

$$\{Z_1^{(1)}, \ldots, Z_{\ell_1}^{(1)}\}, \ldots, \{Z_1^{(u)}, \ldots, Z_{\ell_u}^{(u)}\}$$

be the orbits of $L^\Omega$ on $\{Z_1, \ldots, Z_\ell\}$, where $u > 1$. Then $Q_i := Z_1^{(i)} \perp \cdots \perp Z_{\ell_i}^{(i)}$ is an $L^\Omega$-invariant subspace of $V$, so by Lemma 2.3.12 (iv) we have

$$Q_i = W_1^{(i)} \perp \cdots \perp W_{t_i}^{(i)}$$

for some subset $\{W_1^{(i)}, \ldots, W_{t_i}^{(i)}\}$ of $\{W_1, \ldots, W_t\}$.

Note that the action of $L^\Omega$ on $Q_i$ is contained in $\overline{H^\circ} \wr K_i$, where $K_i$ is the action of $L^\Omega$ on the orbit $\{Z_1^{(i)}, \ldots, Z_{\ell_i}^{(i)}\}$.

We will first prove that $t_i = 1$ for all $1 \leq i \leq u$. To this end, suppose for the sake of contradiction that $t_i > 1$ for some $1 \leq i \leq u$. If there exists $j \neq i$ such that $\left(H_r^{(j)}\right)^\circ \not\leq \Omega(W_r^{(j)}, \kappa)$ for some $1 \leq r \leq t_j$, we find as in the proof of Lemma 2.3.12 (i) that the action of $L^\Omega$ on $Q_i$ is equal to $\left(H_1^{(i)}\right)^\circ \times \cdots \times \left(H_{t_i}^{(i)}\right)^\circ$, which is maximal solvable in $I(Q_i, \kappa)$ by Theorem 7.5.1. Therefore

$$\left(H_1^{(i)}\right)^\circ \times \cdots \times \left(H_{t_i}^{(i)}\right)^\circ = \overline{H^\circ} \wr K_i,$$

which is a contradiction since $t_i > 1$ and $\overline{H^\circ} \wr K_i$ is irreducible (Lemma 2.2.6).

Therefore we can assume that $\left(H_r^{(j)}\right)^\circ \leq \Omega(W_r^{(j)}, \kappa)$ for all $j \neq i$ and $1 \leq r \leq t_j$. In this case the action of $L^\Omega$ on $Z_i$ is equal to $\left(H_1^{(i)} \times \cdots \times H_{t_i}^{(i)}\right)^\Omega$. Furthermore, (ii)–(iv) do not hold for $\left(H_1^{(i)} \times \cdots \times H_{t_i}^{(i)}\right)^\Omega$, so by induction it is maximal solvable in $\Omega(Z_i, \kappa)$. Thus

$$\left(H_1^{(i)} \times \cdots \times H_{t_i}^{(i)}\right)^{\Omega} = \left(\overline{H^\circ} \wr K_i\right)^{\Omega},$$

which is another contradiction, since $t_i > 1$ and $\left(\overline{H^\circ} \wr K_i\right)^{\Omega}$ is irreducible (Lemma 2.2.21).

Hence $t_i = 1$ for all $1 \leq i \leq u$. In this case $u = t$, and after relabeling we can assume that

$$Q_i = W_i = Z_1^{(i)} \perp \cdots \perp Z_{\ell_i}^{(i)}$$

for all $1 \leq i \leq t$.

If Lemma 2.3.12 (i)(a) does not hold, then by Lemma 2.3.12 the action of $L^{\Omega}$ on $W_i$ is equal to $H_i^\circ$ for all $1 \leq i \leq t$. On the other hand the action of $L^{\Omega}$ on $W_i$ is contained in $\overline{H^\circ} \wr K_i$, so $H_i^\circ = \overline{H^\circ} \wr K_i$ because $H_i^\circ$ is maximal solvable in $I(W_i, \kappa)$. Thus

$$L^\circ = \overline{H^\circ} \wr K_1 \times \cdots \times \overline{H^\circ} \wr K_t.$$

Because (iv) does not hold, it follows from Theorem 1.1.2 that the action of $K_1 \times \cdots \times K_t$ on $\{Z_1, \ldots, Z_\ell\}$ is maximal solvable. We have $K_1 \times \cdots \times K_t \leq \overline{K}$, so $K_1 \times \cdots \times K_t = \overline{K}$ by maximality. But this is a contradiction, since $t > 1$ and $\overline{K}$ is transitive.

We consider then the case where Lemma 2.3.12 (i)(a) holds. Without loss of generality, we can assume that $H_1^\circ \not\leq \Omega(W_1, \kappa)$ and $H_j^\circ \leq \Omega(W_j, \kappa)$ for all $j \neq 1$. Then

$$L^{\Omega} = H_1^{\Omega} \times H_2^\circ \times \cdots \times H_t^\circ.$$

We have $H_j^\circ \leq \overline{H^\circ} \wr K_j$ for $j \neq 1$, so $H_j^\circ = \overline{H^\circ} \wr K_j$ since $H_j^\circ$ is maximal solvable. Thus $\overline{H^\circ} = H^{\Omega}$ by Lemma 2.2.20 (iii).

Furthermore $H_1^{\Omega} \leq \overline{H^\circ} \wr K_1$. We have assumed that (ii) does not hold, so $H_1^{\Omega}$ is maximal solvable in $\Omega(W_1, \kappa)$, and thus $H_1^{\Omega} = \left(\overline{H^\circ} \wr K_1\right)^{\Omega} = \overline{H^\circ} \wr K_1$. Hence

$$H_1^{\Omega} \leq \overline{H^\circ} \wr K_1 \leq H_1^\circ.$$

Since $\overline{H^\circ}$ is metrically primitive irreducible solvable and $K_1$ is maximal transitive solvable, it follows from Theorem 7.4.6 that $\overline{H^\circ} \wr K_1$ is maximal solvable in $I(W_1, \kappa)$. Thus $H_1^\circ = \overline{H^\circ} \wr K_1$, but this contradicts the fact that $H_1^\circ \not\leq \Omega(W_1, \kappa)$.

**Case 1.2: $L^{\Omega}$ acts transitively on $\{Z_1, \ldots, Z_\ell\}$**

We can write $H_i^\circ = X_i \wr Y_i$, where $X_i \leq I(W_i', \kappa_i')$ is metrically primitive maximal irreducible solvable, and $Y_i \leq S_{k_i}$ is maximal transitive solvable, where $k_i \geq 1$.

Then $H_i^\circ$ contains $X_1^{(i)} \times \cdots \times X_{k_i}^{(i)}$ with respect to an orthogonal decomposition $W_i = W_1^{(i)} \perp \cdots \perp W_{k_i}^{(i)}$, where $X_j^{(i)}$ is isometric to $X_i$ for all $1 \le j \le k_i$.

Denote

$$R = X_1^{(1)} \times \cdots \times X_{k_1}^{(1)} \times \cdots \times X_1^{(t)} \times \cdots \times X_{k_t}^{(t)}.$$

Because $X_j^{(i)}$ is metrically primitive maximal irreducible solvable in $I(W_j^{(i)}, \kappa)$ for all $i$ and $j$, by Lemma 7.3.9 the orbits of $R^\Omega$ on $\{Z_1, \ldots, Z_\ell\}$ are all of size 1. In other words, each $Z_i$ is $R^\Omega$-invariant.

It follows then from Lemma 2.3.12 that each $Z_i$ is a sum of some $W_j^{(j')}$'s. Thus we can write

$$Z_1 = W_{j_1}^{(j'_1)} \perp \cdots \perp W_{j_r}^{(j'_r)}$$

for some indices $j_1, j'_1, \ldots, j_r, j'_r$. The action of $N_{L^\Omega}(Z_1)$ on $Z_1$ contains

$$\left(X_{j_1}^{(j'_1)} \times \cdots \times X_{j_r}^{(j'_r)}\right)^\Omega$$

and is contained in $\overline{H^\circ}$, so by Theorem 7.2.18 we have $r = 1$. Because $L^\Omega$ acts transitively on $\{Z_1, \ldots, Z_\ell\}$, we have $t = 1$ and $L^\circ = H_1 = X_1 \wr Y_1$, contradicting our assumption $t > 1$.

**Case 2: $G$ is not irreducible**

Because $L^\Omega$ is metrically completely reducible, it follows that the same is true for $G$ as well. By Theorem 2.3.9 we have $G = (\overline{H_1} \times \cdots \times \overline{H_s}) \cap \Delta(V, \kappa)$ with respect to some decomposition $V = Z_1 \perp \cdots \perp Z_s$, where $\overline{H_i} \le \Delta(Z_i, \kappa)$ is maximal irreducible solvable for all $1 \le i \le s$.

Because $L^\Omega$ acts on $Z_i$ for all $1 \le i \le s$, by Lemma 2.3.12 (iv) we have

$$Z_i = W_1^{(i)} \perp \cdots \perp W_{t_i}^{(i)}$$

for some subset $\{W_1^{(i)}, \ldots, W_{t_i}^{(i)}\}$ of $\{W_1, \ldots, W_t\}$. Then $\left(H_1^{(i)} \times \cdots \times H_{t_i}^{(i)}\right)^\Omega \le \overline{H_i^\Omega}$ for all $1 \le i \le t$.

We will prove that $t_i = 1$ for all $1 \le i \le s$. Suppose, for the sake of contradiction that $t_i > 1$. If there exists $j \ne i$ such that $\left(H_r^{(j)}\right)^\circ \not\le \Omega(W_r^{(j)}, \kappa)$ for some $1 \le r \le t_j$, we find as in the proof of Lemma 2.3.12 (i) that the action of $L^\Omega$ on $Z_i$ is equal to $\left(H_1^{(i)}\right)^\circ \times \cdots \times \left(H_{t_i}^{(i)}\right)^\circ$, which is maximal solvable in $I(Z_i, \kappa)$ by Theorem 7.5.1.

Therefore

$$\left(H_1^{(i)}\right)^{\circ} \times \cdots \times \left(H_{t_i}^{(i)}\right)^{\circ} = \overline{H_i^{\circ}},$$

which is a contradiction since $t_i > 1$ and $\overline{H_i^{\circ}}$ is irreducible.

Therefore we can assume that $\left(H_r^{(j)}\right)^{\circ} \leq \Omega(W_r^{(j)}, \kappa)$ for all $j \neq i$ and $1 \leq r \leq t_j$. In this case the action of $L^{\Omega}$ on $Z_i$ is equal to $\left(H_1^{(i)} \times \cdots \times H_{t_i}^{(i)}\right)^{\Omega}$, which is maximal solvable in $\Omega(Z_i, \kappa)$ by induction. Hence

$$\left(H_1^{(i)} \times \cdots \times H_{t_i}^{(i)}\right)^{\Omega} = \overline{H_i^{\Omega}}.$$

Then $\overline{H_i^{\Omega}}$ is not irreducible, so it follows from Lemma 2.2.21 that $\overline{H_i}$ is metrically primitive. It follows from Theorem 5.6.11 that $\overline{H_i^{\Omega}}$ is metrically primitive, so it cannot act on the summands in the decomposition $Z_i = W_1^{(i)} \perp \cdots \perp W_{t_i}^{(i)}$, which is a contradiction.

Therefore $t_i = 1$ and $Z_i = W_1^{(i)}$ for all $1 \leq i \leq s$. In this case $t = s$, and after relabeling the summands we can assume that $Z_i = W_i$ for all $1 \leq i \leq t$.

We will next prove that $L^{\Omega} = \overline{G^{\Omega}}$, a final contradiction which will complete the proof of the theorem. If Lemma 2.3.12 (i)(a) does not hold, then by Lemma 2.3.12 the action of $L^{\Omega}$ on $W_i$ is equal to $H_i^{\circ}$ for all $1 \leq i \leq t$. Because $q$ is even and $H_i$ is maximal solvable, it follows that $H_i^{\circ}$ is maximal solvable in $I(W_i, \kappa)$ for all $1 \leq i \leq t$. Therefore $H_i^{\circ} = \overline{H_i^{\circ}}$ for all $1 \leq i \leq t$, which implies $L^{\circ} = G^{\circ}$, a contradiction.

It remains to consider the case where Lemma 2.3.12 (i)(a) holds, so there exists a unique $i_0$ such that $H_{i_0}^{\circ} \not\leq \Omega(W_{i_0}, \kappa)$. Suppose without loss of generality that $H_1^{\circ} \not\leq \Omega(W_1, \kappa)$ and $H_j^{\circ} \leq \Omega(W_j, \kappa)$ for all $j \neq 1$. We have $H_j^{\Omega} = H_j^{\circ} \leq \overline{H_j^{\circ}}$ for all $j \neq 1$ and $H_j^{\circ}$ is maximal solvable, so $H_j^{\circ} = \overline{H_j^{\circ}}$ for all $j \neq 1$. Therefore

$$L^{\Omega} = H_1^{\Omega} \times H_2^{\circ} \times \cdots \times H_t^{\circ},$$

$$G^{\Omega} = \overline{H_1^{\Omega}} \times H_2^{\circ} \times \cdots \times H_t^{\circ},$$

where $H_1^{\Omega} \leq \overline{H_1^{\Omega}}$. We have assumed that (ii) does not hold, so $H_1^{\Omega}$ is maximal solvable in $\Omega(W_1, \kappa)$. Thus $H_1^{\Omega} = \overline{H_1^{\Omega}}$, giving $L^{\Omega} = G^{\Omega}$, a contradiction. □

## 7.6 Further Results

With the results proved so far, we have essentially completed the classification of maximal irreducible solvable subgroups of $\Delta(V, \kappa)$ and $I(V, \kappa)$, and for $q$ even we have classified maximal irreducible solvable subgroups of $\Omega(V, \kappa)$. Furthermore, for $\kappa \neq 0$ we have a classification of metrically completely reducible maximal solvable subgroups of $\Delta(V, \kappa)$, $I(V, \kappa)$, and ($q$ even) $\Omega(V, \kappa)$.

Some of the results, such as Theorem 7.1.12, involve conditions where $X_i \leq \mathrm{GSp}_{2\ell}(r_i)$ is metrically completely reducible maximal solvable, but $X_i^\circ$ is not maximal solvable in $\mathrm{Sp}_{2\ell}(r_i)$. Furthermore, in Theorem 7.1.13 (iv) we have $X_1 \leq O_{2\ell_1}^{\varepsilon_1}(2)$ metrically completely reducible maximal solvable, but $X_1^\Omega$ is not maximal solvable in $\Omega_{2\ell_1}^{\varepsilon_1}(2)$. Using our main results, we will now give a more precise description of the $X_i$ with these properties. This will then make the application of Theorems 7.1.12, 7.1.13, and 7.1.14 more convenient.

**Proposition 7.6.1** *Assume that $n$ is even, and that $G \leq \mathrm{GSp}_n(q)$ is metrically completely reducible maximal solvable. If $G^\circ$ is not maximal solvable in $\mathrm{Sp}_n(q)$, then 2 is a square in $\mathbb{F}_q$.*

***Proof*** If $q$ is even, then $G = G^\circ Z$ with $Z \leq \mathrm{GL}_n(q)$ the group of scalar matrices, so $G^\circ$ is maximal solvable in $\mathrm{Sp}_n(q)$. Therefore we can assume for the rest of the proof that $q$ is odd. If $n = 2$, then $G$ and $G^\circ$ are metrically primitive and irreducible subgroups of $\mathrm{GSp}_2(q)$. Since $G$ is maximal solvable and $G^\circ$ is not maximal solvable, it follows from Theorems 7.1.12 and 7.1.13 that one of the following cases of Theorem 7.1.13 holds: (iv), (vii) with $q = 9$, (viii) with $q = 7$, or (ix) with $q = 7$. Here (iv) does not apply since $\Omega_2^-(2)$ is maximal solvable, so $q \in \{7, 9\}$ in which case 2 is a square in $\mathbb{F}_q$.

We will assume then that $n > 2$ and proceed by induction on $n$. We can write $G = (H_1 \times \cdots \times H_t) \cap \mathrm{GSp}_n(q)$, where $H_i \leq \mathrm{GSp}_{n_i}(q)$ is maximal irreducible solvable for all $1 \leq i \leq t$, and $n = n_1 + \cdots + n_t$. Now $G$ must be of multiplier 1 since $G^\circ$ is not maximal solvable, so $H_i$ is of multiplier 1 for all $1 \leq i \leq t$.

We consider first the case where $t \geq 2$. If there exists $1 \leq i \leq t$ such that $H_i^\circ$ is not maximal solvable in $\mathrm{Sp}_{n_i}(q)$, by induction 2 is a square in $\mathbb{F}_q$. Therefore we can assume that $H_i^\circ$ is maximal irreducible solvable in $\mathrm{Sp}_{n_i}(q)$ for all $1 \leq i \leq t$. Because $G^\circ$ is not maximal solvable, by Theorem 7.5.1 there exists $i \neq j$ such that $H_i^\circ$ is isometric to $H_j^\circ \wr S_k$, for some $1 \leq k \leq 3$. But then it follows from Theorem 7.5.4 that $G$ is not maximal solvable, which is a contradiction.

Consider then the case $t = 1$, so $G$ is irreducible. Suppose first that $G$ is metrically imprimitive. If $G = \mathrm{semiwr}(H^\circ)$ is semiprimary (Example 2.2.7), then $G^\circ = H_1^\circ \times H_2^\circ$, where $H_i^\circ$ are similar to $H^\circ$, but $H_1^\circ$ and $H_2^\circ$ are not isometric. In this case $G^\circ$ is maximal solvable by Theorem 7.5.4, contrary to what we have assumed. Thus $G = (H \wr K) \cap \mathrm{GSp}_n(q)$ is an isometric imprimitive subgroup, where $H \leq \mathrm{GSp}_{n'}(q)$ is metrically primitive maximal irreducible solvable, and $K \leq S_{n''}$ is maximal transitive solvable with $n'' > 1$. Because $G$ is maximal solvable and $G^\circ$ is not maximal solvable, it follows from Theorems 7.4.7 and 7.4.6 that $H^\circ$ is not maximal solvable in $\mathrm{Sp}_{n'}(q)$. Thus 2 is a square in $\mathbb{F}_q$, by applying induction on $H \leq \mathrm{GSp}_{n'}(q)$.

It remains to consider the case where $G$ is metrically primitive and irreducible, say $G = G^{\mathcal{B}}_{\mu,\nu}(X_1, \ldots, X_k)$. Since $G \leq \mathrm{GSp}_n(q)$, it follows from Corollary 5.6.15 that $G^\circ$ is metrically primitive and irreducible. Then by Theorems 7.1.12 and 7.1.13, one of the following cases of Theorem 7.1.13 holds: (iv), (vii) with $q = 9$, (viii) with $q = 7$, or (ix) with $q = 7$. Now 2 is a square in $\mathbb{F}_q$ for $q \in \{7, 9\}$, so we can assume that (iv) holds. In this case $G = G^{\mathcal{B}_3}_{\mu,1}(X_1)$ is of type $\mathcal{B}_3$, and $X_1^{\Omega}$ is not maximal solvable in $\Omega^{\varepsilon_1}_{2\ell_1}(2)$.

We will see that this assumption leads to a contradiction. Because $G \leq \mathrm{GSp}_n(q)$, we have $\varepsilon_1 = -$. We consider first the case where $X_1$ is irreducible. Because $X_1$ is maximal solvable in $O^-_{2\ell_1}(2)$ and $X_1^{\Omega}$ is not maximal solvable in $\Omega^-_{2\ell_1}(2)$, it follows from Theorem 7.4.8 that $X_1$ is metrically primitive and irreducible. In this case $X_1$ must be of type $\mathcal{B}_2$ since $\varepsilon_1 = -$, so $X_1^{\Omega}$ is metrically primitive and irreducible by Theorem 5.6.11.

We can write $X_1 = G^{\mathcal{B}_2}_{\mu',\nu'}(Y_1, \ldots, Y_s)$, where $Y_i \leq \mathrm{GSp}_{2\ell_i'}(r_i')$ is metrically completely reducible maximal solvable for all $1 \leq i \leq s$. Since $X_1^{\Omega}$ is metrically primitive irreducible and not maximal solvable in $\Omega^-_{2\ell_1}(2)$, it follows from Theorem 7.1.14 that there exists $1 \leq i \leq s$ such that $Y_i^\circ$ is not maximal solvable in $\mathrm{Sp}_{2\ell_i'}(r_i')$. By induction $\left(\frac{2}{r_i}\right) = +$, so $\left(\frac{q}{r_i}\right) = +$. But then Theorem 7.1.13 (iii) holds (case (iii)(d)), so it follows from Theorem 7.1.13 that $X_1$ is not maximal solvable, which is a contradiction.

Consider then the case where $X_1 \leq O^-_{2\ell_1}(2) = I(W, \kappa')$ is not irreducible. Then $X_1 = T_1 \times \cdots \times T_u$ with respect to a decomposition $W = W_1 \perp \cdots \perp W_u$, where $u \geq 2$ and $T_i \leq I(W_i, \kappa')$ is maximal irreducible solvable for all $1 \leq i \leq u$ (Lemma 2.3.7). Because $X_1^{\Omega}$ is not maximal solvable, by Theorem 7.5.7 one of Theorem 7.5.7 (ii)–(iv) holds. Here (iv) is not applicable, since in this case it follows from Theorem 7.5.1 that $X_1$ is not maximal solvable. Thus (ii) or (iii) holds, in which case there exists $1 \leq i \leq u$ such that $T_j \leq \Omega(W_j, \kappa')$ for all $j \neq i$. By Lemma 5.1.4 the form on $(W_j, \kappa')$ is of plus type for all $j \neq i$, so the form on $(W_i, \kappa')$ must be of minus type. Thus Theorem 7.5.7 (iii) does not hold. Then Theorem 7.5.7 (ii) holds, in which case $T_i$ is maximal solvable and $T_i^{\Omega}$ is not maximal solvable. Since the form on $(W_i, \kappa')$ is of minus type, arguing as in the case where $X_1$ is irreducible leads to a contradiction. □

**Proposition 7.6.2** *Assume that $n$ and $q$ are even, and that $G \leq \mathrm{GO}^{\varepsilon}_n(q)$ is metrically completely reducible maximal solvable. If $G^{\Omega}$ is not maximal solvable in $\Omega^{\varepsilon}_n(q)$, then $\varepsilon = +$.*

***Proof*** By arguing as for $X_1$ at the end of the proof of Proposition 7.6.1 (last three paragraphs), the result follows from Proposition 7.6.1. □

**Theorem 7.6.3** *Assume that $n$ and $q$ are even and that $\kappa$ is a quadratic form. Suppose that $G \leq \Delta(V,\kappa)$ is maximal irreducible solvable. Then the following statements hold:*

*(i) $G^\circ$ is maximal irreducible solvable in $I(V,\kappa)$.*
*(ii) If $G^\Omega$ is irreducible, then $G^\Omega$ is maximal irreducible solvable in $\Omega(V,\kappa)$.*
*(iii) $G^\Omega$ is not irreducible if and only if $G$ is metrically primitive of type $\mathcal{B}_1$, with $n/2$ odd and $\varepsilon = +$.*
*(iv) If $G^\Omega$ is not irreducible, then $G^\Omega$ is not maximal solvable, unless $G = \mathrm{GO}_2^+(q)$.*

***Proof*** We have $G = G^\circ Z$, where $Z \leq \mathrm{GL}(V)$ is the group of scalar matrices, so claim (i) holds. Claims (iii) and (iv) follow from Lemmas 2.2.21, 5.2.4, and Remark 5.2.5.

We consider then claim (ii), so assume that $G^\Omega$ is irreducible. In the metrically imprimitive case, it follows from Theorem 7.4.8 that $G^\Omega$ is maximal solvable in $\Omega(V,\kappa)$. Suppose then that $G^\Omega$ is metrically primitive. If $G^\Omega$ is not maximal solvable, it follows from Proposition 7.6.2 that $\varepsilon = +$, so $G$ must be of type $\mathcal{B}_1$. Now $G^\Omega$ is irreducible and not maximal solvable, so Theorem 7.1.14 (iii)(a) holds. But in this case it follows from Theorem 7.1.13 that $G$ is not maximal solvable, which is a contradiction. □

**Theorem 7.6.4** *Assume that $n$ and $q$ are even and that $\kappa$ is a quadratic form. Suppose that $G \leq \Delta(V,\kappa)$ is metrically completely reducible maximal irreducible solvable. Then $G^\circ = H_1^\circ \times \cdots \times H_t^\circ$ with respect to a decomposition $V = W_1 \perp \cdots \perp W_t$, where $H_i^\circ \leq I(W_i,\kappa)$ is maximal irreducible solvable for all $1 \leq i \leq t$. Furthermore, the following statements hold.*

*(i) $G^\circ$ is maximal solvable in $I(V,\kappa)$.*
*(ii) Assume that $t \geq 2$. If $G^\Omega$ is not maximal solvable in $\Omega(V,\kappa)$, then $\varepsilon = +$ and there exists $1 \leq i \leq t$ such that the following hold:*

*(a) $H_j^\circ \leq \Omega(W_j,\kappa)$ for all $j \neq i$;*
*(b) $H_i^\circ = O_2^+(q)$, or $H_i$ is metrically primitive of type $\mathcal{B}_1$ with $\dim W_i/2$ odd.*

***Proof*** Claim (i) follows as in Proposition 7.6.3. Then the fact that $G^\circ = H_1^\circ \times \cdots \times H_t^\circ$ with $H_i^\circ$ maximal irreducible solvable follows from Lemma 2.3.7.

We consider then claim (ii). Suppose that $t \geq 2$ and that $G^\Omega$ is not maximal solvable in $\Omega(V,\kappa)$. It follows from Proposition 7.6.2 that $\varepsilon = +$. Since $G^\circ$ is maximal solvable, it follows from Theorem 7.5.1 that Theorem 7.5.7 (iv) does not apply. Thus Theorem 7.5.7 (ii) or (iii) must hold, which combined with Theorem 7.6.3 gives the result. □

**Theorem 7.6.5** *Assume that $n$ is even, and that $G \leq \mathrm{GSp}_n(q)$ is metrically completely reducible maximal solvable. If $G^\circ$ is not maximal solvable in $\mathrm{Sp}_n(q)$, then $q \in \{7,9\}$ and one of the following holds:*

*(i)* $G = G^{\mathcal{B}}_{\mu,\nu}(X_1, \ldots, X_k)$ *is metrically primitive irreducible, and one of the following holds:*

*(a)* $G$ *is of type* $\mathcal{B}_1$, $q = 9$, $\nu = 1$, *and* $\mu = 1$ *or* $\varepsilon_1 = +$.
*(b)* $G$ *is of type* $\mathcal{B}_1$, *with* $n$ *a power of two,* $q = 7$, $\nu = 1$, *and* $\mu = 1$ *or* $\varepsilon_1 = +$.
*(c)* $G$ *is of type* $\mathcal{B}_2$, $q = 7$, $\nu = 1$, *and* $\mu = 1$ *or* $\varepsilon_1 = +$.

*(ii)* $G = (H \wr K) \cap \mathrm{GSp}_n(q)$ *is an isometric imprimitive subgroup as in Theorem 2.2.14 (i), where* $H \leq \mathrm{GSp}_{n'}(q)$ *is metrically primitive maximal irreducible solvable and* $H^\circ$ *is not maximal solvable in* $\mathrm{Sp}_n(q)$.
*(iii)* $G = (H_1 \times \cdots \times H_t) \cap \mathrm{GSp}_n(q)$ *where* $H_i \leq \mathrm{GSp}_{n_i}(q)$ *is maximal irreducible solvable for all* $1 \leq i \leq t$, $t \geq 2$, $n = n_1 + \cdots + n_t$, *and there exists* $1 \leq j \leq t$ *such that* $H_j^\circ$ *is not maximal solvable in* $\mathrm{Sp}_{n_j}(q)$.

*Furthermore, in cases (i)–(iii) the group* $G^\circ$ *is not maximal solvable in* $\mathrm{Sp}_n(q)$.

***Proof*** If $q$ is even, then $G = G^\circ Z$ where $Z \leq \mathrm{GL}(V)$ is the group of scalar matrices, so $G^\circ$ is also maximal solvable. Therefore we can assume that $q$ is odd. Suppose that $G^\circ$ is not maximal solvable in $\mathrm{Sp}_n(q)$. We will first check that one of (i)–(iii) holds.

Consider first the case where $G$ is metrically primitive and irreducible. Then $G^\circ$ is metrically primitive and irreducible by Corollary 5.6.15. Because $G$ is maximal solvable and $G^\circ$ is not maximal solvable, it follows from Theorems 7.1.12 and 7.1.13 that one of the following cases of Theorem 7.1.13 holds: (iv), (vii) with $q = 9$, (viii) with $q = 7$, or (ix) with $q = 7$. Here (iv) does not apply since 2 is a square in $\mathbb{F}_q$ by Proposition 7.6.1. Therefore one of (i)(a)–(i)(c) holds and $q \in \{7, 9\}$, as required.

Next we consider the case where $G$ is metrically imprimitive and irreducible. As seen in the proof of Proposition 7.6.1 (paragraph 4), it follows in this case that (ii) holds. Since $H$ must be as in case (i), it follows that $q \in \{7, 9\}$. Similarly in the case where $G$ is not irreducible, it follows as in the proof of Proposition 7.6.1 (paragraph 3) that (iii) holds. Because $H_i$ must be as in case (i) or (ii), it follows that $q \in \{7, 9\}$.

It remains to verify that in case (i)–(iii) the group $G^\circ$ is not maximal solvable in $\mathrm{Sp}_n(q)$. In case (i) this follows from Theorem 7.1.13. In case (ii), the fact that $G^\circ$ is not maximal solvable is clear since $G^\circ = H^\circ \wr K$. In case (iii) this is also clear, since $G^\circ = H_1^\circ \times \cdots \times H_t^\circ$. □

**Remark 7.6.6** By Theorem 7.6.5, the exceptions in Theorem 7.1.12 (ii), Theorem 7.1.13 (iii), and Theorem 7.1.14 (iii) can only appear for $r_i = 7$.

**Remark 7.6.7** In the case where $\kappa$ is a quadratic form, we can similarly use our results to give a list of all cases where $G \leq \Delta(V, \kappa)$ is metrically completely reducible maximal solvable and $G^\circ \leq I(V, \kappa)$ is not maximal solvable. We omit the details, but in this case there will be more examples, as the cases in Corollary 5.6.13 must be included (see Proposition 7.2.23).

# Chapter 8
# Examples

In this chapter, we will illustrate the main results in special cases, and provide tables of maximal solvable subgroups in small degrees. Specifically, this chapter contains the following tables:

- Table 8.1: Maximal transitive solvable subgroups of $S_n$ for $5 \leq n \leq 33$.
- Table 8.2: Metrically completely reducible maximal solvable subgroups of $O_n^{\varepsilon}(2)$, where $n$ is even and $2 \leq n \leq 12$.
- Table 8.3: Maximal irreducible solvable subgroups of $\mathrm{GSp}_n(q)$ and $\mathrm{Sp}_n(q)$, where $n$ is even and $2 \leq n \leq 6$.
- Table 8.4: Maximal irreducible solvable subgroups of $\mathrm{GO}_n^{\varepsilon}(q)$ and $O_n^{\varepsilon}(q)$, where $2 \leq n \leq 6$.
- Table 8.5: Maximal irreducible solvable subgroups of $\mathrm{GL}_n(q)$, for $2 \leq n \leq 8$.
- Table 8.6: Maximal irreducible solvable subgroups of $\mathrm{GL}_r(q)$ and $\mathrm{GL}_{2r}(q)$, where $r > 2$ is a prime number.

Similar tables can be generated efficiently for any given degree, using our main results. From the construction of the groups involved, one can also write down generators for the maximal solvable subgroups.

## 8.1 Examples and Summary of Construction

**Example 8.1.1** With Theorems 1.1.1–1.1.2 and Propositions 1.3.7–1.3.9, the construction of maximal solvable subgroups of $S_n$ is reduced to the construction of maximal irreducible solvable subgroups of $\mathrm{GL}_k(p)$, where $p$ is a prime and $p^k \leq n$.

In Table 8.1, we list maximal transitive solvable subgroups of $S_n$ for $5 \leq n \leq 33$, up to conjugacy in $S_n$. This requires maximal irreducible solvable subgroups of $\mathrm{GL}_k(p)$ for $p$ prime and $p^k \leq 33$. For $k = 1$ the only possibility is $\mathrm{GL}_1(p)$. The



M. Korhonen, *Maximal Solvable Subgroups of Finite Classical Groups*,
Lecture Notes in Mathematics 2346, https://doi.org/10.1007/978-3-031-62915-0_8

remaining cases are $p^k \in \{3^2, 3^3, 2^2, 2^3, 2^4, 2^5, 5^2\}$, which are straightforward to deal with—see Example 8.1.2 below.

Note that from the list of maximal transitive solvable subgroups, it is straightforward to generate the list of all maximal solvable subgroups (Theorems 1.1.1, 1.1.2, Proposition 1.3.7.)

**Example 8.1.2** Let $q$ be a prime power and $n \geq 1$ an integer. The recursive construction of all maximal irreducible solvable subgroups of $\mathrm{GL}_n(q)$ (up to conjugacy in $\mathrm{GL}_n(q)$) proceeds as follows. We know that $\mathrm{GL}_1(q)$ is solvable, so suppose that $n > 1$.

First we reduce to the primitive case. By Theorem 2.2.14, an imprimitive maximal irreducible solvable subgroup of $\mathrm{GL}_n(q)$ is of the form $G = H \wr K$, where:

- $n = dk$, with $d$ and $k$ integers and $k > 1$.
- $H \leq \mathrm{GL}_d(q)$ is maximal primitive irreducible solvable, with $(q, d) \neq (2, 1)$.
- $K \leq S_k$ is maximal transitive solvable.

Furthermore, by Theorems 2.2.18 and 7.4.6, we know that any such $G = H \wr K$ is maximal irreducible solvable, with the following exceptions:

- $G = \mathrm{GL}_1(3) \wr S_2 \wr K'$
- $G = \mathrm{GL}_1(5) \wr S_2 \wr K'$
- $G = \mathrm{GL}_1(4) \wr S_3 \wr K'$
- $G = \mathrm{GL}_1(3) \wr S_4 \wr K'$

By Proposition 7.4.10, it suffices to consider $H$ up to conjugacy in $\mathrm{GL}_d(q)$, and $K$ up to conjugacy in $S_k$.

Here $d < n$, so we can assume that the possibilities for $H$ are known by recursion. Furthermore, as noted in Example 8.1.1, we can describe $K$ in terms of maximal irreducible solvable subgroups of $\mathrm{GL}_\ell(r)$, where $r$ is a prime and $r^\ell \leq k$. Here $\ell < r^\ell \leq n$, so by recursion we can assume that the possibilities for $K$ are known as well.

Therefore we can reduce to the primitive case. For $G \leq \mathrm{GL}_n(q)$ primitive maximal irreducible solvable, we have a factorization $n = \mu\nu$ such that after replacing $G$ with a conjugate, we have

$$G = G_{\mu,\nu}^{\mathcal{B}_0}(X_1, \ldots, X_k)$$

for some $X_1, \ldots, X_k$. Here $\mu$ has factorization $\mu = r_1^{\ell_1} \cdots r_k^{\ell_k}$, with $r_1 < \cdots < r_k$ primes such that $q^\nu \equiv 1 \mod r_i$ and $X_i \leq \mathrm{GSp}_{2\ell_i}(r_i)$ is metrically completely reducible maximal solvable for all $1 \leq i \leq k$. If $r_1 = 2$, then $X_1 \leq O_{2\ell_1}^{\varepsilon_1}(2)$.

The conjugacy among groups of the form $G_{\mu,\nu}^{\mathcal{B}_0}(X_1, \ldots, X_k)$ is described by Lemma 5.8.1 and Theorem 5.8.3. It follows that it suffices to consider $G$ with $\mu$ fixed, and $X_i$ up to $\mathrm{Sp}_{2\ell_i}(r_i)$-conjugacy.

If $\mu = 1$, then $G = G^{\mathcal{B}_0}_{1,1} = \Gamma\mathrm{L}_1(q^n)$ is constructed explicitly in Remark 2.9.9, and is unique up to conjugacy in $\mathrm{GL}_n(q)$. Furthermore, in this case it follows from Lemma 5.2.2, Theorems 5.6.9, and 7.1.15 that $G$ is primitive maximal irreducible solvable, except for $(q, \nu) = (3, 2)$. In the exceptional case $G = \Gamma\mathrm{L}_1(3^2) \lneq \mathrm{GL}_2(3)$ is not maximal solvable.

Consider then the case where $\mu > 1$. It follows from Lemma 5.2.2 and Theorem 5.6.9 that every $G$ of the form $G = G^{\mathcal{B}_0}_{\mu,\nu}(X_1, \dots, X_k)$ is primitive irreducible solvable. Furthermore, by Theorem 7.1.12 any such $G$ is maximal solvable, with a few exceptions which are described by Theorem 7.1.12 (ii)–(iii).

Thus we reduce to the construction of metrically completely reducible maximal solvable subgroups of the following groups:

- $\mathrm{GSp}_{2\ell}(r)$ and $\mathrm{Sp}_{2\ell}(r)$, for $r$ prime with $r^\ell \mid n$.
- $O^+_{2\ell}(2)$ and $O^-_{2\ell}(2)$, with $2^\ell \mid n$.

These maximal solvable subgroups are classified by our main results, and again the construction reduces to groups of smaller degree. The base case of the construction consists of the metrically primitive groups with $\mu = 1$, in other words, $G^{\mathcal{B}_1}_{1,\nu}$ and $G^{\mathcal{B}_2}_{1,\nu}$. As a summary, the relevant results for $\mathrm{GSp}_{2\ell}(r)$, $\mathrm{Sp}_{2\ell}(r)$, and $O^\pm_{2\ell}(2)$ are as follows:

- Metrically completely reducible maximal solvable subgroups are described in terms of maximal irreducible solvable subgroups in Theorems 7.5.1, 7.5.3, and 7.5.4.
- In the irreducible case, the metrically imprimitive maximal solvable subgroups are classified in terms of metrically primitive maximal solvable subgroups in Theorems 7.4.6, 7.4.7, and 7.4.9.
- In the metrically primitive case, the maximal solvable subgroups are classified by Theorems 7.1.12 and 7.1.13, in terms of groups of smaller degree. Here it is necessary to also take into account the description of when $G = G^{\mathcal{B}}_{\mu,\nu}(X_1, \dots, X_k)$ is irreducible (Lemmas 5.2.1 and 5.2.2) and metrically primitive (Theorem 5.6.9, Corollary 5.6.15).
- In Theorems 7.1.12 and 7.1.13, one needs to know the multiplier of a metrically completely reducible maximal solvable subgroup $X < \mathrm{GSp}_d(r)$. The cases where $X$ is of multiplier 2 are precisely the following:
  - Non-irreducible: $X = (H_1 \times \cdots \times H_t) \cap \mathrm{GSp}_d(r)$, where some $H_i$ is of multiplier 2.
  - Irreducible metrically imprimitive: $X = (H \wr K) \cap \mathrm{GSp}_d(r)$, where $H$ is metrically primitive maximal irreducible solvable of multiplier 2 (Theorems 2.2.18 and 2.2.19).
  - Metrically primitive: $X$ of type $\mathcal{B}_3$ such that 2 is a square in $\mathbb{F}_r$ (Lemmas 4.3.13, 4.3.22, and 5.1.5).
- In Theorems 7.1.12 and 7.1.13, one also needs information on cases where $X_i < \mathrm{GSp}_{2\ell_i}(r_i)$ is maximal solvable, but $X_i^\circ$ is not maximal solvable in $\mathrm{Sp}_{2\ell_i}(r_i)$. This is described by Theorem 7.6.5.

- In Theorem 7.1.13 we also have a case where $X_1 \leq O_{2\ell_1}^{\varepsilon_1}(2)$ is maximal solvable, but $X_1^{\Omega}$ is not maximal solvable in $\Omega_{2\ell_1}^{\varepsilon_1}(2)$. This is handled by Theorem 7.6.4.
- For conjugacy among the groups constructed, see the following results:
  - Non-irreducible: Lemmas 2.3.13, 2.3.14.
  - Irreducible, metrically imprimitive: Propositions 7.4.10, 7.4.11.
  - Irreducible, metrically primitive: Sect. 5.8, Theorem 5.8.3.
  - For $X < \mathrm{GSp}_d(q)$, the $\mathrm{GSp}_d(q)$-conjugacy class of $X^\circ$ forms a single $\mathrm{Sp}_d(q)$-conjugacy class if $X$ is of multiplier 1, and splits into two classes if $X$ is of multiplier 2 (Lemma 2.1.10).

**Remark 8.1.3** Previously in [77, 21.3, Theorem 6], Suprunenko gave a list of maximal irreducible solvable subgroups of $\mathrm{GL}_r(q)$, for $r$ a prime number. We note that [77, 21.3, Remarks to Theorem 6] is missing the observation that (in the notation of [77, 21.3, Theorem 6]) the groups $\mathcal{G}_{7,3}$ and $\mathcal{G}_{7,4}$ are not maximal solvable in $\mathrm{GL}(7, \Delta)$.

In our notation, we have $\mathcal{G}_{7,3} = G_{7,1}^{\mathcal{B}_0}(G_{1,1}^{\mathcal{B}_1})$ and $\mathcal{G}_{7,4} = G_{7,1}^{\mathcal{B}_0}(G_{1,1}^{\mathcal{B}_2})$. These groups are not maximal solvable in $\mathrm{GL}_7(q)$, since for $X = G_{1,1}^{\mathcal{B}_1} = \mathrm{GL}_1(7) \wr S_2$ and $X = G_{1,1}^{\mathcal{B}_2} = \Gamma\mathrm{L}_1(7^2)$ we have $X$ maximal solvable in $\mathrm{GSp}_2(7)$, but $X^\circ$ is not maximal solvable in $\mathrm{Sp}_2(7)$. (See Lemmas 5.5.12, 5.5.13, and Example 5.5.15.)

## 8.2 Tables of Examples

See Tables 8.1, 8.2, 8.3, 8.4, 8.5, and 8.6.

**Table 8.1** List of maximal transitive solvable subgroups of $S_n$ for $5 \le n \le 33$, up to conjugacy in $S_n$

| $n$ | $X$ |
|---|---|
| 5 | $\mathrm{AGL}_1(5)$ |
| 6 | $S_3 \wr S_2$ |
| | $S_2 \wr S_3$ |
| 7 | $\mathrm{AGL}_1(7)$ |
| 8 | $\mathbb{F}_2^3 \rtimes \Gamma\mathrm{L}_1(2^3)$ |
| | $S_4 \wr S_2$ |
| | $S_2 \wr S_4$ |
| 9 | $\mathrm{AGL}_2(3)$ |
| | $S_3 \wr S_3$ |
| 10 | $\mathrm{AGL}_1(5) \wr S_2$ |
| | $S_2 \wr \mathrm{AGL}_1(5)$ |
| 11 | $\mathrm{AGL}_1(11)$ |
| 12 | $S_4 \wr S_3$ |
| | $S_3 \wr S_4$ |
| | $S_2 \wr S_3 \wr S_2$ |
| 13 | $\mathrm{AGL}_1(13)$ |
| 14 | $\mathrm{AGL}_1(7) \wr S_2$ |
| | $S_2 \wr \mathrm{AGL}_1(7)$ |
| 15 | $\mathrm{AGL}_1(5) \wr S_3$ |
| | $S_3 \wr \mathrm{AGL}_1(5)$ |
| 16 | $\mathbb{F}_2^4 \rtimes \Gamma\mathrm{L}_1(2^4)$ |
| | $\mathbb{F}_2^4 \rtimes (\Gamma\mathrm{L}_1(2^2) \wr S_2)$ |
| | $(\mathbb{F}_2^3 \rtimes \Gamma\mathrm{L}_1(2^3)) \wr S_2$ |
| | $S_2 \wr (\mathbb{F}_2^3 \rtimes \Gamma\mathrm{L}_1(2^3))$ |
| | $S_4 \wr S_4$ |
| | $S_2 \wr S_4 \wr S_2$ |
| 17 | $\mathrm{AGL}_1(17)$ |
| 18 | $S_2 \wr \mathrm{AGL}_2(3)$ |
| | $\mathrm{AGL}_2(3) \wr S_2$ |
| | $S_2 \wr S_3 \wr S_3$ |
| | $S_3 \wr S_2 \wr S_3$ |
| | $S_3 \wr S_3 \wr S_2$ |
| 19 | $\mathrm{AGL}_1(19)$ |
| 20 | $S_4 \wr \mathrm{AGL}_1(5)$ |
| | $\mathrm{AGL}_1(5) \wr S_4$ |
| | $S_2 \wr \mathrm{AGL}_1(5) \wr S_2$ |
| 21 | $\mathrm{AGL}_1(7) \wr S_3$ |
| | $S_3 \wr \mathrm{AGL}_1(7)$ |
| 22 | $\mathrm{AGL}_1(11) \wr S_2$ |
| | $S_2 \wr \mathrm{AGL}_1(11)$ |
| 23 | $\mathrm{AGL}_1(23)$ |
| 24 | $(\mathbb{F}_2^3 \rtimes \Gamma\mathrm{L}_1(2^3)) \wr S_3$ |
| | $S_3 \wr (\mathbb{F}_2^3 \rtimes \Gamma\mathrm{L}_1(2^3))$ |
| | $S_2 \wr S_3 \wr S_4$ |
| | $S_2 \wr S_4 \wr S_3$ |
| | $S_3 \wr S_2 \wr S_4$ |
| | $S_3 \wr S_4 \wr S_2$ |
| | $S_4 \wr S_2 \wr S_3$ |
| | $S_4 \wr S_3 \wr S_2$ |
| 25 | $\mathbb{F}_5^2 \rtimes \Gamma\mathrm{L}_1(5^2)$ |
| | $\mathbb{F}_5^2 \rtimes G_{2,1}^{\mathcal{B}_0}(O_2^-(2))$ |
| | $\mathrm{AGL}_1(5) \wr \mathrm{AGL}_1(5)$ |
| 26 | $S_2 \wr \mathrm{AGL}_1(13)$ |
| | $\mathrm{AGL}_1(13) \wr S_2$ |
| 27 | $\mathrm{AGL}_2(3) \wr S_3$ |
| | $S_3 \wr \mathrm{AGL}_2(3)$ |
| | $S_3 \wr S_3 \wr S_3$ |
| | $\mathbb{F}_3^3 \rtimes (\mathrm{GL}_1(3) \wr S_3)$ |
| | $\mathbb{F}_3^3 \rtimes \Gamma\mathrm{L}_1(3^3)$ |
| 28 | $S_4 \wr \mathrm{AGL}_1(7)$ |
| | $\mathrm{AGL}_1(7) \wr S_4$ |
| | $S_2 \wr \mathrm{AGL}_1(7) \wr S_2$ |
| 29 | $\mathrm{AGL}_1(29)$ |
| 30 | $S_2 \wr S_3 \wr \mathrm{AGL}_1(5)$ |
| | $S_2 \wr \mathrm{AGL}_1(5) \wr S_3$ |
| | $S_3 \wr S_2 \wr \mathrm{AGL}_1(5)$ |
| | $S_3 \wr \mathrm{AGL}_1(5) \wr S_2$ |
| | $\mathrm{AGL}_1(5) \wr S_2 \wr S_3$ |
| | $\mathrm{AGL}_1(5) \wr S_3 \wr S_2$ |
| 31 | $\mathrm{AGL}_1(31)$ |
| 32 | $S_2 \wr S_4 \wr S_4$ |
| | $S_4 \wr S_2 \wr S_4$ |
| | $S_4 \wr S_4 \wr S_2$ |
| | $(\mathbb{F}_2^3 \rtimes \Gamma\mathrm{L}_1(2^3)) \wr S_4$ |
| | $S_4 \wr (\mathbb{F}_2^3 \rtimes \Gamma\mathrm{L}_1(2^3))$ |
| | $S_2 \wr (\mathbb{F}_2^3 \rtimes \Gamma\mathrm{L}_1(2^3)) \wr S_2$ |
| | $(\mathbb{F}_2^4 \rtimes (\Gamma\mathrm{L}_1(2^2) \wr S_2)) \wr S_2$ |
| | $S_2 \wr (\mathbb{F}_2^4 \rtimes (\Gamma\mathrm{L}_1(2^2) \wr S_2))$ |
| | $(\mathbb{F}_2^4 \rtimes \Gamma\mathrm{L}_1(2^4)) \wr S_2$ |
| | $S_2 \wr (\mathbb{F}_2^4 \rtimes \Gamma\mathrm{L}_1(2^4))$ |
| | $\mathbb{F}_2^5 \rtimes \Gamma\mathrm{L}_1(2^5)$ |
| 33 | $S_3 \wr \mathrm{AGL}_1(11)$ |
| | $\mathrm{AGL}_1(11) \wr S_3$ |

**Table 8.2** Metrically completely reducible maximal solvable subgroups of $\Delta(V,\kappa) = O_n^{\varepsilon}(2)$ with $2 \le n \le 12$ even. Note that $O_2^+(2)$ is not metrically completely reducible

| $\Delta(V,\kappa)$ | $X$ | $[X : X^{\Omega}]$ |
|---|---|---|
| $O_2^+(2)$ | None | n/a |
| $O_2^-(2)$ | $O_2^-(2) = G_{1,1}^{\mathcal{B}_2}$ | 2 |
| $O_4^+(2)$ | $O_4^+(2) = O_2^-(2) \wr S_2$ | 2 |
| $O_4^-(2)$ | $G_{1,2}^{\mathcal{B}_2}$ | 2 |
| $O_6^+(2)$ | $G_{1,3}^{\mathcal{B}_1}$ | 2 |
| | $G_{1,2}^{\mathcal{B}_2} \times O_2^-(2)$ | 2 |
| $O_6^-(2)$ | $O_2^-(2) \wr S_3$ | 2 |
| | $G_{3,1}^{\mathcal{B}_2}(\mathrm{GSp}_2(3))$ | 2 |
| $O_8^+(2)$ | $O_2^-(2) \wr S_4$ | 2 |
| | $G_{1,2}^{\mathcal{B}_2} \wr S_2$ | 2 |
| | $G_{1,4}^{\mathcal{B}_1}$ | 1 |
| | $G_{3,1}^{\mathcal{B}_2}(\mathrm{GSp}_2(3)) \times O_2^-(2)$ | 2 |
| $O_8^-(2)$ | $G_{1,4}^{\mathcal{B}_2}$ | 2 |
| | $G_{1,2}^{\mathcal{B}_2} \times (O_2^-(2) \wr S_2)$ | 2 |
| | $G_{1,3}^{\mathcal{B}_1} \times O_2^-(2)$ | 2 |
| $O_{10}^+(2)$ | $G_{1,5}^{\mathcal{B}_1}$ | 2 |
| | $G_{1,4}^{\mathcal{B}_2} \times O_2^-(2)$ | 2 |
| | $G_{1,3}^{\mathcal{B}_1} \times (O_2^-(2) \wr S_2)$ | 2 |
| | $(O_2^-(2) \wr S_3) \times G_{1,2}^{\mathcal{B}_2}$ | 2 |
| | $G_{3,1}^{\mathcal{B}_2}(\mathrm{GSp}_2(3)) \times G_{1,2}^{\mathcal{B}_2}$ | 2 |
| $O_{10}^-(2)$ | $O_2^-(2) \wr \mathrm{AGL}_1(5)$ | 2 |
| | $G_{1,5}^{\mathcal{B}_2}$ | 2 |
| | $(O_2^-(2) \wr S_4) \times O_2^-(2)$ | 2 |
| | $(G_{1,2}^{\mathcal{B}_2} \wr S_2) \times O_2^-(2)$ | 2 |
| | $G_{1,4}^{\mathcal{B}_1} \times O_2^-(2)$ | 2 |
| | $G_{1,3}^{\mathcal{B}_1} \times G_{1,2}^{\mathcal{B}_2}$ | 2 |
| | $(O_2^-(2) \wr S_3) \times (O_2^-(2) \wr S_2)$ | 2 |
| | $G_{3,1}^{\mathcal{B}_2}(\mathrm{GSp}_2(3)) \times (O_2^-(2) \wr S_2)$ | 2 |
| $O_{12}^+(2)$ | $O_2^-(2) \wr S_3 \wr S_2$ | 2 |
| | $O_2^-(2) \wr S_2 \wr S_3$ | 2 |
| | $G_{1,3}^{\mathcal{B}_1} \wr S_2$ | 2 |
| | $G_{3,1}^{\mathcal{B}_2}(\mathrm{GSp}_2(3)) \wr S_2$ | 2 |
| | $G_{1,6}^{\mathcal{B}_1}$ | 1 |
| | $(O_2^-(2) \wr \mathrm{AGL}_1(5)) \times O_2^-(2)$ | 2 |
| | $G_{1,5}^{\mathcal{B}_2} \times O_2^-(2)$ | 2 |
| | $(O_2^-(2) \wr S_4) \times (O_2^-(2) \wr S_2)$ | 2 |
| | $(G_{1,2}^{\mathcal{B}_2} \wr S_2) \times (O_2^-(2) \wr S_2)$ | 2 |
| | $G_{1,4}^{\mathcal{B}_1} \times (O_2^-(2) \wr S_2)$ | 2 |
| | $G_{1,4}^{\mathcal{B}_2} \times G_{1,2}^{\mathcal{B}_2}$ | 2 |
| | $G_{3,1}^{\mathcal{B}_2}(\mathrm{GSp}_2(3)) \times (O_2^-(2) \wr S_3)$ | 2 |
| | $G_{1,3}^{\mathcal{B}_1} \times G_{1,2}^{\mathcal{B}_2} \times O_2^-(2)$ | 2 |
| $O_{12}^-(2)$ | $G_{1,2}^{\mathcal{B}_2} \wr S_3$ | 2 |
| | $G_{1,6}^{\mathcal{B}_2}$ | 2 |
| | $G_{1,5}^{\mathcal{B}_1} \times O_2^-(2)$ | 2 |
| | $(O_2^-(2) \wr S_4) \times G_{1,2}^{\mathcal{B}_2}$ | 2 |
| | $G_{1,4}^{\mathcal{B}_1} \times G_{1,2}^{\mathcal{B}_2}$ | 2 |
| | $G_{1,4}^{\mathcal{B}_2} \times (O_2^-(2) \wr S_2)$ | 2 |
| | $G_{1,3}^{\mathcal{B}_1} \times (O_2^-(2) \wr S_3)$ | 2 |
| | $G_{1,3}^{\mathcal{B}_1} \times G_{3,1}^{\mathcal{B}_2}(\mathrm{GSp}_2(3))$ | 2 |
| | $G_{3,1}^{\mathcal{B}_2}(\mathrm{GSp}_2(3)) \times G_{1,2}^{\mathcal{B}_2} \times O_2^-(2)$ | 2 |

**Table 8.3** Maximal irreducible solvable (MIRS) subgroups of $G = \mathrm{GSp}_n(q)$ and $G^\circ = \mathrm{Sp}_n(q)$, for $q$ odd. In the table, $(*)$ denotes the condition that 2 is a square in $\mathbb{F}_q$

| $G$ | Type of $X$ | $X$ MIRS in $\mathrm{GSp}_n(q)$ | $X^\circ$ MIRS in $\mathrm{Sp}_n(q)$ | Multiplier |
|---|---|---|---|---|
| $\mathrm{GSp}_2(q)$ | $G^{\mathcal{B}_1}_{1,1}$ | $q \neq 3, 5$ | $q \neq 3, 5, 7, 9$ | 1 |
| | $G^{\mathcal{B}_2}_{1,1}$ | $q \neq 3$ | $q \neq 3, 7$ | 1 |
| | $G^{\mathcal{B}_3}_{2,1}(O_2^-(2))$ | | | 1 if not $(*)$<br>2 if $(*)$ |
| $\mathrm{GSp}_4(q)$ | $G^{\mathcal{B}_1}_{1,1} \wr S_2$ | $q \neq 3, 5$ | $q \neq 3, 5, 7, 9$ | 1 |
| | $G^{\mathcal{B}_2}_{1,1} \wr S_2$ | $q \neq 3$ | $q \neq 3, 7$ | 1 |
| | $G^{\mathcal{B}_3}_{2,1}(O_2^-(2)) \wr S_2$ | | | 1 if not $(*)$<br>2 if $(*)$ |
| | $\mathrm{semiwr}(G^{\mathcal{B}_3}_{2,1}(O_2^-(2))^\circ)$ | $(*)$ | $X^\circ$ not irreducible | 1 |
| | $G^{\mathcal{B}_1}_{1,2}$ | $q \neq 3, 5$ | $q \neq 3, 5$ | 1 |
| | $G^{\mathcal{B}_2}_{1,2}$ | | | 1 |
| | $G^{\mathcal{B}_1}_{2,1}(O_2^-(2))$ | $q \neq 3$ | $q \neq 3$ | 1 |
| | $G^{\mathcal{B}_2}_{2,1}(O_2^-(2))$ | | | 1 |
| | $G^{\mathcal{B}_3}_{4,1}\left(G^{\mathcal{B}_2}_{1,2}\right)$ | | | 1 if not $(*)$<br>2 if $(*)$ |
| $\mathrm{GSp}_6(q)$ | $G^{\mathcal{B}_1}_{1,1} \wr S_3$ | $q \neq 3, 5$ | $q \neq 3, 5, 7, 9$ | 1 |
| | $G^{\mathcal{B}_2}_{1,1} \wr S_3$ | $q \neq 3$ | $q \neq 3, 7$ | 1 |
| | $G^{\mathcal{B}_3}_{2,1}(O_2^-(2)) \wr S_3$ | | | 1 if not $(*)$<br>2 if $(*)$ |
| | $G^{\mathcal{B}_1}_{1,3}$ | | | 1 |
| | $G^{\mathcal{B}_2}_{1,3}$ | | | 1 |
| | $G^{\mathcal{B}_1}_{3,1}(\mathrm{GSp}_2(3))$ | $q \equiv 1 \mod 3$ | $q \equiv 1 \mod 3$ | 1 |
| | $G^{\mathcal{B}_2}_{3,1}(\mathrm{GSp}_2(3))$ | $q \equiv 2 \mod 3$ | $q \equiv 2 \mod 3$ | 1 |

Notes: $\mathrm{GSp}_2(q) = \mathrm{GL}_2(q)$
$G^{\mathcal{B}_1}_{1,1} = \mathrm{GL}_1(q) \wr S_2$
$G^{\mathcal{B}_2}_{1,1} = \Gamma\mathrm{L}_1(q^2) = G^{\mathcal{B}_0}_{1,2}$
$G^{\mathcal{B}_3}_{2,1}(O_2^-(2)) = G^{\mathcal{B}_0}_{2,1}(O_2^-(2))$
$\mathrm{GSp}_2(2) = G^{\mathcal{B}_2}_{1,1}$
$\mathrm{GSp}_2(3) = G^{\mathcal{B}_3}_{2,1}(O_2^-(2))$
Notation such as $G^{\mathcal{B}_1}_{1,1} \wr S_2$ in the "type of $X$" column denotes that $X = (G^{\mathcal{B}_1}_{1,1} \wr S_2) \cap \mathrm{GSp}_4(q)$, as in the construction of Example 2.2.5

**Table 8.4** Maximal irreducible solvable (MIRS) subgroups of $G = \mathrm{GO}_n^\varepsilon(q)$ and $G^\circ = O_n^\varepsilon(q)$. In the table, $(*)$ denotes the condition that 2 is a square in $\mathbb{F}_q$

| $G$ | Type of $X$ | $X$ MIRS in $\mathrm{GO}_n^\varepsilon(q)$ | $X^\circ$ MIRS in $O_n^\varepsilon(q)$ | Multiplier |
|---|---|---|---|---|
| $\mathrm{GO}_2^+(q)$ | $\mathrm{GO}_2^+(q)$ | $q \neq 2$ | $q \neq 2, 3$ | 1 |
| $\mathrm{GO}_2^-(q)$ | $\mathrm{GO}_2^-(q)$ | | | 1 |
| $\mathrm{GO}_3(q)$ | $\mathrm{GO}_1(q) \wr S_3$ | $q$ odd | $q$ odd | 2 |
| $\mathrm{GO}_4^+(q)$ | $\mathrm{GO}_2^+(q) \wr S_2$ | $q \neq 2, 3, 5$ | $q \neq 2, 3, 5$ | 1 |
| | $\mathrm{GO}_2^-(q) \wr S_2$ | $q \neq 3$ | $q \neq 3$ | 1 |
| | $G_{1,2}^{\mathcal{B}_1}$ | $q \neq 2, 3, 5$ | $q \neq 2, 3, 5$ | 1 |
| | $G_{2,1}^{\mathcal{B}_1}(O_2^-(2))$ | $q$ odd, $q \neq 3, 5$ | $q$ odd, $q \neq 3, 5, 7, 9$ | 1 |
| | $G_{2,1}^{\mathcal{B}_2}(O_2^-(2))$ | $q$ odd, $q \neq 3$ | $q$ odd, $q \neq 3, 7$ | 1 |
| | $G_{4,1}^{\mathcal{B}_3}(O_4^+(2))$ | $q$ odd | $q$ odd | 1 if not $(*)$<br>2 if $(*)$ |
| $\mathrm{GO}_4^-(q)$ | $G_{1,2}^{\mathcal{B}_2}$ | | | 1 |
| $\mathrm{GO}_5(q)$ | $\mathrm{GO}_1(q) \wr \mathrm{AGL}_1(5)$ | $q$ odd | $q$ odd | 1 |
| $\mathrm{GO}_6^+(q)$ | $\mathrm{GO}_2^+(q) \wr S_3$ | | | 1 |
| | $\mathrm{semiwr}(O_1(q) \wr S_3)$ | $q \equiv 3 \mod 4$ | $X^\circ$ not irreducible | 1 |
| | $G_{1,3}^{\mathcal{B}_1}$ | | | 1 |
| | $G_{3,1}^{\mathcal{B}_1}(\mathrm{GSp}_2(3))$ | $q \equiv 1 \mod 3$ | $q \equiv 1 \mod 3$ | 1 |
| $\mathrm{GO}_6^-(q)$ | $\mathrm{GO}_2^-(q) \wr S_3$ | | | 1 |
| | $\mathrm{semiwr}(O_1(q) \wr S_3)$ | $q \equiv 1 \mod 4$ | $X^\circ$ not irreducible | 1 |
| | $G_{1,3}^{\mathcal{B}_2}$ | | | 1 |
| | $G_{3,1}^{\mathcal{B}_2}(\mathrm{GSp}_2(3))$ | $q \equiv 2 \mod 3$ | $q \equiv 2 \mod 3$ | 1 |

Notes: $O_2^+(2)$ and $O_2^+(3)$ are not irreducible
$\mathrm{GO}_2^+(3) = \mathrm{semiwr}(O_1(3))$
$\mathrm{GO}_2^+(q) = G_{1,1}^{\mathcal{B}_1}$ for $q > 3$
$\mathrm{GO}_2^-(q) = G_{1,1}^{\mathcal{B}_2}$
$O_2^-(3) = O_1(3) \wr S_2$
$O_2^+(5) = O_1(5) \wr S_2$
$\mathrm{GO}_4^+(3) = G_{4,1}^{\mathcal{B}_3}(O_4^+(2))$
Notation such as $\mathrm{GO}_2^+(q) \wr S_2$ in the "type of $X$" column denotes that $X = (\mathrm{GO}_2^+(q) \wr S_2) \cap \mathrm{GO}_4^+(q)$ as in the construction of Example 2.2.5

**Table 8.5** Maximal irreducible solvable subgroups of $G = \mathrm{GL}_n(q)$ for $2 \leq n \leq 8$, up to conjugacy in $G$

| $G$ | Subgroup | $q$ |
|---|---|---|
| $\mathrm{GL}_2(q)$ | $\mathrm{GL}_1(q) \wr S_2$ | $q \neq 2, 3, 5$ |
| | $\Gamma\mathrm{L}_1(q^2)$ | $q \neq 3$ |
| | $G^{\mathcal{B}_0}_{2,1}\left(O_2^-(2)\right)$ | $q$ odd |
| $\mathrm{GL}_3(q)$ | $\mathrm{GL}_1(q) \wr S_3$ | $q \neq 2, 4$ |
| | $\Gamma\mathrm{L}_1(q^3)$ | |
| | $G^{\mathcal{B}_0}_{3,1}(\mathrm{GSp}_2(3))$ | $q \equiv 1 \mod 3$ |
| $\mathrm{GL}_4(q)$ | $\mathrm{GL}_1(q) \wr S_4$ | $q \neq 2, 3$ |
| | $G^{\mathcal{B}_0}_{1,2} \wr S_2$ | $q \neq 3$ |
| | $G^{\mathcal{B}_0}_{2,1}\left(O_2^-(2)\right) \wr S_2$ | $q$ odd |
| | $\Gamma\mathrm{L}_1(q^4)$ | |
| | $G^{\mathcal{B}_0}_{2,2}\left(O_2^-(2)\right)$ | $q$ odd, $q \neq 3$ |
| | $G^{\mathcal{B}_0}_{4,1}\left(O_4^+(2)\right)$ | $q$ odd |
| | $G^{\mathcal{B}_0}_{4,1}\left(G^{\mathcal{B}_2}_{1,2}\right)$ | $q$ odd |
| $\mathrm{GL}_5(q)$ | $\mathrm{GL}_1(q) \wr \mathrm{AGL}_1(5)$ | $q \neq 2$ |
| | $\Gamma\mathrm{L}_1(q^5)$ | |
| | $G^{\mathcal{B}_0}_{5,1}\left(G^{\mathcal{B}_2}_{1,1}\right)$ | $q \equiv 1 \mod 5$ |
| | $G^{\mathcal{B}_0}_{5,1}\left(G^{\mathcal{B}_3}_{2,1}(O_2^-(2))\right)$ | $q \equiv 1 \mod 5$ |
| $\mathrm{GL}_6(q)$ | $\mathrm{GL}_1(q) \wr S_3 \wr S_2$ | $q \neq 2, 4$ |
| | $\mathrm{GL}_1(q) \wr S_2 \wr S_3$ | $q \neq 2, 3, 5$ |
| | $\Gamma\mathrm{L}_1(q^2) \wr S_3$ | $q \neq 3$ |
| | $G^{\mathcal{B}_0}_{2,1}\left(O_2^-(2)\right) \wr S_3$ | $q$ odd |
| | $\Gamma\mathrm{L}_1(q^3) \wr S_2$ | |
| | $G^{\mathcal{B}_0}_{3,1}(\mathrm{GSp}_2(3)) \wr S_2$ | $q \equiv 1 \mod 3$ |
| | $\Gamma\mathrm{L}_1(q^6)$ | |
| | $G^{\mathcal{B}_0}_{2,3}\left(O_2^-(2)\right)$ | $q$ odd |
| | $G^{\mathcal{B}_0}_{3,2}(\mathrm{GSp}_2(3))$ | $q \equiv \pm 1 \mod 3$ |
| | $G^{\mathcal{B}_0}_{6,1}\left(O_2^-(2), \mathrm{GSp}_2(3)\right)$ | $q \equiv 1 \mod 6$ |
| $\mathrm{GL}_7(q)$ | $\mathrm{GL}_1(q) \wr \mathrm{AGL}_1(7)$ | $q \neq 2$ |
| | $\Gamma\mathrm{L}_1(q^7)$ | |
| | $G^{\mathcal{B}_0}_{7,1}\left(G^{\mathcal{B}_3}_{2,1}(O_2^-(2))\right)$ | $q \equiv 1 \mod 7$ |
| | $G^{\mathcal{B}_0}_{7,1}\left(xG^{\mathcal{B}_3}_{2,1}(O_2^-(2))x^{-1}\right)$ | $q \equiv 1 \mod 7$ |
| $\mathrm{GL}_8(q)$ | $\mathrm{GL}_1(q) \wr S_2 \wr S_4$ | $q \neq 2, 3, 5$ |
| | $\mathrm{GL}_1(q) \wr S_4 \wr S_2$ | $q \neq 2, 3$ |
| | $\mathrm{GL}_1(q) \wr \left(\mathbb{F}_2^3 \rtimes \Gamma\mathrm{L}_1(2^3)\right)$ | $q \neq 2$ |
| | $\Gamma\mathrm{L}_1(q^2) \wr S_4$ | $q \neq 3$ |
| | $G^{\mathcal{B}_0}_{2,1}(O_2^-(2)) \wr S_4$ | $q$ odd |
| | $\Gamma\mathrm{L}_1(q^4) \wr S_2$ | |
| | $G^{\mathcal{B}_0}_{2,2}\left(O_2^-(2)\right) \wr S_2$ | $q$ odd, $q \neq 3$ |
| | $G^{\mathcal{B}_0}_{4,1}\left(O_4^+(2)\right) \wr S_2$ | $q$ odd |
| | $G^{\mathcal{B}_0}_{4,1}\left(G^{\mathcal{B}_2}_{1,1}\right) \wr S_2$ | $q$ odd |
| | $\Gamma\mathrm{L}_1(q^8)$ | |
| | $G^{\mathcal{B}_0}_{2,4}\left(O_2^-(2)\right)$ | $q$ odd |
| | $G^{\mathcal{B}_0}_{4,2}\left(O_4^+(2)\right)$ | $q$ odd, $q \neq 3$ |
| | $G^{\mathcal{B}_0}_{4,2}\left(G^{\mathcal{B}_2}_{1,2}\right)$ | $q$ odd, $q \neq 3$ |
| | $G^{\mathcal{B}_0}_{8,1}\left(G^{\mathcal{B}_1}_{1,3}\right)$ | $q$ odd |
| | $G^{\mathcal{B}_0}_{8,1}\left(G^{\mathcal{B}_2}_{1,2} \times O_2^-(2)\right)$ | $q$ odd |
| | $G^{\mathcal{B}_0}_{8,1}\left(G^{\mathcal{B}_2}_{3,1}(\mathrm{GSp}_2(3))\right)$ | $q$ odd |
| | $G^{\mathcal{B}_0}_{8,1}\left(O_2^-(2) \wr S_3\right)$ | $q$ odd |

Notes: $G^{\mathcal{B}_0}_{1,n} = \Gamma\mathrm{L}_1(q^n)$ for all $n$

For $n = 7$, we denote $x = \begin{pmatrix} 3 & 0 \\ 0 & 1 \end{pmatrix}$, so that $x \in \mathrm{GSp}_2(7)$ with $\tau(x) \notin \left(\mathbb{F}_7^\times\right)^2$

**Table 8.6** Maximal irreducible solvable subgroups of $G = \mathrm{GL}_r(q)$ and $G = \mathrm{GL}_{2r}(q)$ for $r > 2$ prime, up to conjugacy in $G$

| $G$ | Subgroup | Conditions |
|---|---|---|
| $\mathrm{GL}_r(q)$ | $\mathrm{GL}_1(q) \wr \mathrm{AGL}_1(r)$ | $q \neq 2$ and $(q, r) \neq (4, 3)$ |
| | $\Gamma\mathrm{L}_1(q^r)$ | |
| | $G_{r,1}^{\mathcal{B}_0}\left(G_{1,1}^{\mathcal{B}_1}\right)$ | $q \equiv 1 \mod r$ and $r \neq 3, 5, 7$ |
| | $G_{r,1}^{\mathcal{B}_0}\left(G_{1,1}^{\mathcal{B}_2}\right)$ | $q \equiv 1 \mod r$ and $r \neq 3, 7$ |
| | $G_{r,1}^{\mathcal{B}_0}\left(G_{2,1}^{\mathcal{B}_3}(O_2^-(2))\right)$ | $q \equiv 1 \mod r$ |
| | $G_{r,1}^{\mathcal{B}_0}\left(xG_{2,1}^{\mathcal{B}_3}(O_2^-(2))x^{-1}\right)$ | $q \equiv 1 \mod r$ and $r \equiv \pm 1 \mod 8$ |
| $\mathrm{GL}_{2r}(q)$ | $\mathrm{GL}_1(q) \wr S_2 \wr \mathrm{AGL}_1(r)$ | $q \neq 2, 3, 5$ |
| | $\mathrm{GL}_1(q) \wr \mathrm{AGL}_1(r) \wr S_2$ | $q \neq 2$ and $(q, r) \neq (4, 3)$ |
| | $\Gamma\mathrm{L}_1(q^2) \wr \mathrm{AGL}_1(r)$ | $q \neq 3$ |
| | $G_{2,1}^{\mathcal{B}_0}\left(O_2^-(2)\right) \wr \mathrm{AGL}_1(r)$ | $q$ odd |
| | $\Gamma\mathrm{L}_1(q^r) \wr S_2$ | |
| | $G_{r,1}^{\mathcal{B}_0}\left(G_{1,1}^{\mathcal{B}_1}\right) \wr S_2$ | $q \equiv 1 \mod r$ and $r \neq 3, 5, 7$ |
| | $G_{r,1}^{\mathcal{B}_0}\left(G_{1,1}^{\mathcal{B}_2}\right) \wr S_2$ | $q \equiv 1 \mod r$ and $r \neq 3, 7$ |
| | $G_{r,1}^{\mathcal{B}_0}\left(G_{2,1}^{\mathcal{B}_3}(O_2^-(2))\right) \wr S_2$ | $q \equiv 1 \mod r$ |
| | $G_{r,1}^{\mathcal{B}_0}\left(xG_{2,1}^{\mathcal{B}_3}(O_2^-(2))x^{-1}\right) \wr S_2$ | $q \equiv 1 \mod r$ and $r \equiv \pm 1 \mod 8$ |
| | $\Gamma\mathrm{L}_1(q^{2r})$ | |
| | $G_{2,r}^{\mathcal{B}_0}\left(O_2^-(2)\right)$ | $q$ odd |
| | $G_{r,2}^{\mathcal{B}_0}\left(G_{1,1}^{\mathcal{B}_1}\right)$ | $q \equiv \pm 1 \mod r$, and $r \neq 3, 5$<br>$q \equiv -1 \mod 7$ if $r = 7$ |
| | $G_{r,2}^{\mathcal{B}_0}\left(G_{1,1}^{\mathcal{B}_2}\right)$ | $q \equiv \pm 1 \mod r$ and $r \neq 3$<br>$q \equiv -1 \mod 7$ if $r = 7$ |
| | $G_{r,2}^{\mathcal{B}_0}\left(G_{2,1}^{\mathcal{B}_3}(O_2^-(2))\right)$ | $q \equiv \pm 1 \mod r$ |
| | $G_{r,2}^{\mathcal{B}_0}\left(xG_{2,1}^{\mathcal{B}_3}(O_2^-(2))x^{-1}\right)$ | $q \equiv \pm 1 \mod r$<br>$r \equiv \pm 1 \mod 8$ and $\left(\frac{q}{r}\right) = +$ |
| | $G_{2r,1}^{\mathcal{B}_0}\left(O_2^-(2), G_{1,1}^{\mathcal{B}_1}\right)$ | $q \equiv 1 \mod 2r$ and $r \neq 3, 5, 7$ |
| | $G_{2r,1}^{\mathcal{B}_0}\left(O_2^-(2), G_{1,1}^{\mathcal{B}_2}\right)$ | $q \equiv 1 \mod 2r$ and $r \neq 3, 7$ |
| | $G_{2r,1}^{\mathcal{B}_0}\left(O_2^-(2), G_{2,1}^{\mathcal{B}_3}(O_2^-(2))\right)$ | $q \equiv 1 \mod 2r$ |
| | $G_{2r,1}^{\mathcal{B}_0}\left(O_2^-(2), xG_{2,1}^{\mathcal{B}_3}(O_2^-(2))x^{-1}\right)$ | $q \equiv 1 \mod 2r$ and $r \equiv \pm 1 \mod 8$ |

Notes: $x = \begin{pmatrix} \zeta & 0 \\ 0 & 1 \end{pmatrix}$, where $\zeta \in \mathbb{F}_r^\times$ is nonsquare. $\left(G_{1,1}^{\mathcal{B}_1}\right)^\circ$ is not maximal solvable in $\mathrm{Sp}_2(7)$. Thus for $G_{r,2}^{\mathcal{B}_0}\left(G_{1,1}^{\mathcal{B}_1}\right)$ with $r = 7$, we need $q \equiv -1 \mod 7$ so that $q$ is nonsquare modulo 7 (Similar remarks apply for $G_{r,2}^{\mathcal{B}_0}\left(G_{1,1}^{\mathcal{B}_2}\right)$ with $r = 7$)

# References


1. Abel, N.H.: Oeuvres complètes de Niels Henrik Abel, vol. 2. Grøndahl, Christiania (1881)
2. Aschbacher, M.: On the maximal subgroups of the finite classical groups. Invent. Math. **76**(3), 469–514 (1984)
3. Aschbacher, M.: Finite Group Theory, vol. 10, 2nd edn. Cambridge Studies in Advanced Mathematics. Cambridge University Press, Cambridge (2000)
4. Aschbacher, M., Scott, L.: Maximal subgroups of finite groups. J. Algebra **92**(1), 44–80 (1985)
5. Beglaryan, G.V., Zalesskiĭ, A.E.: Spectra of $p$-elements in the normalizer of the extraspecial linear group. Mat. Zametki **49**(5), 7–15, 157 (1991)
6. Betti, E.: Sulla risoluzione delle equazioni algebriche. Ann. Sci. Mat. Fisiche **3**, 49–115 (1852)
7. Bray, J.N., Holt, D.F., Roney-Dougal, C.M.: The Maximal Subgroups of the Low-Dimensional Finite Classical Groups, vol. 407. London Mathematical Society Lecture Note Series. Cambridge University Press, Cambridge (2013). With a foreword by M. Liebeck
8. Brechenmacher, F.: Self-portraits with Évariste Galois (and the shadow of Camille Jordan). Rev. Hist. Math. **17**(2), 273–371 (2011)
9. Bucht, G.: Die umfassendsten primitiven metazyklischen Kongruenzgruppen mit drei oder vier Variablen. Ark. Mat. Astron. Fys. **11**(26), 96 (1917)
10. Burness, T.C.: Fixed point ratios in actions in finite classical groups. II. J. Algebra **309**(1), 80–138 (2007)
11. Burness, T.C., Ghandour, S., Marion, C., Testerman, D.M.: Irreducible almost simple subgroups of classical algebraic groups. Mem. Am. Math. Soc. **236**(1114), vi+110 (2015)
12. Cohen, A.M.: Finite complex reflection groups. Ann. Sci. École Norm. Sup. (4) **9**(3), 379–436 (1976)
13. Cox, D.A.: Galois Theory. Pure and Applied Mathematics (Hoboken), 2nd edn. Wiley, Hoboken (2012)
14. Curtis, C.W., Reiner, I.: Representation Theory of Finite Groups and Associative Algebras. Pure and Applied Mathematics, vol. XI. Interscience Publishers (a division of John Wiley & Sons, Inc.), New York-London (1962)
15. Detinko, A.S.: Maximal solvable subgroups of the special linear group over an arbitrary field. Sibirsk. Mat. Zh. **33**(6), 39–46, 229 (1992)
16. Dieudonné, J.: Notes sur les travaux de C. Jordan relatifs a la théorie des groupes finis. Oeuvres de Camille Jordan **1**, xvii–xlii (1961)
17. Dixon, J.D.: The Structure of Linear Groups. Van Nostrand-Reinhold, London (1971)
18. Dixon, J.D., Mortimer, B.: Permutation Groups, vol. 163. Graduate Texts in Mathematics. Springer, New York (1996)


 
M. Korhonen, *Maximal Solvable Subgroups of Finite Classical Groups*,
Lecture Notes in Mathematics 2346, https://doi.org/10.1007/978-3-031-62915-0

19. Doerk, K., Hawkes, T.: Finite Soluble Groups, vol. 4. De Gruyter Expositions in Mathematics. Walter de Gruyter & Co., Berlin (1992)
20. Dye, R.H.: A geometric characterization of the special orthogonal groups and the Dickson invariant. J. Lond. Math. Soc. (2) **15**(3), 472–476 (1977)
21. Dye, R.H.: Interrelations of symplectic and orthogonal groups in characteristic two. J. Algebra **59**(1), 202–221 (1979)
22. Ehrhardt, C.: A social history of the "Galois Affair" at the Paris Academy of Sciences (1831). Sci. Context **23**(1), 91–119 (2010)
23. Eick, B., Höfling, B.: The solvable primitive permutation groups of degree at most 6560. LMS J. Comput. Math. **6**, 29–39 (2003)
24. Fieker, C., Klüners, J.: Computation of Galois groups of rational polynomials. LMS J. Comput. Math. **17**(1), 141–158 (2014)
25. Girstmair, K.: On the computation of resolvents and Galois groups. Manuscr. Math. **43**(2–3), 289–307 (1983)
26. Gluck, D., Manz, O.: Prime factors of character degrees of solvable groups. Bull. Lond. Math. Soc. **19**(5), 431–437 (1987)
27. Goozeff, J.T.: Abelian $p$-subgroups of the general linear group. J. Austral. Math. Soc. **11**, 257–259 (1970)
28. Gorenstein, D.: Finite Groups, 2nd edn. Chelsea Publishing Co., New York (1980)
29. Gow, R.: Real-valued characters of solvable groups. Bull. Lond. Math. Soc. **7**, 132 (1975)
30. Gross, F.: On the uniqueness of wreath products. J. Algebra **147**(1), 147–175 (1992)
31. Guralnick, R.M., Maróti, A.: On the non-coprime $k(GV)$-problem. J. Algebra **385**, 80–101 (2013)
32. Hall, P., Higman, G.: On the $p$-length of $p$-soluble groups and reduction theorems for Burnside's problem. Proc. Lond. Math. Soc. (3) **6**, 1–42 (1956)
33. Hall, J.I., Liebeck, M.W., Seitz, G.M.: Generators for finite simple groups, with applications to linear groups. Quart. J. Math. Oxf. Ser. (2) **43**(172), 441–458 (1992)
34. Herstein, I.N.: A remark on finite groups. Proc. Am. Math. Soc. **9**, 255–257 (1958)
35. Holt, D.F., Roney-Dougal, C.M.: Constructing maximal subgroups of classical groups. LMS J. Comput. Math. **8**, 46–79 (2005)
36. Holt, D.F., Roney-Dougal, C.M.: Constructing maximal subgroups of orthogonal groups. LMS J. Comput. Math. **13**, 164–191 (2010)
37. Hulpke, A.: Techniques for the computation of Galois groups. In: Algorithmic Algebra and Number Theory (Heidelberg, 1997), pp. 65–77. Springer, Berlin (1999)
38. Huppert, B.: Endliche Gruppen. I. Die Grundlehren der Mathematischen Wissenschaften, Band 134. Springer, Berlin-New York (1967)
39. Ireland, K., Rosen, M.: A Classical Introduction to Modern Number Theory, vol. 84, 2nd edn. Graduate Texts in Mathematics. Springer, New York (1990)
40. Isaacs, I.M.: Character degrees and derived length of a solvable group. Can. J. Math. **27**, 146–151 (1975)
41. Jordan, C.: Mémoire sur le nombre des valeurs des fonctions. J. École Polytech. **22**, 113–194 (1861)
42. Jordan, C.: Mémoire sur les groupes des équations résolubles par radicaux. C. R. Math. Acad. Sci. Paris **58**, 963–966 (1864)
43. Jordan, C.: Lettre à M. Liouville sur la résolution algébrique des équations. J. Math. Pures Appl. **12**, 105–108 (1867)
44. Jordan, C.: Mémoire sur la résolution algébrique des équations. C. R. Math. Acad. Sci. Paris **64**, 269–272, 586–590, 1179–1183 (1867)
45. Jordan, C.: Mémoire sur la résolution algébrique des équations. J. Math. Pures Appl. **12**, 109–157 (1867)
46. Jordan, C.: Sur la résolution algébrique des équations primitives de degré $p^2$ ($p$ étant premier impair). J. Math. Pures Appl. **13**, 111–135 (1868)
47. Jordan, C.: Traité des substitutions et des équations algébriques. Gauthier-Villars, Paris (1870)

48. Jordan, C.: Sur la résolution des equations les unes par les autres. C. R. Math. Acad. Sci. Paris **72**, 283–290 (1871)
49. Jordan, C.: Sur la limite du degré des groupes primitifs qui contiennent une substitution donnée. J. Reine Angew. Math. **79**, 248–258 (1875)
50. Jordan, C.: Notice sur les travaux de M. Camille Jordan: ingénieur des mines, professeur à l'École polytechnique: à l'appui de sa candidature à l'Académie des sciences (Section de géométrie). Gauthier-Villars, Paris (1881)
51. Jordan, C.: Recherches sur les groupes résolubles. Memorie della Pontificia Accademia Romana dei Nuovi Lincei **26**, 7–39 (1908)
52. Jordan, C.: Mémoire sur les groupes résolubles. J. Math. Pures Appl. **3**, 263–374 (1917)
53. Kantor, W.M.: Linear groups containing a Singer cycle. J. Algebra **62**(1), 232–234 (1980)
54. Kiernan, B.M.: The development of Galois theory from Lagrange to Artin. Arch. Hist. Exact Sci. **8**(1–2), 40–154 (1971)
55. Kleidman, P., Liebeck, M.: The Subgroup Structure of the Finite Classical Groups, vol. 129. London Mathematical Society Lecture Note Series. Cambridge University Press, Cambridge (1990)
56. Korhonen, M.: Orthogonal irreducible representations of finite solvable groups in odd dimension. Bull. Lond. Math. Soc. **56**(1), 444–448 (2024)
57. Korhonen, M., Li, C.H.: Systems of imprimitivity for wreath products. J. Algebra **587**, 628–637 (2021)
58. Kozel, P.T., Tyškevič, R.I.: Two theorems on solvable groups. Izv. Vysš. Učebn. Zaved. Matematika **1962**(6(31)), 45–50 (1962)
59. Lidl, R., Niederreiter, H.: Finite Fields, vol. 20. Encyclopedia of Mathematics and its Applications, 2nd edn. Cambridge University Press, Cambridge (1997). With a foreword by P. M. Cohn
60. Liebeck, M.W., Seitz, G.M.: Unipotent and Nilpotent Classes in Simple Algebraic Groups and Lie Algebras, vol. 180. Mathematical Surveys and Monographs. American Mathematical Society, Providence (2012)
61. Liebeck, M.W., Shalev, A.: Simple groups, permutation groups, and probability. J. Am. Math. Soc. **12**(2), 497–520 (1999)
62. Liebeck, M.W., Praeger, C.E., Saxl, J.: On the O'Nan-Scott theorem for finite primitive permutation groups. J. Austral. Math. Soc. Ser. A **44**(3), 389–396 (1988)
63. Liouville, J.: Oeuvres Mathématiques d'Évariste Galois. J. Math. Pures Appl. **11**, 381–444 (1846)
64. Manz, O., Wolf, T.R.: Representations of Solvable Groups, vol. 185. London Mathematical Society Lecture Note Series. Cambridge University Press, Cambridge (1993)
65. Mihăilescu, P.: Primary cyclotomic units and a proof of Catalan's conjecture. J. Reine Angew. Math. **572**, 167–195 (2004)
66. Neumann, P.M.: Review of "Traité des substitutions et des équations algébriques" by C. Jordan. Math. Rev. MR1188877 (1989)
67. Neumann, P.M.: The concept of primitivity in group theory and the second memoir of Galois. Arch. Hist. Exact Sci. **60**(4), 379–429 (2006)
68. Neumann, P.M.: The Mathematical Writings of Évariste Galois. Heritage of European Mathematics. European Mathematical Society (EMS), Zürich (2011)
69. Saxl, J., Seitz, G.M.: Subgroups of algebraic groups containing regular unipotent elements. J. Lond. Math. Soc. (2) **55**(2), 370–386 (1997)
70. Scott, L.L.: Representations in characteristic $p$. In: The Santa Cruz Conference on Finite Groups (Univ. California, Santa Cruz, Calif., 1979), vol. 37. Proc. Sympos. Pure Math., pp. 319–331. American Mathematical Society, Providence (1980)
71. Seitz, G.M.: The root subgroups for maximal tori in finite groups of Lie type. Pac. J. Math. **106**(1), 153–244 (1983)
72. Seress, Á.: The minimal base size of primitive solvable permutation groups. J. Lond. Math. Soc. (2) **53**(2), 243–255 (1996)
73. Serret, J.-A.: Cours d'Algèbre Supérieure, 3rd edn. Gauthier-Villars, Paris (1866)

74. Short, M.W.: The Primitive Soluble Permutation Groups of Degree Less Than 256, vol. 1519. Lecture Notes in Mathematics. Springer, Berlin (1992)
75. Stauduhar, R.P.: The determination of Galois groups. Math. Comput. **27**, 981–996 (1973)
76. Suprunenko, D.A.: Soluble and Nilpotent Linear Groups. American Mathematical Society, Providence (1963)
77. Suprunenko, D.A.: Matrix Groups. American Mathematical Society, Providence (1976). Translated from Russian, Translation edited by K. A. Hirsch, Translations of Mathematical Monographs, vol. 45
78. Suprunenko, D.A.: Permutation Groups. Navuka i Tekhnika, Minsk (1996)
79. Taylor, D.E.: The Geometry of the Classical Groups, vol. 9. Sigma Series in Pure Mathematics. Heldermann Verlag, Berlin (1992)
80. Tignol, J.-P.: Galois' Theory of Algebraic Equations. World Scientific Publishing, River Edge (2001)
81. Wielandt, H.: Finite Permutation Groups. Academic, New York-London (1964)
82. Willems, W.: Metrische $G$-Moduln über Körpern der Charakteristik 2. Math. Z. **157**(2), 131–139 (1977)
83. Wussing, H.: The Genesis of the Abstract Group Concept. MIT Press, Cambridge (1984). A contribution to the history of the origin of abstract group theory, Translated from the German by A. Shenitzer and H. Grant
84. Zalesskiĭ, A.E.: Subgroups of the classical groups. Siberian Math. J. **12**, 90–94 (1971)
85. Zalesskiĭ, A.E.: The normalizer of an extraspecial linear group. Vestsī Akad. Navuk BSSR Ser. Fīz.-Mat. Navuk (6), 11–16, 124 (1985)
86. Zalesskiĭ, A.E.: Eigenvalues of matrices of complex representations of finite groups of Lie type. In: Algebra—Some Current Trends (Varna, 1986), vol. 1352. Lecture Notes in Mathematics, pp. 206–218. Springer, Berlin (1988)
87. Zalesskiĭ, A.E., Konjuh, V.S.: Sylow $\pi$-subgroups of the classical groups. Mat. Sb. (N.S.) **101**(143)(2), 231–251 (1976)
88. Zsigmondy, K.: Zur Theorie der Potenzreste. Monatsh. Math. Phys. **3**(1), 265–284 (1892)

# Index





M. Korhonen, *Maximal Solvable Subgroups of Finite Classical Groups*,
Lecture Notes in Mathematics 2346, https://doi.org/10.1007/978-3-031-62915-0

*LECTURE NOTES IN MATHEMATICS* Springer

Editors in Chief: J.-M. Morel, B. Teissier;

**Editorial Policy**

1. Lecture Notes aim to report new developments in all areas of mathematics and their applications – quickly, informally and at a high level. Mathematical texts analysing new developments in modelling and numerical simulation are welcome.

   Manuscripts should be reasonably self-contained and rounded off. Thus they may, and often will, present not only results of the author but also related work by other people. They may be based on specialised lecture courses. Furthermore, the manuscripts should provide sufficient motivation, examples and applications. This clearly distinguishes Lecture Notes from journal articles or technical reports which normally are very concise. Articles intended for a journal but too long to be accepted by most journals, usually do not have this "lecture notes" character. For similar reasons it is unusual for doctoral theses to be accepted for the Lecture Notes series, though habilitation theses may be appropriate.

2. Besides monographs, multi-author manuscripts resulting from SUMMER SCHOOLS or similar INTENSIVE COURSES are welcome, provided their objective was held to present an active mathematical topic to an audience at the beginning or intermediate graduate level (a list of participants should be provided).

   The resulting manuscript should not be just a collection of course notes, but should require advance planning and coordination among the main lecturers. The subject matter should dictate the structure of the book. This structure should be motivated and explained in a scientific introduction, and the notation, references, index and formulation of results should be, if possible, unified by the editors. Each contribution should have an abstract and an introduction referring to the other contributions. In other words, more preparatory work must go into a multi-authored volume than simply assembling a disparate collection of papers, communicated at the event.

3. Manuscripts should be submitted either online at www.editorialmanager.com/lnm to Springer's mathematics editorial in Heidelberg, or electronically to one of the series editors. Authors should be aware that incomplete or insufficiently close-to-final manuscripts almost always result in longer refereeing times and nevertheless unclear referees' recommendations, making further refereeing of a final draft necessary. The strict minimum amount of material that will be considered should include a detailed outline describing the planned contents of each chapter, a bibliography and several sample chapters. Parallel submission of a manuscript to another publisher while under consideration for LNM is not acceptable and can lead to rejection.

4. In general, **monographs** will be sent out to at least 2 external referees for evaluation.

   A final decision to publish can be made only on the basis of the complete manuscript, however a refereeing process leading to a preliminary decision can be based on a pre-final or incomplete manuscript.

   Volume Editors of **multi-author works** are expected to arrange for the refereeing, to the usual scientific standards, of the individual contributions. If the resulting reports can be forwarded to the LNM Editorial Board, this is very helpful. If no reports are forwarded or if other questions remain unclear in respect of homogeneity etc, the series editors may wish to consult external referees for an overall evaluation of the volume.

5. Manuscripts should in general be submitted in English. Final manuscripts should contain at least 100 pages of mathematical text and should always include

   - a table of contents;
   - an informative introduction, with adequate motivation and perhaps some historical remarks: it should be accessible to a reader not intimately familiar with the topic treated;
   - a subject index: as a rule this is genuinely helpful for the reader.
   - For evaluation purposes, manuscripts should be submitted as pdf files.

6. Careful preparation of the manuscripts will help keep production time short besides ensuring satisfactory appearance of the finished book in print and online. After acceptance of the manuscript authors will be asked to prepare the final LaTeX source files (see LaTeX templates online: https://www.springer.com/gb/authors-editors/book-authors-editors/manuscriptpreparation/5636) plus the corresponding pdf- or zipped ps-file. The LaTeX source files are essential for producing the full-text online version of the book, see http://link.springer.com/bookseries/304 for the existing online volumes of LNM). The technical production of a Lecture Notes volume takes approximately 12 weeks. Additional instructions, if necessary, are available on request from lnm@springer.com.

7. Authors receive a total of 30 free copies of their volume and free access to their book on SpringerLink, but no royalties. They are entitled to a discount of 33.3 % on the price of Springer books purchased for their personal use, if ordering directly from Springer.

8. Commitment to publish is made by a *Publishing Agreement*; contributing authors of multiauthor books are requested to sign a *Consent to Publish form*. Springer-Verlag registers the copyright for each volume. Authors are free to reuse material contained in their LNM volumes in later publications: a brief written (or e-mail) request for formal permission is sufficient.

**Addresses:**

Professor Jean-Michel Morel, CMLA, École Normale Supérieure de Cachan, France
E-mail: moreljeanmichel@gmail.com

Professor Bernard Teissier, Equipe Géométrie et Dynamique,
Institut de Mathématiques de Jussieu – Paris Rive Gauche, Paris, France
E-mail: bernard.teissier@imj-prg.fr

Springer: Ute McCrory, Mathematics, Heidelberg, Germany,
E-mail: lnm@springer.com

The manufacturer's authorised representative in the EU is Springer Nature Customer Service Centre GmbH, Europaplatz 3, 69115 Heidelberg, Germany. If you have any concerns regarding our products, please contact ProductSafety@springernature.com

Printed and bound by CPI Group (UK) Ltd, Croydon, CR0 4YY
25/03/2026
02078185-0014